Nanotechnologies for the Life Sciences
Volume 3

Nanosystem Characterization Tools in the Life Sciences

Edited by
Challa S. S. R. Kumar

Related Titles

Kumar, C. S. S. R. (ed.)

Nanotechnologies for the Life Sciences (NtLS)

Book Series

Vol. 1

Biofunctionalization of Nanomaterials

385 pages with 153 figures
Hardcover
ISBN 3-527-31381-8

Vol. 2

Biological and Pharmaceutical Nanomaterials

400 pages with 160 figures
Hardcover
ISBN 3-527-31382-6

Baltes, H., Brand, O., Fedder, G. K., Hierold, C., Korvink, J. G., Tabata, O. (eds.)

Advanced Micro & Nanosystems (AMN)

Book Series

Vol. 1

Enabling Technologies for MEMS and Nanodevices

ISBN 3-527-30746-X

Vol. 2

CMOS-MEMS

ISBN 3-527-31080-0

Kumar, C. S. S. R., Hormes, J., Leuschner, C. (eds.)

Nanofabrication Towards Biomedical Applications

Techniques, Tools, Applications, and Impact

442 pages with 161 figures and 19 tables
Hardcover
ISBN 3-527-31115-7

Fecht, H.-J., Werner, M. (eds.)

The Nano-Micro Interface

Bridging the Micro and Nano Worlds

351 pages with 102 figures and 27 tables
2004
Hardcover
ISBN 3-527-30978-0

Öberg, P. Å., Togawa, T., Spelman, F. A. (eds.)

Sensors in Medicine and Health Care

Sensors Applications Vol. 3

444 pages with 219 figures and 7 tables
2004
Hardcover
ISBN 3-527-29556-9

Niemeyer, C. M., Mirkin, C. A. (eds.)

Nanobiotechnology

Concepts, Applications and Perspectives

491 pages with 193 figures and 9 tables
2004
Hardcover
ISBN 3-527-30658-7

Willner, I., Katz, E. (eds.)

Bioelectronics

From Theory to Applications

492 pages with 241 figures and 4 tables
2005
Hardcover
ISBN 3-527-30690-0

Nanotechnologies for the Life Sciences
Volume 3

Nanosystem Characterization Tools in the Life Sciences

Edited by
Challa S. S. R. Kumar

1st Edition

WILEY-VCH Verlag GmbH & Co. KGaA

The Editor of this Book

Dr. Challa S. S. R. Kumar
The Center for Advanced
Microstructures and Devices
(CAMD)
Louisiana State University
6980 Jefferson Highway
Baton Rouge, LA 70806
USA

Cover
Cover design by G. Schulz based on micrograph courtesy of S. Rilger, Nanostructural Analysis Network Organisation, NANO-MNRF, The University of Sydney, Australia

Library of Congress Card No.: applied for
British Library Cataloging-in-Publication Data:
A catalogue record for this book is available from the British Library.

Bibliographic information published by Die Deutsche Bibliothek
Die Deutsche Bibliothek lists this publication in the Deutsche Nationalbibliografie; detailed bibliographic data is available in the Internet at http://dnb.ddb.de.

Printed in the Federal Republic of Germany.
Printed on acid-free paper.

Typesetting Asco Typesetters, Hong Kong
Printing betz-druck GmbH, Darmstadt
Binding Litges & Dopf Buchbinderei GmbH, Heppenheim

ISBN-13 978-3-527-31383-9
ISBN-10 3-527-31383-4

Contents

Nanotechnologies for the Life Sciences Vol. 3
Nanosystem Characterization Tools in the Life Sciences. Edited by Challa S. S. R. Kumar

ISBN: 3-527-31383-4

Preface

The third volume in this series, *Nanosystem Characterization Tools for the Life Sciences*, is being presented to you sooner than expected thanks to the dedication of all the contributing authors. In a very timely manner, they present here their in-depth reviews on several important characterization tools that are currently being utilized and are in the process of development and refinement for nanotechnological applications in the life sciences.

While the book is certainly not an exhaustive source for all the characterization tools, it is for sure first of its kind to provide, under one umbrella, several of the very important tools. My gratitude goes to all those who have been part of this project, most importantly the authors, my employer, family, friends and Wiley-VCH publishers. I do hope that the information provided will help the readers in their scientific pursuits. I do realize that there is always a scope for improvement and therefore, your comments and suggestions are very valuable.

Traditional fluorescence imaging tools have so far helped life scientists in unraveling several mysteries. However, the advent of quantum dots is poised to take fluorescence imaging to a different level altogether. In the first chapter, *Fluorescence Imaging In Biology using Nanoprobes*, author Daniele Gerion from Lawrence Livermore National Laboratory in California, USA, provides a unique perspective to how various nanoprobes are changing the limits of fluorescence imaging, and shares her views on what future holds in this exciting field.

The second chapter in the book is a journey into imaging of biological samples using scanning probe microscopy techniques (SPM). The authors, Anthony W. Coleman and colleagues from CNRS-UCBL, France, critically analyse the utility of SPM in imaging hitherto unseen biological systems, understanding their mechanical properties and how they are useful in developing sensor systems. The chapter is aptly entitled *Characterization of Nanoscale Systems in Biology using Scanning Probe Microscopy Techniques.*

Continuing on the similar theme of imaging techniques, Rhonda Dzakpasu and Daniel Axelrod from the University of Michigan at Ann Arbor, USA, provide an exceptional account of a novel imaging technique using optical microscopy based on dynamic light scattering in the tenth chapter—*Dynamic Light Scattering Microscopy.* It has an all-embracing account on the theory of DLSM followed by a description of the detection system that uses a slow scan CCD to record the rapid kHz range

Nanotechnologies for the Life Sciences Vol. 3
Nanosystem Characterization Tools in the Life Sciences. Edited by Challa S. S. R. Kumar

ISBN: 3-527-31383-4

fluctuations of DLS, and finally followed by tests on a model system of small polystyrene beads in suspension and a living cell system of macrophages.

While the imaging techniques described in the first two chapters provide direct tools for understanding biological systems, the third chapter is an insight into use of an indirect tool, quartz crystal microbalance (QCM), which is a widely-used, popular and effective tool for evaluation of the assembly of biological structures. David E. Cliffel's group from Vanderbilt University in Nashville, USA, has done a remarkable job in capturing their experience in utilization of QCM techniques in life sciences. In this chapter, *Quartz Crystal Microbalance Characterization of Nanostructure Assemblies in Biosensing*, the authors cover important biological applications of QCM in addition to providing the principles and operational aspects.

Yet another indirect approach for understanding nanoscaled biologically important systems is the use of traditional spectroscopy tools. Nuclear Magnetic Resonance is playing a key role in the development and design of nanoscaled pharmaceutical carriers by providing comprehensive data on the structure and the function of these systems under a large variety of conditions. Christian Mayer from Gerhard Mercator University in Duisburg, Germany, provides an authoritative account of application of NMR tools for nanoscaled systems in the fourth chapter, entitled *NMR Characterization Techniques—Application to Nanoscaled Pharmaceutical Carriers*. The author's own words succinctly summarize the importance of this growing field: Unlike any other analytical technique, it combines a distinctly non-invasive character with the ability to analyze for a chemical composition as well as for local mobility of individual system components.

In the sixth chapter—*In Situ Characterization of Drug Nanoparticles by FTIR Spectroscopy*—Michael Türk and Ruth Signorell from the University of Göttingen, Germany, focus on the *in situ* characterization of nanoparticles by Fourier-transform infrared (FTIR) spectroscopy. *In situ* characterization tools such as these are a must for life scientists as they allow not only investigation, but also provide control of the properties of nanoparticles, especially drug nanoparticles, during their formation.

Electron Spin Resonance (ESR) or synonymously Electron Paramagnetic Resonance (EPR) can be regarded as a sister method for Nuclear Magnetic Resonance (NMR). This is yet another traditional spectroscopy tool that is being developed further to provide understanding of nanoscale drug delivery systems. The seventh chapter contributed by Karsten Mäder from Martin Luther University of Halle, Germany, discusses how ESR can contribute to shedding more light on nanoscaled drug delivery systems, providing examples that show how useful information can be extracted from the spectra and how this information can be used to characterize drug delivery systems. This chapter—*Characterization of Nanoscaled Drug Delivery Systems by Electron Spin Resonance*—is a must not only for practioners of ESR, but also for life scientists interested in nanoscale drug delivery systems.

Synchrotron radiation based spectroscopies have become an indispensable tool in many areas of biological sciences. Of these, soft x-ray absorption spectroscopy (XAS) and soft x-ray emission spectroscopy (XES) are gaining importance in the field of nanoscience. The eighth chapter, entitled *X-ray Absorption and Emission*

Spectroscopy in Nanoscience and Life Science, by Jinghua Guo from Lawrence Berkeley National Laboratory in California, USA, brings out how XAS and XES techniques are sensitive detection tools for nanostructured and molecular materials using a number of examples.

Each characterization tool provides unique information. However, a combination of several techniques is always required in order to obtain complete information about any nanoscaled biological systems. Two chapters, numbers five and nine, provide a broader perspective to the application of multiple characterization tools in nanosystem's characterization. The fifth chapter, *Characterization of Nanofeatures in Biopolymers using Small-angle X-ray Scattering, Electron Microscopy and Modeling*, was written by Angelika Krebs and co-workers from the European Molecular Biology Laboratory (EMBL) at Heidelberg, Germany. This chapter is an indication of how an appropriate combination of characterization tools will be used in future even though such an approach has not yet been well established.

The ninth chapter, *Some New Advances and Challenges in Biological and Biomedical Materials Characterization*, by F. Braet and co-workers from the Australia Key Centre for Microscopy and Microanalysis at Sydney University, Australia, describe recent advances as well as challenges in the microscopy of selected biological and biomedical materials using three very important tools: atom probe tomography, atomic force microscopy, and cryo-transmission electron microscopy.

In the final chapter of the book, *X-ray Characterization Tools for Nanosystems in Life Science*, Cheng K. Saw from Lawrence Livermore National Laboratory in California, USA, provides the basics of using X-ray diffraction techniques to obtain information on the structure and morphology of nanosystems in general.

I do hope that this book will be a useful source of information both for life scientists interested in nanoscale systems and for characterization specialists interested in applying their tools in biological systems.

October 2005, Baton Rouge *Challa S. S. R. Kumar*

List of Contributors

Daniel Axelrod
The University of Michigan
275 West Hall
Randall Laboratory
500 E. University Ave.
Ann Arbor, MI 48109-1120
USA

Bettina Böttcher
European Molecular Biology Laboratory
Structural and Computational Biology
Programme
Meyerhofstrasse 1
69117 Heidelberg
Germany

Filip Braet
The University of Sydney
Nanostructural Analysis Network Organisation
Major National Research Facility, NANO-MNRF
Electron Microscope Unit, Madsen Building, F09
Sydney, NSW, 2006
Australia

Sebastien Cecillon
Laboratoire d'Assemblages Moléculaires
d'Intérêt Biologique
Institut de Biologie et Chimie des Protéines –
CNRS/UCBL UMR 5086
7, passage du Vercors
F69367 Lyon Cedex 07
France

David E. Cliffel
Vanderbilt University
Department of Chemistry
Station B 351822
Nashville, TN 37235
USA

Anthony W. Coleman
Laboratoire d'Assemblages Moléculaires
d'Intérêt Biologique
Institut de Biologie et Chimie des Protéines –
CNRS/UCBL UMR 5086
7, passage du Vercors
F69367 Lyon Cedex 07
France

Rhonda Dzakpasu
The University of Michigan
Department of Physics
Biophysics Research Division
275 West Hall
Randall Laboratory
500 E. University Ave.
Ann Arbor, MI 48109-1120
USA

Aren E. Gerdon
Vanderbilt University
Department of Chemistry
Station B 351822
Nashville, TN 37235
USA

Daniele Gerion
Lawrence Livermore National Laboratory
Physics and Advanced Technology (PAT)
Directorate and Chemistry and Materials
Sciences (CMS) Directorate
7000 East Avenue
Livermore, CA 94551
USA

Jinghua Guo
Advanced Light Source Division
Lawrence Berkeley National Laboratory
One Cyclotron Road, MS 7R0222
Berkeley, CA 94720-8226
USA

Thomas F. Kelly
Imago Scientific Instruments Corporation
6300 Enterprise Lane
Madison, WI 53719
USA

Angelika Krebs
European Molecular Biology Laboratory
Structural and Computational Biology
Programme
Meyerhofstrasse 1
69117 Heidelberg
Germany

David J. Larson
Imago Scientific Instruments Corporation
6300 Enterprise Lane
Madison, WI 53719
USA

Adina N. Lazar
Laboratoire d'Assemblages Moléculaires
d'Intérêt Biologique
Institut de Biologie et Chimie des Protéines –
CNRS/UCBL UMR 5086
7, passage du Vercors
F69367 Lyon Cedex 07
France

Karsten Mäder
Institute of Pharmaceutics & Biopharmacy
Martin Luther University of Halle
Wolfgang-Langenbeck-Str. 4
06120 Halle a. d. Saale
Germany

Christian Mayer
Fachbereich Chemie
University of Duisburg-Essen
Lotharstr. 1
47057 Duisburg
Germany

Simon P. Ringer
The University of Sydney
Australian Key Centre for Microscopy &
Microanalysis
Nanostructural Analysis Network Organisation
Major National Research Facility, NANO-
MNRF
Electron Microscope Unit, Madsen Building,
F09
Sydney, NSW, 2006
Australia

Cecile F. Rousseau
Laboratoire d'Assemblages Moléculaires
d'Intérêt Biologique
Institut de Biologie et Chimie des Protéines –
CNRS/UCBL UMR 5086
7, passage du Vercors
F69367 Lyon Cedex 07
France

Cheng K. Saw
Materials Science & Technology Division
Chemistry & Materials Science Directorate
Lawrence Livermore National Laboratory
P.O. Box 808, L350
Livermore, CA 94551
USA

Patrick Shahgaldian
BCP
UMR 5086 CNRS-UCBL
7, passage du Vercors
69367 Lyon cedex 07
France

Ruth Signorell
University of British Columbia
Department of Chemistry
2036 Main Mall
Vancouver, B. C. V6T 1Z1
Canada

Lilian Soon
The University of Sydney
Nanostructural Analysis Network Organisation
Major National Research Facility, NANO-
MNRF
Electron Microscope Unit, Madsen Blg, F09
Sydney, NSW, 2006
Australia

Michael Türk
University of Karlsruhe
Institut für Technische Thermodynamik
und Kältetchnik
Engler–Bunte–Ring 21
76131 Karlsruhe
Germany

David W. Wright
Vanderbilt University
Department of Chemistry
Station B 351822
Nashville, TN 37235
USA

1
Fluorescence Imaging in Biology using Nanoprobes

Daniele Gerion

1.1
Introduction and Outlook

1.1.1
A New Era in Cell Biology

Fluorescence is ubiquitous in biology. Indeed, any biology textbook contains a multitude of colorful images, most of which would not disfigure an art gallery [1]. For decades, biologists have mastered the use of fluorescently labeled molecules to stain cells in cultures or tissues. They developed the tools to target different compartments inside the cells, such as the nucleus and mitochondria, and subcompartment structures, such as chromosomes and the telomeres using light-emitting organic markers. These traditional approaches permitted the unraveling of a wealth of information on organs and tissues and, to a smaller scale, on the cell organization and its functioning. Images of cellular division captured by fluorescence microscopy are breathtaking, particularly the movement of the duplicated chromosomes along the spindle apparatus towards the two poles of the parent cell. Similar examples abound in the scientific literature and in fluorescent dye catalogs to the point that biology may convey a false sense of completeness. Indeed, an engineer or a physicist like myself may have the impression that everything interesting in biology has been already discovered and that only a few blanks remain to be filled.

Yet, in recent years, biology has witnessed an extraordinary revolution. Modern biology looks beyond responses to stimuli or the morphological description of structures. Modern biology does not satisfy itself with the successful sequencing of the human genome because a list of four repeating letters does not reveal a biological function. Modern biology is much more ambitious. It seeks to understand how biological and chemical processes work together to make cells and living organisms [2]. The journey towards this "Holy Grail" of biology depends on our ability to decipher interconnected cellular networks, mainly by observing molecular pathways of proteins and other metabolites in living organisms. The ultimate goal is to know where and when proteins and metabolites are expressed, how and at

Nanotechnologies for the Life Sciences Vol. 3
Nanosystem Characterization Tools in the Life Sciences. Edited by Challa S. S. R. Kumar

ISBN: 3-527-31383-4

which level they are distributed in time and space and how are signals generated and transmitted to maintain cell vitality [3].

1.1.2
Nanotechnology and its Perspectives for Fluorescence Imaging in Cell Biology

It is instructive to describe a problem of primary interest for biologists that illustrates the importance of knowing the entire pathway of a biological process and exemplifies how nanotechnology may enter the game. The example concerns double-stranded silencing RNA (siRNA [4–6]), a potent agent able to temporarily knockdown selected genes in mysterious fashion. Elucidation of this process is of considerable importance because it may reveal new routes that regulate the expression of genes and lead to the development of drugs with the ability to selectively keep the "good" genes on and turn the "bad" ones off. To study the mechanism by which siRNA works, a simple strategy would be to follow its path when it travels through the cell. This can be done by labeling it with a fluorescent chromophore and following its path with a fluorescence microscope. The double strand nature of the duplex raises the possibility that siRNA opens up at some point and that only one of the strands becomes the active agent that knocks down the gene expression. Where and when does it happen? A partial answer can be gained by labeling each strand of the siRNA duplex with a different fluorophore and following each of them separately. However, the colocalization of both strands with a resolution of a few nanometers far exceeds the diffraction limit of conventional microscopes, and therefore the very idea of observing the siRNA denaturation needs a different strategy. An elegant approach would be to use fluorescence resonance energy transfer or FRET [7] between two fluorophores, each located on a different strand of the siRNA duplex. When the duplex is intact, the fluorescence of the donor is resonantly transferred to the acceptor. Therefore, the donor fluoresce is "low" and the acceptor emission is "high". The opening up of the molecule manifests itself by the switching off of the FRET mechanism with the consequence that the donor fluorescence becomes "high" while the acceptor fluorescence goes to "low". Such a plan to study siRNA cannot rely on organic dyes. They bleach too rapidly under continuous illumination, especially at the single molecule level where the time window to observe them is limited to a few seconds. In addition, the photobleaching is generally associated with the formation of reactive oxygen radicals, which are a major cause of DNA damage and may result in the poisoning and death of the cell [8].

Enter nanotechnology. Over the past ten years, chemists have learned how to make fluorescent colloidal nanocrystals [9], also called quantum dots, and how to solubilize them in water, and to functionalize them with biomolecules [10, 11]. What makes these quantum dots interesting is their remarkable optical properties [12] and the fact that live cells seem to tolerate them quite well [13–15]. Quantum dots could represent an ideal solution to the shortcomings of organic dyes for labeling siRNA in live cells. The fate of siRNA is only one of a handful of problems where quantum dots can play a decisive role in biology or medicine. Other examples are described below. In fact, nanotechnology could represent a platform that would feed biologist's requests for more sophisticated tools to see smaller and smaller units with faster and better resolution. Whether chemists' dreams of using quantum dots to label live cells and solve important problems will come true is an open issue; but as the number of scientists working in nanoscience grows, new horizons open and the role of nanotechnology in biology is likely to expand in directions not yet been foreseen.

This chapter presents one area in which nanotechnology is rapidly establishing itself, namely fluorescence imaging. It seeks to describe different imaging techniques, and how new types of fluorescent probes, i.e., quantum dots, can be used in the context of biolabeling. In Section 1.2, I review some fundamentals of fluorescence by summarily describing the properties of several types of fluorescent probes. A brief section describes a few ways to excite the fluorescent probes and some principles of detection that will illustrate the potential of qdots for bioimaging. Section 1.3 considers how sources and detectors can be assembled to allow the observation of fluorescent signals in live organisms. Then I will take a brief tour and describe different strategies to make images, insisting on the advantages that nanoparticles may offer (Section 1.4). Section 1.5 then presents some example applications where quantum dots may indeed be the probes that biologists have been looking for.

1.2 Fundamentals of Fluorescence

1.2.1 Basic Principles

Fluorescence is the result of molecules emitting photons after an excitation. The fluorescence process is generally illustrated by the Jablonski diagram [16] (Fig. 1.1), a schematic based on the energy levels of the molecules. The molecule can be excited from a ground electronic state to a higher energy electronic state by the absorption of a photon whose energy is in resonance with the transition energy. Very rapidly, internal conversion takes place. During this process the molecule dissipates some excess energy into the vibrational states and falls to the bottom of the electronic excited state. There, the molecule still needs to dissipate some of its excess of energy, which is too large to go only through the vibrational modes. The return to the ground state can use different channels, which are described in details in the caption of Fig. 1.1. One possible relaxation channel proceeds through the emission of a photon (black arrows K_R) and constitutes what is called "fluorescence". The emission energy is always lower than the excitation energy, which translates into the fact that the wavelength of the emission is always red-shifted compared with that of the excitation. This difference in wavelength is called the Stokes shift. The Stokes shift depends on the nature of the molecules and on their environment (i.e., the solvent), but it is in the order of 10–20 nm. A large Stokes

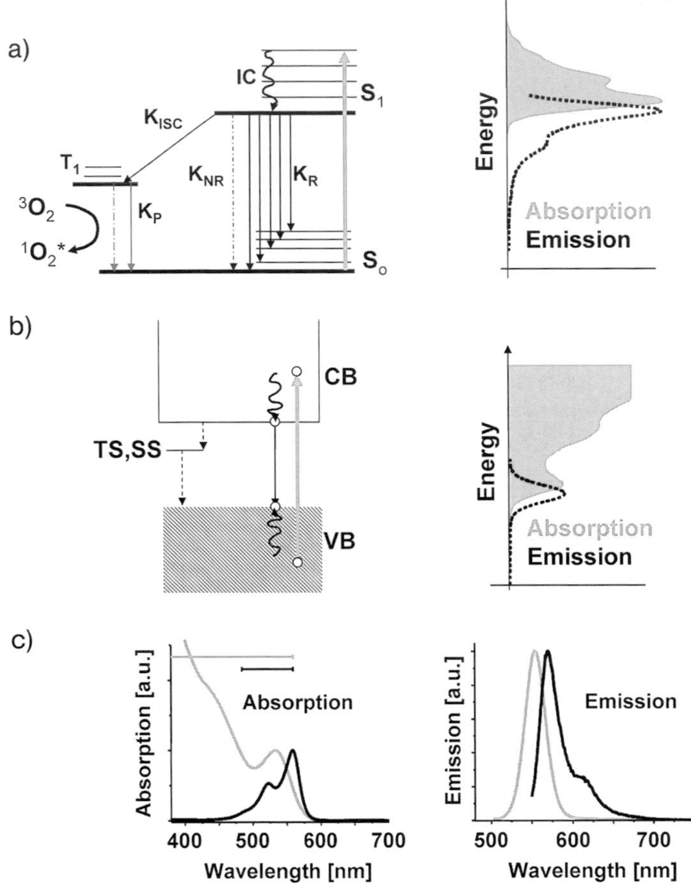

Fig. 1.1. Optical properties of organic dyes and qdots. (A) Jablonski diagram for organic dyes. Upon absorption of a photon of appropriate energy (light gray arrow), a molecule in a vibrational sublevel of the electronic ground state (S_0) is promoted to a vibrational sublevel of the first excited electronic state (S_1). This excess energy supplied by the photon is relaxed partially into vibrational modes by internal conversion (IC). IC is a radiation-less process that brings the molecule to the lowest S_1 sublevel within picoseconds. The remaining excess energy is too large to be dissipated only into vibrational modes. Thus, the dye emits one photon to reach one sublevel of S_0 (black arrows, K_R). Emission takes place a few nanoseconds after excitation. This is the fluorescence mechanism of an organic dye. The energy can also be dissipated through collisions with other molecules in solution and, in this case, there is no need to emit a photon (black dotted arrow, K_{NR}). A third route is possible in organic molecules: intersystem crossing to the first triplet state (T_1, K_{ISC}). The electronic transition from T_1 to S_0 is spin-forbidden and, therefore, requires the assistance of phonons or external collisions. Consequently, relaxation to the ground state is in the microsecond to the millisecond range. It can take place through emission of a photon (phosphorescence) or through a nonradiative process. Note also that oxygen reacts readily with the triplet state (T_1) of the dye, producing quenching of the emission. In this reaction, the ground state oxygen molecule, $^3O_2$ (a triplet), is excited into a singlet state, $^1O_2^*$. This singlet state can initiate chemical reactions that lead to bleaching of the dye and its phototoxicity. The right-hand panel shows the absorption band of an organic dye (Cy3B) as a

shift is desirable for imaging because it facilitates separation between the excitation light and the emission (Section 1.3).

The precedent emission properties are shared by all organic dyes because their electronic structure is composed of discrete electronic states with vibrational envelopes. Nanotechnology has brought along new types of fluorescent systems that differ markedly from dyes. These systems are nanometer-size crystals made of semiconductor materials [12, 17]. In bulk semiconductors, the electronic states merge into bands, a valence band and a conduction band, separated by a bandgap [18]. In the ground state, the valence band is fully occupied and the conduction band is empty. Any photon of energy superior to the bandgap can excite the semiconductor by forming an exciton, i.e., by bringing one electron in the conduction band and leaving a hole in the valence band. Both electrons and holes release part of their energy by relaxing to the bottom of the conduction band and the top of the valence band, respectively. Once there, one possible energy relaxation mechanism is a recombination process where the electron recombines with the hole and the excess of energy is released as a photon. The photon energy corresponds to the value of the bandgap. In confined systems of nanometer size dimension, the same overall picture applies except that the bandgap varies with the diameter of the nanoparticles: the smaller the particle, the larger the bandgap [17, 19]. Therefore, the fluorescence of a quantum dot is size tunable. Similarly to the fluorescence process in organic molecules, the energy of the emitted photon is lower

function of photon energy along with the associated fluorescence emission. Note the limited extension of the absorption band, the redshift of the emission maximum compared to the maximum of absorption and the tail in the fluorescence emission extending towards low energies. The tail represents the contribution of photons relaxing into vibrational levels of S_0.

(B) A diagram similar to the Jablonski plot can be drawn for quantum dots. In a qdot, the electronic energy levels merge into two bands called valence and conduction bands. They are separated by an energy gap. In the ground state, the valence band is full and the conduction band is empty. Upon absorption of a photon with energy larger than the bandgap, an electron is promoted to the conduction band, leaving a hole in the valence band. Both electrons and holes can easily relieve energy by relaxing to the bottom of the conduction band or the top of the valence band, a process analogous to internal conversion in organic dyes (although involving electronic states of qdots and not vibrations). At this stage, there are no more electronic states available and the excess energy would be too large to be dissipated through vibrations of the qdot. Therefore, qdots emit a photon, whose energy corresponds to the bandgap. As in organic dyes, qdots also undergo nonradiative processes. For instance, a nonradiative channel is induced by surface/trap states (defect at the surface of the dots) that create discrete energy levels within the bandgap. The right-hand panel shows the absorption spectrum of qdots along with its emission spectra. Unlike dyes, the absorption is a continuous band not limited to a specific energy range. The fluorescence emission is also redshifted compared with the absorption band. Remarkably, the position of the emission is independent of the excitation energy, although its amplitude depends on it.

(C) Absorption spectra of a qdots solution (light gray) and the organic dye Cy3B (in black). Horizontal lines at the top of the figure represent the range of wavelengths that can effectively excite the chromophores. In the emission spectra, notice the symmetry of the qdot emission and the tail in that of dyes at higher wavelengths.

than that of the absorbed photon and, therefore, the fluorescence light has a longer wavelength than the excitation.

A marked difference between a quantum dot and an organic dye is in their absorption spectra. For a dye, the absorption is centered around bands, conveying the discrete nature of the electronic states. In quantum dots, the absorption is a continuum from the band-edge up, mapping the continuous nature of the conduction band of the material. A continuous excitation spectrum represents clear advantages for imaging purposes. First, a main issue concerns the filtering out of the excitation source. When the excitation and emission are close in wavelength, as with organic dyes, the excitation light can be filtered out but a fair amount of the emission is also blocked. The possibility to excite quantum dots far from their emission greatly improves fluorescent detection. Second, it becomes possible to excite all sizes of quantum dots with just one single wavelength and therefore obtained all colors of emission with one excitation. Similarly, the emission of organic dyes exhibits a red tail resulting from transitions from the bottom of the excited state into the various vibrational levels of the ground state (black arrows K_R in Fig. 1.1). In contrast, quantum dots have symmetric and narrower emission patterns, because the energies of the radiated photons are determined solely by the size of the bandgap. Those properties represent distinct advantages for multicolor labeling, as the rest of this chapter will show.

1.2.2
A Few Types of Fluorescent Probes

Now that the mechanism of fluorescence has been described in general terms, it is worth looking in more detail into the nature of fluorescent molecules encountered in biology.

1.2.2.1 Small Luminescent Units and Autofluorescence of Living Organisms

Fluorescence in organic molecules comes from delocalized π orbital states of conjugated double bonds [16, 20]. Thus, most aromatic molecules exhibit a natural emission. This is the case of tryptophan, an amino acid, NADH (nicotinamide adenine dinucleotide), an important coenzymes found in cells used extensively in glycolysis and in the citric acid cycle of cellular respiration, and other coenzymes such as flavins and some forms of pyridoxyl. Because these fluorophores are natural, cells and tissues fluoresce when excited with a UV source; this is the source of autofluorescence. The dominant autofluorescence from biological samples usually occurs below 500 nm.

Autofluorescence represents a fair amount of light. Unfortunately, these natural fluorophores do not possess properties that are useful for biolabeling. Most proteins contain a few tryptophan amino acids and therefore all proteins emit to a certain extent. Because autofluorescence may easily overshadow other unnatural fluorescent labels, it must be removed by an appropriate combination of filters, or by the implementation of detection techniques that selectively eliminate autofluorescence (Section 1.4.3).

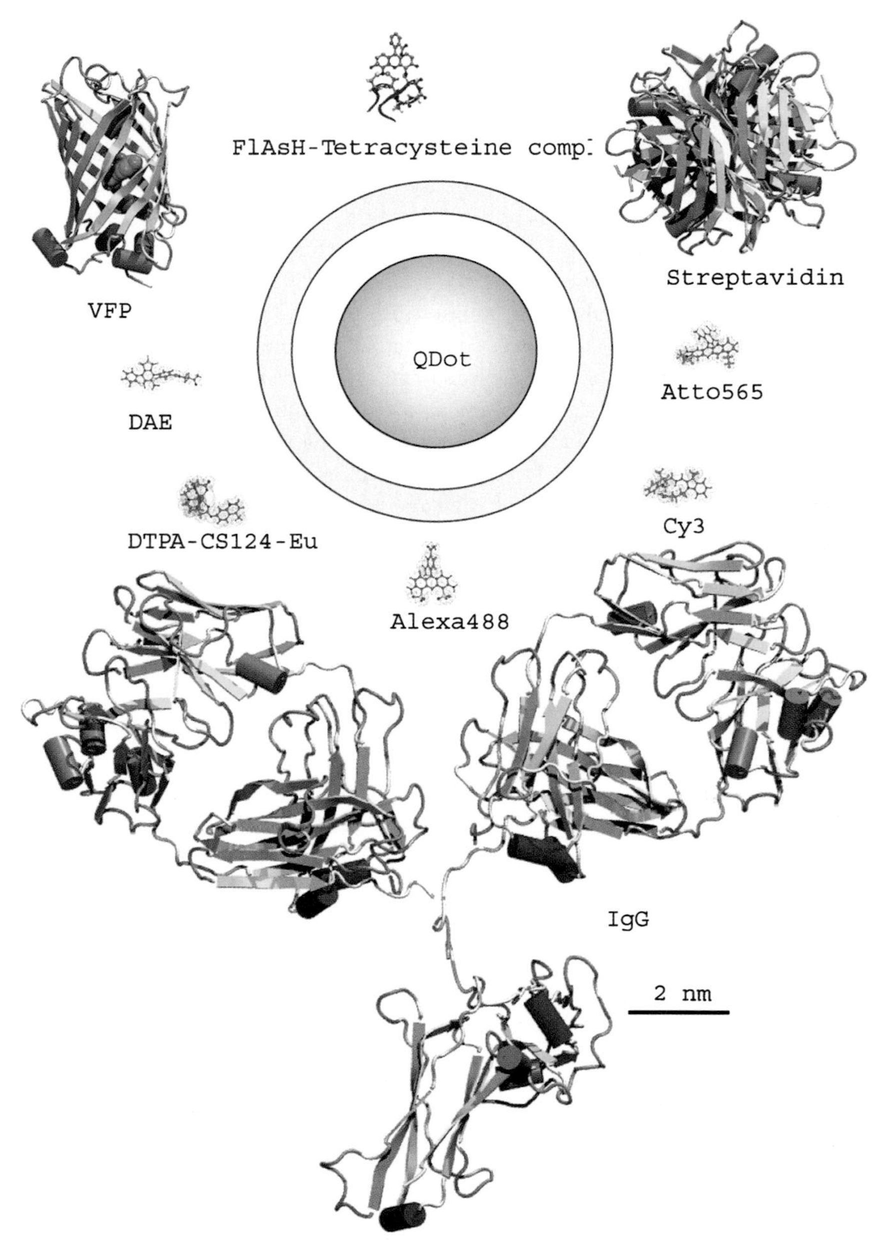

Fig. 1.2. Structure of commonly used organic dyes, fluorescent proteins, qdots and biomolecules, and the comparative sizes of common fluorophores, fluorescent proteins, qdots, some protein carriers and antibodies. The scale bar applies to all molecules. Organic dyes (FLAsh tetracysteine complex, DAE, DTPA-CS124-Eu, alexa, Cy3 and atto) are organic conjugated structures 1–2 nm in size. Qdots passivated to become water-soluble are ~10 nm – comparable to proteins such as streptavidin and VFP (violet fluorescent proteins) – but are smaller than a typical IgG antibody. (Copyright 2004 Nature Publishing Group. Reproduced with permission from Ref. [7].)

1.2.2.2 A few Organic Dyes and their Limitation in Live Cell Labeling

A plethora of unnatural organic dyes cover the whole range of desirable emissions and have unique optical signatures. Fluorescein, cyanine, rhodamine, alexa or atto are among the most popular. Figure 1.2 shows the structure of organic dyes and compares their size with other fluorescent probes and biological units. Organic dyes consist of aromatic conjugated structures of 2 to 10 units providing different degrees of electron delocalization, and hence colors of emission. The emission of an organic dye has a full-width at half-maximum (fwhm) >50 nm and exhibits a red tail (Fig. 1.1C). The quantum yield, the ratio of the number of photons emitted to the number of photons absorbed, can be close to one, but it is quite dependent on the local environment of the dye. Alteration of the media pH, buffer components, solvent polarity, or dissolved oxygen can affect or even quench an organic probe. For instance, in an environment below pH 7, fluorescein is significantly quenched. Similarly, the solubility of the dyes depends on the solvent and not all dyes are soluble in aqueous environment. The fluorescence decay time or lifetime of most organic fluorophores falls in a narrow range of 1 to 5–10 ns, a time window shared by autofluorescence lifetime. This limits the use of time-gated detection methods to distinguish between the fluorescence of an unnatural dye and the autofluorescence (Section 1.4.3). These properties represent an inconvenience for the use of organic dyes as biolabels. But, in fairness, the dye industry has managed to downplay these issues by synthesizing alternative or new types of probes that minimize these effects.

A more serious shortcoming that has not been solved yet is photostability. Dyes bleach after the emission of 10^6–10^8 photons [21]. The exact physical bases for bleaching are not entirely clear in all cases. The dye is not physically destroyed, but rather it is altered so that it can no longer emit. One cause of photobleaching is the oxidative reaction between atmospheric oxygen and the triplet state of the dye (Fig. 1.1 and its caption). Photobleaching of the dye produces radical oxygen species (ROS) that may poison and kill the cell in a short while [8]. One way to markedly reduce photobleaching consists in using antioxidants, an oxygen-free atmosphere or oxygen scavengers (such as the enzyme glucose oxidase along with glucose and catalase). Such approaches work well for studying conformation change of biomolecules, catalytic activity of ribosomes adsorbed on surfaces or fixed (dead) cells [22]. However, they are inadequate for the study of live organisms that need oxygen to survive.

1.2.2.3 Green Fluorescent Protein and its Cousin Mutants

Green fluorescent protein (GFP) is a natural fluorescent protein produced by the jellyfish *Aequorea* victoria [23]. It has 238 amino acids and a molecular weight of 27 or 30 kDa. The gene responsible for this protein was identified in 1961 [24] and cloned in 1992 [23]. GFP has a barrel-shape enclosing a short polypeptide fluorescent unit (Fig. 1.2). The nature of the polypeptide determines the color of emission. Biologists have selectively modified this short amino acid sequence and have been able to clone mutants that emit at different wavelengths. The proteins are called YFP (yellow fluorescent protein), BFP (blue), RFP (red), or CFP (cyan).

All exhibit, however, a fairly large emission spectrum with a fwhm larger than 50 nm and the distinctive red tail common to all organic dyes. Fluorescent proteins are relatively stable in the very diverse environments of the cells, although their fluorescence properties may be affected. GFP for instance remains fully fluorescent at pH above 6, below which the fluorescence decreases dramatically.

By applying the powerful methods of DNA recombinant technology it is possible to link together a specific gene of interest, for instance a therapeutic gene, with the GFP gene, and to transfect a mammalian cell or a whole organism with this engineered DNA. The cell machinery can translate this DNA into the protein of interest tagged with GFP. The protein complex is fluorescent but, amazingly, the tagged protein maintains the same function as the natural protein. Biologists have, therefore, a fluorescent complex whose location in space and time can be monitored. This is a marked improvement over traditional antibody labeling techniques that work on fixed (dead) cells. The ability to fuse the GFP gene to any protein brings biologists very close to what we called the promised land in the introduction.

Yet GFP and its cousin mutants are still cumbersome to construct. GFP fluorescence is the final product of a long, complex pathway involving transcription, translation, and post-translational modifications. Moreover, GFP and its mutants tend to form dimeric structures, and there have been questions on the feasibility of using GFP in certain conditions, for instance for gene expression studies under tumor conditions [25]. As a result, although GFP and its cousins allowed a huge step forward for molecular imaging, there is still room and need for alternative probes.

1.2.2.4 **Quantum Dots**

Quantum dots, often abbreviated qdots, are nanometer-size crystals made of semiconductor materials [12, 17]. The most widely used and studied quantum dots consist of a core of cadmium selenide or CdSe [9]. Colloidal chemistry allows the synthesis of nanoparticles with a very narrow size distribution ($<10\%$) and in (milli)-gram quantities (Fig. 1.3a). The dimension of the core determines the bandgap and hence the color of emission of the nanoparticles. For instance, emission of CdSe quantum dots covers the visible spectrum from green (520 nm, ~2.3 nm dots) to red (650 nm, ~5 nm dots). It is possible to cover the near-IR and IR spectrum by synthesizing quantum dots made from other materials such as CdTe [26], InP [27], InAs [28], PdS [29], PbSe [30, 31] but most of these chemical syntheses do not allow the degree of control reached in the case of CdSe. In principle though, the emission of qdots can be coarse-tuned by the choice of the material and later fine-tuned by playing with the size of the core (Fig. 1.3b).

The emission of the qdots presents many advantages over that of organic dyes and GFP. For instance, the emission of a batch of CdSe qdots is symmetrical and its fwhm is generally around 30–35 nm, although experienced groups can reach an fwhm as low as 22–25 nm. Moreover, the position of the qdot emission does not depend on the excitation wavelength. If qdots are excited at 360, 400 or at 500 nm, the emission will occur at the same wavelength (but the intensity will vary). All together, these relatively narrow and symmetric emissions that can be excited by a single excitation wavelength are ideal candidates for multiplexing detection

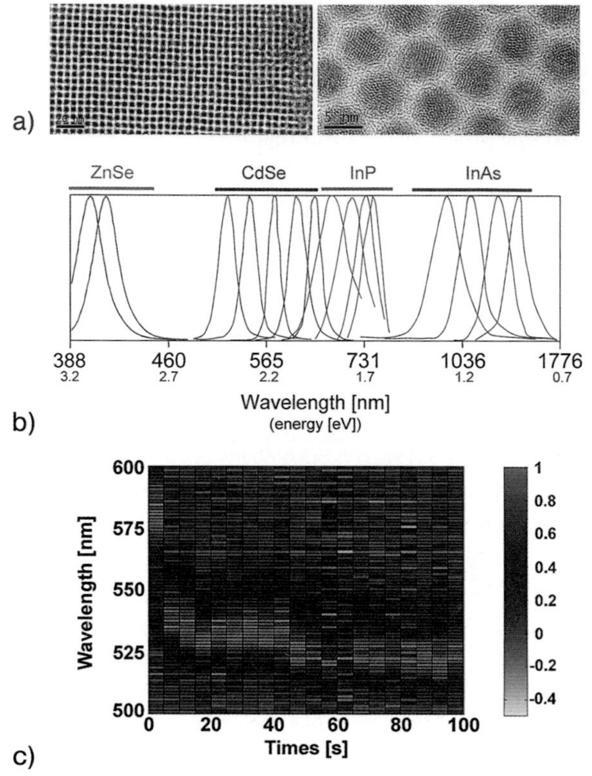

Fig. 1.3. Overview of CdSe qdots and their optical properties. (A) Low-resolution transmission electron microscope (TEM) image of CdSe qdots, showing an ordered lattice of CdSe nanocrystals, called a superlattice. The superlattice extends over a few microns. Scale bar: 20 nm. Right-hand side: high-resolution TEM view of the superlattice, revealing the internal structure of the CdSe qdots. Scale bar: 5 nm. Note the different alignment of qdots and the homogeneous particle size. (Image courtesy of Dr Natalia Zaitseva, Lawrence Livermore National Laboratory.) (B) Fluorescence emission of qdots spans from the UV to the IR. Different materials emit in different bands. For a given material, the emission can be fine-tuned by the size of the qdots. For instance, with CdSe, the maximum emission is at 530, 550, 590, 620 and 640 nm for qdots of 2.1, 2.4, 3.1, 3.6, 4.6 nm respectively. (C) Comparison of the photostability of an organic dye (Cy3B) and a silica-coated CdSe/ZnS qdot in phosphate buffer. The qdot–dye conjugate was synthesized as a model system to study FRET. The conjugate was illuminated continuously with a picosecond laser diode (400 nm) and its emission spectrum collected every 5 s with a CCD camera. The dye emits at 579 nm and the qdot emits at ~530 nm. In the first 5 s of illumination, the qdot is not seen because it transfers its emission energy to the dye through a FRET process (Section 1.4.4). However, after <5 s, the dye photobleaches, the FRET channel closes and the qdot shows up. The qdot emission remains strong throughout the measurement, with only a slight blue-shift. The long-lasting emission of single dots is the basis of long-term monitoring of molecular processes in live cells (Section 1.5.2).

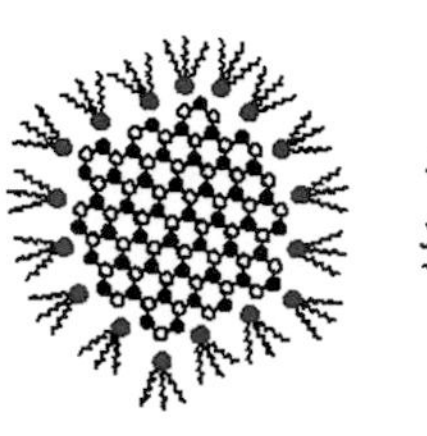

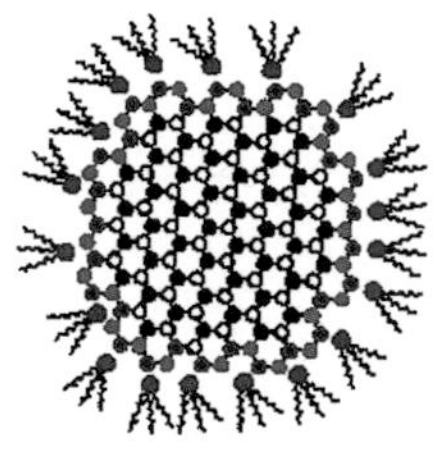

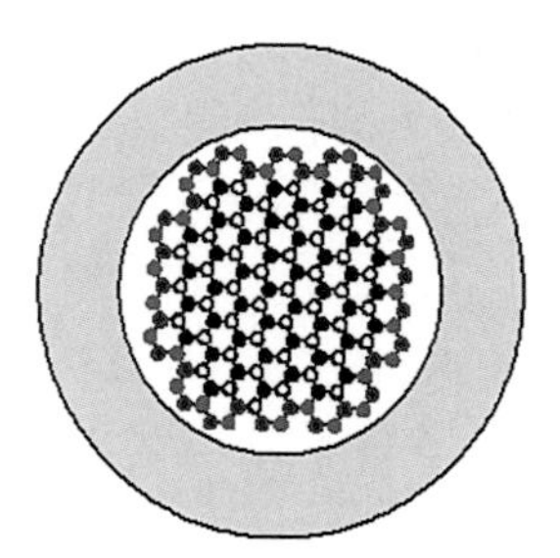

Fig. 1.4. CdSe, CdSe/ZnS and water-soluble qdots. Schematic of qdot cores, core/shell and water-soluble particles used in bioapplications (drawing not to scale). Left: a core of CdSe surrounded by TOPO (tri-octylphosphine oxide). The phosphine oxide head of TOPO (red ball) sticks to the Se atoms of the qdot. Middle: a thin shell of ZnS passivates the CdSe core. The shell increases the fluorescence emission of the core. The quantum yield of core/shell qdots is 30–50%. Since TOPO molecules are hydrophobic, core and core/shell are not soluble in water, where biology takes place. An additional shell makes the interface between the inorganic qdot and the organic/water outside (right-hand side). If the shell is made of glass, functional groups on the glass surface allow bioconjugation of the qdots to biomolecules. Note that, regardless of the solubilization method, the final size of water-soluble qdots is close to or above 10 nm, even if the active CdSe core is only 2–3 nm.

with minimal or no cross-talk between detection channels. In a proof-of principle experiment, five orthogonal channels have been simultaneously detected within the visible range using a single laser excitation [32], but under more realistic conditions, which use ligand–receptor interactions mediated by single-stranded DNA, only four qdot channels could be unambiguously sorted out [33].

In general, qdots grown by colloidal methods have surface trap states that reduce their emission to a few percents. However, these trap states can be removed by passivating the surface with a few layers of a second semiconductor material with a larger bandgap. For instance, a few monolayers of zinc sulfide (ZnS) can be grown on top of a CdSe core (Fig. 1.4) [34]. The core/shell CdSe/ZnS are about 20–50× brighter than CdSe cores and their quantum yield can reach 30–50%, although certain laboratories with more experience in the art of growing core/shell qdots can reach quantum yields above 80–90% by further doping the shell. Core/shell qdots do not bleach as fast as organic dyes. In fact, qdots in solutions can be excited continuously well over 4 h with an Ar^+ laser excitation and their emission remains fairly stable [35]. Eventually, over the long term, photobleaching occurs. It is easier to quantify the resistance to photobleaching at the single molecule level. With organic dyes exposed to air, photobleaching occurs in a matter of seconds, even for Cy3B, one of the most robust dyes (Fig. 1.3C). Under similar excitation conditions, qdots can emit for well over a few minutes and in some cases for over 20 min. The photostability of qdots is the major advantage that could allow the tracking of single molecules over extended periods (see Section 1.5.2).

Qdots are characterized by very high photoabsorption cross-sections. Chemists prefer to use the extinction coefficients, which measure the absorption of a 1 M solution along a 1 cm path. Extinction coefficients scale roughly as the volume of the qdots and range from $\sim 1.5 \times 10^5$ M^{-1} cm^{-1} for CdSe qdots of 2 nm up to over 5×10^5 M^{-1} cm^{-1} for qdots of 5 nm [36]. For comparison, extinction coefficients of organic dyes vary from less than 1×10^5 to 1.5×10^5. High extinction coefficients mean that a lower amount of light is needed to excite them, a particularly beneficiary thing for the study of delicate systems such as live cells. Similarly, two-photon cross sections of CdSe/ZnS qdots exceed by far the values of organic dyes. In fact, two-photon cross-sections of qdots are two to three orders of magnitude larger than those of organic dyes and one order of magnitude larger than those of molecules designed specifically for enhanced two-photon absorption [37]. These high values enable the imaging of organs and tissues at greater depth and lower power than with any other dyes. A further improvement may come from the use of IR-qdots whose emitted photons are more transparent to tissues (Section 1.5.3).

Other optical signatures that distinguish organic dyes and qdots are lifetime and polarization. Qdots have a lifetime of ~10–20 ns, which is significantly larger than the lifetime of organic dyes (1–5 ns) or the lifetime of autofluorescence [38]. This longer lifetime provides an efficient way to reduce the autofluorescence background by time-gating the detection (Section 1.4.3). Conversely, a negative point of qdots vs. dyes is the lack of strong polarization of the emitted photons. In fact, elongated qdots, also called nanorods, do emit photons polarized along the long axis of the rod [39]. But as the ratio between the long axis and the short axis of the rod approaches one, the polarization vanishes. Since chemical procedures to solubilize nanorods are not perfected yet, the present qdot technology does not allow orientational sensing. Further developments in the post-processing of nanorods will likely lift this limitation.

Although qdots have photophysical properties that surpass organic dyes in many aspects, they have one big shortcoming: quantum dots are not water-soluble. Qdots are generally synthesized in organic solvents and are passivated with hydrophobic molecules. It is possible to modify the surface of the qdots to render them hydrophilic. Many methods have been developed and there is probably no consensus on which one works best. The most promising solubilization methods for bioapplications consist in embedding the qdots into a shell or surfactant layer that cannot be further exchanged. The reason is that the shell/layer represents a protection that masks the true hydrophobic nature of the qdots. Examples of solubilization strategies include ligand exchange by thiol-containing molecules [40], oligomeric phosphines [41], dendrons [42], or two-domain peptides [43]. Other solubilizations proceed through encapsulation of the qdots by layers of phospholipids micelles [14], amphiphilic diblock [44], or triblock copolymers [45], polymer [46] or silica shells [35] and amphiphilic polysaccharides [47]. Regardless of the solubilization strategy, the difficulty to interface these inorganic objects with the bioworld is a major concern for biolabeling applications, and is likely to require constant improvement in forthcoming years.

As with organic dyes, water-soluble qdots can be conjugates to biomolecules using standard bioconjugation approaches [48]. Biologists have developed covalent conjugation schemes based on thiol, amine, and carboxylic group chemistries (the natural functional groups found in amino acids) to crosslink proteins together. It turns out that similar schemes can be used to link qdots to any biomolecule. Most biomolecules (DNA, oligonucleotides [14, 33, 49, 50], peptides [13, 43], antibodies [44], virus peptides [13], proteins [45]) have been linked to water-soluble qdots and, surprisingly, the binding has no effect either on the qdot optical signature or on the functionality of the biomolecules. For instance, a single strand of DNA of 24 bases long (~7 nm) linked to a qdot of ~10 nm is able to find its perfect complement within a pool of competing stands. Even a single base mismatch does not produce hybridization [51]. In addition, the DNA strand does not modify the emission of the qdot. Similarly, qdots linked to antibodies for surface receptor proteins indeed find their target on the cell membranes without losing their optical properties [43]. It is remarkable that a 10 nm qdot does not perturb the function of a biomolecule and conversely that a biomolecule does not modify the emission of the qdot, although it does change the local dielectric constant around the dot. It becomes therefore possible to envision studying biomolecules in their native cellular environment (Section 1.5.2) [2].

1.2.2.5 **Toxicity Issues of Nanomaterials**

So far, we have described the benefits of using qdots for *in vivo* labeling applications. In fact, a major issue for biologists is the toxicity of the qdots for live organisms. Because qdots are made of toxic chemicals, they have to be toxic, or so the rationale goes. This has sparked a debate that encompasses the general theme of the environmental impact of nanotechnology [52, 53]. An extreme position calls for a ban on nanoscience until its environmental impact is accessed. A wiser position may consist in taking seriously the potential toxicity of these new nanoprobes and studying in parallel benefits and shortfalls of this technology. There are no benefits in banning nanoscience all together. However, there are strong merits in warning against the potential toxicity of the nanomaterials, while we await the results of long-term cytotoxicity or genotoxicity studies.

Besides its unknown long-term effects, short-term toxicity is an important factor for live cell labeling, because the markers must not interfere with the regular cell cycle. For instance, the presence of qdots in the cell cytoplasm should not impede the cellular division, nor should it switch off vital cell–cell communication channels. So far, the toxicity of qdots towards live cells has been studied only through simple experiments where the survival rate of cells transfected with different types of foreign material was quantified [15, 54, 55]. The major result in all those studies is that surface functionalization plays the key role in nanoparticle toxicity. For instance, an apparently inert molecule like C_{60}, with a well-defined surface and no available dangling bonds, is extremely harmful to cells even at low doses because C_{60} is an excellent electron acceptor that can readily react with available oxygen and water and generate free radicals [56]. These radicals eventually cause oxidative damage to the cellular membrane. C_{60} derivatized with hydroxyl groups showed much lower toxicity because derivatized fullerenes are less efficient in producing oxygen radicals. A similar situation exists for water-soluble CdSe/ZnS. These qdots tend to release Cd^{2+} into the cellular environment, due to photooxidation, which leads to the death of the cells by poisoning. If these qdots are solubilized by ligand exchange with a mercaptoacid [40] the particles present low solubility and a high toxicity in the cells, essentially because the bond between the qdot and the mercapto-surfactant is weak, dynamic and unable to prevent diffusion of dissolved Cd^{2+}. In contrast, qdots embedded in a cross-linked silica shell have greatly reduced toxicity [55]. However, even in such cases, the overall toxicity is a function of surface functionalization. Silica-coated qdots with negatively-charged phosphonate groups at the surface cause a higher rate of cell death than silica-coated qdots with neutral PEG groups. Such experiments tell us a lot about the compatibility of the probes to live organisms, but they lack the ability to discriminate further than the simple toxic/less toxic/non-toxic conclusions. A classification of the effects of nanomaterials on the expression of genes in live cells would help clarify the potential dangers of nanomaterials. Those experiments are likely underway.

1.2.3 Sources and Detectors

With a toolkit of bioconjugated qdots in hand, the doors to biology get closer. But before going there, let us tour the world of sources and detectors that could be used for imaging qdots. There are many ways to excite fluorescent probes and similarly there are many different ways of detecting them. The selection of a particular type of source and detector is dictated by the type of experiments and the information that one wants to access. Imaging time-resolved and dynamic events puts a stronger constraint on the equipment than the imaging of a static field of fluorescent probes. In the next section, we look briefly into the different types of light sources and detectors commonly used in fluorescence microscopy.

1.2.3.1 Light Sources

Light sources commonly used for fluorescent microscopy are arc lamps and lasers. Lasers are essentially used for confocal microscopy where an important aspect is to focus the excitation source (Section 1.3.2). For wide-field (Section 1.3.1) illumination, however, biologists tend to prefer arc lamps because of their lower price, relative flexibility (i.e., any wavelength may be selected – an important factor when using two different dyes) and the lack of interference fringes from reflections inside the optical system.

Arc Lamps These include mercury lamps or xenon lamps. Mercury lamps exhibit a compact arc, with some lines dominating the whole spectrum. They can reach very high flux densities, mainly concentrated in the wavelengths of the mercury lines. When a broader, more homogenous excitation spectrum is needed, xenon lamps are the sources of choice. They have a fairly flat spectrum with only a few lines in the blue and IR region. A main disadvantage of arc lamps is the need of optical filters to block the excitation light from the fluorescence light. Considering the broad excitation that often overlaps with the emission of the probes, complete removal of the excitation source also reduces the fluorescent signal that reaches the detector. This is a problem when one wants to detect a few labels inside a cell. In that case, every photon counts. Arc lamps also lack coherence properties. Because of their wide spectrum, different wavelengths going through an objective are focused at different locations in the sample. This is one source of chromatic aberrations in the excitation arm of the microscope. One consequence is that two different dyes, exited by two different wavelengths, may show up at the same location on the recorded image even though they are physically far apart. As a result, high precision colocalization of two fluorophores cannot be achieved with an arc lamp. A high level of localization requires coherent light sources.

Lasers Lasers are coherent, monochromatic light sources. While it becomes experimentally easier to get rid of the excitation light using interference filters, lasers present the inconvenience that not all organic dyes can be excited by a single wavelength, so a second excitation wavelength is needed; and the story continues. However, qdots have a broad absorption band and therefore a single wavelength can produce the emission of all qdot colors at once. For instance, with a simple Ar^+ laser emitting at 488 nm and a shortpass filter with cutoff below 500 nm, it is easy to observe the glowing of five different colors of CdSe qdots simultaneously and record it with a color consumer-grade CCD camera. A decided advantage of monochromatic excitation is that chromatic aberrations in the excitation path are suppressed altogether. This opens up the possibility of colocalizing multiple qdots with nanometer precision (Section 1.5.1). A disadvantage of a laser source over an arc lamp is the smaller area that can be illuminated. A Xe lamp can cover an area of a few cm^2. In contrast, a defocused laser beam can illuminate an area of a few tens of μm^2 only [33, 57].

1.2.3.2 Detectors

Fluorescence microscopy uses digital imaging due to the high sensitivity that such detectors can reach. The detection of fluorescence signals is often a trivial task, especially if the signals are strong. This is the case when the nucleus of a cell is stained with DAPI dyes or when the cell membrane is stained with fluorescent markers. In these cases, the fluorescence signal is easily observable even to naked eyes, and can be captured by low-tech techniques such as the traditional emulsion-based films. Most of today applications, however, look intensively to fluorescence signals coming from rare events where only a handful of probes emit. One example concerns the expression of proteins in live cells following an external stimulus. Often, these proteins are expressed at a low level (a few hundred to a thousand of copies per cells) and their localization is spread all around the cell cytoplasm. Detecting them is like finding a nail in a haystack. In addition, irreversible photobleaching imposes an additional constraint on the quality of the detector, because not only there are a few emitting probes, but the time window to detect them is limited. The detection of weak signals becomes therefore a central piece in biomedical applications. Fortunately, the field of optics has developed ultrasensitive detection devices for all sorts of single-molecule applications that encompass the field of materials science, chemistry, physics and biology. Broadly defined, there are two classes of detectors: point detectors and two-dimensional arrays. The next two sections discuss their properties.

Point Detectors These detectors include photomultiplier tubes (PMT) and photodiodes (PD). Both devices employ a photosensitive surface that captures the impinging photons and generates electronic charges that are amplified and sensed. Therefore, a common feature of these single-elements detectors is the lack of spatial resolution.

In a PMT, the impinging photon strikes a photocathode that releases secondary electrons that get amplified by a series of dynodes. PMT sensitivity depends on the composition of the photocathode [58]. Photocathodes made of gallium-arsenide-phosphide (GaAsP) are sensitive in the range of 250–650 nm and GaAs photocathodes are sensitive to 300–800 nm. InGaAs photocathodes extend the sensitivity from 900 to 1700 nm in the IR range, but they are less common or available in regular laboratories. The output signal from a dynode is proportional to the number of impinging photons and a gain of up to 10^6 can be obtained. Because PMTs do not store charge, and respond to changes in photon fluxes within a few nanoseconds, they can be use to monitor extremely fast events such as photon bursts. A limiting factor of PMTs is their low quantum efficiency (QE), i.e., the ability to generate a signal from the percentage of incident photons. It ranges from 20% for GaAs up to 40% for GaAsP photocathodes. In contrast, PMTs have a large detection area, on the order of 1×1 cm. Such a large area is a great experimental advantage because it does not require tedious alignment of the microscope on a daily basis.

Avalanches photodiodes (APDs) are an alternative to PMTs with greater efficiencies. They consist for instance in p-n junctions or heterojunctions operated at reverse biases close to avalanche breakdown voltage to enable the multiplication of photogenerated carriers [58]. APDs have generally higher QEs (up to 95% for silicon APDs) and a relatively flat response over the entire visible range (400–850 nm). APDs have been fabricated in many semiconductor materials. While Si APDs have a 90% QE up to ~850 nm, their efficiency is greatly diminished in the 1000–1600 nm range. For this wavelength range, APDs made of Ge or other III–V compounds are used. The temporal response of APDs is comparable to PMTs but the sensing surface area is much smaller (a few hundreds of microns in diameter). Although PMTs offer higher gain, APDs feature better quantum efficiency, lower noise, higher linearity, and compact packaging.

Two-dimensional Array Detectors Two-dimensional detectors consist of a dense array of photodiodes that collect simultaneously the light coming from an area several microns or millimeters in size. These detectors are used for imaging a wide field in a single shot. There are several variations on the same principle, but the most popular 2D detector used in fluorescence microscopy is the CCD (charge couple device). A photosensor called a pixel is coupled to a charge storage region that is connected to an amplifier. The amplifier reads out the quantity of accumulated charge.

The spectral sensitivity of CCD camera is usually lower than that of a simple photodiode, because the CCD surface has channels used for charge transfer that are shielded by polysilicon electrodes. These structures do absorb in the blue region. This reduces the quantum efficiency of the device down to ~40% in the case of a Si chip. A marked improvement consists in using a back-illuminated CCD. In this configuration, the light impinges on the back of the CCD on a surface that has been thin-etched to the point of being transparent (~10–15 μm). This results in quantum efficiencies of ~90% over the entire visible spectrum. In addition, when CCD cameras are cooled to liquid nitrogen temperature, the background noise can be reduced to a few counts per second, making it possible to detect single emitters spread on a surface in a heartbeat.

The limiting factor of CDD cameras, beside their cost, is the transfer rate, i.e., the speed at which the collected frames are read out and transferred to the computer. CCD cameras are often termed slow-scan cameras because their standard frame rate is usually less than that of a video camera. This is because the data readout is performed pixel row by pixel row. The fastest full resolution readout time reaches a few hundreds of milliseconds or about 10 frames per second in the case of frame-transfer cameras. As a result, CCD cameras are not appropriate to monitor time-dependent events occurring on a timescale of millisecond or below. For instance, the dynamics of proteins folding, the lifetime emission of fluorophores and time-correlated events are too fast for CCD detection, but can be caught quite easily with a PMT/APD.

1.3 Microscope Configurations

So far, we have discussed how to excite a fluorescent probe and how to detect its emission. Let us now look at how to assemble these devices into a custom-made microscope to acquire images. The next few sections discuss the advantages and disadvantages of each configuration (Fig. 1.5).

1.3.1 Wide-field Methods: Epi-, and Total Internal Reflection (TIR)

Most fluorescence detection techniques in biology use a wide-field illumination. As the name indicates, a laser or an arc lamp illuminates a large area of interest and the fluorescence from the whole field of view is collected.

1.3.1.1 Epifluorescence Illumination

By far the simplest microscope uses an epifluorescence configuration in which the excitation and the emission pass through the same objective (Fig. 1.5). The excitation light, usually a mercury or xenon source, or in rarer cases a defocused laser beam [33, 57], is directed through an objective by a dichroic mirror and illuminates the sample. The fluorescence emission is captured back through the same objective, and passes through an appropriate long pass filter that further removes any residual excitation light scattered back from the surface. The fluorescent light is then detected. The simplicity of such a setup has made it the most commonly used fluorescent imaging technique.

Epifluorescence microscopy has certain limitations. For instance, the samples are illuminated along the whole path of the excitation light and therefore probes

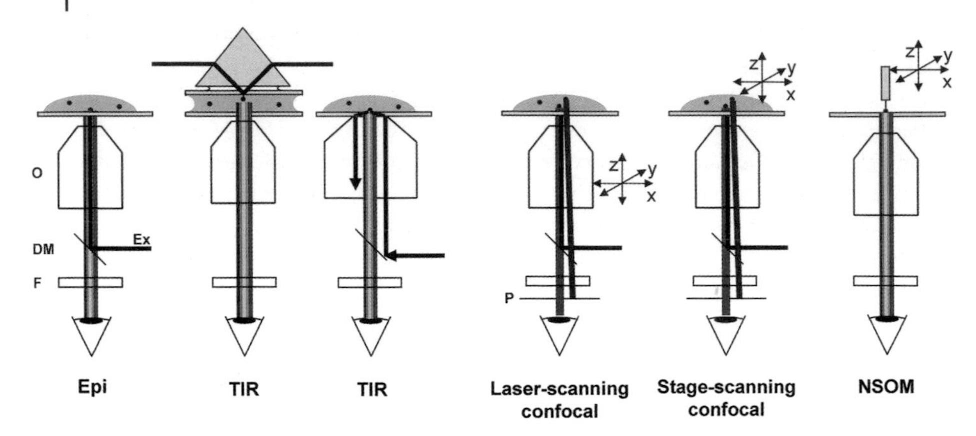

Fig. 1.5. A qdot sample can be illuminated in many ways. Excitation light is represented by the black line. The most popular illumination is *epi*fluorescence, where the excitation source (arc lamp or laser) and the fluorescence emission go through the same objective. A dichroic mirror (DM) is used to direct the excitation in the back of the objective. The back-scattered excitation light is further blocked by a bandpass filter (F) that is transparent to the fluorescence emission that is directed to the detector. Total internal reflection (TIR) uses an evanescent wave at the interface of two phases of different refractive indexes to excite the probes. It is, therefore, used to study structures close to interfaces. TIR illumination can use either a prism or a special objective configuration. In both cases, excitation and emission paths are disconnected. The detection arm of a TIR microscope is similar to that of an epifluorescence microscope. In a scanning *confocal* microscope, a focused light source, i.e., a laser, is scanned over the surface at a fixed height. Although all probes crossing the laser beam will emit, emissions from outside the focal plane of the objective are blocked by a pinhole (P). By varying the focal depth, sections of the samples can be isolated and imaged. In a stage-scanning confocal microscope, the stage is scanned while the objective is kept fixed. This configuration allows the removal of chromatic aberration in the excitation arm. Finally, in a near-field scanning optical microscope (NSOM), the tip of a fiber is brought in close to the sample and is scanned over it.

that are slightly out of the objective focal plane emit light that can reach the detector. This results in images that contain a fair amount of background signal. As a consequence, epifluorescence imaging lacks the sectioning ability and has a limited spatial resolution. It does not allow discrimination between a probe on the membrane of the cell or a probe engulfed in the cytoplasm.

1.3.1.2 Total Internal Reflection (TIR) Illumination

One strategy to decrease the background and increase the spatial resolution is to eliminate fluorescence from outside the focal plane. Various strategies are employed to restrict the excitation and detection of fluorophores to a thin region of the specimen. Among them, total internal reflection (TIR) is one way to reduce the background signal while maintaining the wide-field illumination. Confocal microscopy is a second option, which will be discussed in Section 1.3.2.1.

The principle of TIR is simple and illustrated in Fig. 1.5. A light beam, incident on an interface with different refractive indexes for the two phases, is totally reflected if the incident angle exceeds a critical angle. While the light is totally reflected, a portion of the radiation exists in the distal phase called the evanescent wave. The evanescent wave will continue to travel into the medium of higher refractive index, but its strength will decay exponentially. The TIR illumination system takes advantage of the evanescent wave to specifically illuminate only a range of 100–200 nm from the coverslip. The result is that only probes located within such distance from the surface are excited. This provides an optical sectioning capability of a few hundreds of nanometers, at least 5× thinner than any existing confocal microscope (Section 1.3.2).

In general, TIR illumination has potential benefits in any applications requiring imaging of minute structures or single molecules in specimens having large numbers of fluorophores located outside the optical plane of interest. Since the excitation light is completely reflected away from the detection, one can easily discriminate the fluorescence signal from the excitation light and achieve very high sensitivities and detection limits. This capability is particularly useful in many types of applications. For instance, TIR fluorescence is very helpful in monitoring the interaction between an intracellular protein and the substrate, especially in cases where the intracellular protein is of very high abundance in the cytoplasm, which will inevitably generate very high internal noise using a conventional epi-illumination. Another important application of TIR is in the characterization of forces exerted on the substrate during cell motility. For instance, the observation of points of contacts or focal adhesions between the cell and the substrate holds information on how the traction forces are transmitted [59]. Finally, by varying the illumination incidence angle, and consequently the penetration depth of the evanescent wave, fluorophores can be distinguished by depth on a nanometric scale. If one were able to vary rapidly the evanescent field depth, target vesicles or other structures can be tracked at different depths and their positions accurately determined.

The main limitation of TIR, however, resides in the small penetration depth of the evanescent field. Imaging chromophores inside cells becomes exponentially difficult and requires a different strategy.

1.3.2 Scanning Methods for Microscopy

1.3.2.1 Laser-scanning or Stage-scanning Confocal Microscopy

Traditionally, biologists have physically sliced through specimens to look at internal structures with a conventional light or electron microscope. The *laser-scanning* confocal microscope allows optical sectioning through a whole intact sample. Its principle is illustrated in Fig. 1.5. A computer-controlled laser beam is tightly focused and scanned over the surface at a fixed depth (in fact the objective is scanned). For each position of the light beam, the emission light is collected back through the same objective. By using a pinhole in the path to the detector, the fluorescence light coming from outside the objective focal plane is rejected, thus providing a high-resolution image at that depth. By progressively changing the plane of focus, one can section the entire specimen optically, producing a sharp image of the fluorescently marked components for many different depths. A confocal reconstruction is produced when all these layers are put together to provide a two-dimensional representation of the three-dimensional information. The slicing resolution of a confocal microscope is of the order of ~ 1 μm, i.e., about $5\times$ the z-resolution achievable by TIR illumination (Section 1.3.1.2). However, the ability to probe entire cells or organisms instead of only their contact point with the surface makes confocal microscopy the method of choice for many applications. One inconvenience of a confocal microscope is the time needed to acquire an image. Considering an integration time of a few tens of milliseconds per pixel, it takes a few minutes to capture a high resolution image.

In a laser-scanning confocal microscope, the spatial resolution is limited by the optics and by the size of the focused laser beam. In particular, diffraction limits its size so that a tightly focus beam of wavelength λ still has a spot with extension of roughly $\lambda/2$. Considering that excitation wavelengths are in the 400–500 nm range, the spot of a focused laser beam is at least 200 nm wide. Two point-like sources closer than this distance will be excited simultaneously by the focused beam, and one will not be able to distinguish them. There are two strategies to reduce the resolution limit of a confocal microscope. First, the excitation volume can to be narrowed. This can be achieved through interference methods [60] or by stimulated emission depletion [61]. Resolutions beyond the Rayleigh limit have been achieved, although these methods are cumbersome to implement. The second strategy is useful when one is interested in the relative distance between two point-like sources with different emissions rather than the exact location of each object separately [32]. In this case, one can acquire the diffraction-limited shape of each point-like source in separate channels, also called point spread function or PSF, and determine the position of the center of each PSF [62]. This requires a signal strong enough to accurately fit the PSFs. An additional critical point here is to make sure that both objects are excited identically. Usually, wavelength-dependent properties of the objective produce different z-focuses for different wavelengths. In addition, spherical aberrations and uneven flatness of the back of the objective will produce distortions in beams scanned off-axis. Finding the exact positions of the PSFs in these conditions requires an impractical characterization of the whole system. One way to circumvent these problems is to use a monochromatic excitation and move the stage instead of the laser beam. This constitutes the principle of the *stage-scanning confocal* microscope. By keeping the laser beam fixed with respect to the optical setup, chromatic aberrations in the excitation path are removed. With such a setup, colocalization of two chromophores with distinct emission spectra in the nanometer range has been reported [32, 62, 63].

1.3.2.2 **Near-field Scanning Optical Microscopy (NSOM)**

An alternative way to beat the Rayleigh diffraction limit is to use near-field methods [21, 64]. Near-field optics uses laser spots and tiny fibers with an aperture diameter smaller than the diffraction limit. If the light that leaks through the fiber tip is captured in close proximity of the aperture, the excitation spot size is on the order of the aperture size, i.e., it has a subwavelength dimension. Transmitted and fluorescence light are collected in the far-field through conventional optics and detectors. To make an image, the tip is scanned over the surface and an image is built pixel by pixel.

To obtain a high-resolution image of a specimen, the tip has to be positioned within a few nanometers of the sample surface all along the scanning process. This requires a feedback mechanism, usually based on shear force. Although NSOM has been successful in imaging biological specimens on surfaces, it still presents a tremendous challenge to image unfixed cells under buffered conditions. An additional limitation of NSOM comes from the delicate process of manufacturing the fiber tips. These are brittle and may break easily if in contact with rough surfaces. In addition, the near-field regime disappears quickly in the z-direction, preventing observations in the cell interior. Finally, the experimental challenges in NSOM often overcome by far its benefits over confocal microscopy.

1.4 Strategies for Image Acquisition

There are different strategies to collect information and form an image. Each photon encodes a wealth of information: it has a wavelength, a polarization, and an arrival time. Each of these properties can be used to create an image. It is also possible to use a combination of detectors to measure multiple properties of the photons simultaneously, i.e., the location of multiple chromophores, their spectra, their polarization and their lifetime, provided there are enough photons to generate a signal.

1.4.1 Intensity Imaging

The simplest method to acquire an image is to attribute to each pixel a value proportional to the intensity of light impinging on the detector. In its simplest form, such approach is color-blind since one green photon and one red photon are not discriminated by a PMT/APD or a CCD camera. However, with the appropriate use of bandpass filters, it is possible block all photons outside a spectral window and leave only a few selected photons to reach the detector. By sequentially using different filters with orthogonal spectral windows, images corresponding to each spectral window are collected and then overlaid to produce a false-color image. Figure 1.6(A) illustrates this procedure. It shows a field of HeLa cells whose nucleus is stained with DAPI, an organic dye, and whose cytoplasm is stained with qdots. Cells are observed under epifluorescence illumination using a Hg lamp. Since DAPI emission occurs at ~450–550 nm and the qdots emission is centered around 620 nm, the two channels of emission present minimal spectral overlap. By insert-

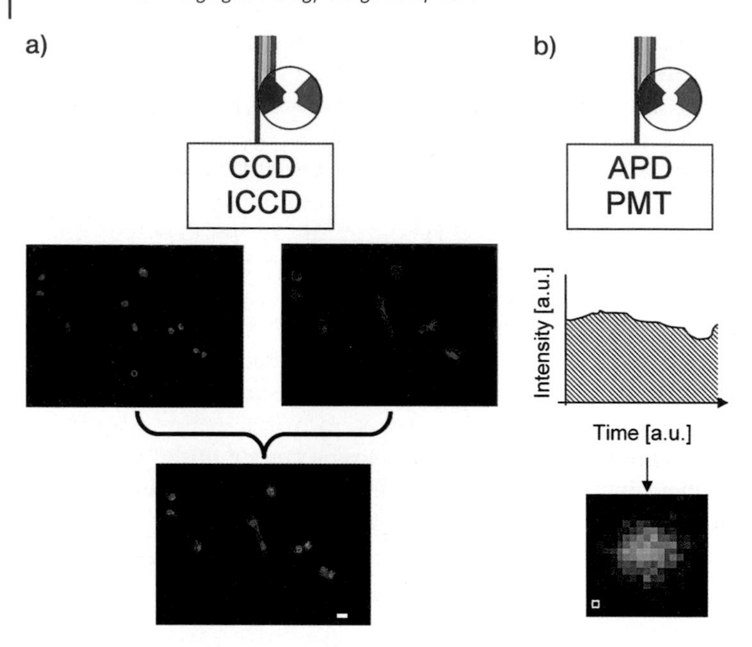

Fig. 1.6. A simple option for making images – Intensity imaging. (A) The simplest method of image making is to collect the fluorescence light in a CCD or intensified CCD camera. To discriminate different colors, filters are inserted in the emission path to block undesired colors. By changing filters, the same CCD camera can record, sequentially, different emissions from the same area. A color image is created by overlaying individual images. In the example shown, HeLa cells are imaged through epifluorescence illumination with an Hg lamp. The nuclei of the cells were stained with a blue dye (DAPI) and the cytoplasm stained with red qdots. Scale bar: 5 nm. (B) If a PMT or an APD were used, one would have measured a time-dependent signal with no encoded spatial information. The integrated intensity over a period of time may be used for imaging. If a focused laser beam is scanned over a surface, each position of the objective can be translated into a pixel. The value of the pixel is proportional to the integrated intensity. This is illustrated by measuring the spot produced by a single qdot. Note that the qdot spot is diffraction-limited even if the emitter is a point source. The white square represents a single pixel. Pixel size: 50 nm, integration time: 100 ms $pixel^{-1}$.

ing the appropriate bandpass filters (for instance a short pass filter with cutoff at 520 nm and a long pass with cutoff at 580 nm), a blue image of the nuclei and a red image of the cell cytoplasm can be recorded separately and then overlaid. When multiple chromophores with closer emission wavelengths are used, a clear distinction between channels is not always possible, and spectral imaging (Section 1.4.2) can represent one way to minimize the overlaps. This is especially true for organic dyes where the emissions are broad (>50 nm) and tailed to the red; but it also applies to qdots with narrower and more symmetric emission spectra.

If the CCD camera were replaced by a PMT or APD detector, one would acquire a trace similar to that shown in Fig. 1.6(B). There is no spatial resolution in this case but a time dependence of the signal, which represents the amount of photons emitted from all over the illuminated area that reach the detector. For imaging purposes, epifluorescence illumination using an arc lamp and PMT detection is of limited use. However, things are different if a cw or pulsed laser is used. Here, the coherence of the laser allows its focusing down to a tiny spot. The fluorescence intensity from the illuminated spot is proportional to the integrated signal reaching the PMT/APD detector. If the laser spot is scanned over the surface, and the out-of-focus light blocked by a pinhole (Section 1.3.2.1), an image of the emitters can be constructed pixel by pixel. For example, Fig. 1.6(B) shows the confocal image of a single CdSe/ZnS qdot. Although the qdot is less than 10 nm in diameter, its image is diffraction-limited because it maps the point-spread function of the excitation. With a sufficient signal-to-noise ratio, the PSF can be accurately fitted and its center determined (Sections 1.3.2.1 and 1.5.1) [32, 62]. Because of the scanning nature of the acquisition process and because typical integration times of a few tens of millisecond per pixel are needed, the acquisition of a 256 × 256 pixels image necessitates a few minutes. The use of a pulsed excitation affords additional imaging modalities. It allows correlation of the arrival time of the photon onto the PMT/APD detector to an external clock (i.e., the pulse of the laser), which constitutes the basis for measuring the lifetime of a chromophore (Fig. 1.8 and Section 1.4.3 below).

1.4.2 Spectral Imaging

A general problem with organic chromophores is their relatively broad emission spectra. This makes the spectral separation by bandpass filters difficult and impractical, especially when multiple chromophores with close emissions are used. It is often impossible to avoid the leaking of one color channel into another one. It is then advantageous to use spectral imaging. These methods collect all photons together and record an emission spectrum. Because fluorescence is incoherent, the total signal is simply the sum of every individual photon. The knowledge of the emission properties of each single fluorophore permits the deconvolution of the signal into diverse contributions.

Spectral imaging is illustrated in Fig. 1.7. Green-emitting qdots (emission at 540 nm) are covalently linked to several Cy3B, an organic dye with emission at about 575 nm. These conjugates are model systems for FRET imaging that will be discussed in Section 1.4.4. A tightly focused laser excites the conjugates. The fluorescence emission is spectrally resolved using a grating and collected by a CCD camera. The spectrum clearly shows two components, corresponding to qdots and dyes, respectively. Notice the narrower contribution of the qdot. A deconvolution of the spectrum into qdot and dye components allows the encoding of this pixel into several colors. If the same procedure is repeated for other pixels, a multicolor map can be constructed. In the simple case of Fig. 1.7, the deconvolution is a trivial task because the chromophores have almost non-overlapping emissions. However, spectral imaging may become quite complex as the number of chromophores increases and their spectra start to overlap. With CdSe/ZnS qdots, four colors of

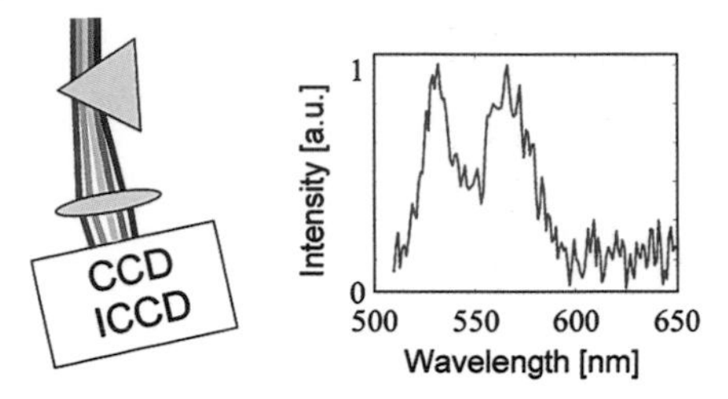

Fig. 1.7. Another simple option for making images is spectral imaging. Instead of using filters to separate the different emissions, the entire beam can be split through a prism or a grating and detected with a CCD camera or an intensified CCD camera. Here, the wavelength of the photon is encoded into the area of the CCD chip. The spectrum on the right shows the emission of a complex made of a single CdSe/ZnS qdot (emission ~530 nm) linked to several Cy3B, an organic dye (emission ~580 nm). By scanning the surface and deconvoluting the spectrum into these two components for each pixel, a multicolored map can be created with a single scan.

emission in the visible range could be specifically coded and detected by linking single-stranded DNA to each color of emission [33]. The use of qdots with emissions in the IR (Fig. 1.3B) can provide additional detection channels while preserving the simplicity of the excitation/detection schemes.

1.4.3
Lifetime and Time-gated Imaging

Lifetime is another property of fluorophores that can be used for imaging because it allows high-sensitivity and background-free imaging (Fig. 1.8). After excitation by a laser pulse, most organic fluorophores emit one photon within a characteristic time of a few nanoseconds. Two distinct fluorophores are likely to have different

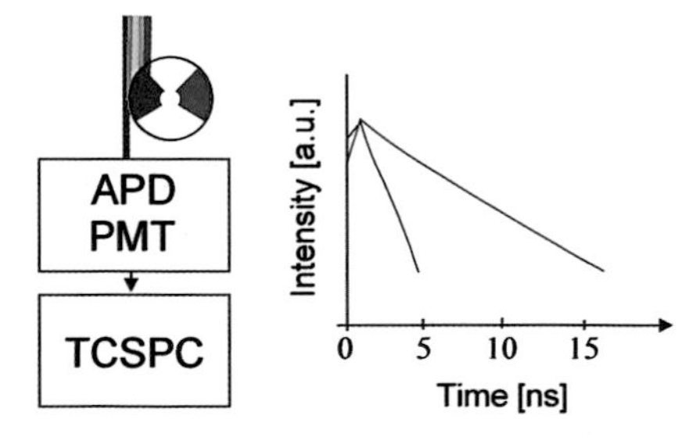

Fig. 1.8. Lifetime imaging is a further simple option for making images. A PMT/APD can also be used to map the lifetime of the emission. The arrival time is correlated with the firing of the laser by a time correlated single-photon counter (TCSPC). The example shown here compares the lifetimes of Cy3B, an organic dye (blue curve, ~2 ns), and qdots (red curve, ~10–17 ns). The large difference allows electronic gating as a way of increasing the signal-to-noise ratio.

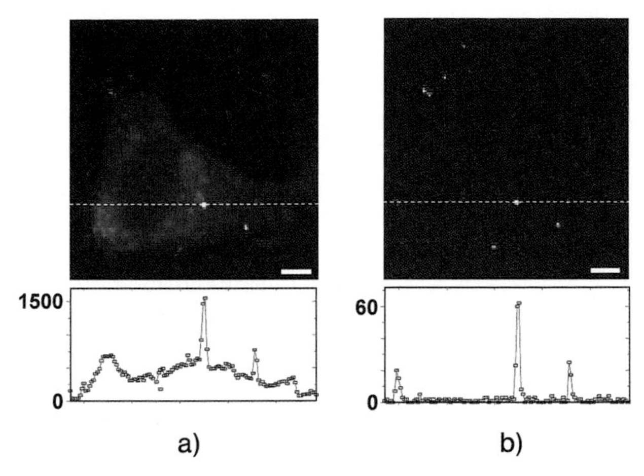

Fig. 1.9. Time-gate detection. The long lifetime of qdot permits suppression of the fluorescence of short-lived species, for instance autofluorescence and emission from other organic dyes. In this example, 3T3 mouse fibroblasts were incubated with red qdots. Imaging was performed using a confocal microscope (10 ms integration per pixel, lifetime window of 0–150 ns after the laser pulse, 5 MHz repetition rate). (A) All photons have been used to create the image. (B) Only photons arriving after the disappearance of the autofluorescence background were retained (30–65 ns after the laser pulse). Even though the total signal decreases markedly, the signal-to-noise ratio increases from 3 to 45. (Adapted with permission from Ref. [38]. Copyright 2001, Optical Society of America.)

lifetimes and a simple idea is to harness this difference to encode the image. Often, however, the difference in lifetime is too small to allow a proper discrimination between two distinct fluorophores. For instance moieties responsible for the autofluorescence background of cells and conventional organic dyes have very close lifetimes and it is difficult to isolate and remove only the background contribution. Other types of fluorophores have lifetimes of the orders of milliseconds (chelated lanthanide, metal ligand complexes), but this slow emission process results in a low turnover rate, i.e., a few photons emitted per second. Qdots combine the advantage of a high turnover rate with a long lifetime (10–20 ns) that seems more amenable for high-sensitive or background-free imaging. Lifetime imaging can be implemented by using either a time correlated single-photon counter (TCSPC) [65] or a time-gated ICCD camera [38]. In this latter case, the camera is electronically gated to detect only photons emitted after a certain time window after the laser pulse and reject all other photons.

Figure 1.9 shows an implementation of lifetime imaging of a cell using qdots to increase the signal-to-noise ratio [38]. 3T3 epithelial cells were transfected with qdots and imaged with a stage-scanning confocal microscope (Section 1.3.2). Figure 1.9(A) shows the image of the cell constructed from all photons reaching the APD detector. A major contribution comes from autofluorescence background that masks partially the qdot contribution. The electronically gated image constructed from photons that arrived 35–65 ns after the laser pulse is shown in Fig. 1.9(B). The autofluorescence background is suppressed but the emission of the qdots is still visible. The linescan shows that the time-gated imaging, allowed by the long lifetime of the qdots, considerably increases the signal-to-noise ratio.

1.4.4 Other Imaging Modalities: Polarization and FRET Imaging

Polarization imaging can be achieved by splitting the emission according to the polarizations of the photons. This can be achieved through the use of a polarizing beam-splitter or through the use of a 50/50 beam-splitter and the insertion of polarizers in front of the detector. Similarly, imaging based on fluorescence resonance energy transfer (FRET) can also be performed [66]. In a FRET process, a fluorescent donor molecule is excited but, instead of relaxing through the emission of a photon, it transfers its energy into an acceptor molecule. Emission comes from the acceptor although excitation was provided by the donor. The FRET process efficiency is very sensitive to the distance between the donor and the acceptor and vanishes for distances larger than ~10 nm. Therefore FRET can resolve molecular interactions, because even though both donor and acceptor probes cannot be localized below the diffraction-limit individually the fact that FRET occurs is an indication that donor and acceptor are less than 10 nm apart. The detection of FRET usually proceeds through the detection in parallel of two different channels, corresponding to the donor and acceptor probes. Several protein interactions have been demonstrated in cells by FRET microscopy [67]. In fact, FRET imaging has been performed only with pairs of organic dyes or pairs of fluorescent proteins (CFP and YFP) [68] because the use of qdots as FRET pairs has not yet been established. CdSe/ZnS qdots are excellent donors even at the single molecule level, as Fig. 1.3(C) indicated. Whether they also can act as acceptors is an open issue. It may well be that, because of their size and optical properties (Section 1.2.2.4), qdots will not be efficient probes neither as FRET pairs nor for polarization imaging.

1.5 Qdots in Biology: A Few Selected Examples

So far, we have focused on generic aspects of qdot fluorescence, by insisting on the properties that distinguish them from organic dyes. We have described several ways to excite and detect them, and several modalities for imaging them. It is time to consider their use in life sciences applications. In fact, qdot technology has generated a great deal of excitement in chemistry and material sciences communities for all their wonderful properties, but biologists have yet to embrace it. One reason is that the interface between the bioworld and the inorganic world is still a complex field. For example, live organisms have developed defense mechanisms to fight foreign intruders. Forcing them to tolerate unnatural inorganic probes may require tricks such as masking the surface of qdots with natural moieties. This section goes through some examples that illustrate how qdots can provide unique detection signatures in biology. The examples will give an overview of the possibilities of qdots for biolabeling. We will start by describing briefly the problematic that biologists are facing and how qdot technology may help solve those issues.

1.5.1
Ultra-high Colocalization of Qdots for Genetic Mapping

An important issue in genetic disease screening and diagnostics is the ability to detect DNA abnormalities in high-throughput. Many diseases result from deletions, additions or rearrangements of short segments of DNA, some of them only a few bases long. Although single nucleotide polymorphisms can be detected by DNA microarray methods using either dyes [69] or qdots [51], when the length of the abnormalities gets bigger, so do the challenges to detect and map them. A new approach was proposed in the mid-1990s by combining a technique to stretch the full genome on a glass slide with a technique to hybridize short fluorescently labeled oligonucleotides along the DNA backbone at selected locations [70]. This new approach was named dynamic molecular combing, or DMC. It features a homogenous stretching of the DNA (2 μm per kilobase) that allows the conversion of physical distances into genetic units. DMC measures the gap between loci labeled with two different dyes. Loci can be chosen to be contiguous of a specific gene and, therefore, the measured gap between loci holds information on the length of the gene. By comparing the gap measured on the normal allele with the gap measured on the abnormal allele, the size of the deletion/addition can be determined. DMC proved to be successful in observing kilobase deletions in *tuberous sclerosis 2* gene on patients' DNA [70].

Detection of smaller abnormalities using DMC requires a method that can localize two distinct sets of probes distance of 10 to 200 nm. This range of distances lies between the range of FRET detection (1–10 nm) and the Rayleigh resolution limit of conventional optical microscopes. Two recent methods promised to reach this goal. In the first approach, resolutions down to the 5–10 nm range have been reported using organic dyes, TIR illumination and a CCD camera [60]. The method makes use of the photobleaching behavior of one of the two probes. By fitting images collected both before and after photobleaching of one of the dyes, it is possible to localize both dyes separately and compute their distance. The method has been validated by using two identical Cy3 dyes linked by a double stranded DNA [60] or multicolor dyes that are more amenable for DNA mapping [71]. Colocalization by photobleaching requires that all random and individual photobleaching events are resolved. Besides, the numerical methods used to compute the relative distances may affect the precision of the localization.

The second approach does not rely on such constraints but builds on the strengths of fluorescent probes: it uses the photostability of qdots. The approach is based on stage-scanning confocal microscopy with a single excitation source that allows the imaging of qdots free of chromatic aberrations (Section 1.3.2.1)

[32]. A laser beam is focused onto a sample sitting on a stage that can be positioned and moved with nanometer precision. For each position of the stage, the intensity of fluorescence is collected, directed into an APD, and converted into a pixel value. Pixel by pixel, an image of the emitting object is constructed as illustrated in Fig. 1.6(B). By using two APDs/PMTs, each recording the light in an orthogonal fashion, it is possible to image simultaneously two point sources emitting at different wavelengths. Because the excitation arm is fixed, it does not produce chromatic aberrations. Comparison of the positions of the two PSFs allows the determination of their relative distances. Using such a stage-scanning confocal setup and qdots, two qdots have been localized with better than 10 nm resolution. Advances in nanometer-localization imaging are likely to open new doors for DNA mapping. It may be envisioned that such fluorescence techniques would yield physical maps of single nucleotide polymorphisms of the whole genome adsorbed on a surface a few inches wide.

1.5.2
Dynamics of Biomolecules in a Cellular Environment

The ability to track molecules directly in live cells is an important issue in biology that may lead to an integrative view of cellular function [2, 3]. Indeed, seeing the whole pathway of a molecule in its native environment, seeing how it works, where it goes and what it does may reveal precious information on its function and its communication with the rest of the interconnected cellular network. One way to study the motion of a biomolecule is to label it with a chromophore and acquire a succession of fluorescence images that are recombined in a movie. The first experiments of fluorescence tracking of lipids on cell surfaces were reported in 2000, using TIR fluorescence microscopy [72]. Since then, ion channels, cell-adhesion proteins [73], virus infection pathways [74] and other biological processes have been visualized at the single molecule level in live cells. Most of the measurements provide a low contrast and, above all, a time window for observation limited to a few seconds before the dye or GFP bleaches out [3, 57, 75].

The photostability of qdots provides a bright and long-lasting photon source that can be used to monitor biomolecular motion either on cell membranes [11, 76, 77] or directly inside cells [13] with a high signal-to-noise ratio for many minutes. The use of an epifluorescence illumination with a CCD detector is a good compromise between the necessity to monitor an area of a few tens/hundreds of micron square, a reasonable signal-to-noise ratio, and the time resolution of $\sim$100 ms required to monitor moving molecules. In fact, the monitoring of molecular motion through fluorescence only reveals changes in the location of a fluorescent spot. It is then necessary to analyze its position as a function of time and relate it to a physical or biological process. Traditionally, this has been analyzed in terms of diffusion constants. An example of a physical process is Brownian motion, which is characterized by a single diffusion coefficient and a mean square travel distance that depends linearly on time. If linearity is not observed, it may be a sign that an underlying biological process is at play.

1.5.2.1 Trafficking of Glycine Receptors in Neural Membranes of Live Cells

The first example of molecule tracking using qdots concerns the study of the membrane structure of spinal neurons. This is achieved by tracking the dynamics of qdot-labeled glycine receptors (GlyR) for long periods [76, 77]. In general, membrane neurotransmitter receptors, such as GlyR, are free to diffuse in the membrane. However, they can accumulate at synaptic sites by interacting with subsynaptic scaffold proteins. The number of neurotransmitter receptors at synaptic sites is an important parameter because it determines synapse plasticity and the strength of synaptic transmission. For example, removal of glutamate receptors from synapses is involved in long-term depression while their insertion contributes to long-standing synaptic efficiency. Generally, immunochemistry methods used to study variations in receptors distribution provide a static view of the neurotransmitter receptors and give no information on their dynamics. To study the dynamics of neurotransmitter receptors, a simple strategy consists in labeling them with fluorescent latex beads and tracking their motions with a simple epifluorescence illumination and a CCD detector. The size of the latex beads (~500 nm) prevented them from accessing the synaptic cleft. Things change by replacing beads with qdots. Because of their much smaller size (~10–20 nm), qdots can be used to study the lateral movement of individual GlyR in great detail.

Figure 1.10 shows an example of QD-GlyR motion over the neural surface. Synaptic buttons are labeled in red using an organic dye (FM4-64) while GlyR are labeled with green qdots using a bridge of biotinylated antibodies and streptavidin coated qdots. The sequence of images shows one QD-GlyR (in the white circle) moving from one synaptic button to another one about 4 μm away. In the first 30 s (top row), QD-GlyR diffuses freely in the synaptic membrane, it then docks at the synaptic button (bottom row). Further quantitative information is obtained by analysis of QD-GlyR motions over more than 20 min. The diffusion coefficients reveal three types of behavior: some receptors diffuse freely and cover large areas

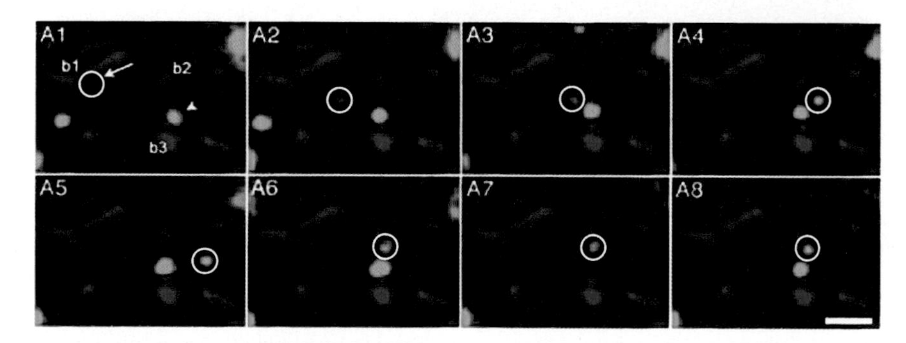

Fig. 1.10. Dynamics of glycine receptors revealed by qdots tracking: The example here shows the movement of a single membrane glycine receptor (circle). The receptor is labeled by a single green qdot moving from one synaptic button, labeled b1, to another labeled b2. The synaptic buttons are labeled with a red dye. Images on the panel are selected snapshots from a 20 min sequence. (Reproduced with permission from Ref. [76]. Copyright 2003, The American Association for the Advancement of Science.)

on the membrane, other receptors remain at the synapse, whereas a final pool of receptors move in a confined space around the synapse. This leads to the classification of the neural membrane in three domains (extrasynaptic, synaptic and perisynaptic, respectively) with distinct diffusion properties. This beautiful work of Dahan and co-workers underlies the advantages of using robust qdot probes for tracking small biomolecules and illustrates the potential of qdots for the study of membrane dynamics.

1.5.2.2 Dynamics of Labeled Nuclear Localization Sequences Inside Living Cells

The observation of dynamical events directly inside live cells adds one layer of complexity because the cytoplasmic environment is vastly different from conditions found outside the cell. Live cells have various kinds of inner membranes that maintain compartmental organization and they have developed various defense mechanisms to protect themselves against foreign intruders [1]. Whether exogenous qdot labels are compatible with the cell cytoplasm environment is largely unknown.

Only a handful of preliminary targeting studies using qdots in live cells have been performed. Perhaps, the simplest approach is to target the largest and most distinctive organelle of a cell: its nucleus. Nature provides the drive on how to do that. For example, viruses are hollow particles containing infectious RNA. Their

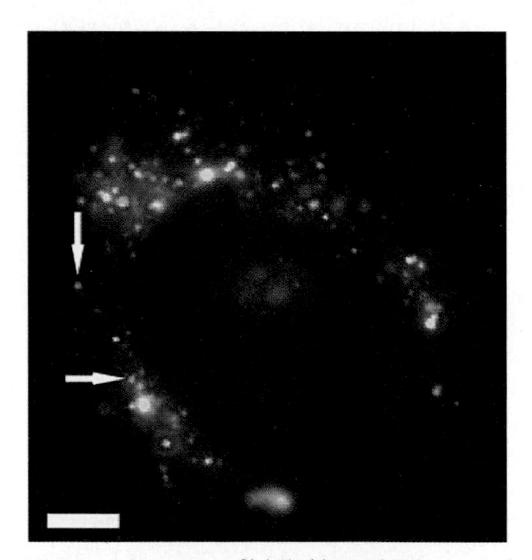

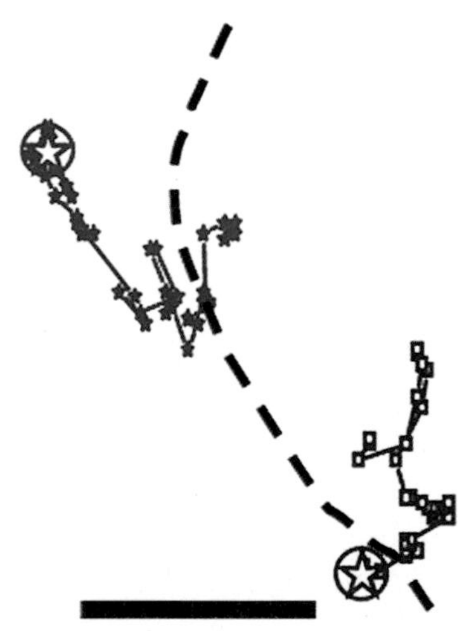

Fig. 1.11. Dynamics of labeled biomolecules inside live cells. HeLa cells were transfected with qdots linked to a nuclear localization signal (NLS). The NLS–qdots localize around the nucleus of the cell, and attempt to penetrate the nuclear envelope. The nucleus of a single cell is visible in the image. The image is an overlay of two frames taken 10 s apart that show how the qdots move during these 10 s (green is at time t, red at $t + 10$ s, yellow is the overlay, 700 ms integration time). Scale bar: 5 nm. The motions of the two NLS–qdots indicated by the arrows are shown on the right of the image. Positions at the beginning of the imaging process are indicated by stars. The nuclear membrane is shown as a dashed line. The NLS–qdot conjugates jiggle around the membrane and eventually cross it. Scale bar: 5 nm.

envelope contains peptide sequences, called nuclear localization signals (NLSs), that are recognized by the cell machinery and ferried towards the nucleus by motor proteins. It is possible to link a NLS sequence derived from a virus SV40 [78] to a silanized qdot [13]. Figure 1.11 shows the end stage of the infection process where NLS-qdots attempt to penetrate through the nuclear pores of a live cell. In a movie sequence that lasts well over 15 min, the somewhat chaotic jiggling of qdots can be observed. Some qdots get stuck at the nuclear membrane, while others manage to cross the nuclear membrane. Two examples of such NLS-qdots are highlighted on the right of the figure.

Although fluorescence imaging shows qdots concentrated around the nucleus, the fate of the NLS peptide is not known. The presence of the qdot exogenous agent is likely to activate the defense mechanisms of the cells. This may result in the cleavage of the qdot from the NLS peptide with the obvious consequence that the tracking of the fluorescent probe does not mimic the molecular pathway of the biomolecule. This is an obvious concern that needs to be addressed in forthcoming years when, no doubt, people will start tracking biomolecules inside live cells using qdots.

1.5.3 *In Vivo* and Non-invasive Detection Using Qdot Reporters

A final example concerns the observation of fluorescent qdots inside small animals through non-invasive imaging. In general, biomedical imaging in living tissues requires the use of exogenous contrast agents. These include MRI contrast agents, radioactive PET reporters, or infrared photons that can penetrate deeper into tissues than visible photons. In particular, the use of IR or near-IR probes would provide submicron spatial resolution and high sensitivity with simple and cost-effective detector technologies. Unfortunately, the requirements of absorption cross section, quantum yield and stability are not met by conventional organic fluorophores emitting in the IR. However, infra-red qdots may be efficient alternatives.

The fact that IR core/shell CdTe/CdSe qdots may be useful probes for non-invasive detection was demonstrated in early 2004 by mapping the sentinel lymph nodes in mice [79]. This allowed a major cancer surgery to be performed in an animal under complete image guidance. A subsequent step forward was achieved when near-IR CdSe/ZnS dots functionalized with a cancer targeting antibody were used to detect non-invasively the location of cancerous metastasis in live mice [45]. These two preliminary reports open new windows of opportunities for nanotechnology in medicine. The area of non-invasive imaging using qdots is likely to grow, as chemists get better at synthesizing IR probes.

1.6 Outlook: Is there a Role for Nanoscience in Cellular Biology and in Medicine?

In this chapter we have focused on fluorescence detection, a central diagnostics in biology and medicine. Fluorescence imaging has reached an incredible level of sensitivity, mainly fueled by the desire of some pioneers to watch single molecules. Today, researchers want to see more than single molecules, they want to see them rock'n roll (as beautifully phrased by Xavier Michalet [80]), and possibly in an *in vivo* environment. Moreover, as other examples discussed in this chapter have shown, biology and medicine have very ambitious plans to observe molecular events directly in their native environment. What limits such experiments today is the availability of appropriate probes, probes that are non-toxic, probes that do not bleach, probes that can be functionalized. Organic dyes, and to a lesser extent engineered fluorescent proteins, have certain limitations. Their shortcomings will likely not be compensated by an upgrade of current technologies. Future breakthroughs will rely on the development of advanced technologies based on a radically different platform. Nanotechnology can provide this platform.

Biology and nanoscience have come to a crossroads after a long courtship. Driven by the intellectual challenge of addressing the properties of particles with a finite size (first enounced by Feynman [81] and later addressed theoretically by Ryogo Kubo [82] and Terrell L. Hill [83] in the early 1960s), experimentalists started working on the production of small particles in the late 1960s and early 1970s. Over the course of two decades, synthetic methods have been perfected and the range of material synthesized has been expanded. In the 1990s, the use of chemical routes permitted the synthesis of semiconductor, metallic and oxide nanoparticles with excellent crystallinity, and with controlled size and shape [17]. In particular, semiconductor nanocrystals, whose flag carrier is cadmium selenide, exhibited remarkable optical properties, described in an earlier section, that attracted the interest of biologists as possible biolabels. Eight years after the first two reports [40, 84], the use of fluorescent nanocrystals in biology is still limited. This is largely due to the difficulty of interfacing inorganic nanoobjects with the bioworld. Yet, chemists and physicists are learning how to do it at a fast pace, and it is likely that fluorescent particles will emerge as useful tools for biology.

We can dream that qdot technology will provide highly integrated probes that can be customized to perform many tasks at once. For instance, qdots would be synthesized to carry diagnostics (they fluoresce), functional (they may be programmed to recognize tumoral cells), and therapeutics properties (they may kill tumoral cells). Qdots could be linked to MRI or PET active compounds to allow non-invasive imaging. Because qdots are made of heavy elements they are easily observable in electron microscopy and through synchrotron radiation, which will provide access to high-resolution cellular localization. We may imagine that further developments will allow the tagging of stem cells with qdots, their transplantation in live animal models, and the remote detection of their differentiation *in vivo*. One may argue that the view expressed here is much too optimistic, and even wonder whether nanotechnology will have a real impact in biology and medicine. Only time can tell.

Acknowledgments

I would like to thank Fanqing Chen (Lawrence Berkeley Laboratory), Balaji Kannan (UC Berkeley) and Xavier Michalet (UCLA) for many discussions and collaborations over the years. I want to thank Natalia Zaitseva, Giulia Galli and Sirine Fakra for their support and Maxime Dahan, Thomas Jobin and Elizabeth Jares-Erijman for permission to reproduce their figures. This work was performed under the auspices of the US Department of Energy by the University of California at the LLNL under contract no W-7405-Eng-48.

References

1 H. Lodish, D. Baltimore, A. Berk, S. L. Zipurski, P. Matsudaira, J. Darnell. *Molecular Cell Biology* (Scientific American Books, New York, NY, **1995**).

2 R. Y. Tsien. Imagining imaging's future. *Nat. Cell Biol.* **2003**, SS16–SS21.

3 Y. Sako, T. Yanagida. Single-molecule visualization in cell biology. *Nat. Rev. Mol. Cell Biol.* **2003**, *4*, SS1–SS5.

4 R. C. C. Ryther, A. S. Flynt, J. A. I. Phillips, J. C. Patton. siRNA therapeutics: big potential from small RNAs. *Gene Therap.* **2004**, 1–7.

5 D. Semizarov, P. Kroeger, S. Fesik. siRNA-mediated gene silencing: a global genome view. *Nucl. Acids Res.* **2004**, *32*, 3836–3845.

6 G. Meister, T. Tuschl. Mechanisms of gene silencing by double-stranded RNA. *Nature* **2004**, *431*, 343–349.

7 E. A. Jares-Erijman, T. M. Jovin. FRET imaging. *Nat. Biotechnol.* **2003**, *21*, 1387–1395.

8 E. C. Friedberg, G. C. Walker, W. Siede. *DNA Repair and Mutagenesis* (American Society for Microbiology, Washington, D.C., **1995**).

9 C. B. Murray, D. J. Norris, M. G. Bawendi. Synthesis and characterization of nearly monodisperse CdE (E = S, Se, Te) semiconductor nanocrystallites. *J. Am. Chem. Soc.* **1993**, *115*, 8706–8715.

10 P. Alivisatos. The use of nanocrystals in biological detection. *Nat. Biotechnol.* **2004**, *22*, 47–52.

11 X. Michalet, F. F. Pinaud, L. A. Bentolila, J. M. Tsay, S. Doose, J. J. Li, G. Sundaresan, A. M. Wu, S. S. Gambhir, S. Weiss. Quantum dots for live cells, in vivo imaging, and diagnostics. *Science* **2005**, *307*, 538–544.

12 A. P. Alivisatos. Perspectives on the physical chemistry of semiconductor nanocrystals. *J. Phys. Chem.* **1996**, *100*, 13226–13239.

13 F. Q. Chen, D. Gerion. Fluorescent CdSe/ZnS nanocrystal-peptide conjugates for long-term, nontoxic imaging and nuclear targeting in living cells. *Nano Lett.* **2004**, *4*, 1827–1832.

14 B. Dubertret, P. Skourides, D. J. Norris, V. Noireaux, A. H. Brivanlou, A. Libchaber. In vivo imaging of quantum dots encapsulated in phospholipid micelles. *Science* **2002**, *298*, 1759–1762.

15 J. K. Jaiswal, H. Mattoussi, J. M. Mauro, S. M. Simon. Long-term multiple color imaging of live cells using quantum dot bioconjugates. *Nat. Biotechnol.* **2003**, *21*, 47–51.

16 J. R. Lakowicz (ed.). *Principles of Fluorescence Spectroscopy*, P. Press, (New York, NY, 1986).

17 W. J. Parak, L. Manna, F. C. Simmel, D. Gerion, A. P. Alivisatos. In *Nanaoparticles: From Theory to Application*, ed. G. Schmid, pp 4–49 (Wiley-VCH Verlag GmbH & Co KGaA, Weinheim, **2004**).

18 C. Kittel. *Introduction to Solid State Physics* (Wiley and Sons, Inc., Ney York, NY, **1996**).

19 A. L. Erfos, M. Rosen. The electronic structure of semiconductor nanocrystals. *Annu. Rev. Mater. Sci.* **2000**, *30*, 475–521.

20 J. R. Lakowicz. Fluorescent analogues in biological research. *Encyclopedia of Life Sciences* **2001**, 1–11.

21 W. E. Moerner, D. P. Fromm. Methods of single-molecule fluores-

cence spectroscopy and microscopy. *Rev. Sci. Instrum.* **2003**, *74*, 3597–3619.

22 S. C. Blanchard, H. D. Kim, R. L. Gonzalez, J. D. Puglisi, S. Chu. tRNA dynamics on the ribosome during translation. *Proc. Natl. Acad. Sci. U.S.A.* **2004**, *101*, 12893–12898.

23 D. C. Prasher, V. K. Eckenrode, W. W. Ward, F. G. Prendergast, M. J. Cormier. Primary structure of the Aequorea victoria green fluorescent protein. *Gene* **1992**, *111*, 229–233.

24 O. Shimomura, F. H. Johnson, Y. Saiga. Extraction, purification and properties of aequorin, a bioluminescent protein from the luminous hydromedusan, Aequarea. *J. Cell. Comparative Physiol.* **1962**, *59*, 223–239.

25 C. Coralli, M. Cemazar, C. Kanthou, G. M. Tozer, G. U. Dachs. Limitations of the reporter green fluorescent protein under simulated tumor conditions. *Cancer Res.* **2001**, *61*, 4784–4790.

26 W. W. Yu, Y. A. Wang, X. Peng. Formation and stability of size-, shape-, and structure-controlled CdTe nanocrystals: Ligand effects on monomers and nanocrystals. *Chem. Mater.* **2003**, *15*, 4300–4308.

27 A. A. Guzelian, J. E. B. Katari, A. V. Kadavanich, U. Banin, K. Hamad, E. Juban, A. P. Alivisatos, R. H. Wolters, C. C. Arnold, J. R. Heath. Synthesis of size-selected, surface-passivated InP nanocrystals. *J. Phys. Chem.* **1996**, *100*, 7212–7219.

28 A. A. Guzelian, U. Banin, A. V. Kadavanich, X. Peng, A. P. Alivisatos. Colloidal chemical synthesis and characterization of InAs nanocrystal quantum dots. *Appl. Phys. Lett.* **1996**, *69*, 1432–1434.

29 L. Bakueva, I. Gorelikov, S. Musikhin, X. S. Zhao, E. H. Sargent, E. Kumacheva. PbS quantum dots with stable efficient luminescence in the near-IR spectral range. *Adv. Mater.* **2004**, *16*, 926–929.

30 J. M. Pietryga, R. D. Schaller, D. Werder, M. H. Stewart, V. I. Klimov, J. A. Hollingsworth. Pushing the band gap envelope: Mid-infrared emitting colloidal PbSe quantum dots. *J. Am. Chem. Soc.* **2004**, *126*, 11752–11753.

31 R. B. Vasiliev, S. G. Dorofeev, D. N. Dirin, D. A. Belov, T. A. Kuznetsov. Synthesis and optical properties of PbSe and CdSe colloidal quantum dots capped with oleic acid. *Mendeleev Commun.* **2004**, 169–171.

32 T. D. Lacoste, X. Michalet, F. Pinaud, D. S. Chemla, A. P. Alivisatos, S. Weiss. Ultrahigh-resolution multicolor colocalization of single fluorescent probes. *Proc. Natl. Acad. Sci. U.S.A.* **2000**, *97*, 9461–9466.

33 D. Gerion, W. J. Parak, S. C. Williams, D. Zanchet, C. M. Micheel, A. P. Alivisatos. Sorting fluorescent nanocrystals with DNA. *J. Am. Chem. Soc.* **2002**, *124*, 7070–7074.

34 M. A. Hines, P. Guyotsionnest. Synthesis and characterization of strongly luminescing ZnS-capped CdSe nanocrystals. *J. Phys. Chem.* **1996**, *100*, 468–471.

35 D. Gerion, F. Pinaud, S. C. Williams, W. J. Parak, D. Zanchet, S. Weiss, A. P. Alivisatos. Synthesis and properties of biocompatible water-soluble silica-coated CdSe/ZnS semiconductor quantum dots. *J. Phys. Chem. B* **2001**, *105*, 8861–8871.

36 W. W. Yu, L. H. Qu, W. Z. Guo, X. G. Peng. Experimental determination of the extinction coefficient of CdTe, CdSe, and CdS nanocrystals. *Chem. Mater.* **2003**, 15, 2854–2860.

37 D. R. Larson, W. R. Zipfel, R. M. Williams, S. W. Clark, M. P. Bruchez, F. W. Wise, W. W. Webb. Water-soluble quantum dots for multiphoton fluorescence imaging in vivo. *Science* **2003**, *300*, 1434–1436.

38 M. Dahan, T. Laurence, F. Pinaud, D. S. Chemla, A. P. Alivisatos, M. Sauer, S. Weiss. Time-gated biological imaging by use of colloidal quantum dots. *Opt. Lett.* **2001**, *26*, 825–827.

39 J. T. Hu, L. S. Li, W. D. Yang, L. Manna, L. W. Wang, A. P. Alivisatos. Linearly polarized emission from

colloidal semiconductor quantum rods. *Science* **2001**, *292*, 2060–2063.
40 W. C. W. Chan, S. M. Nie. Quantum dot bioconjugates for ultrasensitive nonisotopic detection. *Science* **1998**, *281*, 2016–2018.
41 S. Kim, M. G. Bawendi. Oligomeric ligands for luminescent and stable nanocrystal quantum dots. *J. Am. Chem. Soc.* **2003**, *125*, 14652–14653.
42 W. Z. Guo, J. J. Li, Y. A. Wang, X. G. Peng. Conjugation chemistry and bioapplications of semiconductor box nanocrystals prepared via dendrimer bridging. *Chem. Mater.* **2003**, *15*, 3125–3133.
43 F. Pinaud, D. King, H. P. Moore, S. Weiss. Bioactivation and cell targeting of semiconductor CdSe/ZnS nanocrystals with phytochelatin-related peptides. *J. Am. Chem. Soc.* **2004**, *126*, 6115–6123.
44 X. Y. Wu, H. J. Liu, J. Q. Liu, K. N. Haley, J. A. Treadway, J. P. Larson, N. F. Ge, F. Peale, M. P. Bruchez. Immunofluorescent labeling of cancer marker Her2 and other cellular targets with semiconductor quantum dots. *Nat. Biotechnol.* **2003**, *21*, 41–46.
45 X. H. Gao, Y. Y. Cui, R. M. Levenson, L. W. K. Chung, S. M. Nie. In vivo cancer targeting and imaging with semiconductor quantum dots. *Nat. Biotechnol.* **2004**, *22*, 969–976.
46 T. Pellegrino, L. Manna, S. Kudera, T. Liedl, D. Koktysh, A. L. Rogach, S. Keller, J. Radler, G. Natile, W. J. Parak. Hydrophobic nanocrystals coated with an amphiphilic polymer shell: A general route to water soluble nanocrystals. *Nano Lett.* **2004**, *4*, 703–707.
47 F. Osaki, T. Kanamori, S. Sando, T. Sera, Y. Aoyama. A quantum dot conjugated sugar ball and its cellular uptake on the size effects of endocytosis in the subviral region. *J. Am. Chem. Soc.* **2004**, *126*, 6520–6521.
48 G. T. Hermanson. *Bioconjugate Techniques* (Academic Press, Inc., San Diego, CA, **1996**).
49 W. J. Parak, D. Gerion, D. Zanchet, A. S. Woerz, T. Pellegrino, C. Micheel, S. C. Williams, M. Seitz, R. E. Bruehl, Z. Bryant, C. Bustamante, C. R. Bertozzi, A. P. Alivisatos. Conjugation of DNA to silanized colloidal semiconductor nanocrystalline quantum dots. *Chem. Mater.* **2002**, *14*, 2113–2119.
50 S. Pathak, S. K. Choi, N. Arnheim, M. E. Thompson. Hydroxylated quantum dots as luminescent probes for in situ hybridization. *J. Am. Chem. Soc.* **2001**, *123*, 4103–4104.
51 D. Gerion, F. Q. Chen, B. Kannan, A. H. Fu, W. J. Parak, D. J. Chen, A. Majumdar, A. P. Alivisatos. Room-temperature single-nucleotide polymorphism and multiallele DNA detection using fluorescent nanocrystals and microarrays. *Anal. Chem.* **2003**, *75*, 4766–4772.
52 V. L. Colvin. Sustainability for nanotechnology. *Scientist* **2004**, *18*, 26–27.
53 V. L. Colvin. The potential environmental impact of engineered nanomaterials (vol. 21, p. 1166, 2003). *Nat. Biotechnol.* **2004**, *22*, 760–760.
54 A. M. Derfus, W. C. W. Chan, S. N. Bhatia. Probing the cytotoxicity of semiconductor quantum dots. *Nano Lett.* **2004**, *4*, 11–18.
55 C. Kirchner, T. Liedl, S. Kudera, T. Pellegrino, A. Muñoz Javier, H. E. Gaub, S. Stölzle, N. Fertig, W. J. Parak. Cytotoxicity of colloidal CdSe and CdSe/ZnS nanoparticles. *Nano Lett.* **2005**, *5*, 331–338.
56 C. M. Sayes, J. D. Fortner, W. Guo, D. Lyon, A. M. Boyd, K. D. Ausman, Y. J. Tao, B. Sitharaman, L. J. Wilson, J. B. Hughes, J. L. West, V. L. Colvin. The differential cytotoxicity of water-soluble fullerenes. *Nano Lett.* **2004**, *4*, 1881–1887.
57 J. Deich, E. M. Judd, H. H. McAdams, W. E. Moerner. Visualization of the movement of single histidine kinase molecules in live Caulobacter cells. *Proc. Natl. Acad. Sci. U.S.A.* **2004**, *101*, 15921–15926.
58 A. S. Sedra, K. C. Smith. *Microelectronic Circuits* (Oxford University Press, New York, NY, **1998**).
59 P. Roy, Z. Rajfur, P. Pomorski,

K. Jacobson. Microscope-based techniques to study cell adhesion and migration. *Nat. Cell Biol.* **2002**, *4*, E91–E96.

60 J. T. Frohn, H. F. Knapp, A. Stemmer. True optical resolution beyond the Rayleigh limit achieved by standing wave illumination. *Proc. Natl. Acad. Sci. U.S.A.* **2000**, *97*, 7232–7236.

61 T. A. Klar, S. Jakobs, M. Dyba, A. Egner, S. W. Hell. Fluorescence microscopy with diffraction resolution barrier broken by stimulated emission. *Proc. Natl. Acad. Sci. U.S.A.* **2000**, *97*, 8206–8210.

62 X. Michalet, T. D. Lacoste, S. Weiss. Ultrahigh-resolution colocalization of spectrally separable point-like fluorescent probes. *Methods* **2001**, *25*, 87–102.

63 X. Michalet, F. Pinaud, T. D. Lacoste, M. Dahan, M. P. Bruchez, A. P. Alivisatos, S. Weiss. Properties of fluorescent semiconductor nanocrystals and their application to biological labeling. *Single Mol.* **2001**, *2*, 261–276.

64 R. C. Dunn. Near-field scanning optical microscopy. *Chem. Rev.* **1999**, *99*, 2891–2928.

65 W. Becker, A. Bergmann, K. Koenig, U. Tirlapur. Picosecond fluorescence lifetime microscopy by TCSPC imaging. *Proc. SPIE* **2001**, *4431*, 414–419.

66 W. Becker, K. Benndorf, A. Bergmann, C. Biskup, K. Konig, U. Tirlapur, T. Zimmer. FRET measurements by TCSPC laser scanning microscopy. *Proc. SPIE* **2001**, *4431*, 94–98.

67 F. S. Wouters, P. J. Verveer, P. I. H. Bastiaens. Imaging biochemistry inside cells. *Trends Cell Biol.* **2001**, *11*, 203–211.

68 A. G. Harpur, F. S. Wouters, P. I. H. Bastiaens. Imaging FRET between spectrally similar GFP molecules in single cells. *Nat. Biotechnol.* **2001**, *019*, 167–169.

69 P. O. Brown, D. Botstein. Exploring the new world of the genome with DNA microarrays. *Nat. Genet.* **1999**, *21*, 33–37.

70 X. Michalet, R. Ekong, F. Fougerousse, S. Rousseaux, C. Schurra, N. Hornigold, M. vanSlegtenhorst, J. Wolfe, S. Povey, J. S. Beckmann, A. Bensimon. Dynamic molecular combing: Stretching the whole human genome for high-resolution studies. *Science* **1997**, *277*, 1518–1523.

71 X. H. Qu, D. Wu, L. Mets, N. F. Scherer. Nanometer-localized multiple single-molecule fluorescence microscopy. *Proc. Natl. Acad. Sci. U.S.A.* **2004**, *101*, 11298–11303.

72 G. J. Schutz, G. Kada, V. P. Pastushenko, H. Schindler. Properties of lipid microdomains in a muscle cell membrane visualized by single molecule microscopy. *Embo J.* **2000**, *19*, 892–901.

73 R. Iino, I. Koyama, A. Kusumi. Single molecule imaging of green fluorescent proteins in living cells: E-cadherin forms oligomers on the free cell surface. *Biophys. J.* **2001**, *80*, 2667–2677.

74 G. Seisenberger, M. U. Ried, T. Endress, H. Buning, M. Hallek, C. Brauchle. Real-time single-molecule imaging of the infection pathway of an adeno-associated virus. *Science* **2001**, *294*, 1929–1932.

75 M. Goulian, S. M. Simon. Tracking single proteins within cells. *Biophys. J.* **2000**, *79*, 2188–2198.

76 M. Dahan, S. Levi, C. Luccardini, P. Rostaing, B. Riveau, A. Triller. Diffusion dynamics of glycine receptors revealed by single-quantum dot tracking. *Science* **2003**, *302*, 442–445.

77 M. Dahan. Watching the dynamics of individual proteins in live cells using quantum dots. *J. Histochem. Cytochem.* **2004**, *52*, S18–S18.

78 D. A. Jackson, R. H. Symons, P. Berg. Biochemical method for inserting new genetic information into DNA of Simian Virus 40: Circular SV40 DNA molecules containing lambda phage genes and the galactose operon of *Escherichia coli*. *Proc. Natl. Acad. Sci. U.S.A.* **1972**, *69*, 2904–2909.

79 S. Kim, Y. T. Lim, E. G. Soltesz, A. M. De Grand, J. Lee, A. Nakayama, J. A.

Parker, T. Mihaljevic, R. G. Laurence, D. M. Dor, L. H. Cohn, M. G. Bawendi, J. V. Frangioni. Near-infrared fluorescent type II quantum dots for sentinel lymph node mapping. *Nat. Biotechnol.* **2004**, *22*, 93–97.

80 X. Michalet, A. N. Kapanidis, T. Laurence, F. Pinaud, S. Doose, M. Pflughoefft, S. Weiss. The power and prospects of fluorescence microscopies and spectroscopies. *Annu. Rev. Biophys. Biomol. Struct.* **2003**, *32*, 161–182.

81 R. P. Feynman. There's plenty of room at the bottom. http://www.zyvex.com/nanotech/feynman.html **1959**.

82 R. Kubo. Electronic properties of metallic fine particles. *J. Phys. Soc. Jpn.* **1962**, *17*, 975–986.

83 T. L. Hill. Thermodynamics of small systems. *J. Chem. Phys.* **1962**, *36*, 3182–3197.

84 M. Bruchez, M. Moronne, P. Gin, S. Weiss, A. P. Alivisatos. Semiconductor nanocrystals as fluorescent biological labels. *Science* **1998**, *281*, 2013–2016.

2
Characterization of Nanoscale Systems in Biology using Scanning Probe Microscopy Techniques

Anthony W. Coleman, Adina N. Lazar, Cecile F. Rousseau, Sebastien Cecillon, and Patrick Shahgaldian

2.1
Introduction

The numerous variants of Scanning Probe Microscopies (SPM) that have appeared since the invention of Scanning Tunneling Microscopy (STM) in 1982 [1] have opened up a new world for the visualization and characterization of surfaces. Their application to the biological sciences has allowed imaging of a vast range of systems hitherto unseen and the spin-offs of SPM have allowed biologists to probe the forces [2], mechanical properties of biological entities [3] and, with the use of SPM derived cantilever arrays, develop novel bio-sensors [4–8].

However, this meeting of physics, biology, chemistry and micro-technology was and still remains, in my opinion, not quite the universal solution to the imaging of the nano-world of biology. The problems would appear to stem from the near immiscibility of the scientific rigor of biology and physics, and this chapter has been approached from the point of view of a tool user not a tool maker, but one who has been over the last ten years confronted by the need to characterize and understand complex structures interacting with biological systems while using physical methods. As a sideline, I have also spent the last ten years translating between the scientific languages and habits of biologists, physicists, chemists and microsystems scientists.

Thus this chapter is oriented not to the theory of Scanning Probe Microscopies but rather to how to obtain, using SPM, useful and valid information about biological systems at all levels. The physics of SPM has been widely treated in several excellent texts and readers are referred to the list of books given in Appendix 1. Similarly, an extensive body of reviews exists on the imaging of biological systems (Appendix 2).

With regard to the use of SPM on biological samples, the questions to ask are really: what information can one obtain from SPM; how can one best set up the bio-SPM experiment to maximize the chances of success; how can one avoid obtaining skewed information from a bio-SPM experiment and finally, and most importantly, does the information obtained have a real biological sense?

Nanotechnologies for the Life Sciences Vol. 3
Nanosystem Characterization Tools in the Life Sciences. Edited by Challa S. S. R. Kumar

ISBN: 3-527-31383-4

Tab. 2.1. The various microscopies available to the bio-sciences, and their characteristics.

	Optical microscope	*SEM*	*SPM*	*SNOM-NSOM*	*Confocal microscopy [9, 10]*	*Environmental electron microscopy*
Sample operating environment	Air, liquid or vacuum	Vacuum	Gas, liquid or vacuum	Gas, liquid or vacuum	Gas, liquid or vacuum	Gas[a] (up to 10 Torr)
Depth of field	Small	Large	Medium	Medium	Small	Large
Depth of focus	Medium	Large	Small	Small	Medium	Large
Resolution: x, y (nm)	1000	5	1–10 AFM; atomic for STM	20–80	200	5–7
Resolution: z (nm)	N/A	N/A	0.1 AFM 0.01 STM	1	200	N/A
Magnification	$1–2 \times 10^3$X	$10–10^8$X	$5 \times 10^2–10^8$			$15–5 \times 10^5$X
Sample preparation requirement	Little	Little to substantial	Little or none	Little to substantial	Little to substantial	Little
Characteristics required for sample	Sample opaque to light	Sample must be vacuum compatible	Local variations in surface height < 10 μm		Incorporation of a fluorescent dye	–

[a] Possible in wet environments; suitable for hydrated specimens (cells, plant samples, tissue, etc.).

In view of these questions we will treat the various SPM methods in terms of the information they generate and also their limitations, some general thoughts on sample preparation and how this may affect the experiment, the pitfalls that may arise from various artifacts, before looking at selected examples of imaging, probing and analyzing biological systems. [Evidently, rarely are artifacts published but I know that the image files of anyone involved in the field are full of various types of artifacts, but with biological samples sometimes the seeming artifact may in fact be real and just seem to be an artifact.] There remains one final question; why carry out a bio-SPM experiment? To start to answer that question, Table 2.1 presents various microscopy methods with their experimental conditions, sample treatments, resolutions, etc.

2.2 The Scanning Probe Microscopy Experiment

The basis of all SPM experiments is simple: a probe is scanned across a surface to generate an image of some type. To obtain the image there must exist a localized interaction between the probe and the surface, in this way the signal obtained will be dominated only by the small part of the probe closest to the surface. Hence it is implicit that there is a strong distance dependence of the interaction so that

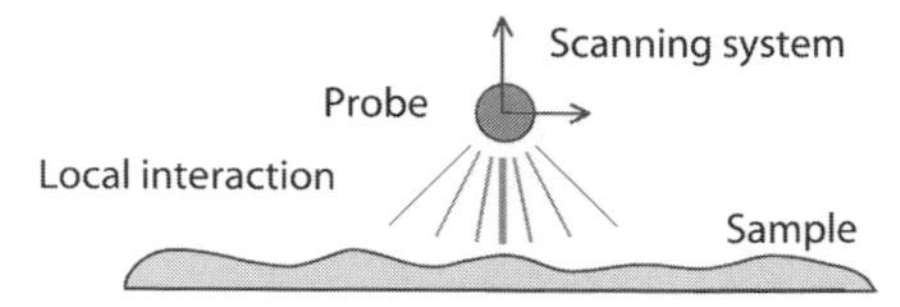

Fig. 2.1. Principle of probe–sample interaction in SPM. (Courtesy of JPK GmbH.)

only those nearest parts of the probe can contribute to the measured signal (Fig. 2.1).

Thus the SPM experiment can be defined as requiring:

$$\text{Signal} = f(d)^n$$

and effectively the resolution will be governed both by the size of the probe and the distance dependence, allowing, for example in STM where both tip size is very small and the distance dependence very strong, imaging at the atomic level.

While the physical basis of the family of Scanning Probe Microscopies is relatively straightforward, and indeed the basis of Scanning Near-Field Optical Microscopy was proposed by Synge in 1926 in a letter to Einstein [11], the technical problems involved in the construction of a working high-resolution SPM system were the real challenge. It was thus only in the 1980s, with the development of working piezoelectric materials [12] and the computing power to implement control over the system, that working systems became available. Even then, their application to biological materials was at first hesitant.

In the following sections the basic modes of imaging, use of force–distance measurements and other SPM derived methods will be discussed with particular emphasis on the information that may be obtained and on the limitations of the various techniques towards imaging biological systems.

2.3 Scanning Tunneling Microscopy Imaging

Scanning Tunneling Microscopy is based on the principle of "tunneling current" flowing between a metallic tip and a conducting material (when the distance is small enough, typically around 1 nm) (Fig. 2.2). This current is the result of the overlapping wave functions between the tip atom and surface atom – electrons can tunnel across the vacuum barrier separating the tip and sample in the presence of a small bias voltage. This current is amplified (7 to 10 orders of magnitude) and allows the distance between the tip and the sample to be kept constant. It is mandatory to do this amplification as close as possible to the tip because any noise incorporated before the amplifier would be amplified with the relevant signal. If the tunneling current goes over its set-point the distance between tip and sample is increased, if it falls below this value then the distance is decreased.

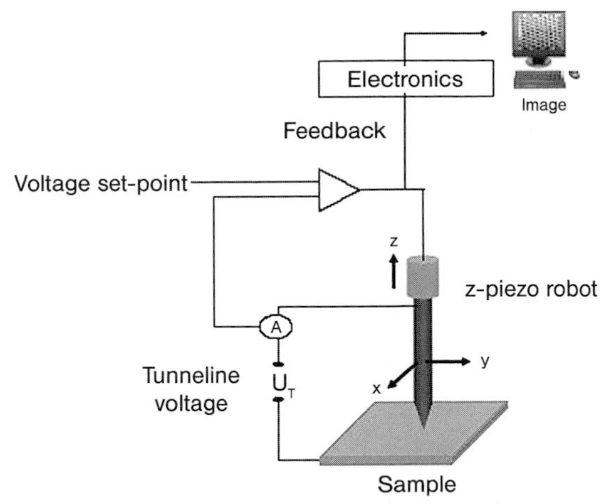

Fig. 2.2. Principle of scanning tunneling microscopy (STM) imaging.

Tunneling current originates from the wavelike properties of particles (electrons, in this case) in quantum mechanics. When a small voltage, *V*, is applied between the tip and the sample, the overlapped electron wave function permits quantum mechanical tunneling and a current flows across the vacuum gap (Fig. 2.3).

2.4 Atomic Force Microscopy

2.4.1 Generalities

Atomic Force Microscopy (AFM) and its various imaging modes is the most widely used member of the SPM family in the study of biological systems (Fig. 2.4). The interactions used in AFM are the local attractive and repulsive forces occurring between a tip attached to a flexible cantilever and the sample surface. A typical curve of tip–sample interaction is given below (Fig. 2.5) along with one of the more generally used scanning conformations. As the tip approaches the surface from an infinite distance, firstly attractive forces are present and due to the flexible nature of the cantilever it deflects toward the surface; at close ranges repulsive forces will dominate and the cantilever will deflect away from the surface. By use of a laser beam reflected from the surface of the cantilever (where the tip is mounted) onto a four quadrant Position Sensitive Device (PSD) the deflection of the cantilever can be measured and a feedback loop can be activated to position the cantilever with regard to the surface (Fig. 2.4). Obviously, the choice of the deflection at a set point to set in place the feedback loop along with the spring constant of the cantilever allows the forces applied to the surface to be determined.

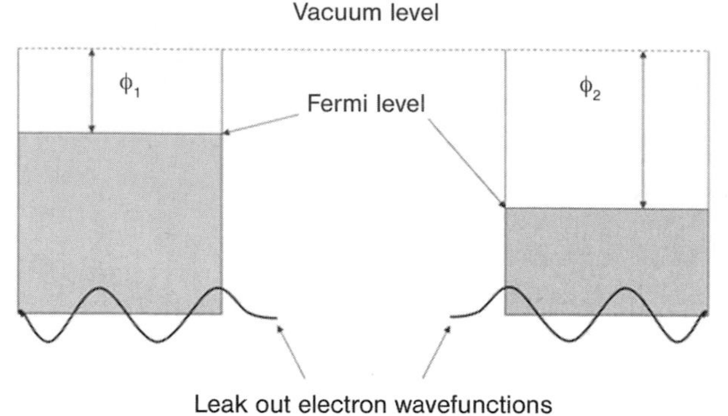

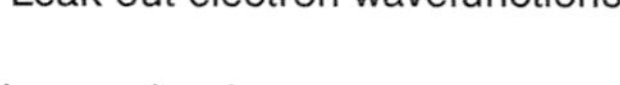

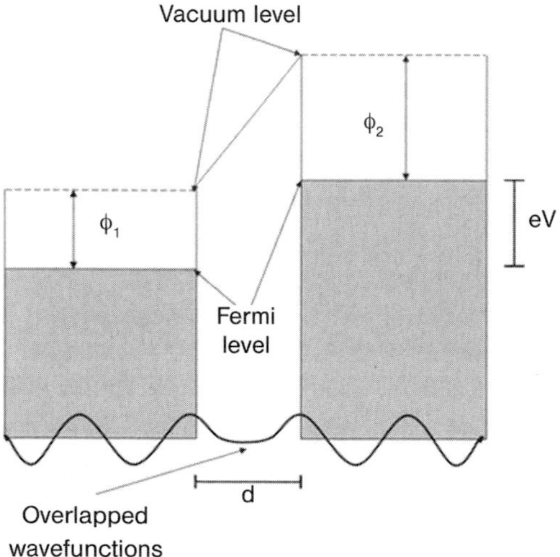

Fig. 2.3. Metal–vacuum–metal tunneling junction.

The schematic shown in Fig. 2.4 is one of two possible configurations of the microscope set up; here the piezo-ceramic scanners used to move in the x, y and z dimensions are positioned to move the sample with the cantilever and the tip is held in a constant position. The second configuration has the piezo-ceramic scanners positioned above the sample to move the cantilever and the tip. Evidently both configurations present advantages and disadvantages; for a moving sample stage the rigidity and, hence, stability of the tip positioning can be maximized (however, access from below to the sample is blocked); this configuration also requires sealed fluid cells to prevent leakage into the piezo-ceramics when scanning in liquid systems. The use of a moving cantilever assembly above the sample will be less rigid and so more prone to "noise" in the images; however, for biological samples the advantage of being able to couple the AFM to a conventional optical microscope far outweighs the possible noise problem. There is also in this configuration much more experimental flexibility with regard to working in liquids.

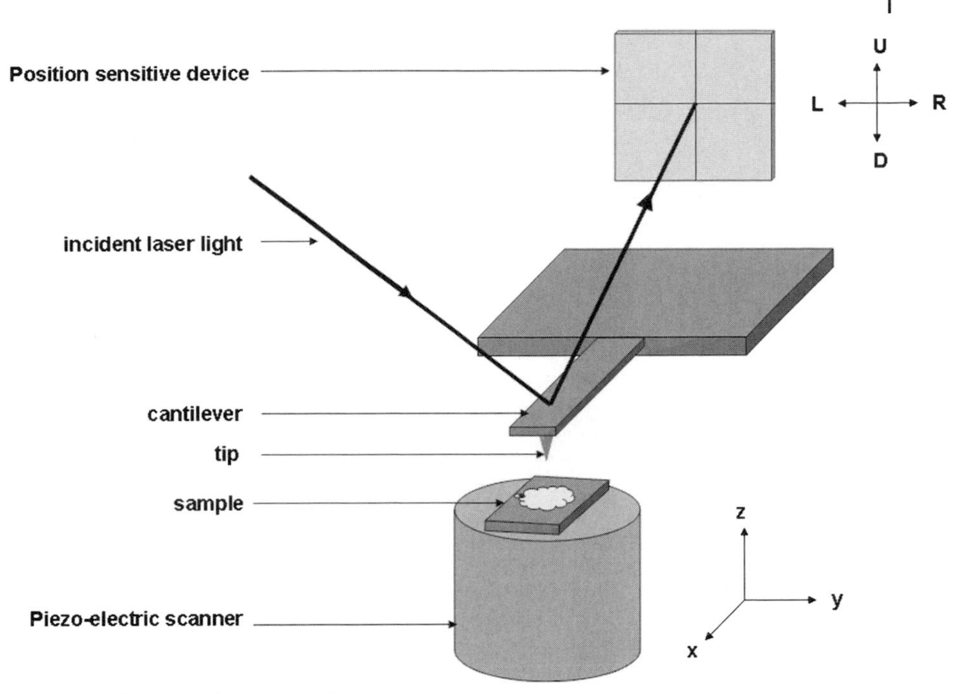

Fig. 2.4. Schematic of the atomic force microscope measurement system using the light deflection mode.

The most used imaging modes in AFM are the contact, force modulation, non-contact and intermittent contact modes. They may be best defined with regard to their positioning along the tip–sample interaction curve (Fig. 2.5).

For both contact and force modulation modes the interactions remain entirely in the repulsive zone of the curve and hence there is no part of the experiment in

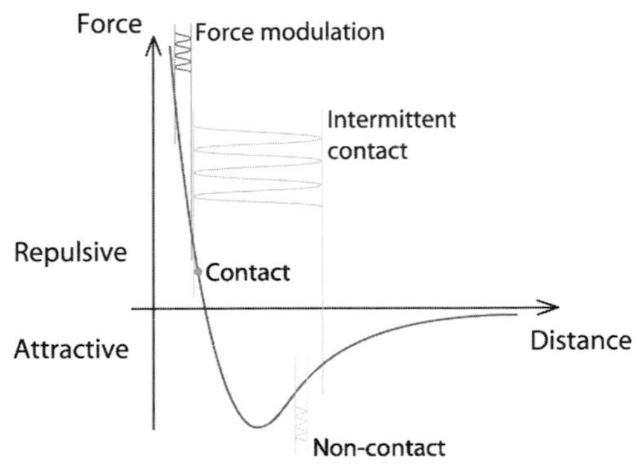

Fig. 2.5. Force curve as function of the distance between the tip and the surface. (Courtesy of JPK GmbH.)

which the tip is not touching the surface. In non-contact mode the interactions remain entirely in the attractive part of the curve, while for intermittent contact (often termed the "tapping" mode) the point will oscillate between attractive and repulsive parts of the curve.

At this point it is necessary to consider the hydration of a surface in air. Effectively, all surfaces, even highly hydrophobic ones, have a thin "contamination" layer of water associated with them, which is generally of the order of 1–5 nm high. Hence, experimentally, contact and force modulation modes of SPM imaging will be with the tip within the aqueous layer, while intermittent contact will involve the tip moving in and out of the water layer. However, for the non-contact mode it is possible to position the tip either within the water layer or when there is high cohesion of this layer to come into feedback at the surface of the aqueous layer; such positioning can modify the images obtained.

The water layer has a second effect. By capillarity, water will move up the tip to form a meniscus, and hence generate secondary drag effects when the tip scans the surface. These capillary forces may often be greater than the applied tip–sample forces and are often a cause of sample distortion or damage (cf. Section 2.6 on artifacts).

2.4.2 Tips and Cantilevers

The SPM experiment is, in reality, simply a mean to position, move and observe the probe across a surface – hence the key to the experiment is the probe or tip.

SPM probes generally consist of a sharp (or ultra-sharp) tip mounted on the end of small flexible cantilevers [13]. Obviously, the shape and the geometry (and also the quality) of the tip will determine the resolution of the SPM image. They are generally produced from Si or Si_3N_4 by micro-fabrication techniques.

Two basic shapes of cantilevers are commercially available, the V-shaped and the beam cantilever (Fig. 2.6). Because of higher mechanical stability, V-shaped cantilevers are generally preferred for contact-mode imaging (Fig. 2.7). When mechanical properties (torsion or friction forces) are studied, beam cantilevers would be preferred. For each type of cantilever, a force (spring) constant could be determined and depends on its geometry, dimensions and the material from which it is fabricated.

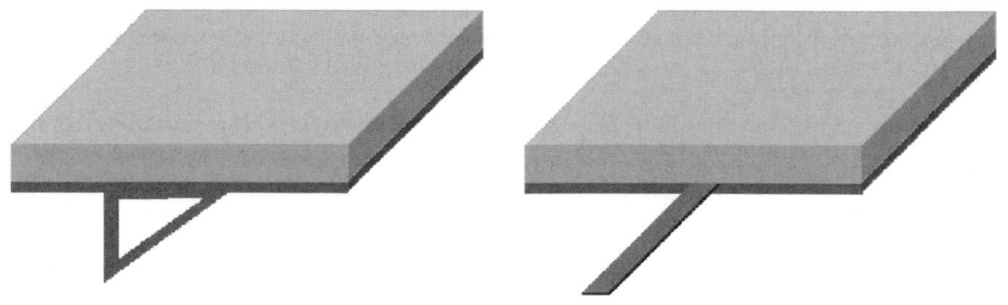

Fig. 2.6. V-shaped and beam SPM cantilevers.

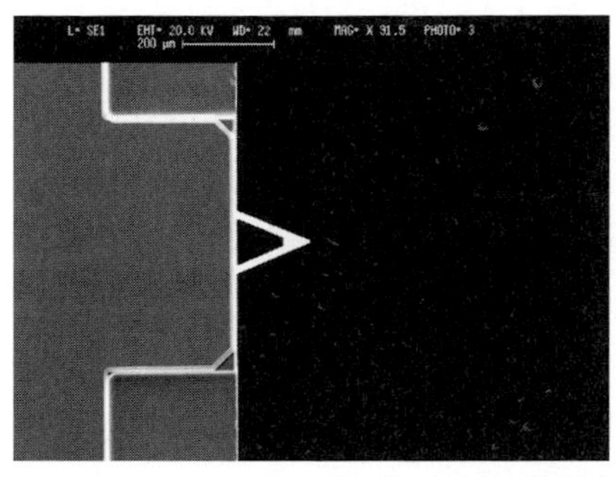

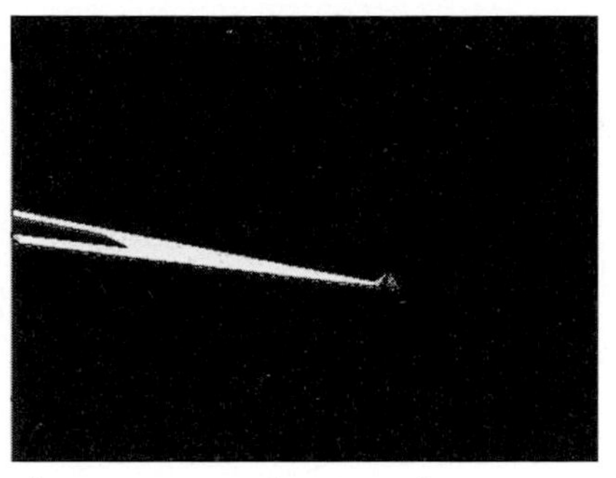

Fig. 2.7. SEM images of V-shaped pyramidal cantilevers. (Courtesy of Dr A. Wirth.)

Table 2.2 summarizes typical physical data for various commercially available cantilevers for contact and dynamic modes of SPM.

AFM tips are usually made of the same material as the cantilever as both are usually fabricated in the same time. The geometry of the tip depends on the micro-fabrication technique [14]; two main tip geometries exist: pyramidal and cone-shaped (Fig. 2.8). These tips may be submitted to additional treatments to sharpen [15] or chemically modify them [16].

Very high aspect ratios may be obtained with electron beam deposited (EBD) tips, which are AFM tips modified to grow a narrow tip-on-the-tip (Fig. 2.9). Normally, shaped silicon intermittent-contact cantilevers are modified in scanning electron microscopes. The electron beam focused on top of the tip deposits a small column of carbon, leading to a tip with narrow radius and high aspect ratio. The

Tab. 2.2. Physical characteristics of commercial cantilevers used in different modes of operation in SPM.

Contact mode cantilevers		
Technical data	***Typical value***	***Range***
Thickness (µm)	2	1.5–2.5
Mean width (µm)	50	45–55
Length (µm)	450	445–455
Force constant (N m^{-1})	0.2	0.07–0.4
Resonant frequency (kHz)	13	9–17
Non-contact and intermittent contact mode cantilevers		
Technical data	***Typical value***	***Range***
Thickness (µm)	4	3.5–4.5
Mean width (µm)	30	25–35
Length (µm)	125	120–130
Force constant (N m^{-1})	42	21–78
Resonant frequency (kHz)	320	250–390

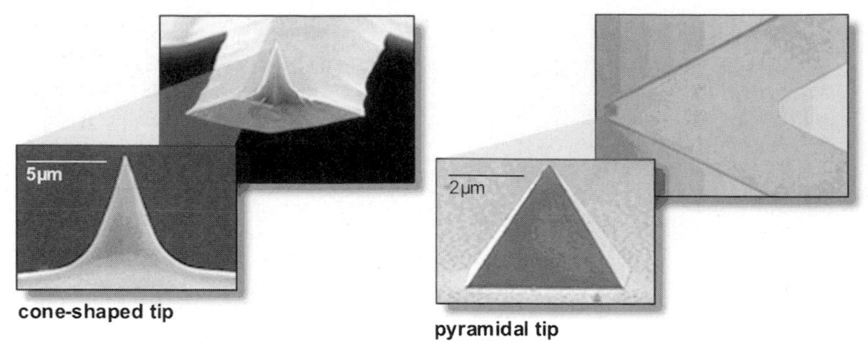

Fig. 2.8. SEM images of pyramidal and cone-shaped SPM tips. (Courtesy of JPK GmbH.)

carbon arises from hydrocarbon contamination in the vacuum chamber of the SEM. The high aspect ratio of the tip allows imaging of structures such as grooves or troughs, which are not properly imaged by tips with conventional shapes and aspect ratios [17].

2.4.3
Contact Mode AFM

In the contact mode the tip is in continuous contact with the surface (Fig. 2.10); consequently, very high resolution imaging becomes possible. On crystalline surfaces atomic level resolution is routine for mineral systems. With well-organized two-dimensional crystals molecular imaging of proteins with sufficient resolution to show sub-units can be achieved (see Section 2.7.2 on protein imaging).

The force exerted by the tip on the sample can readily be controlled by modification of the set point value; given that this is generated by the deflection of the cantilever a lower value will evidently lead to lower interaction forces.

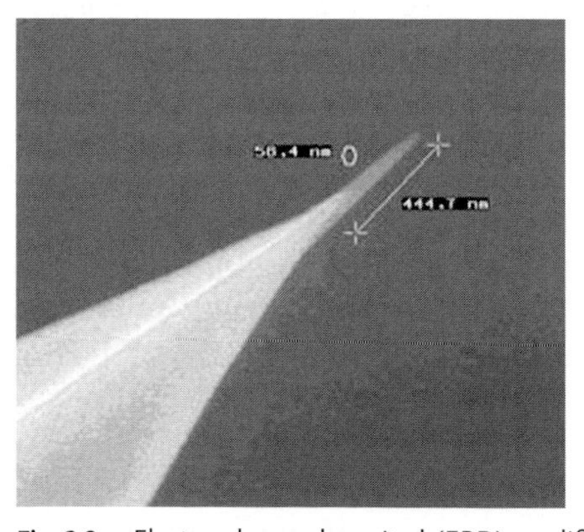

Fig. 2.9. Electron beam deposited (EBD) modified tip. (Courtesy of JPK GmbH.)

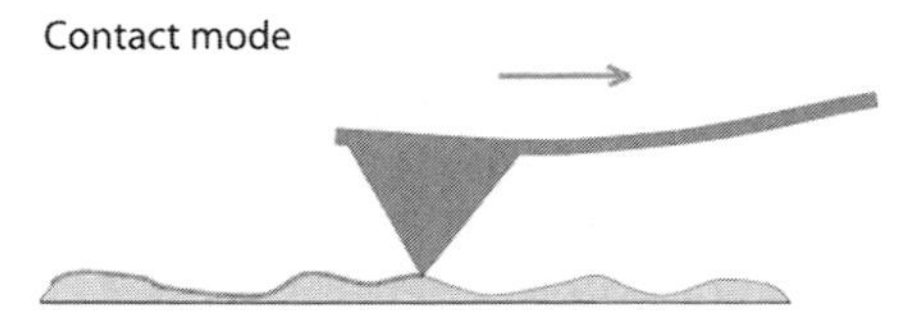

Fig. 2.10. Contact mode imaging; physical contact between the tip and the sample. (Courtesy of JPK GmbH.)

However, lateral forces will also be present as the tip is effectively dragged across the surface and these will, for imaging in air, be combined with effects from the capillary forces of the water contamination layer. This may generate, in systems having low adhesion to the surface on which the sample is placed, movement of all or parts of the sample; similarly, samples having low cohesion may be moved. The problem of capillary forces may be removed by carrying out the imaging in water.

Cantilevers used in contact mode imaging have, in general, low spring constants with resonance frequencies of about 10 kHz.

In nanolithography this effect is taken to its extreme and in fact becomes a useful tool, as by increasing the set point and hence the forces applied to the surface material may be voluntarily removed from the surface. This has been used to measure the heights of Langmuir–Blodgett films and also to remove self-assembled monolayers to allow deposition of other molecular layers.

2.4.4 Dynamic Modes

2.4.4.1 **Generalities**

In the various non-contact or dynamic modes the cantilever vibrates and it is this oscillation of the cantilever that is measured to generate the image rather than direct tip deflection. Various methods can be used to cause oscillation of the cantilever, for imaging in air these are usually mechanical, magnetic or piezoelectric, but in liquid imaging the cantilever is driven acoustically.

The resonant frequency f of the cantilever is defined by Eq. (1).

$$\text{Res. Freq.} \quad f = \frac{1}{2\pi}\sqrt{\frac{k}{m}} \tag{1}$$

where k is the spring constant of the cantilever and m is the mass. Evidently, as the frequency passes through a resonant condition the amplitude of the oscillation increases to a maximum value before decreasing. This peak in the oscillation of the amplitude is accompanied by a switch in phase of the detected oscillation. Hence two modes of detection are possible: either via the amplitude directly or via measurement of the phase shift of the signal.

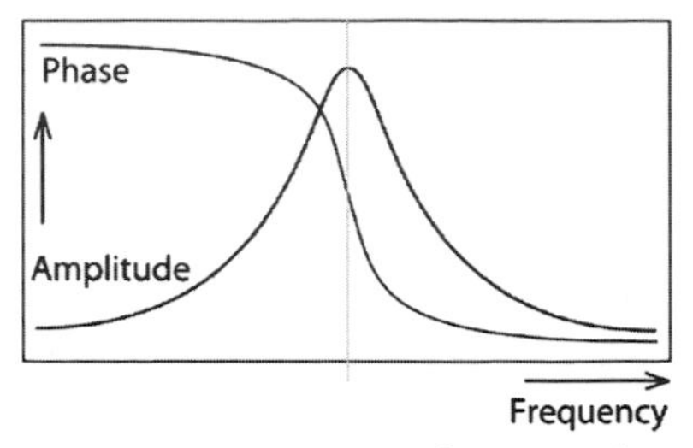

Fig. 2.11. First resonance frequency of a cantilever measured with optical detection. (Courtesy of JPK GmbH.)

Experimentally the situation is seldom as clear as presented in Fig. 2.11 and, often, several resonance peaks are available for use. It is also often not clear which of the resonances will generate the best image!

As the tip approaches the surface the effective mass of the cantilever will change due to the attractive forces acting on the point and hence the resonant frequency will vary. By setting the signal acquisition close to the resonant frequency and defining the set point in terms of the signal amplitude the feedback loop is engaged.

2.4.4.2 **Non-contact Mode**

In non-contact mode (Fig. 2.12) the tip remains at all times in the attractive part of the interaction curve and the tip is scanned at a small distance (a few nanometers) above the surface with a relatively small amplitude. A clear problem with this mode is that the tip may jump into contact with the surface if the attractive forces exerted are greater than the spring constant of the cantilever, and thus much stiffer cantilevers are required. Generally, cantilevers with resonance frequencies in the range 150–300 kHz are used. True non-contact mode is almost unusable in liquid systems as the damping of the small cantilever oscillations by water or other liquids is too large and the signal disappears.

Resolution is also reduced in non-contact imaging, with a minimum value of around 1 nm. However, this may be compensated in the case of very soft samples by the lack of any contact between the tip and the sample, which may allow more accurate information in the z-axis, i.e., on the height.

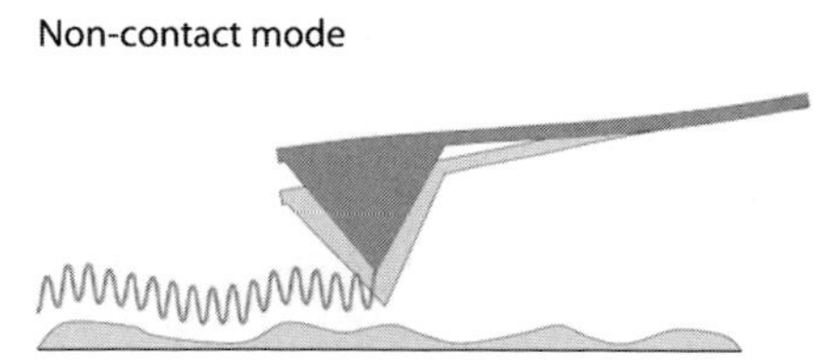

Fig. 2.12. Non-contact mode imaging based on oscillation of the cantilever due to interaction of the tip with the surface. (Courtesy of JPK GmbH.)

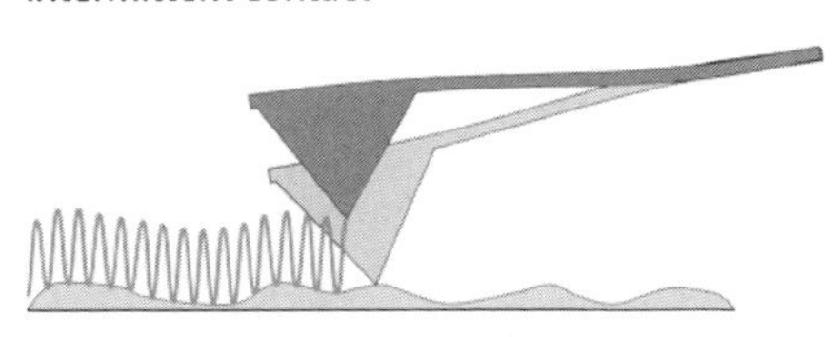

Fig. 2.13. Intermittent contact mode imaging characterized by oscillation with high amplitudes. (Courtesy of JPK GmbH.)

2.4.4.3 Intermittent Contact Mode

In intermittent contact mode, or as it is often termed the tapping mode, the cantilever moves rapidly with a large oscillation between the repulsive and attractive regimes of the interaction curve (Fig. 2.13). Here the maximum orthogonal forces applied to the surface may be lower or higher than those experienced in the contact mode but such forces are not applied constantly, thus lowering drag forces on the sample and reducing to some extent the possibility of sample damage. However, care must be taken with regard to very soft samples in terms of compression and possible distortion of height information.

Again stiff cantilevers, generally with resonant frequencies in the range 200–400 kHz, are used to allow the tip to break free of the water contamination layer. With regard to imaging using intermittent contact mode in water the problem of capillary forces is removed and much softer cantilevers, often contact mode, are generally used. However, again severe damping will occur. Technically this problem is overcome by driving the cantilever at the resonant frequency of the liquid cell or of the acoustic cell used.

As noted above, there will be a shift between the phase of the drive signal and that observed by the lock-in amplifier of the detector. This phase signal is highly sensitive to the tip–sample interaction and can generate information on the mechanical properties of the sample. Such phase shifting may occur via adhesion between the tip and the sample or by a viscoelastic response of the sample. This can be particularly useful to analyze samples in which a phase separation has occurred between two components of a sample having the same height.

2.4.4.4 Force Modulation Mode

This mode combines the oscillation of the cantilever with scanning in the contact mode (Fig. 2.14). The cantilever is now oscillated at very low frequencies, normally between 1 and 5 kHz, and thus much below the resonant frequency of even the cantilevers used in contact mode imaging.

The information extracted concerns the mechanical and viscoelastic properties of the sample, and again is useful for extracting information on samples containing composite materials but which have no topographical differentiation between the components.

However, the forces applied to the sample can be very high and this mode is often damaging to both sample and tip.

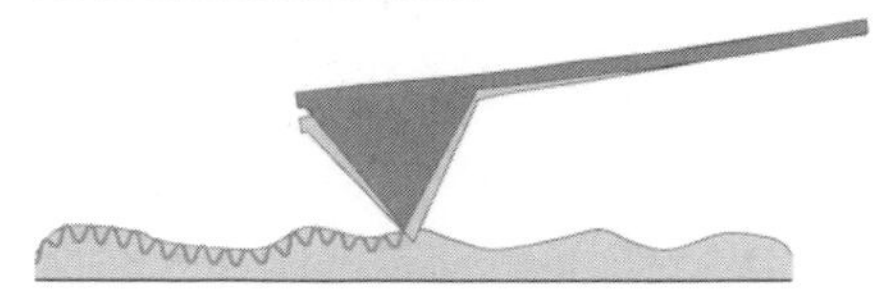

Fig. 2.14. Force modulation mode imaging based on the vertical motion of the scanner undergoing a "feeling out" of the surface by the cantilever. (Courtesy of JPK GmbH.)

2.4.5
Friction Force Mode or Lateral Force Mode

As noted in Section 2.4.1, the experimental setup of the AFM experiment uses a four quadrant detector. Thus information is not only available about the topography of the sample via the cantilever deflection in the vertical axis but also concerning twisting of the cantilever in the horizontal axis. This information concerns the friction forces occurring between the tip and the surface.

Gel surfaces that do not contribute to topographical information can be observed in the lateral mode image.

2.4.6
Force–Distance Analysis

If, in place of scanning, the tip is placed at a fixed point on the sample and now moved in the vertical direction to the surface and then retracted from the surface the deflection of the cantilever as it moves can be measured as shown in Fig. 2.15. The deflection of the cantilever will provide information on the mechanical properties of the material during the part of the approach and the retraction when it is in the repulsive, contact region of the cycle, and also on the adhesion interactions between the tip and the surface as it tries to disengage from the sample. This is termed force–distance measurement.

When the sample is hard and incompressible, as would be seen with glass, ceramics or metallic surfaces (Fig. 2.16a and b), the tip will simply approach the surface, jump into contact and then bend; the retraction curve will be the same, although scanning in air will induce adhesion of the tip to the sample by the water layer. To remove these water adhesion effects, which can be of the order of several to hundreds of nano-Newtons, force–distance measurements are normally carried out in a fluid system.

For more compressible samples the curve will be expected to resemble that in Fig. 2.16(c) and information on the mechanical properties of the sample may be extracted. However, care must be taken as effects of the sample substrate may also be observed. Often, during retraction hysteresis will be seen, particularly if the sample is not perfectly elastic.

However, now suppose that a molecule has been coupled to the tip and a recep-

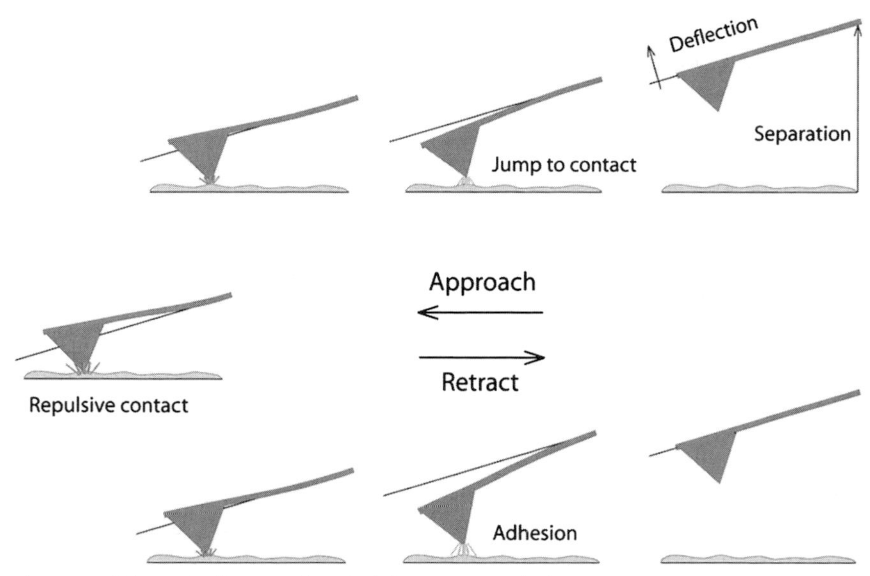

Fig. 2.15. Schematic diagram of vertical tip movement during the approach and retract parts of a force spectroscopy experiment. (Courtesy of JPK GmbH.)

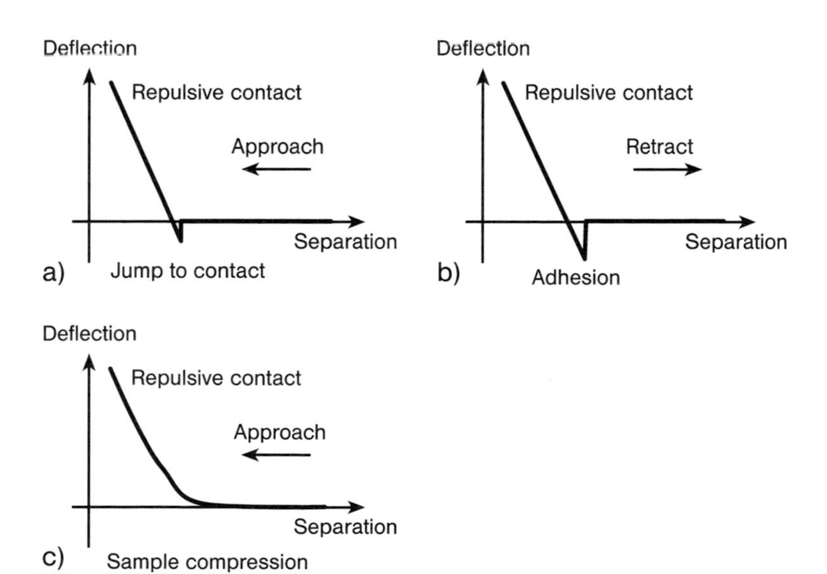

Fig. 2.16. Typical force–distance curves, showing, left to right, no adhesion on a hard surface, adhesion on a hard surface and non-adhesion on a soft surface in a liquid. (Courtesy of JPK GmbH.)

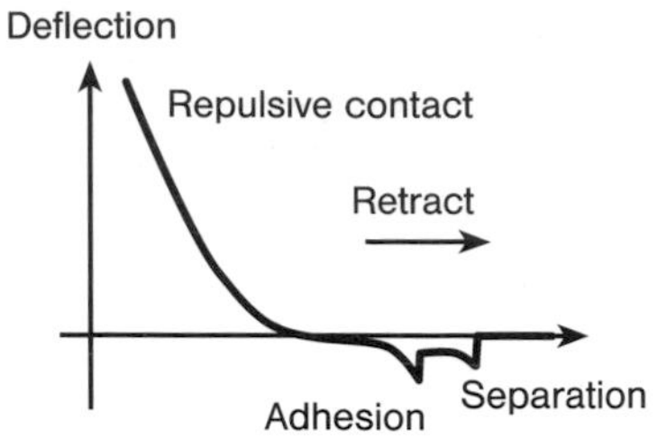

Fig. 2.17. Force–distance measurement in water between a ligand fixed to a coated AFM tip and the receptor present on the sample. (Courtesy of JPK GmbH.)

tor is present on the sample. When the tip is in contact with the surface there will be ligand–receptor coupling; as the tip is retracted, force will be applied to the complex until, eventually, the applied force is greater than the interaction force. At this point the tip will snap off. As the area of the tip is considerably greater than that of a single molecule one cannot be certain that the rupture of the complex is due to a single molecular event, and so multiple analysis and deconvolution of the force–distance curves (Fig. 2.17) are required to calculate a single molecule interaction force. However, where knowledge of the forces involved in single molecular complexes is not required, assuming that a set of tips are roughly the same size and using receptors of the same size, e.g., antibodies, the experiment may be used to analyze whether there is recognition between a ligand and a biomolecule [18].

Interaction forces have been calculated for a wide range of molecular interactions, varying from the streptavidin–biotin complex, which has the strongest non-covalent interaction known for biological systems at 340 ± 120 pN [26], down to carboxylic acid–carboxylic acid interactions where forces are in the range of 17 pN [20]. Table 2.3 gives typical values of molecular interactions forces for biological molecules and interaction forces for cells.

AFM has also been used to evaluate the strength of a covalent bond (Si–C) under an external load [36].

For biological molecules other events may be associated with force–distance curves, including unraveling of the molecule, as has been observed for polysaccharides [37, 38], DNA [39–42] and proteins [43–49].

2.4.7
Chemical Force Imaging

In Chemical Force Microscopy the tip is modified chemically, either directly by using organo-silane chemistry or by alkylthiol chemistry after first coating the tips with a thin layer of gold. In this way charged (ammonium or carboxylate), hydrogen bonding (hydroxyl) or non-interacting hydrophobic (alkane) surfaces may be grafted onto the tips, Fig. 2.18 [50].

Tab. 2.3. Intermolecular forces derived from force spectroscopy.

Receptor	*Ligand*	*Force (pN)*	*Ref.*
Thymine	Adenine	54	19
Carboxylic acid	Carboxylic acid	16.6	20
Biotin	Avidin	160	21
Iminobiotin	Avidin	85	21
Adhesion glycoprotein CsA	Adhesion glycoprotein CsA	23	22
Concavalin-A (lectin)	Mannose	47 ± 9	23
Meromyosin	Actin	15–25	24
Biotin	Streptavidin	200–257	25, 26
		340 ± 120	
20 Base pairs DNA	Complementary strand	1520	27
16		1110	
12		830	
Cell-adhesion proteoglycans	Cell-adhesion proteoglycans	40–125	28
Biotin	Antibiotin antibody	60	29
Anti-HAS, other antibody	HAS, other antigen	49–244	30, 31
P-selectin	Glycoprotein ligand-1	165	32
Cell receptor	RGD	35–120	33
Uterine epithelium	Trophoblast	1000–16000	34
Uncoated surface	Cell-adhesion proteoglycans	19000–100000	35
Coated surface	Cell	100000–220000	35

Obviously, in the case of gold-coated tips the resolution of the experiment will reduce convolution effects due to the tip.

Scanning in the lateral force mode allows chemical imaging of structures present on the surface due to differences in interactions between the tip and the sample. Figure 2.19 shows the results of scanning with alkyl carboxylic acid or alkane

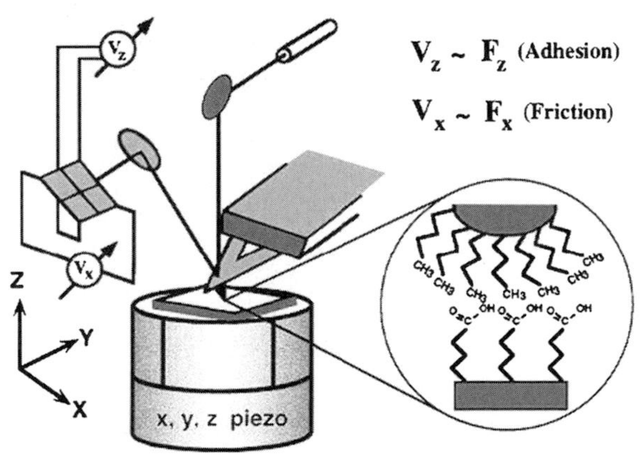

Fig. 2.18. Schematic of the principle of chemical force microscopy (CFM). (Reprinted with permission from Ref. [50]. Copyright (1997) Annual Review of Materials Science.)

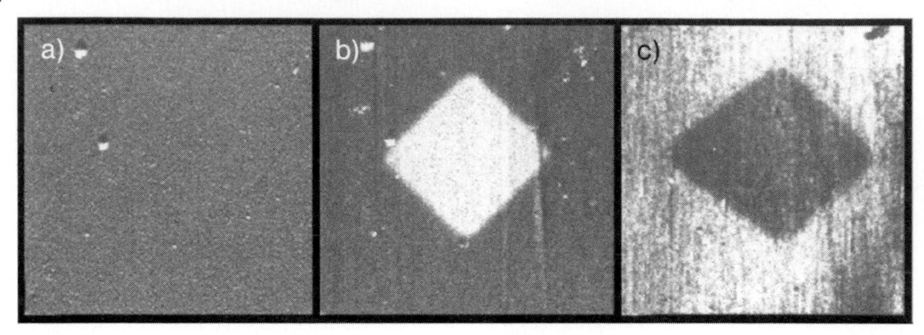

Fig. 2.19. Force microscopy images of a photo-patterned SAM sample. The 10 × 10 μ square region terminates in COOH, and the surrounding region terminates in CH_3. The images are of (A) topography, (B) friction force using a tip modified with a COOH-terminated SAM, and (C) friction force using a tip modified with a CH_3-terminated SAM. (Reprinted with permission from Ref. [50]. Copyright (1997) Annual Review of Materials Science.)

modified points on patterned surfaces in which a square pattern of an alkylcarboxylic acid has been deposited. The topographic image in Fig. 2.19(A) clearly shows that there are no variations in layer heights. Figure 2.19(B) presents a friction force image using a tip modified with a COOH-terminated SAM, and Fig. 2.19(C) a friction force image using a tip modified with a CH_3-terminated SAM. Light regions in Figs. 2.19(B) and (C) indicate high friction, and dark regions indicate low friction [50].

2.4.8
Dip-pen Lithography

In a further extension of SPM derived techniques, the use of AFM tips to directly "write" structures on surfaces has been demonstrated [51–53]. Based on the soft lithography of Whitesides [54] a suitable molecule is adsorbed by capillary forces onto the AFM tip. This is then used in the contact-mode to scan pre-determined patterns on a given surface (Fig. 2.20). The use of suitable programming allows features of any shape and form to transfer to the surface with a resolution of 30 to 100 nm [53]. The choice of molecular ink to be transferred depends on the particular surface chemistry of the substrate, the application of alkane thiols onto gold surfaces being the most generally used chemistry.

2.4.9
Cantilever Array Sensors

In contrast to all SPM derived methods, cantilever sensors [8] are based not on probe–sample interactions but rather on direct interaction of the sample with tipless cantilevers. Signal detection is based on measurement of the mechanical response of the thin Si beams, arranged in a micro-fabricated array (Fig. 2.21).

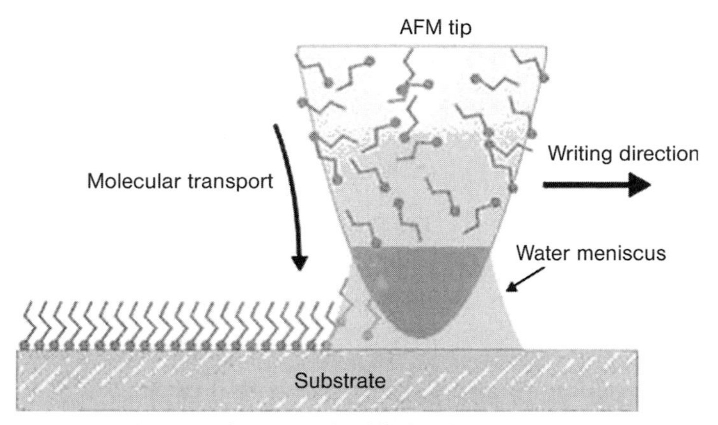

Fig. 2.20. Schematic of the principle of dip-pen lithography. (Reprinted with permission from Ref. [53]. Copyright (1999) Science.)

Fig. 2.21. Microfabricated Si cantilever array. (Reprinted with permission from Ref. [8]. Copyright (2000) Elsevier.)

Cantilever sensor arrays experiments can be performed in either static or dynamic mode. In the static mode, only one side of the cantilever is chemically modified (with a receptor molecule); contact with the analyte is responsible for surface stress that causes a bending of the cantilever. In dynamic mode, both sides of the cantilever may be modified and detection is based on mass-dependant resonance frequency (or phase) shifts of the cantilever.

Detection is either by deflection of the laser on the standard AFM photo-detector or by resonance shifts as detected with dynamic AFM modes (Fig. 2.22).

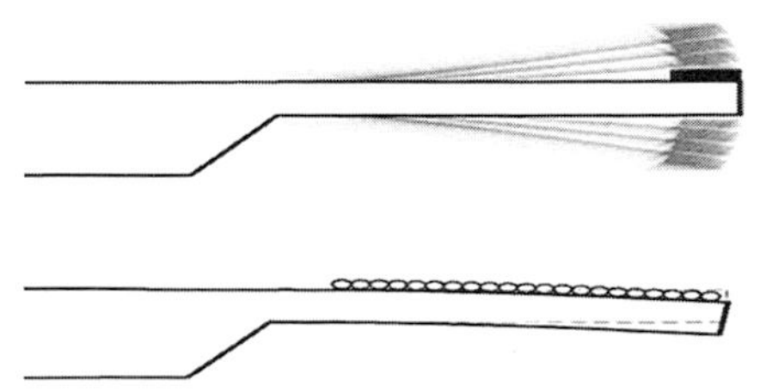

Fig. 2.22. Schematic of the detection principle. (Reprinted with permission from Ref. [5]. Copyright (2001) Elsevier.) AQ3

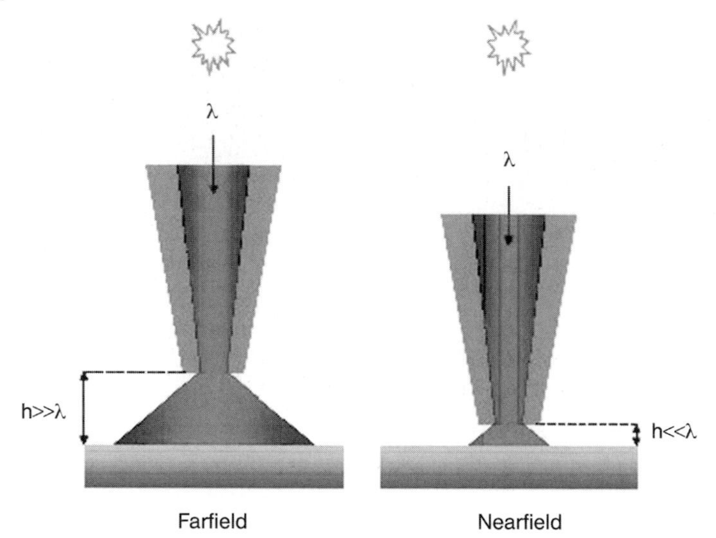

Fig. 2.23. Schematic of far-field and near-field microscopies, and probe positioning.

2.5
Near-field Scanning Optical Microscopy

Near-field illumination of a sample with visible light can resolve features well beyond the resolution of conventional, far-field microscopes [55–58]. Near-field Scanning Optical Microscopy (NSOM) thus has the potential of extending the resolution of techniques such as fluorescent labeling, yielding images of cell structures and molecules on the nanoscale (Fig. 2.23).

The spatial resolution of ordinary light microscopes is limited to about 250 nm, the objective lens or equivalent, at a distance from the object observed that is several times the wavelength, λ, of the illuminating light. At this distance the high spatial frequencies in the image components giving information about the smallest features of the object are not collected by the imaging optics and do not contribute to image formation. It is this phenomenon, rather than the physics of light itself, that sets the $\lambda/2$ diffraction limit. If the collecting or the illuminating optics can be brought closer to the object, to a distance less $\ll \lambda$ from its surface (Figure 23, right-hand side), then the high spatial frequencies can be resolved, at the price of reconstructing the image from the scan of the optics across the sample. The most efficient way to do this is to bring an aperture with diameter $\ll \lambda$ to within a few nm of the object surface. Positioning at this distance allows collection of light from the sample that otherwise would be lost to far-field collection. Conversely, illuminating through the aperture means that the incident light reaches the sample and interacts with it before it can be diffracted and lost. Thus, NSOM, like scanning confocal microscopy or AFM, builds an image by registering intensity for each point of the scan. Like confocal microscopy, NSOM uses visible light. Like AFM, it achieves super-resolution by means of a very small probe. All of the

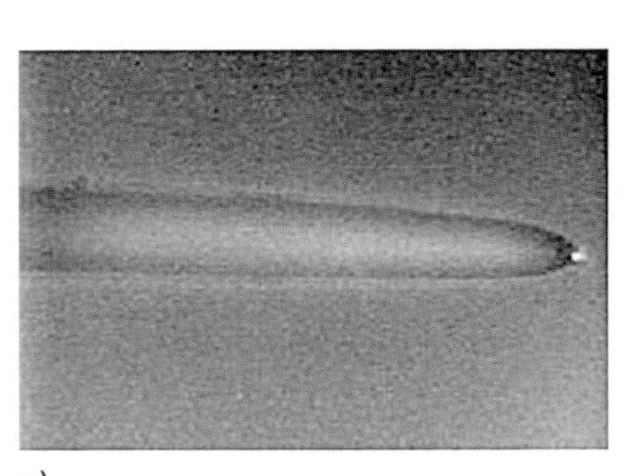

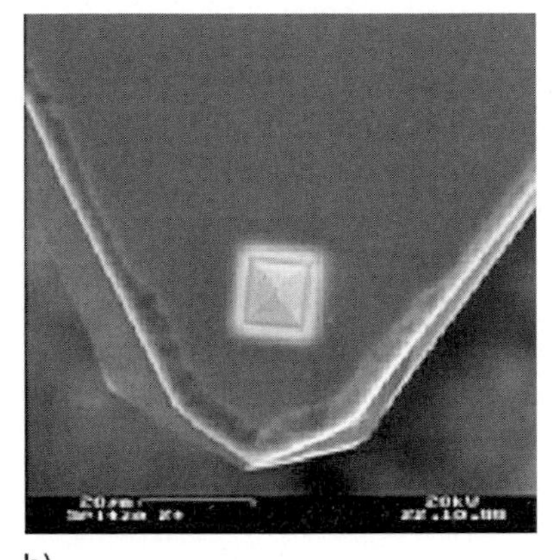

a) b)

Fig. 2.24. The two forms of NSOM tips: (a) a pulled optical fiber with aluminum cladding and (b) an etched aperture in a standard AFM cantilever.

problems and the promise of NSOM then turn on the problems of fabricating the aperture, and scanning across the sample surface while keeping it within near-field distance of the surface.

Two main forms of NSOM tips exist (Fig. 2.24): pulled optical fibers covered with an aluminum layer and AFM tips into which the aperture is etched. Fiber tips have a better theoretical resolution, below 20 nm, while that for etched AFM tips is higher, around 50 nm; however, such resolution is not generally achieved.

In practice, fiber tips are much more fragile than AFM cantilever based tips. With regard to scanning in liquids, often a condition for biological samples, feedback control is easily obtained for the AFM tips in either contact or dynamic modes; however, fiber tips use either shear-force or an attached tuning fork for feedback, and damping of the signal is generally very high, making the experiment difficult.

2.6 Artifacts

2.6.1 Artifacts Related to Tip Size and Geometry

The shape of an AFM tip can modify considerably, with both the geometry and the radius of the tip causing artifacts. Figure 2.25 shows the effects of tip geometry on the form of the observed image of a vertical step.

None of these tip geometries can accurately give a true image of the step, indeed only a truly two-dimensional tip could image correctly such a feature. The image will always contain elements arising from a combination of tip shape and the true topography of the surface. While long thin tips will clearly best reproduce the

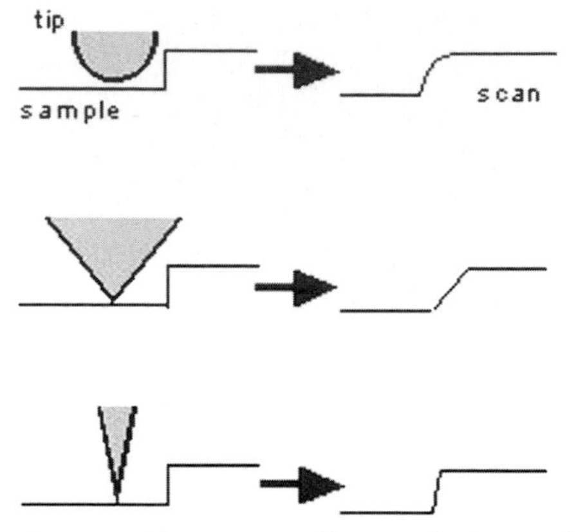

Fig. 2.25. Tip geometry effects on the image of a sharply stepped surface.

topography, there is a trade-off between the mechanical strength of the tip and the quality of the image, with sharp narrow tips being much more fragile and apt to break, which can then lead to other artifacts (see below).

Tip artifacts arising from the geometry can be modeled in terms of the cone angle of the main pyramid forming the tip and the radius of the tip end. Thus images of small sharp objects will be dominated by the tip radius, leading to sizes in the x and y dimensions that effectively measure the tip size. However, the information in the z-axis is correct and may be used to obtain the correct size of spherical or cylindrical objects. For larger objects the artifacts will be dominated by the cone angle, as noted above.

The relationship between the observed width W of a feature and the diameter of the probe tip can be calculated for an idealized tip shape, such as the one shown in Fig. 2.26, where $x^2 = R^2 - (R - d)^2$. For $R \gg d$, $W = \sqrt{(8dR)}$ and $d = (W^2/8R)$. For $R = 10$ nm and $d = 5$ nm, the observed width W would be 20 nm.

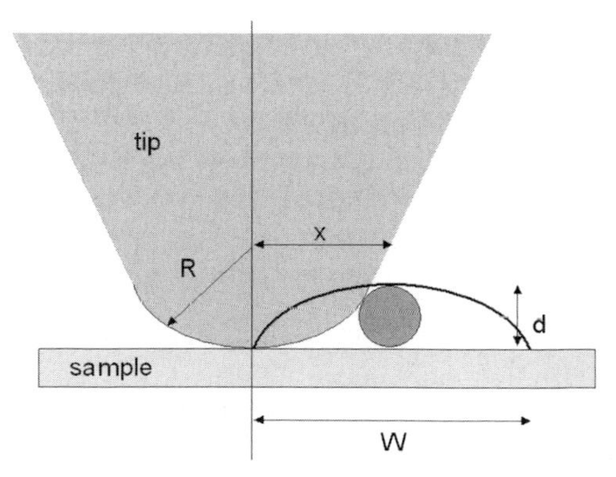

Fig. 2.26. Geometries used in investigating tip deconvolution effects.

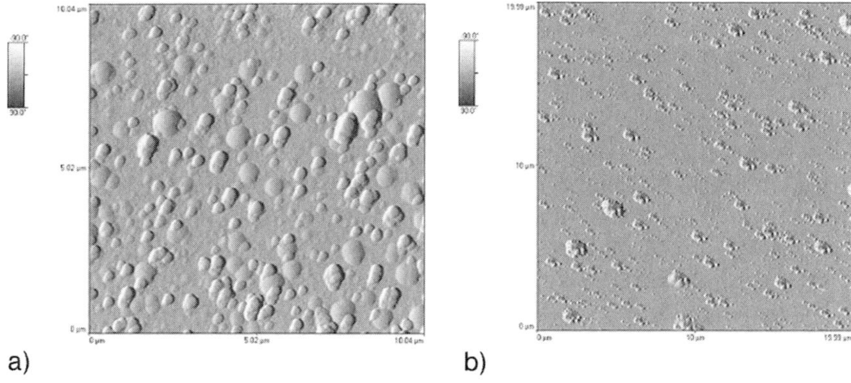

Fig. 2.27. Artifacts arising from (a) a blunt tip with triangular geometry and (b) a broken tip with at least double tips.

2.6.2
Artifacts from Damaged Tips

Sometimes the tip will be damaged during use or during the fabrication process, leading to the occurrence of double or multiple tips. Such tips will generate images in which features will be present as pairs or larger multiples. These are normally relatively easy to identify (Fig. 2.27). Other damaged tips may simply have non-pyramidal shapes – again these can be identified by the presence of repeating geometries of all the features.

2.6.3
Artifacts from Tip–Sample Interactions

Artifacts may also occur when tip–sample interactions are greater than the mechanical strength of the sample being imaged. This is often encountered in the contact mode imaging of soft surfaces, or when the response times of the feedback loop are incorrectly set. This often occurs when there are rapid changes in the height of the sample in certain areas of the sample and the feedback loop values have been set on a flat part of the surface.

2.6.4
Sample Artifacts

Biological samples scanned in air are prone to sample artifacts arising from the presence of buffers in the solution used to prepare the sample. On drying, the buffer may crystallize, yielding dendritic patterns (Fig. 2.28a). By careful washing the buffer can often be selectively removed. However, some care may be taken to determine whether the patterns observed arise from the buffer, the sample or from highly complex structures arising from self-assembly (Fig. 2.28b). Here the

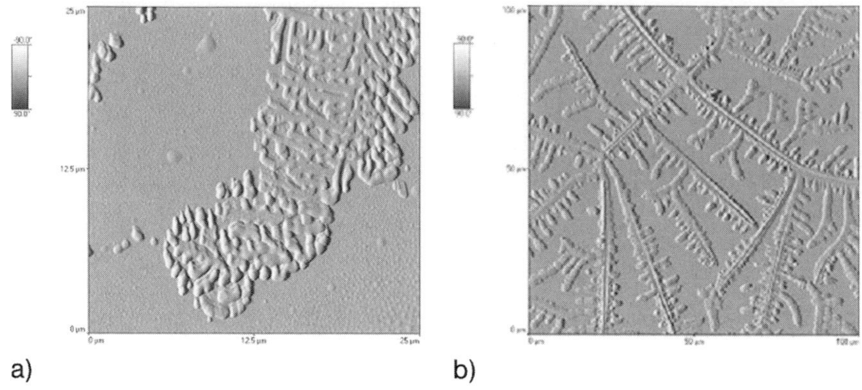

Fig. 2.28. (a) Artifact caused by phosphate buffer; dendritic crystallization is evident in the upper left of the image, (b) Non-artifactual patterns in a bovine serum albumin–sodium chloride mesocomposite.

apparently dendritic patterns are not "true artifacts" but arise from the protein, in this case bovine serum albumin, controlling crystallization of composite structures containing salt–protein complexes [59]. A similar problem was observed in the self-assembly of cowpea mosaic virus particles into highly ordered comb-structures with a spacing of 6 μm [60].

2.7
Using the Tools

In this section we attempt to illustrate how the various methods of SPM can be applied to the study of biological systems, using selected examples for each type of system. Evidently, due to space requirements, the presentations will not be exhaustive, but rather represent my personal choice (A.W.C.). So, in advance, I apologize to any author who feels left out in the examples presented.

2.7.1
DNA

DNA has been imaged with a wide range of SPMs since the very debut of such experiments, firstly by STM [61] and then by AFM [62]. Subsequent experiments have involved various types of manipulation of DNA and determination of the recognition properties [63] and the mechanical unraveling of DNA strands [64]. Application of SPM to DNA has been widely reviewed (see Appendix 2).

2.7.1.1 Topographic Imaging of DNA
In a series of single molecule experiments with different methods of AFM, Anselmetti et al. [65] observed topographically various DNA topologies (Fig. 2.29). Linear

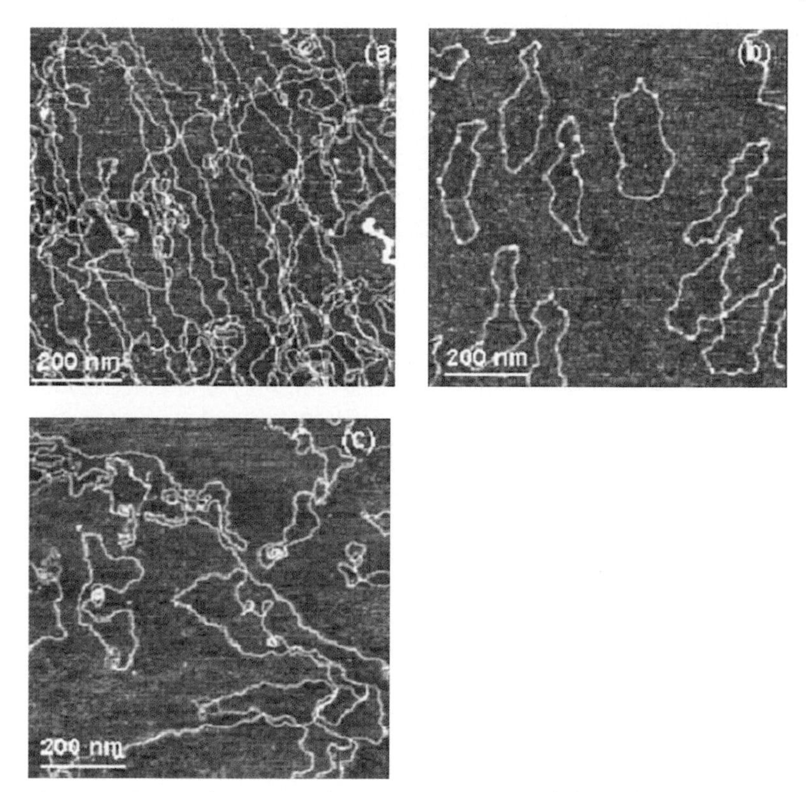

Fig. 2.29. Topographic images of various DNA topologies. (a) Linear l-DNA (48.5 kbp), (b) non-twisted circular DNA plasmids (vector 3.2 kbp) and (c) circular supercoiled DNA with twists and writhes due to internal supercoiling (supercoiled DNA ladder 2–16 kbp). Intermittent contact mode, vertical scale 3 nm. (Reprinted with permission from Ref. [65]. Copyright (2000) Wiley.)

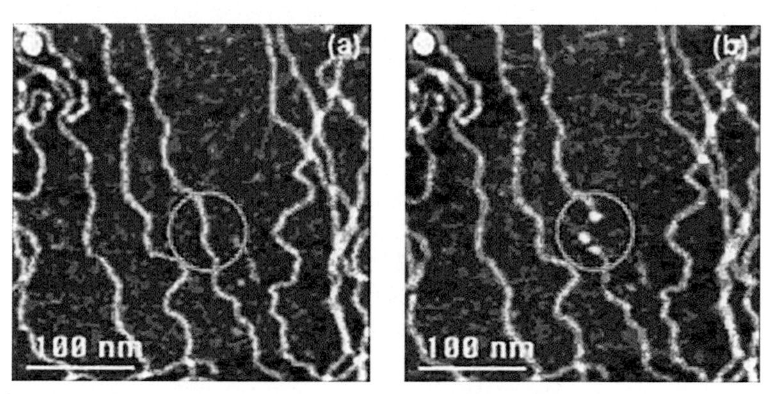

Fig. 2.30. Topographic images of λ-DNA: (a) Before and (b) after cutting a dsDNA strand. Intermittent contact mode, vertical scale 3 nm. (Reprinted with permission from Ref. [65]. Copyright (2000) Wiley.)

double strand DNA, circular plasmids and supercoiled DNA were observed. The measured widths of the strands were in the range 3–7 nm, which can be compared to the known radius of 2 nm [66]; thus the size of the imaged DNA is controlled by the tip convolution effects cited in Section 2.6.1. In comparison the observed height was 1 nm.

The experiments were further expanded to use the AFM tip as a cutting tool to locally cut the DNA strands. By using much higher tip-surface forces than normally used for scanning of soft material and repeatedly scanning at a predefined point on the DNA strand, the authors were able to use the AFM tip as a mechanical restriction endonuclease (Fig. 2.30).

As discussed above in Section 3.4.4.3, both topographic and phase difference information are available from the intermittent contact imaging AFM mode. By using carefully controlled humidity conditions Lee et al. [67] were able to image very short oligonucleotide strands (a biotinylated 21 mer 5′-biotinGAGGAGTTGGGG-GAGCACATT-3′) attached to streptavidin on a gold surface. Under normal conditions such short strands of DNA-oligonucleotides are impossible to image and even here at low humidity little information is obtained. However, as shown in Fig. 6.31, localized signals arising from higher hydration of the oligonucleotide were observed. Normal phase differences are of the order of 10°; however, phase differences in the range of 70–100° were obtained (Fig. 2.31c). These show a typical adsorption curve as a function of concentration, implying saturation of available streptavidin–biotin binding sites.

2.7.1.2 **Imaging DNA Translocation**

Atomic force microscopy imaging in liquids allows observation in almost real time (scan speeds are now such that an image can be acquired in about 1 min) of biological processes. This recognition was applied by Firman et al. [68] to translocation of DNA by Type 1 restriction-modification enzymes. Here large DNA fragments (2364 and 724 bp) both containing a single recognition site at 175 bp were used. Figure 2.32 shows the images obtained after fixed incubation times. Both DNA and the protein are clearly present. By using a large body of images the amount of translocated DNA associated with loop formation was analyzed. Single loops observed for the R1–DNA complex were determined. Evidently, the size of the loop increased with time. The authors were able to determine the velocity of the translocation process, which proved to be much lower than that previously measured [69, 70]. Prudently, the authors note that this may be due to sample surface interactions leading to an artificially show translocation process. With the R2–DNA complex, two loops were formed (Fig. 2.32) – the difference in size between the loops was postulated to arise from the switch from the recognition to translocation process occurring independently in the two sub-units of the enzyme.

2.7.1.3 **DNA Interactions and Stretching**

The double stranded DNA may interact with drugs, specific pharmacological functions being strongly dependent on the nature of the binding of the two systems.

Single-molecule force spectroscopy can discriminate between different interaction modes that modulate the binding of the small molecules to the DNA, by measuring the mechanical properties of DNA [71]. Several structure transitions are characterized upon stretching of double stranded DNA; it was observed, at a level of a single molecule, that the binding of a small molecule on to the DNA strongly affects these transitions (Fig. 2.33).

DNA occurs in solution as an entropic coil. When a DNA molecule attached between an AFM-tip and a surface is stretched it exerts a restoring force (see black curve in the Fig. 2.33). Up to a certain limit, these force–extension characteristics may be described by the worm-like chain model. On further increasing the applied force, the molecule undergoes an important structural transition (corresponding to a plateau in the force–extension profile). Initially, this transition plateau was thought to correspond to a structural modification of B-DNA to an overstretched S-DNA structure. One study [72] explains the plateau in the force–extension curves by a mere melting of the double helix, implying also the breakage of hydrogen bonds in the overstretching transition. After stretching beyond this melting transition the double stranded molecule is finally separated into two single strands, one of which remains tethered between tip and surface. The relaxation trace does not resemble the extension trace and single stranded DNA mechanical properties prevail. At lower forces (<150 pN), partial melting of the DNA molecule can occur and is observable as a deviation of the relaxation trace from the stretching traces (melting hysteresis). This model, however, predicts that the overstretching force should be related to the thermal melting temperature and that the force range over which overstretching occurs should be related to the width of the thermal melting transition.

Ethidium bromide is a well-characterized dye that intercalates into DNA without sequence specificity. Insertion of a single dye molecule increases the base pair rise by 3.4 Å and unwinds the double helix by 26°. Typical force–extension traces obtained on DNA in the presence of ethidium bromide show that low concentrations (0.44 μg mL^{-1} or ca. 1 molecule of ethidium per 10 bp) (Fig. 2.33 curve b) markedly affect the overstretching plateau. The increased slope of the transition regime is indicative of a reduction in cooperativity as compared to the curve obtained on the very same molecule before ethidium was added (Fig. 2.33 curve a). The force is reduced in the beginning of the transition while at the end it rises above 110 pN, almost twice the force at the end of the overstretching transition in non-complexed molecules. The melting transition, however, is still observed and hysteresis between stretching and relaxation remains. At elevated concentrations (Fig. 2.33 curve c), the plateau vanishes and a steadily rising force is observed that begins at larger extensions. Furthermore, the sharpness of the end of the overstretching plateau is lost, and it is difficult to precisely identify the beginning or end of the overstretching transition. The mechanical energy that can be deposited in the DNA double helix before force-induced melting occurs was found to decrease with increasing temperature. This energy correlates with the base-pairing free enthalpy $\Delta G(bp)(T)$ of DNA. Experiments with pure poly(dG-dC) and poly(dA-dT) DNA

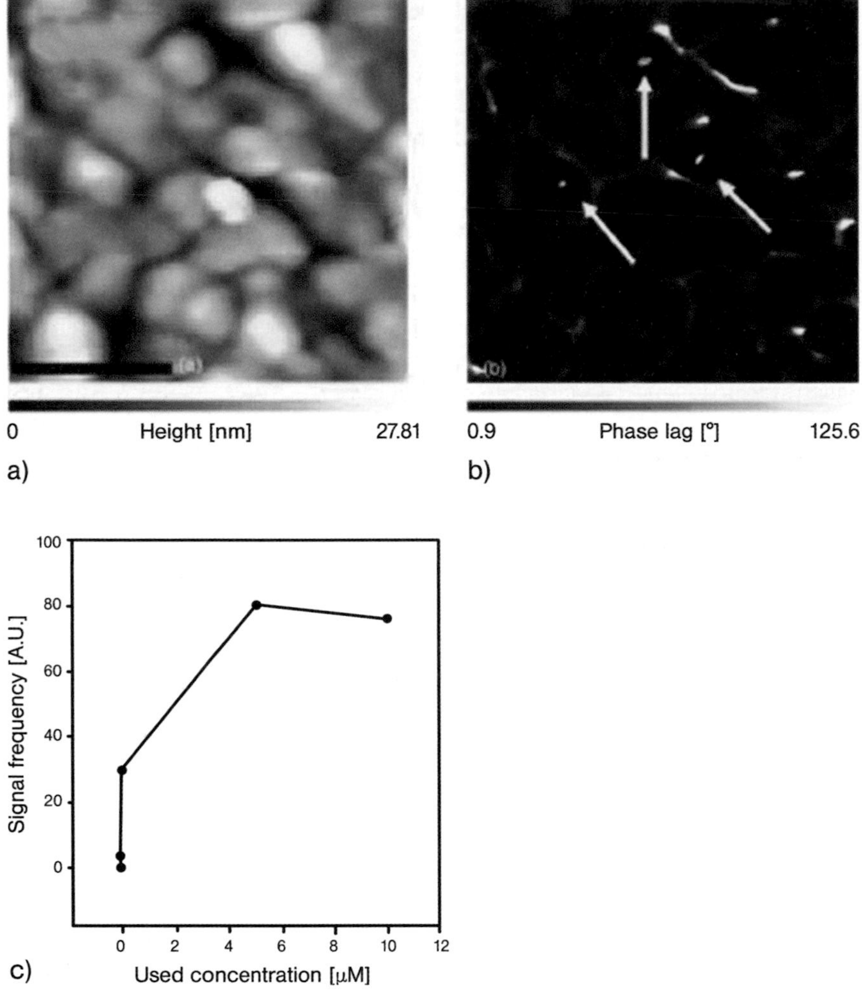

Fig. 2.31. (a) Topographic image of oligonucleotide complex, (b) phase shift image, (c) phase shift variation as a function of oligonucleotide concentration. Image size 250 × 250 nm. (Reprinted with permission from Ref. [67]. Copyright (2004) Elsevier.)

sequences again revealed a close correlation between the mechanical energies at which these sequences melt with base pairing free enthalpies ΔG(bp)(sequence) (Fig. 2.34): while the melting transition occurs between 65 and 200 pN in λ-phage DNA, depending on the loading rate, the melting transition is shifted to ~300 pN for poly(dG-dC) DNA, whereas poly(dA-dT) DNA melts at a force of 35 pN [73].

The force at which the melting transition occurs depends on the pulling velocity, ranging from 68 pN at 0.15 $\mu m\ s^{-1}$ to 300 pN at 3 $\mu m\ s^{-1}$.

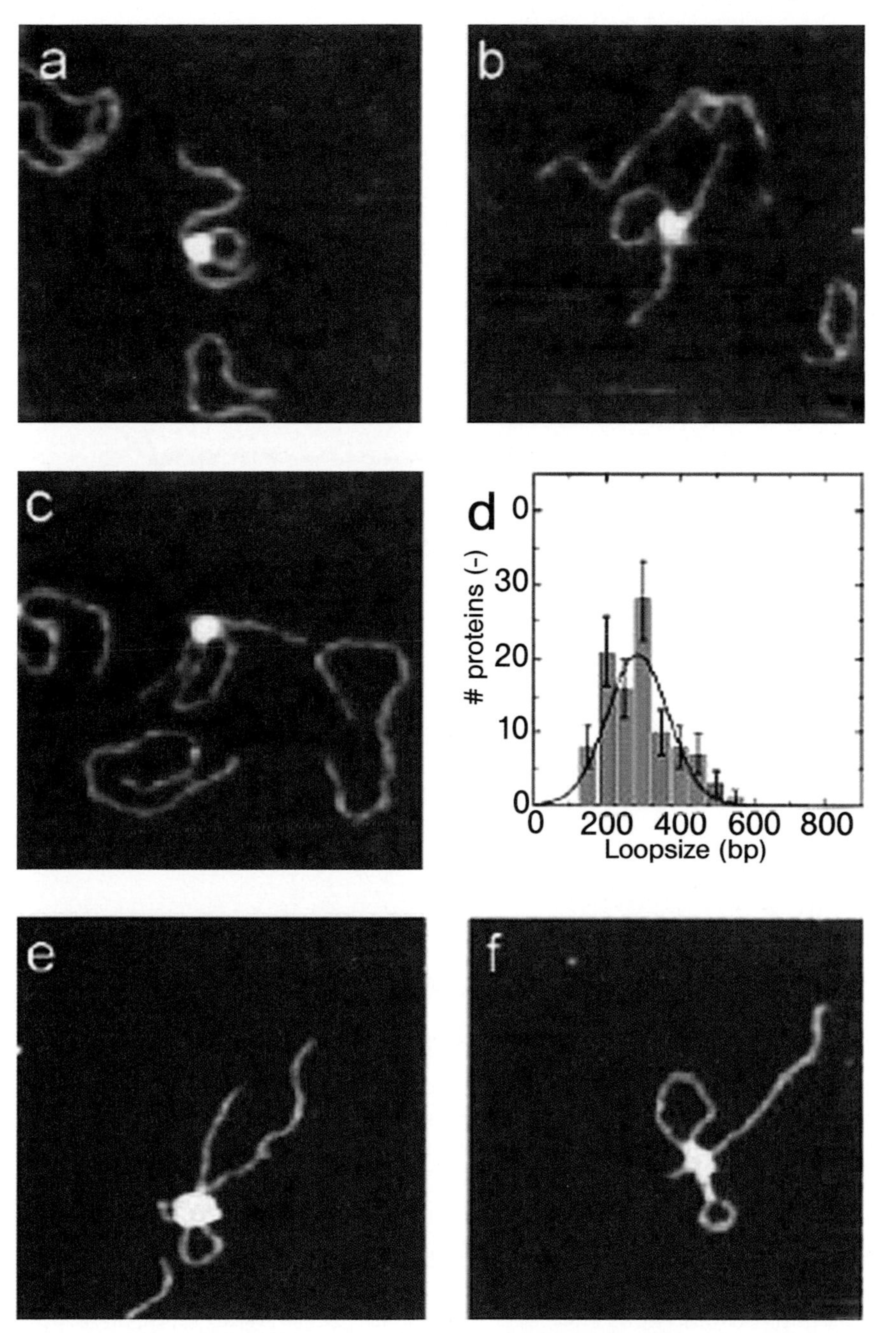

Fig. 2.32. AFM images of DNA translocation. (a)–(c) Scans taken after 10, 30 and 60 s incubation, respectively, with R_1-complex. Scan 250 × 250 nm, height scale 3 nm. (d) Loop size distribution for images taken after 10 s. (e, f) Double loops of DNA translocated by R_2-complexes. (Reprinted with permission from Ref. [68]. Copyright (2004) Oxford Journals.)

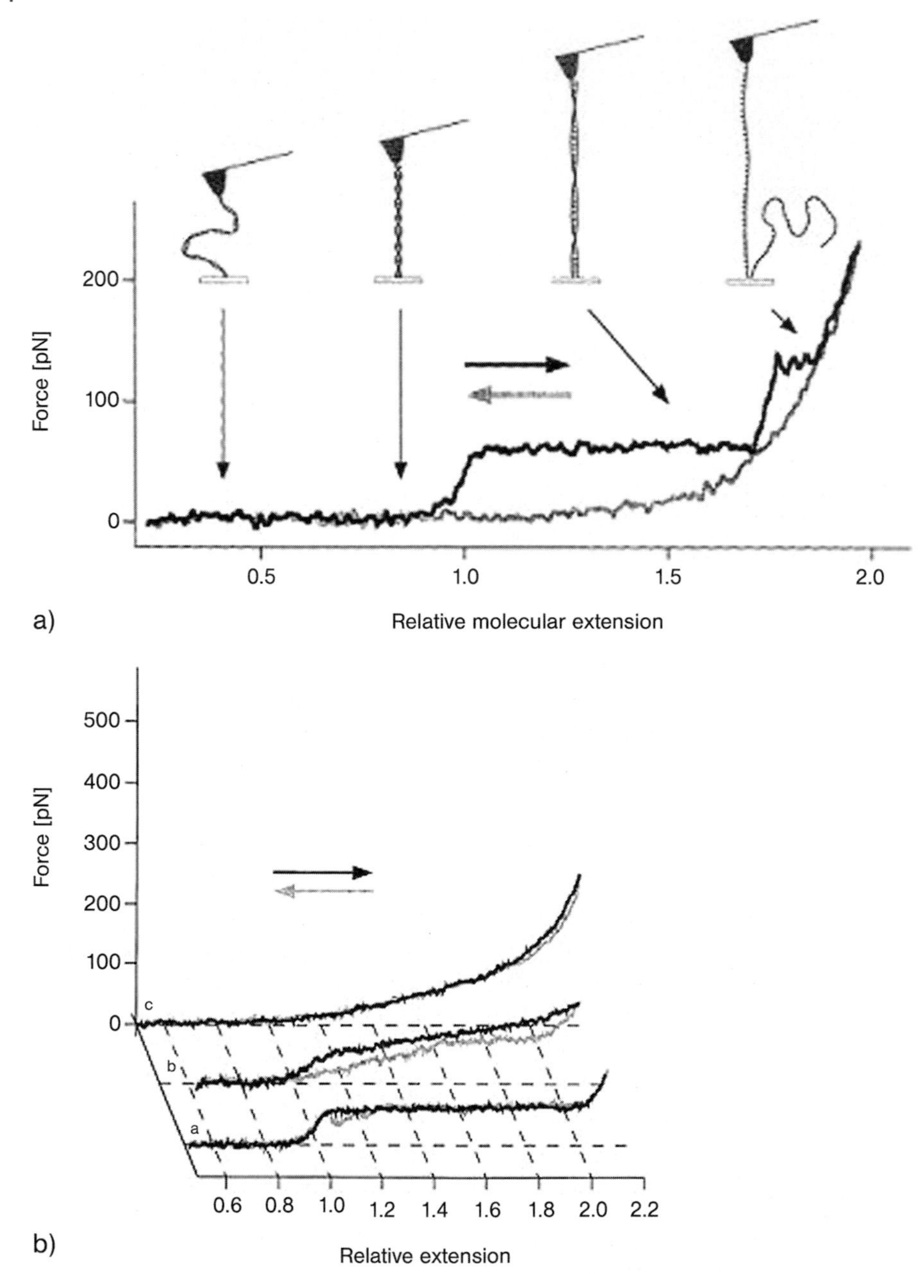

Fig. 2.33. Schematic of the force–extension characteristics of DNA (A): at 65 pN the molecule is overstretched to about 1.7× its contour length; at 150 pN the double strand separates into two single strands, one of which remains attached between tip and surface. (B) Force vs. extension curve of a single molecule of DNA in pure buffer (a) and in the presence of (b) 0.44 and (c) 2.2 μg mL^{-1} ethidium bromide. With increasing ethidium bromide concentration the overstretching plateau shortens while the force at the end of the transition increases to 110 pN. At high concentrations, hysteresis between stretching and relaxation is drastically reduced. (Reprinted with permission from Ref. [71]. Copyright (2002) Elsevier.)

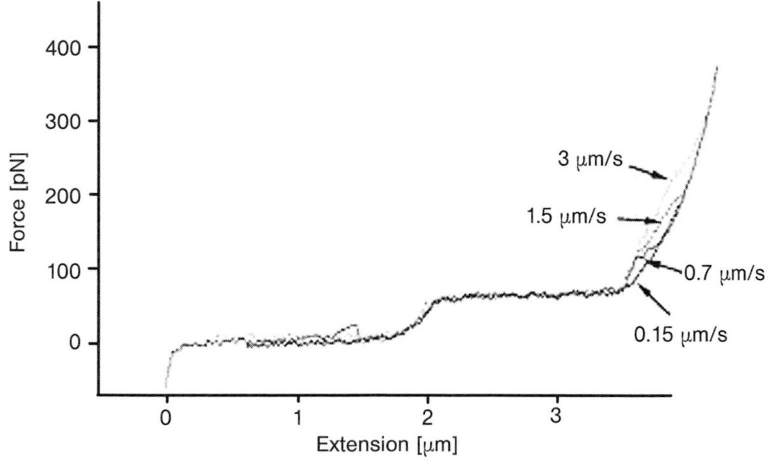

Fig. 2.34. Superposition of four extension traces of the same piece of λ-BstE II digest at different pulling velocities: 3, 1.5, 0.7 and 0.15 μm s^{-1}. (Reprinted with permission from Ref. [73]. Copyright (2002) Biophysical Society.)

Apparently the flexibility of DNA and RNA molecules depends on the D+C content. More than that, it was observed [74] that the stacking forces are stronger in RNA than in DNA, as an increased transition force is noted for RNA molecules.

2.7.2
Proteins

2.7.2.1 Topographic Imaging of Proteins

Collagen presents the archetypal protein for AFM imaging first observed by Lees et al. in 1994 [75]; the bands of collagen spaced at 67 nm is often the first experiment encountered on entering the field of bio-AFM.

AFM high resolution imaging of membrane, reconstituted in high concentrations in lipid domains, is proving to be one of the major tools for obtaining structural information on these proteins, which represent a major difficulty for classical methods, such as X-ray diffraction, of structural analysis.

Contact mode AFM in liquids allows resolution at a level of individual sub-units of assembled arrays of proteins (Fig. 2.35) for the case of chloroplast ATP-synthase [76]. The image shows the protein to be formed of 14 sub-units arranged cylindrically. There are alternating cylinder diameters of 5.9 ± 0.3 and 7.4 ± 0.3 nm, with both having the same internal diameter of 3.5 ± 0.3 nm. Given that SDS-PAGE reveals only a single sharp band at 100 kDa, the authors propose that there is alternating presentation of the extra- and intra-cellular faces of the protein. Furthermore the sub-units project by differing amounts from the lipid bilayer, 1.7 ± 0.3 nm for the wider face and 1.5 ± 0.3 nm for the narrower face.

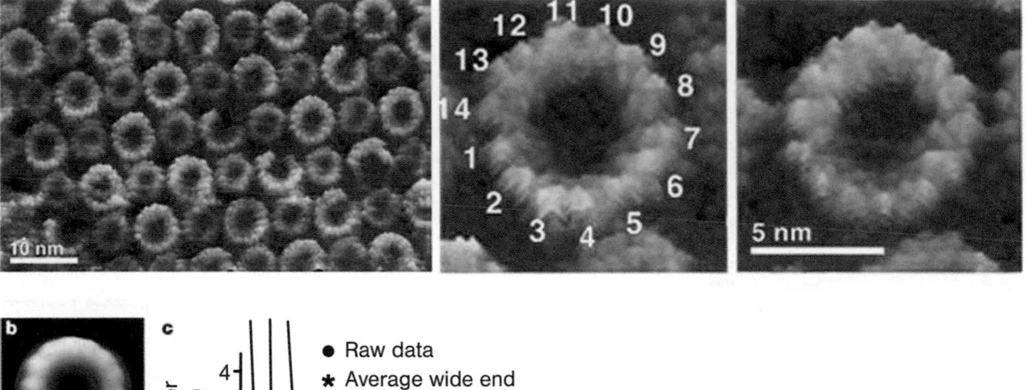

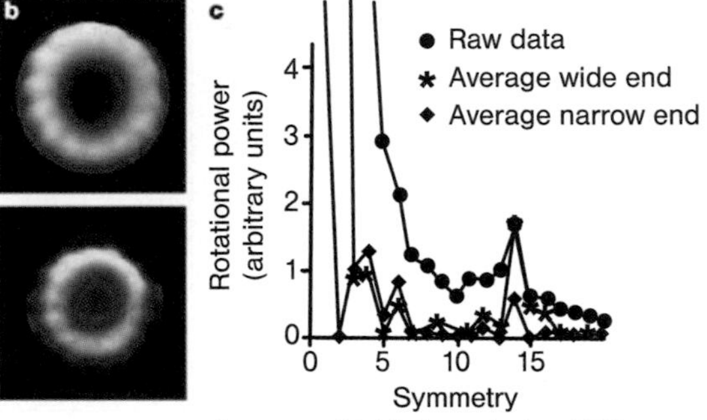

Fig. 2.35. (a) High-resolution AFM contact-mode images of images of chloroplast ATP-synthase, showing the fourteen-fold symmetry; (b) wide and narrow faces; (c) power spectrum analysis, showing the peak for fourteen-fold symmetry. (Reprinted with permission from Ref. [76]. Copyright (2000) Nature.)

Combining AFM with cryo-electron crystallography provides a means to enhance the structural data that may be obtained from molecular resolution AFM topographic imaging of proteins.

The lens major intrinsic protein (MIP) has been studied by Engel et al. [77], using reconstituted purified ovine lens MIP and exogenous lipids to obtain crystalline sheets. The sheets show two surfaces, one smooth and the other rough, Fig. 2.36(A) and (B). From the high level of organization, Fourier transform analysis allowed determination of the unit cell and the presence of a tetrameric structure. The rough surface was imaged by mechanical removal of the smooth part using high tip-surface forces.

Careful analysis of the geometry of the protrusions present on the surfaces of the rough and smooth faces, along with analysis of carboxypeptidase Y digestion, which permitted identification of the smooth face as being cytoplasmic, allowed the authors to propose a tongue and groove fit between the two surfaces, showing MIP assembles as a two-layer system (Fig. 2.36). Use of cryo-electron microscopy projection maps allowed confirmation of a two-layer structure of the crystalline sheets.

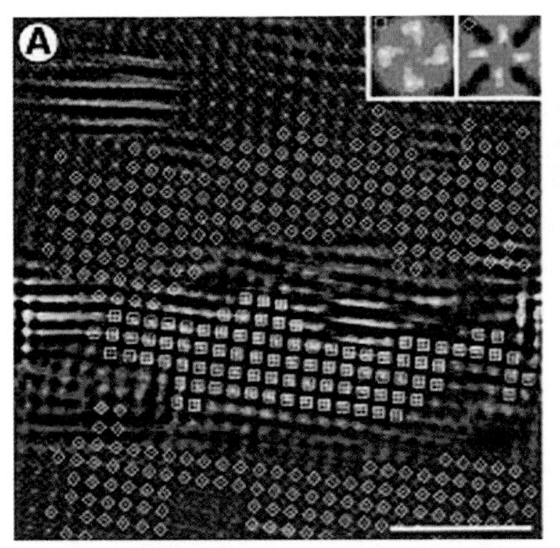

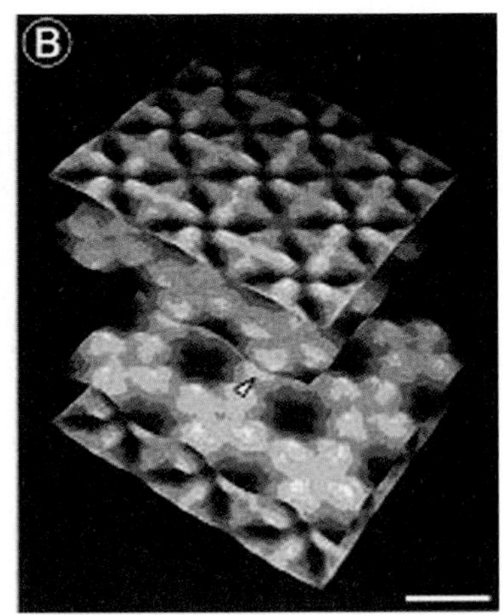

Fig. 2.36. Treated images of MIP from lens fiber cells. (A) Both rough and smooth surfaces are visible after using the point to remove the upper layer. (B) Model of the tongue and groove fit of the two layers. (Reprinted with permission from Ref. [77]. Copyright (2000) Elsevier.)

2.7.2.2 Dip-pen Nanolithography Patterning of Proteins

The use of Dip-pen nanolithography to allow deposition of various geometries of protein arrays on negatively charged SiO_2 surfaces has been demonstrated by Mirkin (Fig. 2.37) [78].

Arrays of 450 nm spots of anti-rabbit IgG with 350 nm spacing were generated in under 5 min, using a deposition time of 3 s. While not as rapid as conventional robotic systems, the spatial resolution is much higher. However, this is counterbalanced by the maximum standard AFM displacement of about 100×100 μm. Individual spots of size as low as 55 nm can be generated by using shorter contact times (0.5 s) with low contact forces (0.5 nN). Such arrays were used to validate immuno-recognition with suitably labeled fluorescent antibodies.

2.7.2.3 Protein–Protein and Protein–Ligand Interactions

In the classical first experiment on force–distance measurement between a ligand and a protein, avidin was coupled to a bovine serum albumin coated tip and reacted with biotin or iminobiotin attached to agarose beads. The force required to separate a single avidin–biotin interaction was 160 pN, and 85 pN was required for a single avidin–iminobiotin complex (Figs. 2.38 and 2.39) [79].

The function of a protein derives from the specific folding of the polypeptide chain. In the case of muscles or cytoskeletal proteins, the folding is designed to maintain a certain structure under varying load conditions. Thus, single-molecule force spectroscopy, in which one end of the protein is attached to the AFM tip and

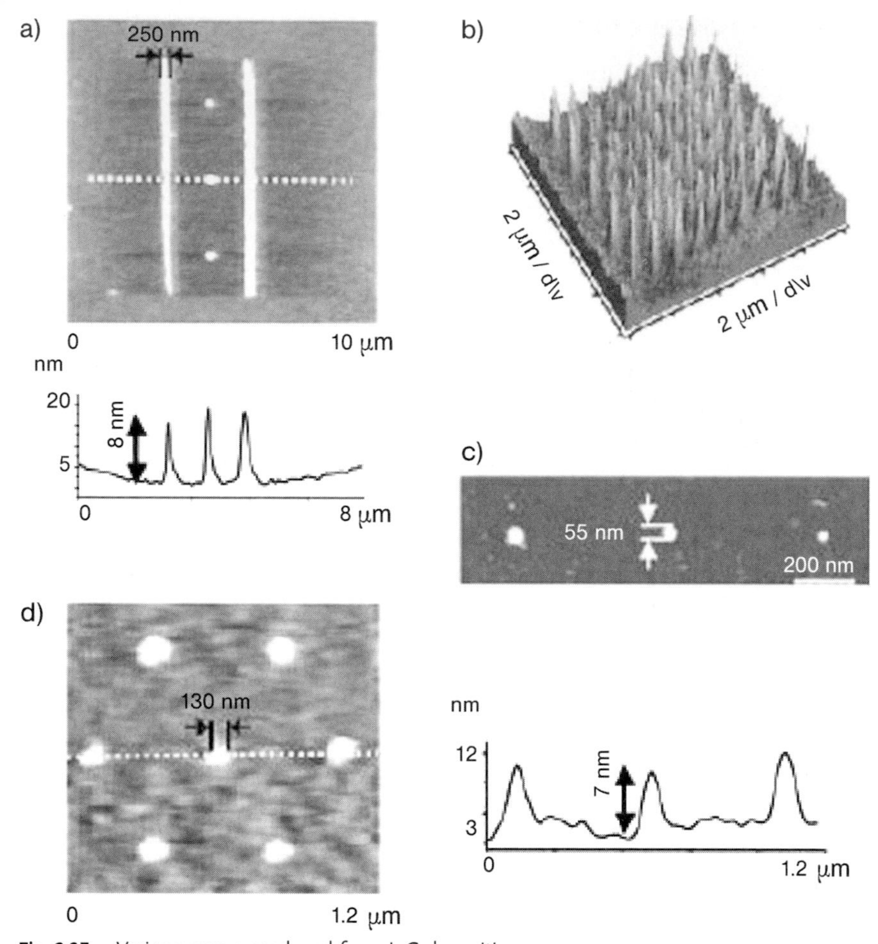

Fig. 2.37. Various arrays produced from IgG deposition on negatively charged Si surfaces. (Reprinted with permission from Ref. [78]. Copyright (2003) Wiley.)

the other end to the surface, provides a unique method of studying single molecular unfolding and folding under controlled loading.

The study by Gaub of reversible unfolding of titin immunoglobulin domains is a textbook example [80]. Titin was first chemisorbed onto a gold surface that was then adsorbed by contact onto the AFM tip. Retraction of the tip from the surface produced, as shown in Fig. 2.40, force–distance curves that show a saw tooth form extending to over 1 μm from the surface.

The initial high force peak may arise from tip–gold interactions. Periodicity between the peaks is between 25 and 28 nm. Using a retraction velocity of 1 μm s^{-1}, maximum forces observed in the saw tooth curves are between 100 and 300 pN. The periodicity is close to the 31 nm expected for unraveling of an Ig domain.

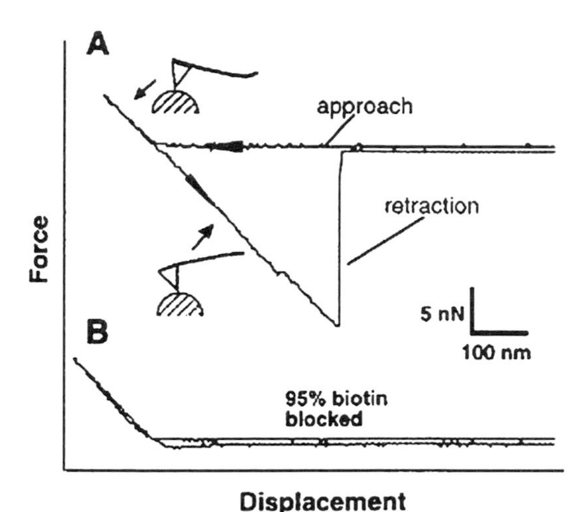

Fig. 2.38. Force–distance curves for avidin–biotin interaction. (Reprinted with permission from Ref. [79]. Copyright (1999) Elsevier.)

This was confirmed using constructs of four or eight Ig domains (Fig. 2.41), which exhibited a strict 25 nm periodicity measured at 100 pN.

Interestingly, the slope of the rising half of the curves, representing the stiffness of the protein, decreases from peak to peak. This implies that the stiffness of the stretched protein decreases, and that the stiffness is dominated by a spring that softens on unfolding of the Ig domains. The data fits to a worm-like chain model. Repeated stretching–relaxation cycles showed that refolding occurs when the construct is allowed to fully retract (Fig. 2.42).

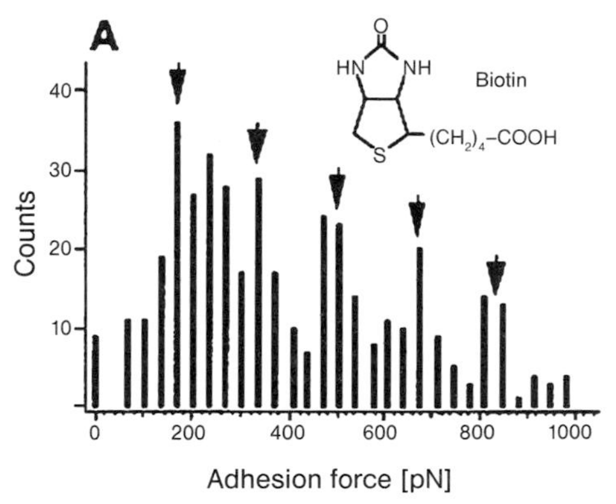

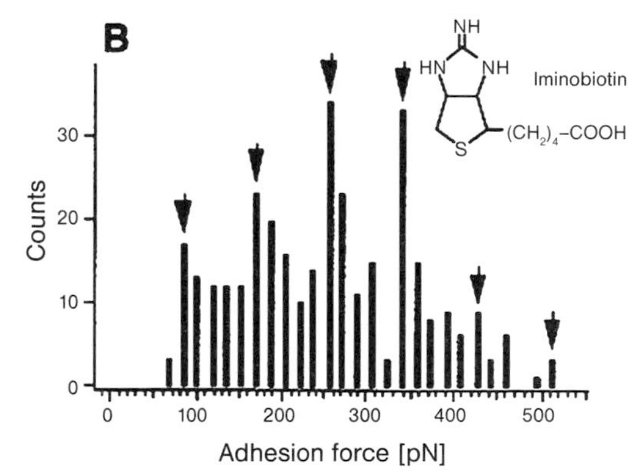

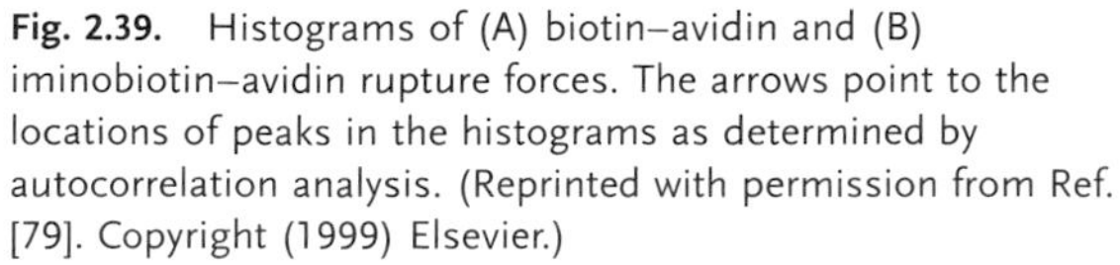
Fig. 2.39. Histograms of (A) biotin–avidin and (B) iminobiotin–avidin rupture forces. The arrows point to the locations of peaks in the histograms as determined by autocorrelation analysis. (Reprinted with permission from Ref. [79]. Copyright (1999) Elsevier.)

Tab. 2.5. Atomic force microscopy study to investigate the interaction of MRP8/MRP14 with cytochrome b_{558} liposomes; measurements of cytochrome b_{558} liposome height in the presence or absence of MRP8/MRP14.

Systems	*Liposome height (nm)*	
	−AA	*+AA*
Cyt b_{558} PMN	5 ± 2	6 ± 2
Cyt b_{558} EBV-BL	7 ± 2	6 ± 4
Cyt b_{558} PMN + MRP8/MRP14 (Ca^{2+})	3 ± 2	5 ± 2
Cyt b_{558} PMN + MRP8/MRP14 (±AA)	8 ± 3	10 ± 2
Cyt b_{558} EBV-BL + MRP8/MRP14 (Ca^{2+})	5 ± 1	15 ± 5

a bimodal form with an interval control, as imprecision in the positioning often generates an antibody–mica contact (Fig. 2.45B). Again work was carried out using native and deglycosylated cytochrome b_{558}.

The results present several interesting points (Table 2.6). Firstly, interaction between the antibody Ab54.1 and the protein are significant even though the epitope is on the intracellular surface of the protein; this implies that orientation of the protein with regard to the membrane is statistical on reconstruction. Secondly, in the case of the mAb11C12 antibody, directed against neutrophil associated cyt-b_{558} recognition occurs for both native and deglycosylated protein, while for EBV-BL–cyt b_{558} for the native protein no recognition is observed; deglycosylation leads to a strong recognition event, implying that here the epitope is present but hidden by the carbohydrate chains. The glycosylation is also related to the biological activity of the complex in the EBV-BL cell lines.

AFM is widely used to investigate the mode of liposomes formation, their size, the transition of attached liposomes to bilayers patches, and the complexes they form with biological molecules such as DNA or proteins. The following is the

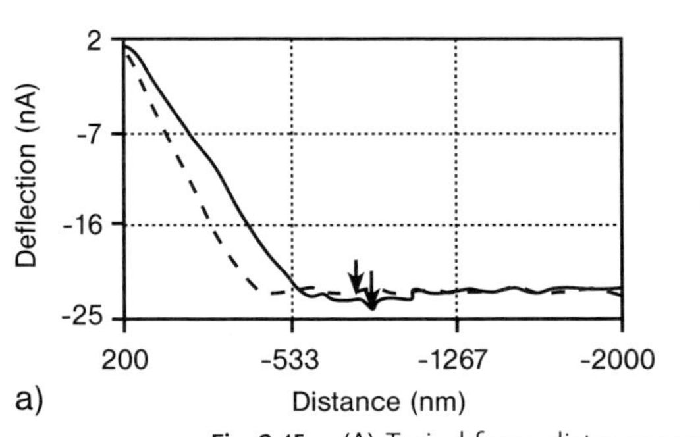

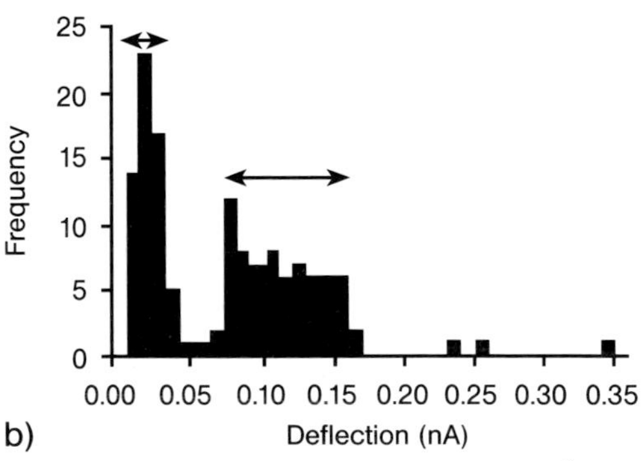

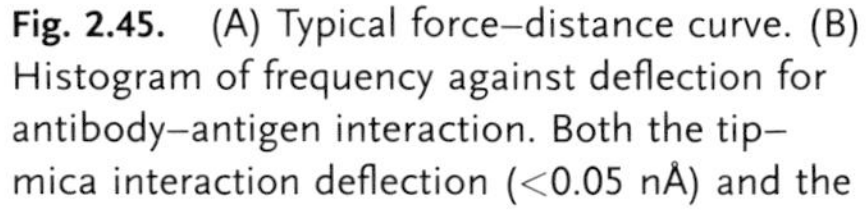

Fig. 2.45. (A) Typical force–distance curve. (B) Histogram of frequency against deflection for antibody–antigen interaction. Both the tip–mica interaction deflection (<0.05 nÅ) and the antibody–antigen interaction (0.1–0.15 nÅ) are visible. (Reprinted with permission from Ref. [18]. Copyright (2001) FEBS.)

Tab. 2.6. Measures of interaction forces between specific antibodies and purified native (N) or deglycosylated (D) cytochrome b_{558}.

Antibody	*Interaction (nN)*				
	Mica	*Neutrophil cyt.* b_{558}		*EBV-BL cyt.* B_{558}	
		N	*D*	*N*	*D*
Negative control	9 ± 5	16 ± 5	8 ± 4	9 ± 5	9 ± 4
54.1	10 ± 5	39 ± 11	26 ± 10	12 ± 2	9 ± 4
11C12	4 ± 1	25 ± 21	23 ± 7	16 ± 12	28 ± 11
(562–569)	5 ± 2	16 ± 2	–	8 ± 3	–
44.1	9 ± 3	13 ± 5	–	5 ± 4	–

work of Vermette et al. [85] on interaction forces between a silica particle and immobilized liposomes layers. The experiments were performed according to the colloid probe method developed by Ducker et al. [86]. Thus, a spherical colloidal particle of pure silica (diameter 4–5 μm) is attached to the microfabricated, gold-coated AFM cantilever spring via an epoxy adhesive. The liposomes had biotinylated PEG lipids. The resultant biotin moieties on the outer surface of the liposomes were capable of docking onto NeutrAvidin molecules immobilized onto the hydrogel interlayers. The intention was to provide a platform for binding NeutrAvidin molecules, at the same time minimizing nonspecific adsorption events as well as the disruption of liposomes upon contact with the surface. Compression of liposome layers by an approaching silica colloid sphere, mounted on the AFM cantilever, can be quantified by comparing the force vs. distance curves measured with a spherical silica particle approaching NeutrAvidin (on hydrogel interlayers) surfaces and putative immobilized liposome surfaces. Comparison of the release characteristics of fluorescent dye from liposomes in solution and liposomes bound to the three different hydrogel interlayers showed that the release characteristics were little affected, indicating that there was minimal disruption of the liposomes upon surface binding. By blocking surface-immobilized NeutrAvidin is was shown that there exists a physicochemical attractive force of non-negligible magnitude between PEGylated liposomes and hydrogel surfaces decorated with NeutrAvidin, and this physicochemical binding mechanism can surface-attach liposomes when the biologically specific biotin–avidin mechanism is absent.

The utility of AFM for measuring the mechanical properties of adsorbed small liposomes and stability changes in liposomes has been demonstrated by Liang et al. [87]. They investigated the effects of cholesterol concentration on the micromechanical properties of the small unilamellar EggPC liposomes adsorbed to mica. These liposomes are stable and maintain their vesicular form without fusion when deposited on a mica surface. The elastic properties are revealed by force curves between an AFM tip and an adsorbed vesicle obtained in the contact mode. Figure 2.46 presents the types of tip–sample interactions.

AFM analyses quantified the Young's modulus (E) and bending modulus (K_c)

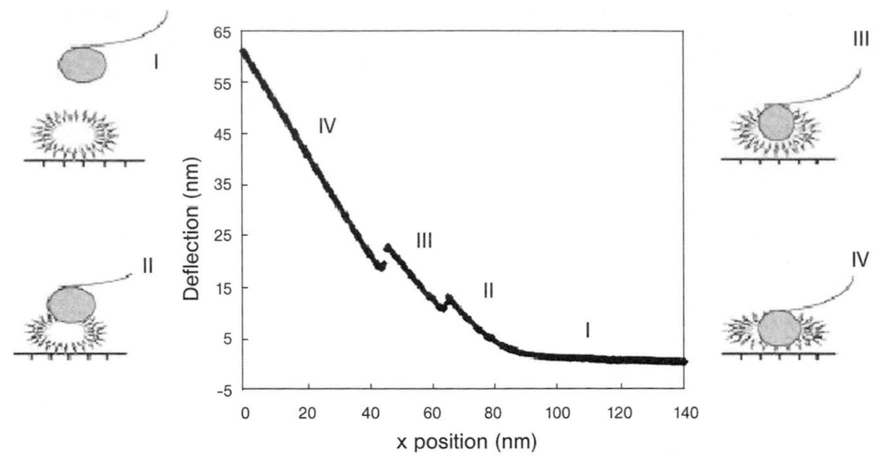

Fig. 2.46. Deflection versus z position approaching curve. Zero separation is not defined as described previously. The force curve was obtained on the EggPC/cholesterol. Zone I is non-contact. Zones II–IV illustrate different on-contact stages of tip and sample interaction. Region II illustrates elastic deformation of the vesicle under tip compression. Region III corresponds to further tip compression after the tip penetrates the top bilayer. Region IV: tip at the top of the mica substrate; the steep slope (0.9946) indicates mica is an infinitely hard surface without deformation. (Reprinted with permission from Ref. [87]. Copyright (2004) Elsevier.)

for those unilamellar vesicles (Table 2.7). The analyses showed an increase in both E and K_c with increasing quantity of cholesterol incorporation into pure EggPC. This may be explained by the fact that the presence of cholesterol molecules enhances the rigidity of the EggPC vesicles.

2.7.4.2 Solid Lipid Nanoparticles (SLNs)

Atomic force microscopy is the most used technique for the size characterization of nanoparticles, generally in conjunction with Dynamic Light Scattering (DLS).

Atomic force microscopy (FMM and LFM) has been shown to be a key tool for imaging nanoparticles incorporated in gels of biological interest [88]. It was seen that SLNs do not aggregate in gels and the use of lateral force and force modulation modes in addition to the classical topographic mode provides additional information on the gel structuring (Figure 2.47).

Tab. 2.7. Comparison of Young's modulus (E) and bending modulus (K_c) of liposome with different cholesterol and lipid ratios.

	Cholesterol:lipid ratio				
	Pure eggPC	***85:15***	***80:20***	***70:30***	***50:50***
$E \times 10^6$ Pa	1.97 ± 0.75	12.07 ± 1.53	10.77 ± 0.64	10.4 ± 4.06	13.0 ± 2.97
$K_c \times 10^6$ Pa	0.27 ± 0.10	1.68 ± 0.21	1.49 ± 0.09	1.44 ± 0.56	1.81 ± 0.41

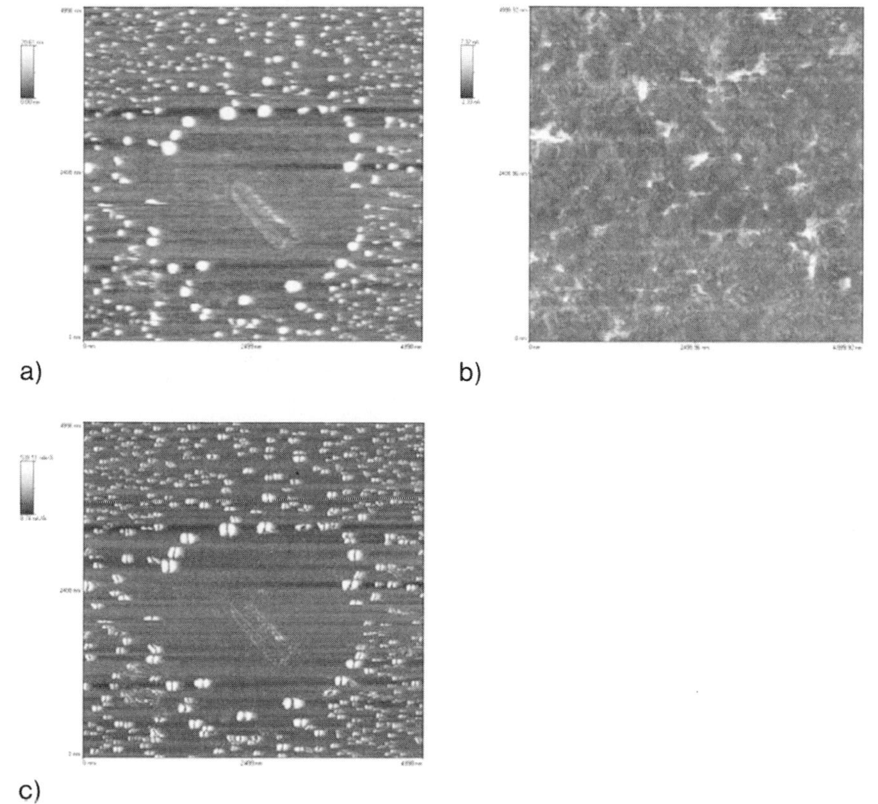

Fig. 2.47. Images of carbopol 2020 containing nanoparticles in topographic (a), lateral force (b) and force modulation (c) modes; average particle diameter is 150 nm and the average height is 120 nm. (Reprinted with permission from Ref. [88]. Copyright (2003) Elsevier.)

To proceed to the use of SLNs in intravenal administration, it is important to verify that they do not aggregate in the presence of albumins. Gualbert et al. performed AFM studies on the interaction of SLNs based on amphiphilic calix[4]arene with one of the major circulatory proteins, serum albumin (Fig. 2.48) [89].

The results show that no aggregation occurs between the SLNs and the serum albumin. The difference in diameter and height of the particles (155 and 15 nm, respectively) with respect to the size of nanoparticles without protein (250 nm in diameter and 55 nm in height) may be explained by the fact that the nanoparticles are included in the protein gel (only the protruding parts being measured). The more circular shape proves a protective action of BSA, i.e., capping the particles, preventing their flattening.

AFM and SEM (Scanning Electron Microscopy) analyses were performed by Dubes et al. [83] on SLNs derived from β-CD21C6 (Fig. 2.49). Comparison of the

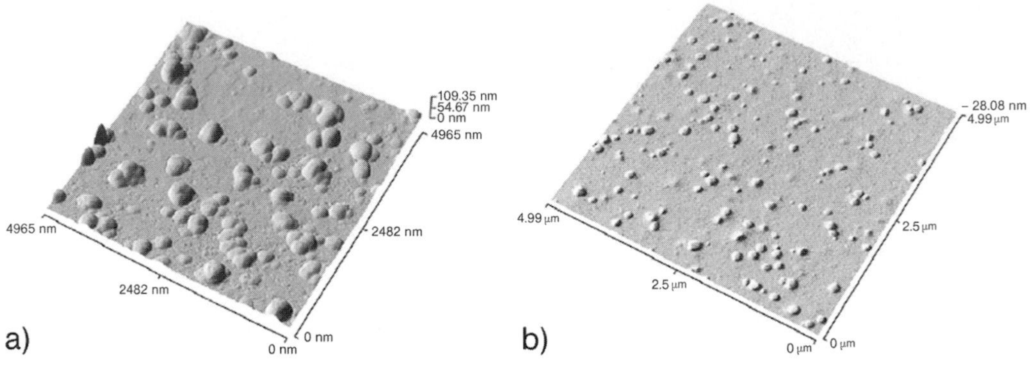

Fig. 2.48. Non-contact mode AFM images of *para*-decanoyl-calix[4]arene-based solid lipid nanoparticles without (a) and with BSA (20 g L^{-1}) (b) on a mica surface. (Reprinted with permission from Ref. [89]. Copyright (2003) Elsevier.)

results shows that the vacuum drying technique used in sample preparation for SEM causes shrinkage in the size of the SLNs, whereas the deposition method used for AFM causes the nanoparticles to form small clusters.

Both techniques confirm that the SLNs are circular. With images obtained from AFM, the SLNs tend to be organized as clusters of 15–30 nanoparticles. This is probably due to the sample preparation method, where the colloidal suspension is slowly dried. At the same time, shrinkage of the nanoparticles analyzed by SEM

Fig. 2.49. Non-contact mode AFM images of β-CD21C6-derived solid lipid nanoparticles (SLNs) at a scan range of 5 mm (a). SEM images of the β-CD21C6-derived SLNs, scale 0.2 mm (b). (Reprinted with permission from Ref. [83]. Copyright (2003) Elsevier.)

Tab. 2.8. Measured diameters on images provided by two analysis techniques.

	Diameter (nm)
SEM	212 ± 12
AFM	
Diameter	359 ± 50
Height	140 ± 27

was observed (Table 2.8). This might have been caused by the freeze-drying treatment prior to analyzing.

The advantage of AFM is the simple sample preparation as no vacuum is needed during operation and the sample does not undergo high vacuum treatment leading to dehydration and shrinkage.

2.7.4.3 **Supported Lipid Bilayers and Monolayers**

Supported lipid bilayers (SLBs) are formed by the spreading and unfolding of liposomes on surfaces. Their formation was first demonstrated by Brian and McConnell [90] in 1984. The mechanism of their formation was proposed from AFM imaging by Jass et al. [91] (Fig. 2.50).

More recently, Richter et al. investigated the adsorption and conformational changes of unilamellar vesicles adsorbed on silica by AFM in real time studies [92]. They proved that the process of vesicle deposition onto surfaces is more complicated and depends strongly on the nature of the component lipids. Zwitterionic, negatively charged and positively charged lipids were used for this investigation.

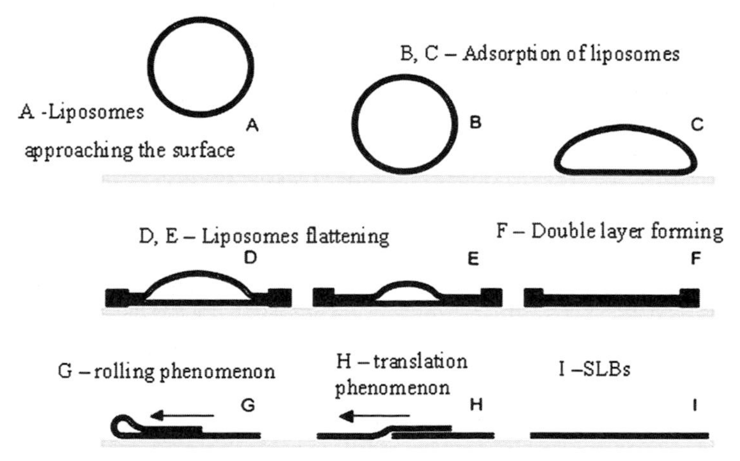

Fig. 2.50. Schematic representation of supported lipid bilayer formation. (Reprinted with permission from Ref. [91]. Copyright (2000) Biophysical Society.)

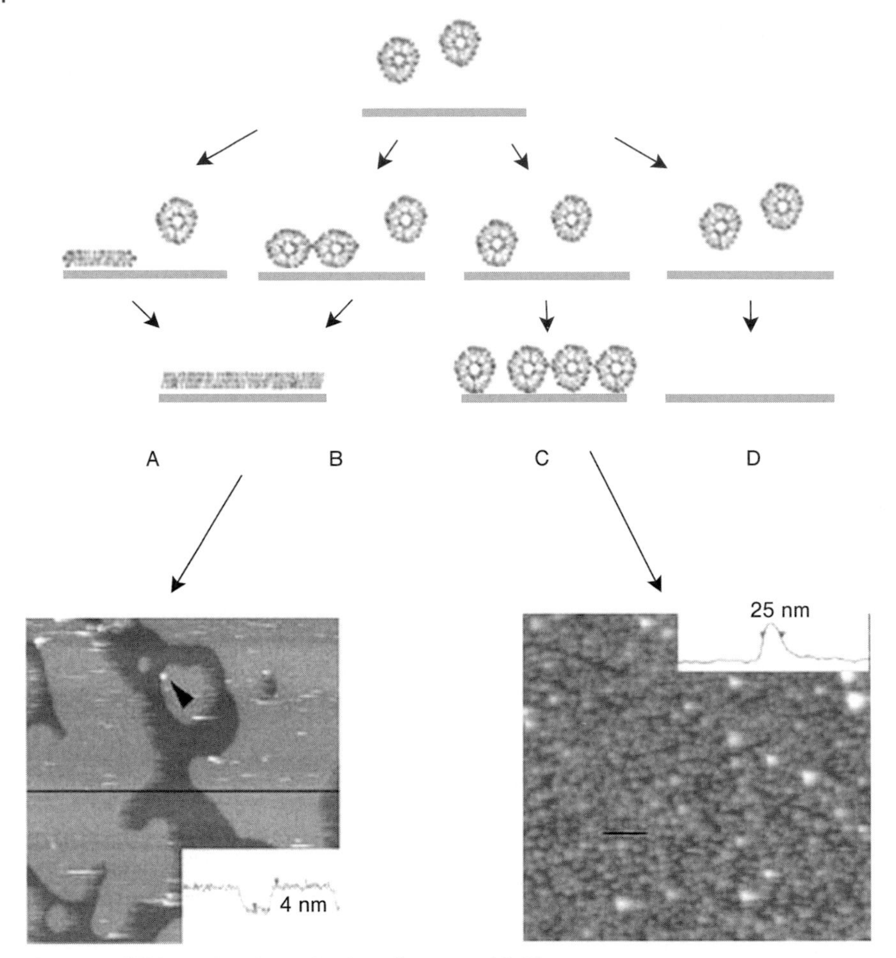

Fig. 2.51. AFM imaging determination of supported lipid bilayer (SLB) formation; (left) image shows SLBs, (right) image of liposomes. Insets: heights of the structure. (Reprinted with permission from Ref. [92]. Copyright (2003) Biophysical Society.)

Based on AFM analysis, a schema of the pathway of vesicles deposition was proposed (Fig. 2.51).

Table 2.9 exemplifies the types of vesicle deposition, as function of the lipid ratio and composition.

The height of the SLBs did vary evidently with lipid composition. Bilayers about 4 nm high described the SLBs film.

The reported existence of microdomains in plasma membranes has prompted considerable recent interest. Microdomains of a similar composition to lipid rafts

Tab. 2.9. Dependence of vesicle deposition pathway on the lipid ratio.

Lipid ratio			*Vesicle deposition pathway*
DOTAP	*DOPC*	*DOPS*	
1	–	–	SLB (I)
1	4	–	SLB (II)
–	1	–	SLB (II)
–	4	1	SLB (II)
–	2	1	SLB (II)
–	1	1	SVL
–	1	2	No adsorption
1	–	–	SLB (I)
1	4	–	SLB (II)
–	1	–	SLB (II)
–	4	1	SLB (II)
–	2	1	SLB (II)
–	1	1	SLB (II), restructuration
–	1	2	SLB (II), restructuration

have been shown to form in lipid bilayers, providing a model to investigate raft formation [93] (Fig. 2.52).

Many studies led to the conclusion that mixed lipid bilayers generate lipid rafts that are enriched in sphingomyelin and cholesterol. These two lipid components are thought to exist in a liquid-ordered phase or gel phase.

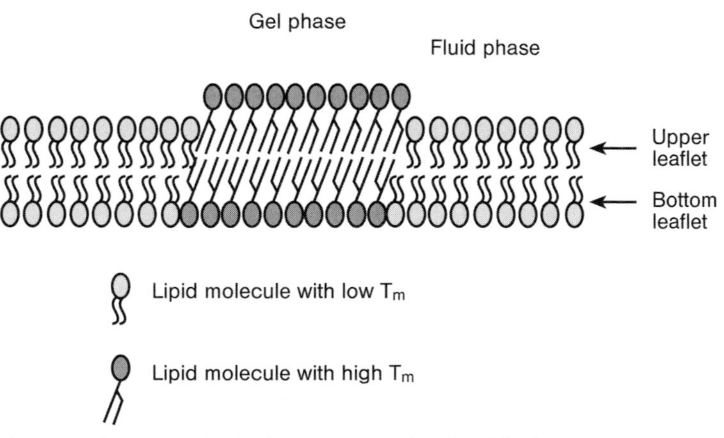

Fig. 2.52. Schematic of a lipid membrane with gel and fluid phase separation. The height difference due to lipid packing is such that AFM topographic imaging can distinguish the zones. (Reprinted with permission from Ref. [93]. Copyright (2003) acsmb@bmb.leeds.ac.uk.)

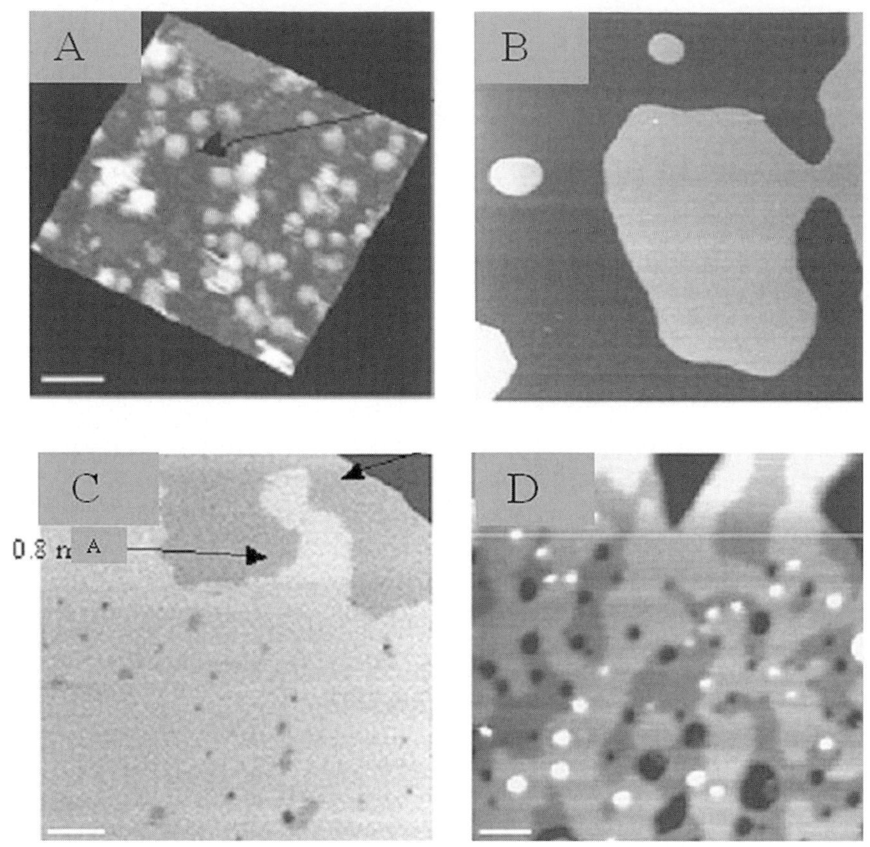

Fig. 2.53. (A) Protein film; (B) SLB containing sphingomyelin and cholesterol; (C) SLB containing sphingomyelin, cholesterol and phosphatidylcholine; (D) protein containing lipid rafts (sphingomyelin, cholesterol and phosphatidylcholine). (Reprinted with permission from Ref. [97]. Copyright (2004) Biochemical Society.)

More evidence about lipids rafts existing in biological membranes is brought by AFM studies that provide high-resolution imaging under physiological conditions [94]. Apparently, crucial cellular functions depend on preferential association of proteins with rafts [95, 96].

Helicobacter pylori vacuolating toxin–SLBs interaction and the preferential association of this protein with rafts [97] are presented below (Fig. 2.53).

The VacA film shows particles 7 nm high with an outer diameter of 30 nm (Fig. 2.53a). Supported lipid bilayers, formed of sphingomyelin and cholesterol, and raft formation (when phosphatidylcholine is added) are observed in Fig. 2.53(b) and (c). The film based on cholesterol and sphingomyelin is about 4 nm high. The presence of phosphatidylcholine in the lipids induces the formation of rafts. Differences of 0.8 nm in height protrude from the lipid bilayer. Analyzing the film of mixed lipids (all three components) in the presence of VacA (Fig. 2.53d), particles

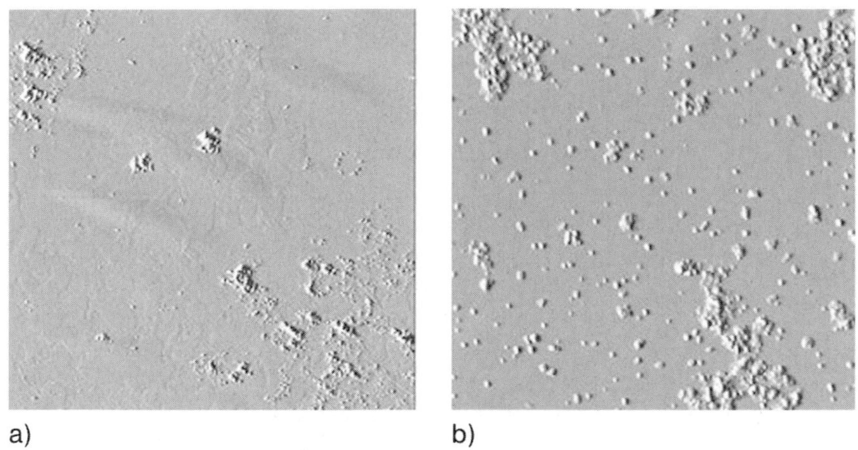

Fig. 2.54. AFM imaging of DRM-enriched membrane bilayers. (A) 60% cholesterol depleted; (B) after repletion with cholesterol. (Reprinted with permission from Ref. [98]. Copyright (2005) Biophysical Society.)

corresponding to the protein, 2 nm higher than the SLB and of much smaller diameter (25 nm), prove that the protein is inserted and traverses the lipid bilayer. The VacA obviously associates preferentially with rafts.

Barakat et al. proved [98], in a similar study on P-glycoprotein in a DRM-enriched membrane bilayer, that the depletion of cholesterol from the rafts induces protein translocation and aggregation outside rafts (Fig. 2.54). Approximately the same height differences (about 0.7 nm) were found for the protruding rafts. Repletion of the sample with cholesterol allows the reinsertion of P-gp in the rafts, probably recovering its P-gp ATPase activity.

Thus, cholesterol appears to be an essential component of DRMs for maintaining a structure of P-gp that is compatible with its optimal activity.

2.7.5 SNOM Imaging

Scanning near-field optical microscopy is generally used complementary to AFM to enrich the structural details of supported lipid monolayers and bilayers (Fig. 2.55). The two techniques can detect the presence of domains on the submicron scale on the supported lipid bilayers. The initial supposition was that a defect in the lipid layer occurred. Further analysis of these structures simultaneously by high-resolution fluorescence and topography measurements confirmed that they arise from coexisting liquid condensed and liquid expanded lipid phases. NSOM is the only reliable technique able to observe the lipid domains. Hollars et al., using NSOM fluorescence measurements, showed that lipid domains are quantitatively similar when formed in lipid bilayers or monolayers [99].

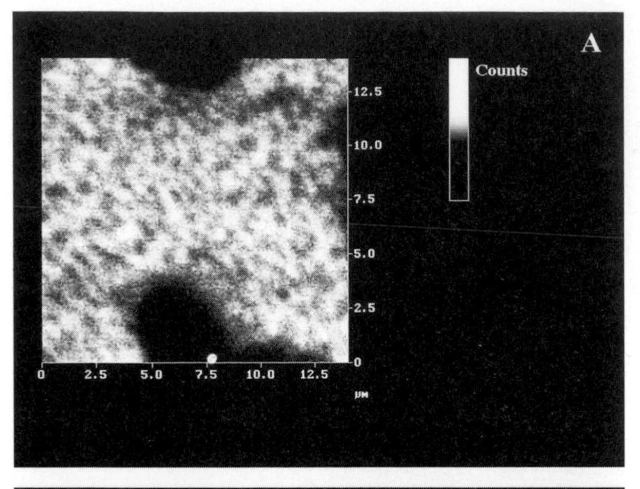

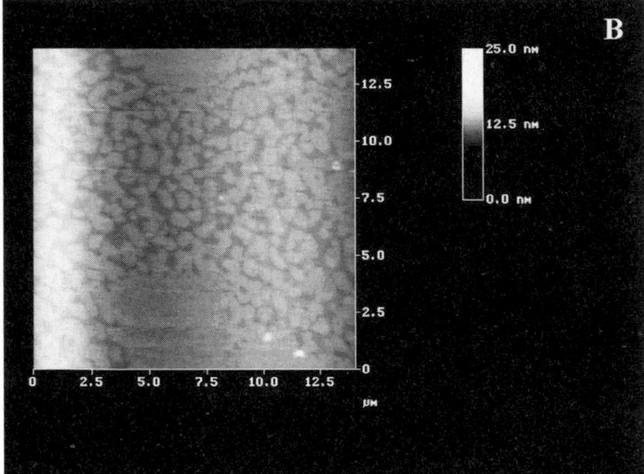

Fig. 2.55. (A) NSOM images and (B) AFM topographic images of monolayers, showing phase separation. The fluorescence is situated in the liquid expanded zones, as shown by the correlation with lower height in the AFM image. (Reprinted with permission from Ref. [99]. Copyright (1998) Biophysical Society.)

Apparently these domains are common features of the sample, regardless of the nature of the substrate.

The formation of phase domains can be influenced by the deposition conditions. As exemplified in Fig. 2.56, the surface pressure at the moment of deposition influences the distribution of the phase domains on the lipidic film [100].

Taken together, these results prove that SNOM might be a unique technique sensitive enough to observe phase structures present in lipid bilayers at the nanometer scale and may provide new clues to the existence of lipid phase domains in natural biomembranes.

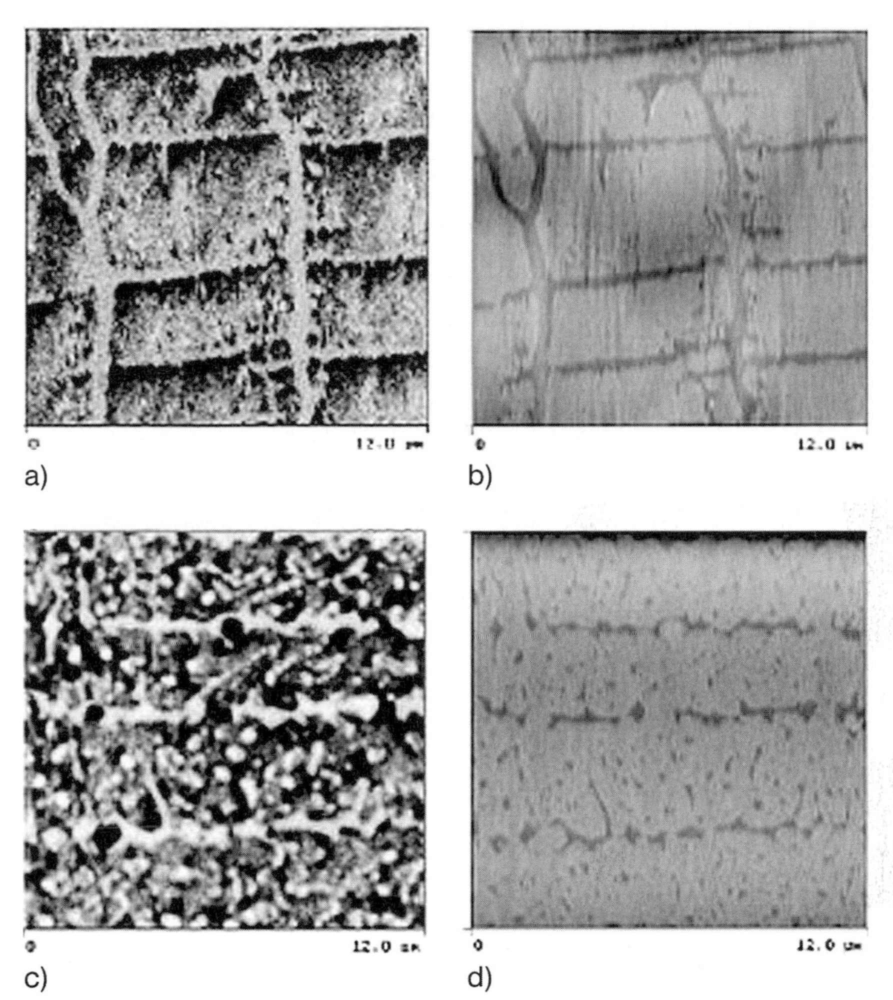

Fig. 2.56. (A, C) NSOM fluorescence images of mixed DPPC, diIC18 monolayers; (B, D) corresponding AFM topographic images. (Reprinted with permission from Ref. [100]. Copyright (1999) American Chemical Society.)

2.7.6
Viruses

Figure 2.57 presents typical images of the Moloney murine Leukemia virus emerging from the surfaces of infected NIH 3T3 cells [101]. The viral particles have an average diameter of 145 nm, but with a degree of polydispersity. The surfaces of the virions are studded with projections (tufts) of proteins, which display a range of packing densities. The tufts are generally 11–12 nm in size and often show a

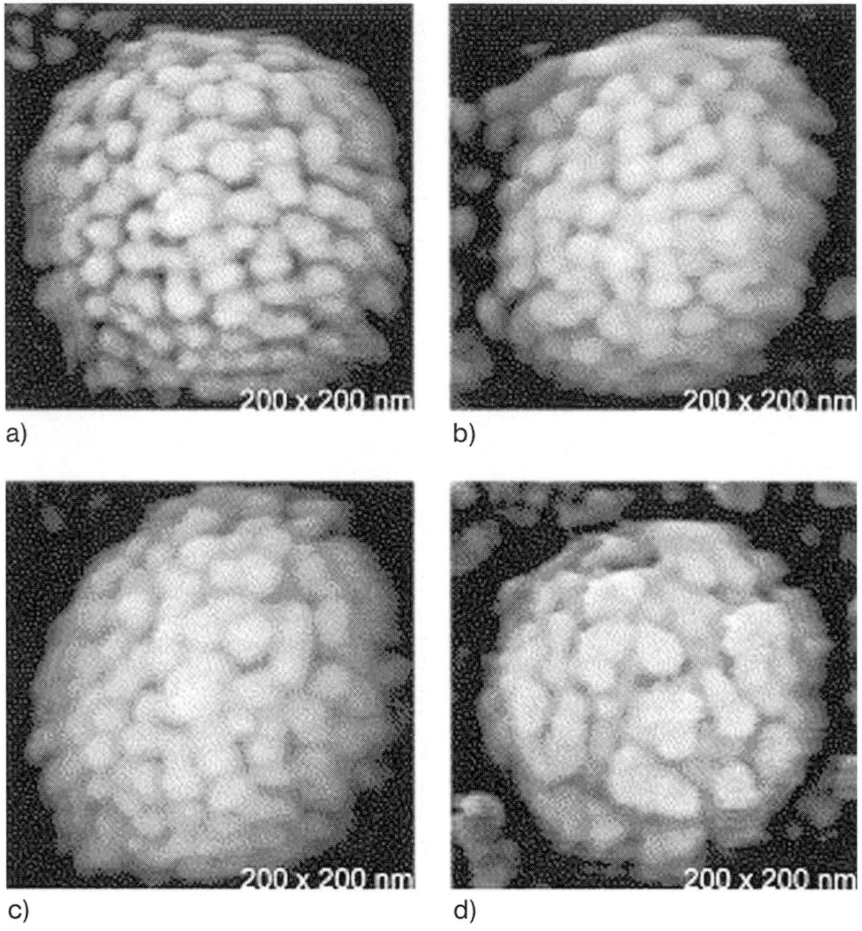

Fig. 2.57. (a–c) Typical AFM images of the surfaces of Moloney murine leukemia virion, showing the dense packing of protein projections. (d) AFM image of a virus particle, showing a low density of protein projection. (Reprinted with permission from Ref. [101]. Copyright (2002) Blackwell.)

high density of surface packing. The authors postulated that these projections are aggregates of SU protein. In view of the varying aggregation, the variation in size and geometry can be explained. Moloney murine leukemia virus (M-MuLV) lacking the gene for the envelope glycoprotein (env^-) was produced in NIH 3T3 cells and investigated using atomic force microscopy (AFM). The particles were compared with similarly produced wild-type virions, some of which had been exposed to a monoclonal antibody against the surface component of the envelope protein (SU protein).

cells and cellular organelles. Enderle et al. have described the use of near-field scanning optical microscopy (NSOM) for mapping and detection of colocalized proteins within a cell, simultaneously and at high resolution [103]. The system of interest was that of human red blood cells invaded by the human malaria parasite *Plasmodium falciparum.* This parasite expresses proteins, during intra-erythrocytic growth, that are transported across the erythrocyte cell membrane and associate with the host skeletal proteins. Specifically, these colocalization studies of the protein pairs of mature parasite-infected erythrocyte surface antigen (MESA)(parasite)/protein 4.1 (host) and *P. falciparum* histidine-rich protein (PfHRP1)(parasite)/protein 4.1 (host) have been analyzed by dual color excitation and NSOM fluorescence detection. A thin blood smear of the trophozoite infected erythrocytes was fixed, then reacted with antibodies against PfHRP1, and labeled with tetramethylrhodamine (Fig. 2.63).

While all three cells are imaged in topography, only the one in the lower right corner was infected and is visible in the fluorescence image. The results imply that the malarial protein MESA interacts with erythrocyte protein in membranes of infected red blood cells, whereas PfHRP1 does not interact with protein 4.1.

2.7.8 Cantilever Arrays as Biosensors

The application of cantilever arrays to the detection of biomolecules may open the way to a new range of biosensors capable of detecting extremely low concentrations of the order of 10 nM of analytes. Figure 2.64 shows three sets of experiments [7].

Firstly, the effect of a base pair mismatch in oligonucleotide–oligonucleotide recognition was demonstrated (Fig. 2.64A). Here the difference between full complementarity and the single mismatch is clearly seen. The detection limit is demonstrated in Fig. 2.64(B), where decreasing concentrations of oligonucleotide were used. Given the differences between 12 mer oligonucleotides and proteins, it may be expected that even lower concentration limits can be attained with protein. Figure 2.64(C) demonstrates the use of a cantilever array for immunodetection. Here differential signals between a cantilever coated with protein A and a second one coated with BSA to allow removal of signals from non-specific binding were used. Some caution must be taken over the use of cantilever arrays as the microfabrication of such systems generally leads to arrays having different cantilever thicknesses and hence differing spring constants. Thus, much care is required to validate the reproducibility of the biosensors.

2.8 Conclusion

This chapter has illustrated the various types of scanning probe microscopy available to the biological community. The examples presented in the section on using

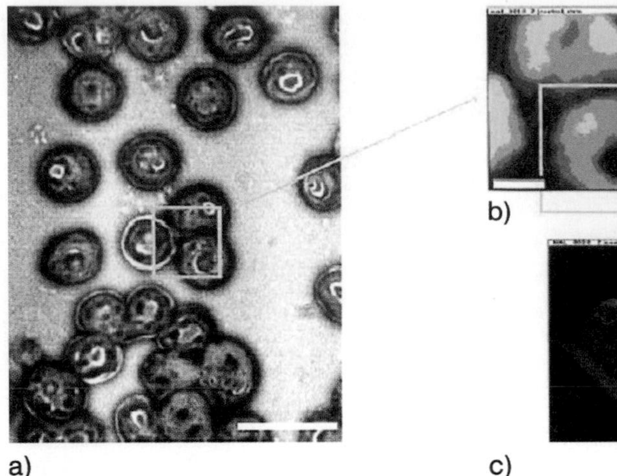

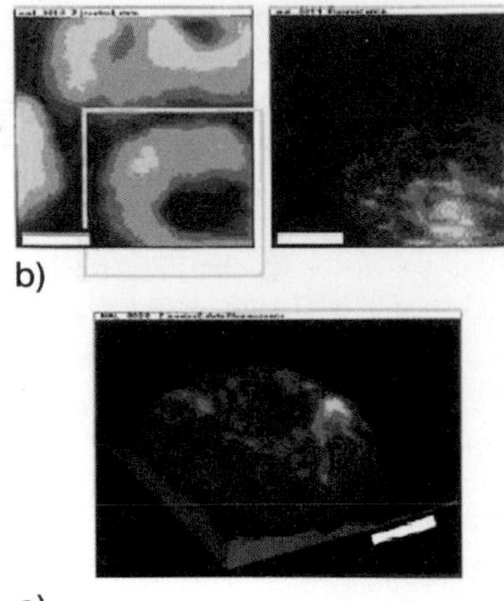

Fig. 2.63. (a) Bright-field image of a blood smear; (b) result of NSOM imaging in topography (left) and fluorescence (right) (128 $\times$ 128 pixels, 60 nm pixel^{-1}); (c) Zoom around the infected cell seen in the lower right corner of (b) (128 $\times$ 128 pixels, 52 nm pixel^{-1}). The fluorescence signal is superimposed as a color value (brightness scale goes from black via red to yellow) to the topography to show the distribution of PfHRP1 in the erythrocyte membrane. (Reprinted with permission from Ref. [103]. Copyright (1997) PNAS.)

the tools are there to stimulate interest in imaging biological systems, and to try and answer the question posed in the introduction. Why use scanning probe microscopies in biology?

The final question is where do we go from here? The reply would seem to be towards integrated studies using SPM as one tool along with the wide panoply of other physical techniques to probe biological activity in its environment.

Acknowledgments

We gratefully acknowledge JPK GmbH of Berlin for allowing us to use many figures from their handbook. We wish to thank the Fondation pour la Recherche Medicale for funding our AFM. S.C. thanks BioMerieux and A.N.L. thanks Delta Proteomics for financial support.

References

1 G. Binning, H. Rohrer, C. Gerber, E. Weibel, Surface studies by scanning tunneling microscopy *Phys. Rev. Lett.* **1982**, *49*, 57–61.

2 J. K. H. Hoerber, M. J. Miles, Scanning probe evolution in biology *Science* **2003**, *302*, 1002–1005.

3 K. D. Jandt, Atomic force microscopy

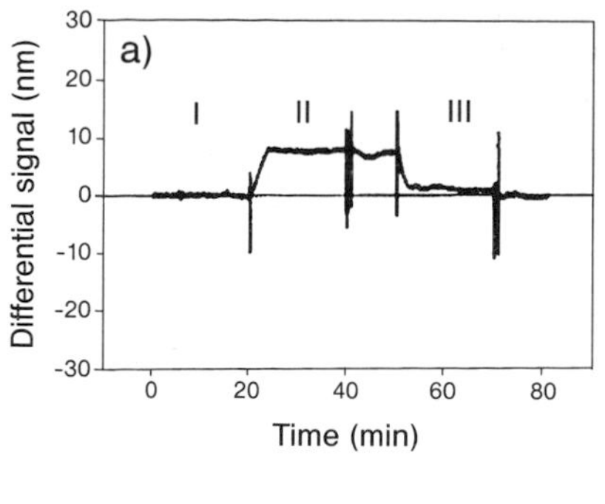

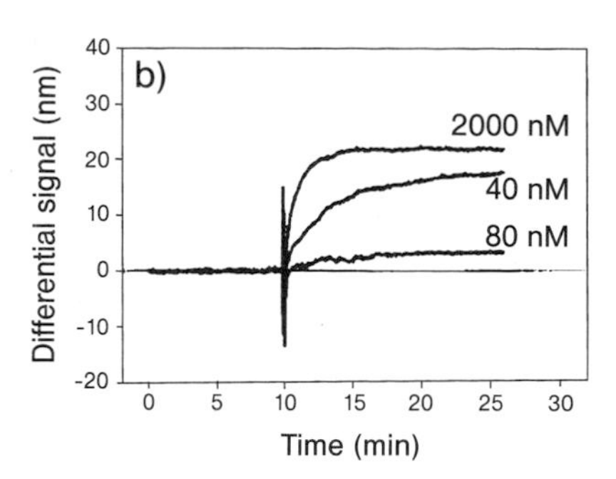

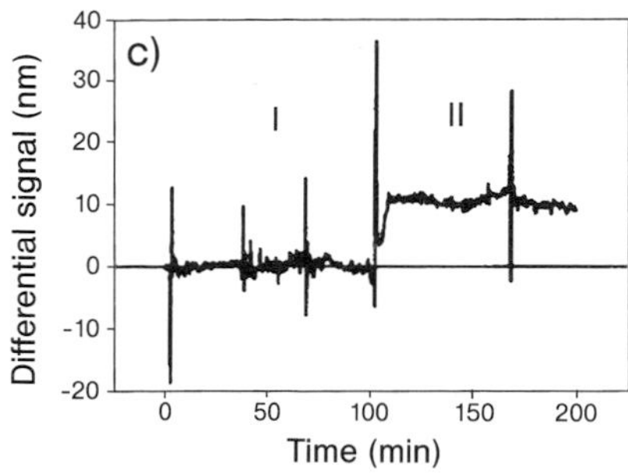

Fig. 2.64. (A) Differential signal of a hybridization experiment, showing detection of a single base mismatch in 12 mer oligonucleotides. Two cantilevers were functionalized with sequences differing only in one base: 59-CTATGTCAGCAC-39 (first oligonucleotide), 59-CTATGTAAGCAC-39 (second oligonucleotide). Injection of the first complementary oligonucleotide increases the differential signal (interval II); injection of the second complementary oligonucleotide decreases the differential signal (interval III). (B) Three successive hybridization experiments with different 12 mer oligonucleotide concentrations, using one array. The concentration detection limit was calculated to be 10 nM, on the basis of a deflection noise of 0.5 nm. (C) Differential signal from a protein–protein interaction. One cantilever was functionalized with protein A and the other with BSA as a reference. First, as a negative control, buffer and then buffer containing goat IgG were injected. In both cases (interval I), no change of the differential signal was detected. In contrast, injection of rabbit IgG increased the differential signal by 12 nm within minutes (interval II). The signal persisted after the liquid cell was purged again with buffer. (Reprinted with permission from Ref. [7]. Copyright (2000) Science.)

of biomaterials surfaces and interfaces *Surf. Sci.* **2001**, *491*, 303–332.

4 R. McKendry, J. Zhang, Y. Arntz, T. Strunz, M. Hegner, H. P. Lang, M. K. Baller, U. Certa, E. Meyer, H.-J. Guntherodt, C. Gerber, Multiple label-free biodetection and quantitative DNA-binding assays on a nanomechanical cantilever array *Proc. Natl. Acad. Sci. U.S.A.* **2002**, *99*, 9783–9788.

5 F. M. Battiston, J. P. Ramseyer, H. P. Lang, M. K. Baller, C. Gerber, J. K. Gimzewski, E. Meyer, H. J. Guntherodt, A chemical sensor based on a microfabricated cantilever array with simultaneous resonance-frequency and bending readout *Sens. Actuators, B* **2001**, *B77*, 122–131.

6 J. Fritz, M. K. Baller, H. P. Lang, T. Strunz, E. Meyer, H. J. Guentherodt, E. Delamarche, C. Gerber, J. K. Gimzewski, Stress at the solid-liquid interface of self-assembled monolayers on gold investigated with a nanomechanical sensor *Langmuir* **2000**, *16*, 9694–9696.

7 J. Fritz, M. K. Baller, H. P. Lang, H. Rothuizen, P. Vettiger, E. Meyer, H. J. Guntherodt, C. Gerber, J. K. Gimzewski, Translating biomolecular recognition into nanomechanics *Science* **2000**, *288*, 316–318.

8 M. K. Baller, H. P. Lang, J. Fritz, C. Gerber, J. K. Gimzewski, U. Drechsler, H. Rothuizen, M. Despont, P. Vettiger, F. M. Battiston, J. P. Ramseyer, P. Fornaro, E. Meyer, H. J. Gunthe-rodt, A cantilever array-based artificial nose *Ultramicroscopy* **2000**, *82*, 1–9.

9 S. W. Paddock (ed.), *Confocal Microscopy Methods and Protocols*, Methods Mol. Biol., vol. 122, Humana, Totowa, **1999**.

10 S. Paddock, Tech sight: Optical sectioning-slices of life *Science* **2002**, *295*, 1319–1321.

11 E. H. Synge, A suggested method for extending microscopic resolution into the ultra-microscopic region *Phil. Mag.* **1928**, *6*, 356–62.

12 Z. Ounaies, Piezoelectric materials, in *Encyclopedia of Biomaterials and Biomedical Engineering*, vol. 2, eds. G. E. Wnek, G. L. Bowlin, Marcel Dekker, New York, **2004**.

13 R. L. Smith, G. S. Rohrer, The preparation of tip and sample surfaces for scanning probe experiments, in *Scanning Probe Microscopy and Spectroscopy. Theory, Techniques and Applications*, 2nd edn, ed. D. A. Bonnel, Wiley & Sons, New York, **1993**.

14 M. J. Madou, *Fundamentals of Microfabrication: The Science of Miniaturization*, 2nd edition, CRC Press, Boca Raton, **2002**.

15 M. Tomitori, T. Arai, Tip cleaning and sharpening processes for non-contact atomic force microscope in ultrahigh vacuum *Appl. Surf. Sci.* **1999**, *140*, 432–438.

16 D. Wouters, U. S. Schubert, Nanolithography and nanochemistry: probe-related patterning techniques and chemical modification for nanometer-sized devices *Angew. Chem. Int. Ed.* **2004**, *43*(19), 2480–2495.

17 H. F. Knapp, W. Wiegräbe, M. Heim, R. Eschrich, R. Guckenberger, Atomic force measurements and manipulation of Langmuir-Blodgett films with modified tips *Biophys. J.* **1995**, *69*, 708–715.

18 M.-H. Paclet, A. W. Coleman, J. Burritt, F. Morel, NADPH oxidase of Epstein-Barr-virus immortalized B lymphocytes. Effect of cytochrome b558 glycosylation *Eur. J. Biochem.* **2001**, *268*, 5197–5208.

19 T. Boland, B. D. Ratner, Direct measurement of hydrogen bonding in DNA nucleotide bases by atomic force microscopy *Proc. Natl. Acad. Sci. U.S.A.* **1995**, *92*(12), 5297–301.

20 T. Han, J. M. Williams, T. P. Jr. Beebe, Chemical bonds studied with functionalized atomic force microscopy tips *Anal. Chim. Acta* **1995**, *307*(2–3), 365–76.

21 E. L. Florin, V. T. Moy, H. E. Gaub, Adhesion forces between individual ligand-receptor pairs *Science* **1994**, *264*, 415–417.

22 M. Benoit, D. Gabriel, G. Gerisch, H. E. Gaub, Discrete interactions in cell adhesion measured by single-molecule force spectroscopy *Nat. Cell Biol.* **2000**, *2*(6), 313–317.

23 T. V. Ratto, K. C. Langry Kevin, R. E. Rudd, R. L. Balhorn, M. J. Allan, M. W. McElfresh, Force spectroscopy of the double-tethered concanavalin-A mannose bond *Biophys. J.* **2004**, *86*(4), 2430–2437.

24 H. Nakajima et al., Scanning force microscopy of the interaction events between a single molecule of heavy meromyosin and actin *Biochem. Biophys. Res. Commun.* **1997**, *234*, 178–182.

25 V. T. Moy, E. L. Florin, H. E. Gaub, Intermolecular forces and energies between ligands and receptors *Science* **1994**, *266*, 257–259.

26 G. U. Lee, D. A. Kidwell, R. J. Colton, Sensing discrete streptavidin-biotin interactions with atomic force microscopy *Langmuir* **1994**, *10*, 354–357.

27 G. U. Lee, L. A. Chrisey, R. J. Colton, Direct measurement of the forces between complementary strands of DNA *Science* **1994**, *266*, 771–773.

28 U. Dammer, O. Popescu, P. Wagner, D. Anselmetti, H. J. Guntherodt, G. N. Misevic, Binding strength between cell adhesion proteoglycans measured by atomic force microscopy *Science* **1995**, *267*, 1173–1175.

29 U. Dammer, M. Hegner, D. Anselmetti, P. Wagner, M. Dreier, W. Huber, H. J. Guntherodt, Specific antigen/antibody interactions measured by force microscopy *Biophys. J.* **1996**, *70*, 2437–2441.

30 P. Hinterdorfer, W. Baumgartner, H. J. Gruber, K. Schilcher, H. Schindler, Detection and localization of individual antibody-antigen recognition events by atomic force microscopy *Proc. Natl. Acad. Sci. U.S.A.* **1996**, *93*, 3477–3481.

31 S. Allen, X. Chen, J. Davies, M. C. Davies, A. C. Dawkes, J. C. Edwards, C. J. Roberts, J. Sefton, S. J. Tendler, P. M. Williams, Detection of antigen-antibody binding events with the atomic force microscope *Biochemistry* **1997**, *36*, 7457–7463.

32 J. Fritz, A. G. Katopodis, F. Kolbinger, D. Anselmetti, Force-mediated kinetics of single P-selectin/ligand complexes observed by atomic force microscopy *Proc. Natl. Acad. Sci. U.S.A.* **1998**, *95*, 12283–12288.

33 P. P. Lehenkari, M. A. Horton, Single integrin molecule adhesion forces in intact cells measured by atomic force microscopy *Biochem. Biophys. Res. Commun.* **1999**, *259*, 645–650.

34 M. Thie, R. Rospel, W. Dettmann, M. Benoit, M. Ludwig, H. E. Gaub, H. W. Denker, Interactions between trophoblast and uterine epithelium: monitoring of adhesives forces *Hum. Reprod.* **1998**, *13*, 3211–3219.

35 G. Sagvolden, I. Giaever, E. O. Pettersen, J. Feder, Cell adhesion force microscopy *Proc. Natl. Acad. Sci. U.S.A.* **1999**, *96*, 471–476.

36 C. Bustamante, D. A. Erie, D. Keller, Biochemical and structural applications of scanning force microscopy *Curr. Opin. Struct. Biol.* **1994**, *4*, 750–760.

37 M. Grandbois, M. Beyer, M. Rief, H. Clausen-Schaumann, H. E. Gaub, How strong is a covalent bond? *Science* **1999**, *283*, 1727–1730.

38 H. Li, M. Rief, F. Oesterhelt, H. E. Gaub, Force spectroscopy on single xanthan molecules *Appl. Phys. A: Mater. Sci. Process.* **1999**, *68*, 407–410.

39 C. Bustamante, Z. Bryant, S. B. Smith, Ten years of tension: single-molecule DNA mechanics *Nature* **2003**, *421*, 423–427.

40 C. Bustamante, S. B. Smith, J. Liphardt, D. Smith, Single-molecule studies of DNA mechanics *Curr. Opin. Struct. Biol.* **2000**, *10*, 279–285.

41 U. Bockelmann, Single-molecule manipulation of nucleic acids *Curr. Opin. Struct. Biol.* **2004**, *14*, 368–373.

42 M. Hegner, W. Grange, Mechanics and imaging of single DNA molecules *J. Muscle Res. Cell Motility* **2002**, *23*, 367–375.

43 J. Van Noort, Unraveling bacteriorhodopsin *Biophys. J.* **2005**, *88*, 763–764.

44 R. Rounsevell, J. R. Forman, J. Clarke, Atomic force microscopy: mechanical unfolding of proteins *Methods* **2004**, *34*, 100–111.

45 R. B. Best, D. J. Brockwell, J. L. Toca-Herrera, A. W. Blake, D. A. Smith, S. E. Radford, J. Clarke, Force mode atomic force microscopy as a tool for protein folding studies *Anal. Chim. Acta* **2003**, *479*, 87–105.

46 R. B. Best, J. Clarke, What can atomic force microscopy tell us about protein folding? *Chem. Commun.* **2002**, 183–192.

47 M. Ludwig, M. Rief, L. Schmidt, H. Li, F. Oesterhelt, M. Gautel, H. E. Gaub, AFM, a tool for single-molecule experiments *Appl. Phys. A: Mater. Sci. Process.* **1999**, *A68*, 173–176.

48 D. P. Allison, P. Hinterdorfer, W. Han, Biomolecular force measurements and the atomic force microscope *Curr. Opin. Biotechnol.* **2002**, *13*, 47–51.

49 T. Basche, S. Nie, J. M. Fernandez, Single molecules *Proc. Natl. Acad. Sci. U.S.A.* **2001**, *98*, 10527–10528.

50 A. Noy, D. V. Vezenov, C. M. Lieber, Chemical force microscopy *Annu. Rev. Mater. Sci.* **1997**, *27*, 381–421.

51 M. Su, V. P. Dravid, C. A. Mirkin, Direct patterning of solid-state and organic materials by dip-pen nanolithography *Abstracts of Papers, 224th ACS National Meeting, Boston, MA, United States, August 18–22,* **2002**, COLL-092.

52 S. Hong, J. Zhu, C. A. Mirkin, Multiple ink nanolithography: Toward a multiple-pen nano-plotter *Science* **1999**, *286*, 523–525.

53 R. D. Piner, J. Zhu, F. Xu, S. Hong, C. A. Mirkin, "Dip-pen" nanolithography *Science* **1999**, *283*, 661–663.

54 B. Gates, Q. Xu, J. C. Love, D. B. Wolfe, G. M. Whitesides, Unconventional nanofabrication. *Annu. Rev. Mater. Res.* **2004**, *34*, 339–372.

55 E. A. Ash, G. Nicholls, Super-resolution aperture scanning microscope *Nature* **1972**, *237*, 510–512.

56 E. Betzig, R. J. Chichester, Near-field optics: microscopy, spectroscopy, and surface modification beyond the diffraction limit *Science* **1992**, *257*, 189–195.

57 E. Betzig, R. J. Chichester, Single molecules observed by near-field scanning optical microscopy *Science* **1993**, *262*, 1422–1425.

58 R. C. Dunn, Near-field optical microscopy *Chem. Rev. Wash DC.* **1999**, *99*(10), 2891–2927.

59 A. N. Lazar, P. Shahgaldian, A. W. Coleman, Anion recognition effects in the structuring of bovine serum albumin films. *J. Supramol. Chem.* **2001**, *1*, 193–199.

60 J. Fang, C. M. Soto, T. Lin, J. E. Johnson, B. Ratna, Complex pattern formation by cowpea mosaic virus nanoparticles *Langmuir* **2002**, *18*(2), 308–310.

61 T. P. Beebe, T. E. Wilson, D. F. Ogletree, J. E. Katz, R. Balhorn, M. B. Salmeron, W. J. Siekhaus, Direct observation of native DNA structures with the scanning tunneling microscope *Science* **1989**, *243*, 370–372.

62 S. Gould, B. Drake, C. B. Prater, A. L. Weisenhorn, S. M. Lindsay, P. K. Hansama, Imaging polymers, proteins and DNA in aqueous solutions with atomic force microscope *Proc. of the 47th Electron Mycroscopy Society of America* **1989**, 32–33.

63 W. Grange, T. Strunz, I. Schumakovitch, H.-J. Guntherodt, M. Hegner, Molecular recognition and adhesion of individual DNA strands studied by dynamic force microscopy, *Single Molecules* **2001**, *2*, 75–78.

64 K. Umemura, T. Okada, R. Kuroda, Cooperativity and intermediate structures of single-stranded DNA binding-assisted RecA-single-stranded DNA complex formation studied by atomic force microscopy *Scanning* **2005**, *27*, 35–43.

65 D. Anselmetti, J. Fritz, B. Smith, X. Fernandez-Busquets, Single molecule DNA biophysics with atomic force microscopy *Single Mol.* **2000**, *1*, 53–58.

66 T. Thundat, X.-Z. Zheng, S. L. Sharp, D. P. Allison, R. J. Warmack, D. C. Joy, T. L. Ferrell, *Scanning. Microsc.* **1992**, *6*, 903–910.

67 J. M. Kim, H. S. Jung, J. Wan Park, H. Y. Lee, T. Kawai, AFM phase lag mapping for protein–DNA oligonucleotide complexes *Anal. Chem. Acta* **2004**, *525*, 151–157.

68 J. van Noort, T. van der Heijden, C. F. Dutta, K. Firman, C. Dekker, Initiation of translocation by Type I restriction modification enzymes is associated with a short DNA extrusion *Nucl. Acids Res.* **2004**, *32*, 6540–6547.

69 K. Firman, M. Szczelkun, Measuring motion on DNA by the type I restriction endonuclease EcoR124I using triplex dissociation *EMBO J.* **2000**, *19*, 2094–2102.

70 R. Seidel, J. van Noort, C. van der Scheer, J. G. P. Bloom, N. H. Dekker, C. F. Dutta, A. Blundell, T. Robinson, K. Firman, C. Dekker, Real-time observation of DNA translocation by the type I restriction-modification enzyme EcoR124I *Nat. Struct. Mol. Biol* **2004**, *11*, 838–843.

71 R. Krautbauer, L. H. Pope, T. E. Schrader, S. Allen, H. E. Gaub, Discriminating small molecule DNA binding modes by single molecule

force spectroscopy *F.E.B.S. Letters* **2002**, *510*, *7*(3), 154–158.

72 H. Clausen-Schaumann, M. Rief, C. Tolksdorf, H. E. Gaub, Mechanical stability of single DNA molecules *Biophys. J.* **2000**, *78*, 1997–2007.

73 M. Bonin, R. Zhu, Y. Klaue, J. Oberstrass, E. Oesterschulze, W. Nellen, Analysis of RNA flexibility by scanning force spectroscopy *Nucl. Acids Res.* **2002**, *30*(16), e81.

74 M. Rief, H. Clausen-Schaumann, H. E. Gaub, Sequence-dependent mechanics of single DNA molecules *Nat. Struct. Biol.* **1999**, *6*, 346–349.

75 S. Lees, K. S. Prostak, V. K. Ingle, K. Kjoller, The loci of mineral in turkey leg tendon as seen by atomic force microscope and electron microscopy *Calcified Tissue International* **1994**, *55*(3), 180–189.

76 H. Seelert, A. Poetsch, N. A. Dencher, A. Engel, H. Stahlberg, D. J. Müller, Proton-powered turbine of a plant motor *Nature* **2000**, *405*, 418–419.

77 D. Fotiadis, L. Hasler, D. J. Muller, H. Stahlberg, J. Kistler, A. Engel, Surface tongue-and-groove contours on lens MIP facilitate cell-to-cell adherence *J. Mol. Biol.* **2000**, *300*, 779–789.

78 J.-H. Lim, D. S. Ginger, K.-B. Lee, J. Heo, J.-M. Nam, C. A. Mirkin, Direct-write dip-pen nanolithography of proteins on modified silicon oxide surfaces, *Angew. Chem. Int. Ed.* **2003**, *42*, 2309–2312.

79 J. Wong, A. Chilkoti, V. T. Moy, Direct force measurements of the streptavidin–biotin interaction *Biomol. Eng.* **1999**, *16*, 45–55.

80 M. Rief, M. Gautel, F. Oesterhelt, J. M. Fernandez, H. E. Gaub, Reversible unfolding of individual titin immunoglobulin domains by AFM *Science* **1997**, *276*(5315), 1109–1112.

81 L. Ng, A. J. Grodzinsky, P. Patwari, J. Sandy, A. Plaas, C. Ortiz, Individual cartilage aggrecan macromolecules and their constituent glycosaminoglycans visualized via atomic force microscopy *J. Struct. Biol.* **2003**, *14*, 242–257.

82 F. Tokumasu, A. J. Jin, G. W. Feigenson, J. A. Dvorak, Atomic force microscopy of nanometric liposome adsorption and nanoscopic membrane domain formation, *Ultramicroscopy* **2003**, *97*, 217–227.

83 A. Dubes, H. Parrot-Lopez, A. Wassim, G. Degobert, H. Fessi, P. Shahgaldian, A. W. Coleman, Scanning electron microscopy and atomic force microscopy imaging of solid lipid nanoparticles derived from amphiphilic cyclodextrins, *Eur. J. Pharm. Biopharm.* **2003**, *55*, 279–282.

84 a: M. H. Paclet, A. W. Coleman, S. Vergnaud, F. Morel, P67-phox-mediated NADPH oxidase assembly: imaging of cytochrome b558 liposomes by atomic force microscopy, *Biochemistry* **2000**, *39*, 9302–9310.

85 H. Brochu, A. Polidori, B. Pucci, P. Vermette, Drug delivery systems using immobilized intact liposomes: A comparative and critical review, *Curr. Drug Delivery* **2004**, *1*, 299–312.

86 W. A. Ducker, T. J. Senden, R. M. Pashley, Direct measurement of colloidal forces using an atomic force microscope *Nature* **1991**, *353*, 239–241.

87 X. Liang, G. Mao, K. Y. S. Ng, Mechanical properties and stability measurement of cholesterol-containing liposome on mica by atomic force microscopy *J. Colloid Interface Sci.* **2004**, *278*, 53–62.

88 P. Shahgaldian, L. Quattrocchi, J. Gualbert, A. W. Coleman, P. Goreloff, AFM imaging of calixarene based solid lipid nanoparticles in gel matrices *Eur. J. Pharm. Biopharm.* **2003**, *55*, 107–113.

89 J. Gualbert, P. Shahgaldian, A. W. Coleman, Interactions of amphiphilic calix[4]arene based solid-lipid-nanoparticles with bovine serum albumin *Int. J. Pharm.* **2003**, *257*(1–2), 69–73.

90 A. A. Brian, H. M. McConnell, Allogenic stimulation of cytotoxic tcelles by supported planar membranes, *Proc. Natl. Acad. Sci. U.S.A.* **1984**, *81*, 6159–6163.

91 J. Jass, T. Tjärnhage, G. Puu, From liposomes to supported, planar bilayer structures on hydrophilic and hydrophobic surfaces: An atomic force microscopy study *Biophys J.* **2000**, *79*, 3153–3163.

92 R. Richter, A. Mukhopadhyay, A. Brisson, Pathways of lipid vesicle deposition on solid surfaces: A combined QCM-D and AFM study *Biophys. J.* **2003**, *85*, 3035–3047.

93 A. Garner, S. Connell, G. Li, A. Smith, J. Colyer, N. Hooper, Atomic force microscopy of microdomains in lipid bilayers, http://www.astbury.leeds.ac.uk/Report/2003/Report/46smith-AFM.pdf, **2003**.

94 T. Berge, D. J. Ellis, D. T. F. Dryden, J. M. Edwardson, R. M. Henderson, Translocation-independent dimerization of the EcoKI endonuclease visualized by atomic force microscopy *Biophys. J.* **2000**, *79*(1), 479–484.

95 D. E. Saslowsky, J. Lawrence, X. Ren, D. A. Brown, R. M. Henderson, J. M. Edwardson, Placental alkaline phosphatase is efficiently targeted to rafts in supported lipid bilayers *J. Biol. Chem.* **2002**, *277*, 26966–26970.

96 H. Ha, H. B. Kwak, S. K. Lee, D. S. Na, C. E. Rudd, Z. H. Lee, H. H. Kim, Membrane rafts play a crucial role in receptor activator of nuclear factor Kappa B (Rank) signaling and osteoclast function *J. Biol. Chem.* **2003**, *278*, (20), 18573–18580.

97 N. A. Geisse, T. L. Cover, R. M. Henderson, T. M. Edwardson, Targeting of *Helicobacter pylori* vacuolating toxin to lipid raft membrane domains analyzed by atomic force microscopy *Biochem. J.* **2004**, *381*, 911–917.

98 S. Barakat, L. Gayet, G. Dayan, S. Labialle, A. Lazar, V. Oleinikov, A. W. Coleman, L. G. Baggetto, Multidrug-resistant cancer cells contain two populations of P-glycoprotein with differently stimulated P-gp ATPase activities. Evidence from atomic force microscopy and biochemical analysis, *Biochem. J.* **2005**, *388*, 563–571.

99 C. W. Hollars, R. C. Dunn, Submicron structure in L-α-dipalmitoylphosphatidylcholine monolayers and bilayers probed with confocal, atomic force, and near-field microscopy *Biophys. J.* **1998**, *75*, 342–353.

100 H. Shiku, R. C. Dunn, Domain formation in thin lipid films probed with near-field scanning optical microscopy *J. Microsc.* **1999**, *194*, 455–460.

101 Y. G. Kuznetsov, A. Low, H. Fan, A. McPherson, Atomic force microscopy investigation of wild-type Moloney murine leukaemia virus particles and virus particles lacking the envelope protein *Virology* **2004**, *323*, 189–196.

102 M. Micic, D. Hu, Y. D. Suh, G. Newton, M. Romine, H. P. Lu, Correlated atomic force microscopy and fluorescence lifetime imaging of live bacterial cells, *Colloids Surf., B: Biointerfaces* **2004**, *34*(4), 205–212.

103 T. H. Enderle, T. Ha, D. F. Ogletree, D. S. Chemla, C. Magowan, S. Weiss, Membrane specific mapping and colocalization of malarial and host skeletal proteins in the *Plasmodium falciparum* infected erythrocyte by dual-color near-field scanning optical microscopy *Proc. Natl. Acad. Sci. U.S.A.* **1997**, *94*, 520–525.

Appendix 1 Books on Scanning Probe Microsopies Reviews on Scanning Probe Microsopies in Biology

M. Allegrini, N. Garcia, O. Marti (**2001**). *Nanometer Scale Science and Technology.* Amsterdam, IOS Press.

G. Attard, C. Barnes (**1998**). *Surfaces.* Oxford, Oxford University Press.

C. Bai (**2000**). *Scanning Tunneling Microscopy*

and Its Application. Heidelberg, Springer Verlag.

J. D. Batteas, G. C. Michales (**2005**). *Applications of Scanned Probe Microscopy to Polymers.* Oxford, Oxford University Press.

R. J. Behm, H. Rohrer, N. Garcia (**1990**). *Scanning Tunneling Microscopy and Related Methods Proceedings.* New York, Kluwer Academic Publishers.

G. Benedek (**1992**). *Surface Properties of Layered Structures,* New York, Kluwer Academic Publishers.

B. Bhushan (**2003**). *Applied Scanning Probe Methods.* Heidelberg, Springer Verlag.

K. S. Birdi (**2003**). *Scanning Probe Microscopes Applications in Science and Technology.* Boca Raton, FL, CRC Press.

P. C. Braga, D. Ricci (**2003**). *Atomic Force Microscopy Biomedical Methods and Applications.* Totowa, Humana Press.

H. Bubert, B. Jenett (**2002**). *Surface and Thin Film Analysis Compendium of Principles, Instrumentation and Applications.* New York, Wiley-VCH.

S. H. Cohen (**1995**). *Atomic Force Microscopy/ Scanning Tunneling Microscopy.* New York, Kluwer Academic Publishers.

D. A. Bonnell (**2000**). *Scanning Probe Microscopy and Spectroscopy: Theory, Techniques, and Applications.* New York, Wiley-VCH.

A. De Stefanis, A. A. G. Tomlinson (**2001**). *Scanning Probe Microscopies From Surface Structure to Nano-scale Engineering.* Zurich, Trans Tech Publications Ltd.

N. J. Dinardo (**2004**). *Nanoscale Characterization of Surfaces and Interfaces.* New York, Wiley-VCH.

G. Doyen, D. Drakova (**2004**). *The Physical Principles of STM and AFM Operation.* New York, Wiley-VCH.

P. L. Gai (**1997**). *In-Situ Microscopy in Materials Research Leading International Research in Electron and Scanning Probe.* New York, Kluwer Academic Publishers.

M. Gentili, C. Giovannella (**1994**). *Nanolithography A Borderland Between STM, EB, IB and X-Ray Lithographies.* New York, Kluwer Academic Publishers.

A. A. Gewirth, H. Siegenthaler (**1995**). *Forces in Scanning Probe Methods.* New York, Kluwer Academic Publishers.

H.-J. Guntherodt, R. Wiesendanger (**1995**). *Scanning Tunneling Microscopy.* Berlin and Heidelberg, Springer Verlag.

H. Neddermeyer (**1993**). *Scanning Tunneling Microscopy.* New York, Kluwer Academic Publishers.

B. P. Jena, J. K. Heinrich Hober (**2002**). *Atomic Force Microscopy in Cell Biology* (Methods in Cell Biology, vol 68). London, Elsevier.

S. Kawata, M. Ohtsu, M. Irie (**2002**). *Nano-Optics.* Heidelberg, Springer Verlag.

D. Lange, O. Brand, H. Baltes (**2002**). *Cantilever-Based CMOS Nano-Electro-Mechanical Systems Atomic-Force Microscopy and Gas Sensing Applications.* New York, Springer-Verlag.

X. Y. Liu, J. J. De Yoreo (**2004**). *Nanoscale Structure and Assembly at Solid-Fluid Interfaces.* New York, Kluwer Academic Publishers.

W. J. Lorenz, W. Pleith (**1999**). *Electrochemical Nanotechnology: In Situ Local Probe Techniques at Electrochemical Interfaces.* New York, Wiley-VCH.

O. Marti, M. Amrein (**1993**). *STM and SFM in Biology.* London, Elsevier Academic Press.

F. Mayer, M. Hopert (**2003**). *Microscopic Techniques in Biotechnology.* New York, Wiley-VCH.

S. Morita, R. Wiesendanger, E. Meyer (**2002**). *Non Contact Atomic Force Microscopy.* Heidelberg, Springer-Verlag.

V. J. Morris, A. R. Kirby, A. P. Gunnin (**2000**). *Atomic Force Microscopy for Biologists.* London, Imperial College Press.

M. Nieto-Vesperinas, N. Garcia (**1996**). *Optics at the Nanometer Scale Imaging and Storing with Photonic Near Fields.* New York, Kluwer Academic Publishers.

A. Periasamy (**2001**). *Methods in Cellular Imaging.* Oxford, Oxford University Press.

R. H. Templer, R. Leatherbarrow (**2003**). *Biophysical Chemistry: Membrane and Proteins.* Cambridge, The Royal Society of Chemistry.

R. Vanselow (**1990**). *Chemistry and Physics of Solid Surfaces.* Heidelberg, Springer Verlag.

T. Vo-Dinh (**2004**). *Protein Nanotechnology Protocols, Instrumentation, and Applications.* Totowa, Humana Press.

R. Wiesendanger (**1994**). *Scanning Probe Microscopy and Spectroscopy Methods and Applications.* Cambridge, Cambridge University Press.

Appendix 2 Reviews on Scanning Probe Microsopies in Biology

N. I. Abu-Lail, T. A. Camesan, Polysaccharide properties probed with atomic force microscopy *J. Microsc.* **2003**, *212*(3), 217–238.

W. M. Albers, I. Vikholm, et al., Interfacial and materials aspects of the immobilization of biomolecules onto solid surfaces *Handbook Surf. Interfaces Mater.* **2001**, *5*, 1–31.

S. Allen, S. M. Rigby-Singleton, et al., Measuring and visualizing single molecular interactions in biology *Biochem. Soc. Trans.* **2003**, *31*(5), 1052–1057.

D. P. Allison, P. Hinterdorfer, et al., Biomolecular force measurements and the atomic force microscope *Curr. Opin. Biotechnol.* **2002**, *13*(1), 47–51.

M. Amrein, H. Gross, Scanning tunneling microscopy of biological macromolecular structures coated with a conducting film *Scanning Microsc.* **1992**, *6*(2), 335–344.

M. Amrein, H. Gross, et al., STM of proteins and membranes *STM SFM Biol.* **1993**, 127–175.

J. E. T. Andersen, J. Ulstrup, et al., Imaging of electrochemical processes and biological macromolecular adsorbates by in-situ scanning tunneling microscopy *Electrochem. Nanotechnol.* **1998**, 27–44.

T. Ando, Atomic force microscope: Application to life science *Seikagaku* **2002**, *74*(11), 1329–1342.

T. Ando, Real-time AFM imaging of protein motions *Farumashia* **2002**, *38*(6), 508–512.

T. Ando, High-speed atomic force microscope: High-speed imaging of nanometer-scale dynamics of biological molecules *Oyo Butsuri* **2003**, *72*(10), 1304–1308.

T. Ando, High speed AFM observes motor protein movement *Kagaku* **2004**, *59*(1), 28–29.

T. Ando, N. Kodera, et al., A high-speed atomic force microscope for studying biological macromolecules in action *ChemPhysChem* **2003**, *4*(11), 1196–1202.

O. N. Antzutkin, Amyloidosis of Alzheimer's Ab peptides: Solid-state nuclear magnetic resonance, electron paramagnetic resonance, transmission electron microscopy, scanning transmission electron microscopy and atomic force microscopy studies *Magn. Resonan. Chem.* **2004**, *42*(2), 231–246.

M. Aoki, M. Itoh, Methods for kinetic analysis of protein-protein interaction *Tanpakushitsu Kakusan Koso* **2004**, *49*(17, Zokan), 2780–2785.

H. Arakawa, Mechanical property of protein measured with AFM *Denshi Kenbikyo* **2003**, *38*(2), 86–89.

P. G. Arscott, V. A. Bloomfield, Scanning tunneling microscopy in biotechnology *Trends Biotechnol.* **1990**, *8*(6), 151–156.

Y. Baba, Micro- and nanobiochip technology *Oyo Butsuri* **2002**, *71*(12), 1481–1487.

K. Balashev, T. R. Jensen, et al., Novel methods for studying lipids and lipases and their mutual interaction at interfaces. Part I. Atomic force microscopy *Biochimie* **2001**, *83*(5), 387–397.

M. L. Bennink, D. N. Nikova, et al., Dynamic imaging of single DNA-protein interactions using atomic force microscopy *Anal. Chim. Acta* **2003**, *479*(1), 3–15.

M. Benoit, Cell adhesion measured by force spectroscopy on living cells *Methods Cell Biol.* **2002**, *68*, 91–114.

P. Berlin, D. Klemm, et al., Film-forming aminocellulose derivatives as enzyme-compatible support matrices for biosensor developments *Cellulose* **2003**, *10*(4), 343–367.

J. A. Blach-Watson, G. S. Watson, et al., Characterization of biological materials on the nano/meso-scale by force microscopy *Mater. Technol.* **2004**, *19*(1), 12–16.

E. Buzaneva, A. Gorchynskyy, et al., Nanotechnology of DNA/NANO – Si and DNA/carbon nanotubes/nano – Si chips *NATO Sci. Ser., II: Mathematics, Physics and Chemistry* **2002**, *57*, 191–212.

G. Charras, P. Lehenkari, et al., Biotechnological applications of atomic force microscopy *Methods Cell Biol.* **2002**, *68*, 171–191.

A. A. Chernov, P. N. Segre, et al., Crystallization physics in biomacromolecular solutions *Crystal Growth: From Fundamentals to Technology, [International Summer School of Crystal Growth], 12th, Berlin, Germany* **2004**, 95–113.

C. R. Clemmer, T. P. Beebe, Jr., A review of graphite and gold surface studies for use as substrates in biological scanning tunneling microscopy studies *Scanning Microsc.* **1992**, *6*(2), 319–333.

L. T. Costa, S. Thalhammer, et al., Atomic force microscopy as a tool in nanobiology. Part II: Force spectroscopy in genomics and proteomics *Cancer Genom. Proteom.* **2004**, *1*(1), 71–76.

W. V. Dashek, Methods for atomic force and scanning tunneling microscopies *Methods Plant Electron Microsc. Cytochem.* **2000**, 215–221.

J. J. Davis, D. A. Morgan, et al., Scanning probe technology in metalloprotein and biomolecular electronics *IEE Proc.: Nanobiotechnol.* **2004**, *151*(2), 37–47.

J. M. de la Fuente, S. Penades, Understanding carbohydrate-carbohydrate interactions by means of glyconanotechnology *Glycoconj. J.* **2004**, *21*(3–4), 149–163.

F. De Lange, A. Cambi, et al., Cell biology beyond the diffraction limit: near-field scanning optical microscopy *J. Cell Sci.* **2001**, *114*(23), 4153–4160.

V. Deckert, Near-field imaging in biological and biomedical applications *Biomed. Photonics Handbook* **2003**, 12/1–12/19.

E. Delain, D. Michel, et al., Near-field microscopy: from the isolated molecule to the living cell *Morphologie: Bull. Assoc. Anatomistes* **2000**, *84*(265), 25–30.

D. E. Discher, P. Carl, New insights into red cell network structure, elasticity, and spectrin unfolding – a current review *Cell. Mol. Biol. Lett.* **2001**, *6*(3), 593–606.

C. M. Drain, J. D. Batteas, et al., Designing supramolecular porphyrin arrays that self-organize into nanoscale optical and magnetic materials *Proc. Natl. Acad. Sci. U.S.A.* **2002**, *99*(9, Suppl. 2), 6498–6502.

R. C. Dunn, Near-field scanning optical microscopy *Chem. Rev. (Washington, D.C.)* **1999**, *99*(10), 2891–2927.

J. A. Dvorak, The application of atomic force microscopy to the study of living vertebrate cells in culture *Methods (San Diego, CA)* **2003**, *29*(1), 86–96.

A. Ebner, F. Kienberger, et al., Molecular look at the cell nucleus *Bioforum* **2004**, *27*(12), 48–49.

M. Edidin, Near-field scanning optical microscopy, a siren call to biology *Traffic (Copenhagen)* **2001**, *2*(11), 797–803.

M. Firtel, T. J. Beveridge, Scanning probe microscopy in microbiology *Micron (Oxford, England)* **1995**, *26*(4), 347–362.

C. S. Fokas, Scanning near-field optical microscopy (SNOM) *Nachrichten Chem., Technik Lab.* **1999**, *47*(6), 648–652.

D. Fotiadis, S. Scheuring, et al., Imaging and manipulation of biological structures with the AFM *Micron* **2002**, *33*(4), 385–397.

P. L. T. M. Frederix, T. Akiyama, et al., Atomic force bio-analytics *Curr. Opin. Chem. Biol.* **2003**, *7*(5), 641–647.

J. K. Gimzewski, Scanning tunneling microscopy of surface structures *NATO ASI Ser., Ser. C: Mathematical Phys. Sci.* **1991**, *328*, 203–215.

D. G. Gray, Investigations of cellulose and pulp fiber surfaces by atomic force microscopy *Preprints – International Paper and Coating Chemistry Symposium, 5th, Montreal, QC, Canada* **2003**, 47–49.

J.-B. D. Green, A. Idowu, et al., Modified tips: molecules to cells *Mater. Today* **2003**, *6*(2), 22–29.

L. Guo, T. Zhao, et al., Carbon nanotube tip for AFM and its application in the biology *Gaojishu Tongxu*, **2002**, *12*(4), 36–41.

L. Haggerty, A. M. Lenhoff, STM and AFM in biotechnology *Biotechnol. Progr.* **1993**, *9*(1), 1–11.

P. K. Hansma, V. B. Elings, et al., Scanning tunneling microscopy and atomic force microscopy: application to biology and technology *Science* **1988**, *242*(4876), 209–216.

H. G. Hansma, Surface biology of DNA by atomic force microscopy *Annu. Rev. Phys. Chem.* **2001**, *52*, 71–92.

H. G. Hansma, K. Kasuya, et al., Atomic force microscopy imaging and pulling of nucleic acids *Curr. Opin. Struct. Biol.* **2004**, *14*(3), 380–385.

M. Hara, New stage in STM/AFM for use in organic and biological studies *Kagaku* **1993**, *48*(2), 142–143.

M. Hara, K. Nakajima, et al., Single molecular measurement using a scanning probe microscope: for nanofishing *Hikari Nanotekunoroji: Seimei Kagaku e no Tenkai, Nippon Bunko Ga kkai Igaku Seibutsugaku Kenkyu Bukai Shinpojumu, Chiba, Japan, Sept. 5*, **2002**, 8–12.

M. Hegner, A. Engel, Single molecule imaging and manipulation *Chimia* **2003**, *56*(10), 506–514.

M. Hibino, I. Hatta, STM observation of organic molecules *Nippon Butsuri Gakkaishi* **1998**, *53*(1), 32–36.

K. Hizume, S. Yoshimura, et al., The connection between biochemistry and morphology: imaging of DNA and proteins *Denshi Kenbikyo* **2003**, *38*(2), 74–77.

J. K. H. Horber, Local probe techniques in biology *Proc. Int. School of Physics \"Enrico Fermi"* **2001**, *144*, 221–246.

M. Horton, G. Charras, et al., Analysis of ligand-receptor interactions in cells by atomic force microscopy, *J. Receptors Signal Transduct.* **2002**, *22*(1–4), 169–190.

M. A. Horton, P. P. Lehenkari, et al., Probing cellular structure and function by atomic force microscopy, *Special Publication – Royal Society of Chemistry* **2002**, *283*, 31–49.

O. Hoshi, T. Ushiki, Three-dimensional structure of human chromosomes by image analysis using atomic force microscopy, *Denshi Kenbiky* **2003**, *38*(2), 78–82.

J. Hu, J.-H. Lu, et al., Nano-manipulation of single DNA molecules *Nucl. Sci. Tech.* **2004**, *15*(3), 140–143.

J. L. Huff, M. P. Lynch, et al., Label-free protein and pathogen detection using the atomic force microscope *J. Biomol. Screen.* **2004**, *9*(6), 491–497.

A. Ikai, Biological applications. Studies on DNA structure by STM/AFM *Nippon Kessho Gakkaishi* **1993**, *35*(2), 157–158.

A. Ikai, Study of organic compounds and biological samples by STM/AFM *Gendai Kagaku* **1995**, *289*, 24–30.

A. Ikai, STM and AFM of bio/organic molecules and structures *Surf. Sci. Rep.* **1997**, *26*(8), 261–332.

A. Ikai, Nanomechanics of surface immobilized protein molecules *Hyomen Kagaku* **2001**, *22*(9), 620–626.

A. Ikai, Methods for the analysis of surface adsorbed proteins *Tanpakushitsu Kakusan Koso* **2004**, *49*(11, Zokan), 1740–1744.

A. Ikai, R. Afrin, Toward mechanical manipulations of cell membranes and membrane proteins using an atomic force microscope: An invited review *Cell Biochem. Biophys.* **2003**, *39*(3), 257–277.

A. Ikai, R. Afrin, et al., Nano-mechanics of protein molecules and cellular structures: A review of recent work done in the laboratory of biodynamics, Tokyo Institute of Technology *Recent Res. Develop. Appl. Phys.* **2003**, *6*(Pt. 2), 717–732.

A. Ikai, R. Afrin, et al., Nano-mechanical methods in biochemistry using atomic force microscopy *Curr. Protein Peptide Sci.* **2003**, *4*(3), 181–193.

A. Ikai, R. Afrin, et al., Measurement of mechanical properties and adsorption of protein molecules *Oyo Butsuri* **2003**, *72*(10), 1300–1303.

A. Ikai, T. Okajima, et al., Mechanical stretching and rheological study of single protein molecules *Toraiborojisuto* **2004**, *49*(1), 49–55.

S. Iwabuchi, E. Tamiya, Bioimagings using a scanning near-field optical/atomic-force microscope (SNOAM) *Maku* **1998**, *23*(4), 169–176.

K. D. Jandt, Atomic force microscopy of biomaterials surfaces and interfaces *Surf. Sci.* **2001**, *491*(3), 303–332.

A. Janicijevic, D. Ristic, et al., The molecular machines of DNA repair: scanning force microscopy analysis of their architecture *J. Microsc.* **2003**, *212*(3), 265–272.

B. P. Jena, S.-J. Cho, The atomic force microscope in the study of membrane fusion and exocytosis *Methods Cell Biol.* **2002**, *68*(Atomic force microscopy in cell biology), 33–50.

N. Kataoka, K. Hashimoto, et al., Nano/micromechanics analysis of vascular endothelial cells related to genesis of arteriosclerosis *Okayama Igakkai Zasshi* **2004**, *116*(2), 97–101.

A. Kawai, Adhesion and coagulation interpretation of micro-solids by atomic force microscope *Setchaku no Gijutsu* **2003**, *23*(2), 11–15.

H. Kawakami, Nano-structure control of protein in microarray fabrication by dip-pen nanolithography *Kagaku* **2004**, *59*(6), 62–63.

S. Kidoaki, T. Matsuda, Mechanistic aspects of protein/material interactions probed by atomic force microscopy *Colloids Surf., B: Biointerfaces* **2002**, *23*(2–3), 153–163.

J. H. Kindt, J. C. Sitko, et al., Methods for biological probe microscopy in aqueous fluids *Methods Cell Biol.* **2002**, *68*, 213–229.

O. I. Kiselyova, I. V. Yaminsky, Scanning

probe microscopy of proteins and protein-membrane complexes *Colloid J.* **1999**, *61*(1), 1–19.

K. Kitagawa, T. Morita, et al., Single molecule observation of helix peptide, and its electron mediator function *Kyoto Daigaku Nippon Kagaku Sen'i Kenkyusho Koenshu* **2004**, *61*, 27–32.

T. Kobori, M. Yokokawa, et al., Atomic force microscopy, passport to nanobiology *Seibutsu Butsuri* **2004**, *44*(6), 255–259.

T. S. Koffas, E. Amitay-Sadovsky, et al., Molecular composition and mechanical properties of biopolymer interfaces studied by sum frequency generation vibrational spectroscopy and atomic force microscopy *J. Biomater. Sci., Polym. Edn.* **2004**, *15*(4), 475–509.

K. Kogure, Application of AFM to marine microorganisms *Denshi Kenbikyo* **2003**, *38*(2), 83–85.

H. Kondo, Scanning tunneling microscopic observation of biorelated substances *Idemitsu Giho* **1992**, *35*(2), 224–227.

S. i. Kuroda, T. Okajima, et al., A new pinpoint gene delivery system using genetically engineered hepatitis B virus envelope L particles *Mater. Integration* **2002**, *15*(7), 12–17.

R. M. Leblanc, V. Konka, Langmuir and Langmuir-Blodgett films of chlorophyll a and photosystem II complex *Surfactant Sci. Ser.* **2001**, *95*, 641–648.

J. Legleiter, T. Kowalewski, Tapping, pulling, probing: atomic force microscopy in drug discovery *Drug Discovery Today: Technol.* **2004**, *1*(2), 163–169.

S. H. Leuba, J. Zlatanova, Single-molecule studies of chromatin fibers: A personal report *Arch. Histol. Cytol.* **2002**, *65*(5), 391–403.

A. Lewis, The optical near-field and cell biology *Seminars Cell Biol.* **1991**, *2*(3), 187–192.

A. Lewis, A. Radko, et al., Near-field scanning optical microscopy in cell biology *Trends Cell Biol.* **1999**, *9*(2), 70–73.

G. Li, W. Fudickar, et al., Rigid lipid membranes and nanometer clefts: Motifs for the creation of molecular landscapes *Angew. Chem., Int. Ed.* **2002**, *41*(11), 1828–1852.

M. Q. Li, Scanning probe microscopy (STM/AFM) and applications in biology *Appl. Phys. A: Mater. Sci. Processing* **1999**, *A68*(2), 255–258.

A. P. Limanskii, Atomic force microscopy: from DNA and protein imaging to measurement of intermolecular interaction force *Uspekhi Sovremennoi Biol.* **2003**, *123*(6), 531–542.

Y. L. Lyubchenko, DNA structure and dynamics: An atomic force microscopy study *Cell Biochem. Biophys.* **2004**, *41*(1), 75–98.

T. Majima, Soft X-ray imaging of living cells in water: flash contact soft X-ray microscope *Trends Anal. Chem.* **2004**, *23*(7), 520–526.

R. E. Marchant, I. Kang, et al., Molecular views and measurements of hemostatic processes using atomic force microscopy *Curr. Protein Peptide Sci.* **2002**, *3*(3), 249–274.

S. J. Marshall, M. Balooch, et al., The dentin-enamel junction – a natural, multilevel interface *J. Eur. Ceram. Soc.* **2003**, *23*(15), 2897–2904.

J. Masai, T. Shibata, et al., Observation of biological materials by STM/AFM – principles and present status of applications *Tanpakushitsu Kakusan Koso. Protein, Nucleic Acid, Enzyme* **1993**, *38*(4), 741–752.

L. Matyus, A. Jenei, et al., Atomic force microscopy in cell biology, in *Fluorescence Microscopy and Fluorescent Probes, [Based on the Proceedings of the Conference on Fluorescence Microscopy and Fluorescent Probes], 2nd, Prague, Apr. 9–12*, **1998**, 41–46.

A. J. Meixner, H. Kneppe, Scanning near-field optical microscopy in cell biology and microbiology *Cell. Mol. Biol.* **1998**, *44*(5), 673–688.

B. Michel, STM in biology *NATO ASI Ser., Ser. B: Phys.* **1991**, *285*, 549–572.

P. E. Milhiet, M.-C. Giocondi, et al., AFM imaging of lipid domains in model membranes *Sci. World* **2003**, *3*(3), 59–74.

V. J. Morris, Probing molecular interactions in foods *Trends Food Sci. Technol.* **2004**, *15*(6), 291–297.

D. J. Mueller, A. Engel, Conformations, flexibility, and interactions observed on individual membrane proteins by atomic force microscopy *Methods Cell Biol.* **2002**, *68*, 257–299.

D. J. Mueller, H. Janovjak, et al., Observing

structure, function and assembly of single proteins by AFM *Progr. Biophys. Mol. Biol.* **2002**, *79*(1–3), 1–43.

D. J. Mueller, H. Janovjak, et al., Folding, structure and function of biological nanomachines examined by AFM *AIP Conference Proc.* **2003**, *696*, 158–165.

R. Mukhopadhyay, Molecular level structural studies of metalloproteins/metalloenzymes by scanning tunneling microscopy: Scopes and promises *Curr. Sci.* **2003**, *84*(9), 1202–1210.

H. Muramatsu, J.-M. Kim, DNA-immobilization on surface of mica and image analysis by scanning near-field optical microscopy *Hyomen Gijutsu* **2002**, *53*(12), 909–911.

K. Nagai, K. Suzuki, Observation of microorganisms and genes using atomic force microscope *Farumashia* **2003**, *39*(2), 133–136.

C. Nakamura, Application of AFM to the lithographic processing of biochips and the measurement of molecular interaction on the biochip *Baiochippu Saishin Gijutsu Oyo* **2004**, 72–80.

H. Oberleithner, H. Schillers, et al., Nanoarchitecture of plasma membrane visualized with atomic force microscopy *Ion Channel Localization* **2001**, 405–424.

T. Ohtani, Nano-measurement of genome by atomic force microscope *Kagaku Seibutsu* **2003**, *41*(2), 129–135.

A. L. Oliveira, I. B. Leonor, et al., Surface treatments and pre-calcification routes to enhance cell adhesion and proliferation *NATO Sci. Ser., II: Mathematics, Phys. Chem.* **2002**, *86*, 183–217.

E. Oroudjev, S. Danielsen, et al., Surface biology: analysis of biomolecular structure by atomic force microscopy and molecular pulling *Nanobiotechnology* **2004**, 387–403.

T. Osada, H. Uehara, et al., Clinical laboratory implications of single living cell mRNA analysis *Adv. Clinical Chem.* **2004**, *38*, 239–257.

R. E. Palmer, Q. Guo, Imaging thin films of organic molecules with the scanning tunneling microscope *PhysChemChemPhys* **2002**, *4*(18), 4275–4284.

L. Paulino Da Silva, Atomic force microscopy and proteins *Protein Peptide Lett.* **2002**, *9*(2), 117–125.

R. D. S. Pereira, Atomic force microscopy as a novel pharmacological tool *Biochem. Pharmacol.* **2001**, *62*(8), 975–983.

P. Perfetti, A. Cricenti, et al., Scanning probe microscopy applied to materials science and biology *Surf. Rev. Lett.* **2000**, *7*(4), 411–422.

O. Popescu, I. Checiu, et al., Quantitative and qualitative approach of glycan-glycan interactions in marine sponges *Biochimie* **2003**, *85*(1/2), 181–188.

L. A. Pugnaloni, E. Dickinson, et al., Competitive adsorption of proteins and low-molecular-weight surfactants: computer simulation and microscopic imaging *Adv. Colloid Interface Sci.* **2004**, *107*(1), 27–49.

J. P. Rabe, Molecules at interfaces: STM in materials and life sciences *Ultramicroscopy* **1992**, 42–44 (Pt A), 41–54.

A. Razatos, Application of atomic force microscopy to study initial events of bacterial adhesion *Methods Enzymol.* **2001**, *337*, 276–285.

A. P. Razatos, G. Georgiou, Evaluating bacterial adhesion using atomic force microscopy *Handbook Bacterial Adhesion* **2000**, 285–296.

C. K. Riener, G. Kada, et al., Bioconjugation for biospecific detection of single molecules in atomic force microscopy (AFM) and in single dye tracing (SDT) *Recent Res. Develop. Bioconj. Chem.* **2002**, *1*, 133–149.

C. J. Roberts, S. Allen, et al., Quantification and mapping of protein-ligand interactions at the single molecule level by atomic force microscopy *Protein–Ligand Interactions: Structure Spectrosc.* **2001**, 407–423.

D. B. Rodriguez-Amaya, M. Sabino, Mycotoxin research in Brazil: the last decade in review *Braz. J. Microbiol.* **2002**, *33*(1), 1–11.

M. Salmeron, T. Beebe, et al., Imaging of biomolecules with the scanning tunneling microscope: problems and prospects *J. Vac. Sci. Technol., A: Vacuum, Surf. Films* **1990**, *8*(1), 635–641.

M. Salmeron, D. F. Ogletree, et al., Imaging of biological material with STM/AFM *Proc. SPIE – Int. Soc. Optical Eng.* **1992**, *1556*, 40–54.

N. C. Santos, M. A. R. B. Castanho, An overview of the biophysical applications of atomic force microscopy *Biophys. Chem.* **2004**, *107*(2), 133–149.

H. Sasabe, Modification, characterization and

handling of protein molecules as the first step to bioelectronic devices *Electron. Biotechno. Adv. (EL.B.A.) Forum Ser.* **1996**, *2*, 157–174.

C. Shannon, Y. Dong, Bioanalytical applications of the atomic force microscope *Genomic/Proteomic Technol.* **2002**, *2*(5), 36–39.

X. Sheng, M. D. Ward, et al., Adhesion between molecules and calcium oxalate crystals: Critical interactions in kidney stone formation *J. Am. Chem. Soc.* **2003**, *125*(10), 2854–2855.

J. M. Sloan, W. F. Smith, Contrast mechanisms for scanning tunneling microscopy of biological molecules *Probe Microsc.* **1997**, *1*(1), 11–21.

D. A. Smith, S. D. Connell, et al., Chemical force microscopy: applications in surface characterization of natural hydroxyapatite *Anal. Chim. Acta* **2003**, *479*(1), 39–57.

H. Stahlberg, A. Engel, et al., Assessing the structure of membrane proteins: combining different methods gives the full picture *Biochem. Cell Biol.* **2002**, *80*(5), 563–568.

O. M. Stukalov, Application of atomic force microscopy in protein and DNA biochips development *NATO Sci. Ser., II: Mathematics, Phys. Chem.* **2002**, *57*, 331–340.

V. Subramaniam, A. K. Kirsch, et al., Scanning near-field optical imaging and spectroscopy in cell biology *Emerging Tools Single-Cell Anal.* **2000**, 271–290.

H. Sumi, Y. Hori, Direct measurement of redox-reorganization energies of electron-transfer proteins by V-I characteristics of photoinduced scanning tunneling microscopy currents *Hyomen Kagaku* **2003**, *24*(1), 8–12.

O. Takeuchi, S. Yasuda, et al., Analysis of single molecular interaction by using scanning probe microscope *Toraiborojisuto* **2004**, *49*(1), 42–48.

I. Taniguchi, Design of functional surfaces of electrodes for bioelectrochemical sensing *Chem. Sensors* **1998**, *14*, 161–164.

H. Tatsumi, Near-field microscope *Byori Rinsho* **1998**, *16*(6), 717–721.

S. Thalhammer, W. M. Heckl, Atomic force microscopy as a tool in nanobiology. Part I: Imaging and manipulation in cytogenetics *Cancer Genomics Proteomics* **2004**, *1*(1), 59–70.

G. Travaglini, M. Amrein, et al., Imaging and conductivity of biological and organic material *NATO ASI Ser., Ser. E: Appl. Sci.* **1990**, *184*, 335–347.

K. Tsunoda, H. Yajima, et al., Excimer laser ablation of collagen film *Kokagaku* **2002**, *33*(1), 18–25.

K. Uosaki, Scanning tunneling microscopy *Bunseki* **1990**, *9*, 675–681.

K. Uosaki, Scanning tunneling microscope and related techniques *Kagaku Kogaku* **1991**, *55*(10), 785–786.

D. W. Urry, T. M. Parker, Mechanics of elastin: molecular mechanism of biological elasticity and its relationship to contraction *J. Muscle Res. Cell Motility* **2002**, *23*(5–6), 543–559.

T. Ushiki, O. Hoshi, et al., The structure of human metaphase chromosomes: Its histological perspective and new horizons by atomic force microscopy *Arch. Histol. Cytol.* **2002**, *65*(5), 377–390.

H. C. Van der Mei, P. Kiers, et al., Measurements of softness of microbial cell surfaces *Methods Enzymol.* **2001**, *337*, 270–276.

J. Van Noort, Unraveling bacteriorhodopsin *Biophys. J.* **2005**, *88*(2), 763–764.

P. G. Vekilov, Microscopic, mesoscopic, and macroscopic lengthscales in the kinetics of phase transformations with proteins *Nanoscale Struct. Assembly Solid-Fluid Interf.* **2004**, *2*, 145–199.

J. A. Veerman, O. Willemsen, Scanning microscopes. Looking with other eyes *Natuur & Techniek (Beek, Netherlands)* **1997**, *65*(9), 66–73.

S. Wang, S.-x. Cai, et al., Mechanism and progress on protein crystal growth by atomic force microscopy *Chongqing Daxue Xuebao, Ziran Kexueban* **2004**, *27*(7), 36–39.

H.-S. Xia, Q. Wang, Advances in nanotechnology *Gaofenzi Cailiao Kexue Yu Gongcheng* **2001**, *17*(4), 1–6.

Y. Yang, H. Wang, et al., Quantitative characterization of biomolecular assemblies and interactions using atomic force microscopy *Methods (San Diego, CA)* **2003**, *29*(2), 175–187.

M. Yokokawa, S. H. Yoshimura, et al., The observation of biological samples by atomic force microscopy *Tanpakushitsu Kakusan Koso* **2004**, *49*(11, Zokan), 1607–1614.

E. Yonemochi, Assessment of drug characteristics by atomic force microscope (AFM) *Yakuzaigaku* **2003**, *63*(3), 129–132.

T. Yoshio, T. Ohtani, Application of SNOW/AFM to biomaterials *Denshi Kenbikyo* **2003**, *38*(2), 94–97.

L. Yu, A. Wu, et al., Deoxyribonucleic acid sample preparation in scanning probe microscope *Fenxi Huaxue* **2001**, *29*(12), 1470–1477.

J. A. Zasadzinski, Scanning tunneling microscopy with applications to biological surfaces *BioTechniques* **1989**, *7*(2), 174–187.

A. H. Zewail, Femtochemistry. Atomic-scale dynamics of the chemical bond using ultrafast lasers *Prix Nobel* **2000**, 110–203.

J. Zhang, Q. Chi, et al., Organization and control of nanoscale structures on Au(111) *Probe Microsc.* **2001**, *2*(2), 151–167.

J. Zhang, Q. Chi, et al., Electronic properties of functional biomolecules at metal/aqueous solution interfaces *J. Phys. Chem. B* **2002**, *106*(6), 1131–1152.

J. Zhang, Q. Chi, et al., Organized monolayers of biological macromolecules on Au(111) surfaces *Russ. J. Electrochem. (Transl. of Elektrokhimiya)* **2002**, *38*(1), 68–76.

J. Zhang, M. Grubb, et al., Electron transfer behavior of biological macromolecules towards the single-molecule level *J. Phys.: Condensed Matter* **2003**, *15*(18), S1873–S1890.

J. Zhang, Y. Wang, et al., Atomic force microscopy of actin *Shengwu Huaxue Shengwu Wuli Xuebao* **2003**, *35*(6), 489–494.

X. Zhang, A. Chen, et al., Probing ligand-receptor interactions with atomic force microscopy *Protein–Protein Interactions* **2002**, 241–254.

T. J. Zieziulewicz, D. W. Unfricht, et al., Shrinking the biologic world – nanobiotechnologies for toxicology. *Toxicol. Sci.* **2003**, *74*(2), 235–244.

J. Zlatanova, S. H. Leuba, Chromatin structure and dynamics: lessons from single molecule approaches *New Comprehensive Biochem.* **2004**, *39*, 369–396.

G. Zuccheri, A. Bergia, et al., Scanning force microscopy studies of the structure of biological macromolecules *Seminars Org. Synth. Summer School \"A. Corbella\", 27th, Gargnano, Italy, June 17–21* **2002**, 7–15.

3
Quartz Crystal Microbalance Characterization of Nanostructure Assemblies in Biosensing

Aren E. Gerdon, David W. Wright, and David E. Cliffel

3.1
Introduction

The characterization of functional nanostructures is crucial in determining their efficacy for life sciences. Nanostructure design requires careful consideration of recognition units and material properties, while synthesis can be meticulous, but evaluation and characterization of the system is equally challenging and important. Nanostructure creation can not easily be deemed successful without analytical evaluation. Furthermore, it may not be used to its potential if it is not well defined and well understood. The design and development of analytical tools and techniques for nanocharacterization is, therefore, imperative. A widely-used and effective tool for structure assembly evaluation is the quartz crystal microbalance (QCM). QCM is a sensitive technique based on the propagation of evanescent acoustic waves, which are affected by adsorption processes as well as local changes in density and viscosity of the contact liquid. Real-time analyte adsorption is monitored through simultaneous measurement of the quartz crystal's resonant frequency of oscillation and the damping resistance caused by liquid loading. One of the paramount applications of QCM has been the study of adsorption of biomolecules on functionalized surfaces. It has also been used in various other applications and is particularly suited to the study of nanoscale materials [1]. This chapter addresses the principles and operational aspects of QCM (Section 3.1.1), followed by important QCM applications to the life sciences (Section 3.1.5). The importance of interfaces between biology and nanoscale materials will be discussed (Section 3.2) and will precede examples of QCM-nanoparticle sensors for chemical (Section 3.3), biological (Section 3.4), and immunological systems (Section 3.5).

3.1.1
Principles of QCM

The quartz crystal microbalance relies on the converse-piezoelectric effect, reported by the Curie brothers in 1880. Piezoelectricity describes the electrical charge produced by pressure applied to solids having certain geometries, while the converse

Nanotechnologies for the Life Sciences Vol. 3
Nanosystem Characterization Tools in the Life Sciences. Edited by Challa S. S. R. Kumar

ISBN: 3-527-31383-4

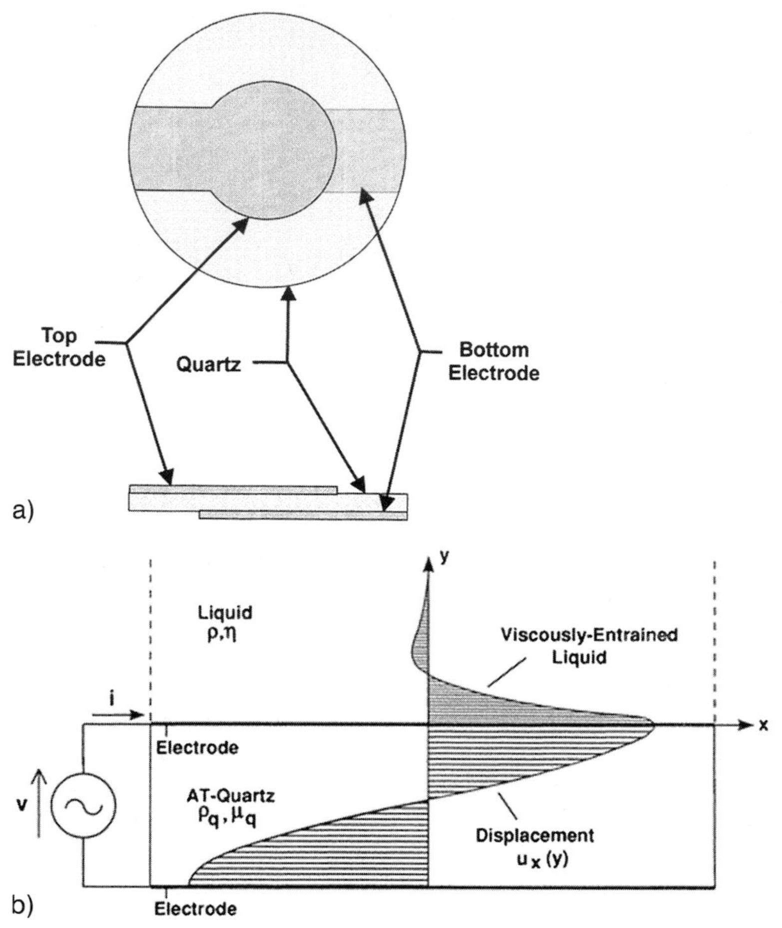

Fig. 3.1. Top and side view of a QCM resonator (a and b) and cross sectional view of a QCM resonator in contact with a liquid (bottom). Wave propagation is relatively lossless in the piezoelectric material, but becomes evanescent in the viscous liquid [7]. (Reprinted with permission the American Chemical Society.)

effect produces a strain on the crystal when an electric field is applied [2]. Coupling the crystal to an oscillating circuit provides crystal oscillation at a resonant frequency (Fig. 3.1).

These quartz resonators have been widely used in communication devices, commercial acoustic electronics, and sonar [2, 3]. Sauerbrey first demonstrated the usefulness of this effect for analytical chemistry by showing a linear relationship between mass deposited on the crystal and the frequency of oscillation [Eq. (1)] [1, 2].

$$\Delta f = -C_f \Delta m \tag{1}$$

Initially, QCM was used in the gas phase, where oscillator resistance is low and mass adsorption is rigid. The device was improved in 1982 by the advent of circuitry capable of crystal operation in liquids. While this advancement introduced QCM to the world of bioanalytical chemistry, it also introduced new difficulties. The first is that operation in liquid slows the mass transport of analyte to the sensor surface. The second is that many binding events in liquid, especially those biological in nature, are not static but tend to be in equilibrium. The third is resonator damping and viscous loading (Fig. 3.1). The density and viscosity of the contact solution affects crystal frequency response, convolutes mass measurements, and requires modification of the Sauerbrey equation [Eq. (1)] [2, 4]. Careful consideration of simultaneous mass and liquid loading by Kanazawa [5] and Martin [6] has shown that Δf relates to adsorbed mass and the density and viscosity of the contact liquid according to Eq. (2) [6].

$$\Delta f \approx C_{\mathrm{f}}\Delta m - C_{\mathrm{f}}(\Delta\rho\eta/4\pi f_{\mathrm{o}})^{1/2} \tag{2}$$

It is, therefore, impossible to distinguish mass loading from liquid loading by frequency measurements alone. Fortunately, the independent term of liquid loading resistance, R_{L}, depends solely on density and viscosity [Eq. (3)] [6].

$$\Delta R_{\mathrm{L}} \approx (\eta_{\mathrm{q}}/[c_{66}C_1]) + (N\pi C_1)^{-1}(\Delta\rho\eta/\pi f_{\mathrm{o}}c_{66}\rho_{\mathrm{q}})^{1/2} \tag{3}$$

Measuring both changes in frequency and resistance provides a means for decoupling mass and liquid loading effects, though different methods have been described [1, 2, 5–9]. Applying Eqs. (2) and (3) to frequency and resistance measurements to reliably calculate adsorbed mass has proven to be non-trivial. A simple approach to a reliable mass measurement has been to create a calibration curve of change in frequency as a function of change in resistance for a system in which mass loading is minimal. In one case, sucrose, a hydrophilic molecule that will not independently adsorb to the QCM electrode, was dissolved in DI water at 0, 5, 10, 20, and 40% by weight. Sequential addition of these five solutions to the QCM sensor provided simultaneous changes in frequency and resistance that were due to density and viscosity effects only (Fig. 3.2) [1]. The relationship between frequency and resistance within this range is linear and provides an accurate calibration. Above ~40% sucrose, the relationship deviates from linearity, making this approach only applicable to systems in which the resistance changes by less than 500 Ω [5, 10]. Problems also occur with this calibration, as in any QCM measurement, when coupling to the sensor is non-rigid and the analyte has its own viscous properties. Calibration allows for the calculation of Δm through simultaneously solving Eqs. (2) and (3) and using a sensitivity factor, C_{f}, of 56.6 Hz cm^2 μg^{-1}, which is known for a 5 MHz crystal [11]. Decreasing the crystal thickness increases the resonant frequency and can provide enhanced sensitivity.

Since his study of simultaneous QCM mass and liquid loading in 1991 [6], Stephen Martin from Sandia National Laboratory and collaborators at Leicester University have continued to contribute experimental and theoretical-based insight

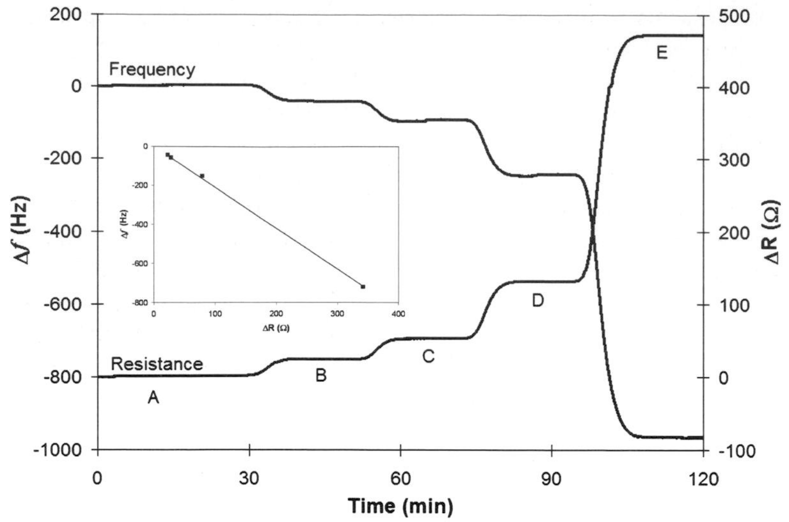

Fig. 3.2. Practical QCM mass calibration using a non-mass loading system. Frequency and resistance changes mirror each other upon the addition of deionized (DI) water with (A) 0, (B) 5, (C) 10, (D) 20 and (E) 40% sucrose. Inset: linear calibration curve of change in frequency vs. change in resistance [1].

into the details of QCM measurement in liquids. Four contributions, since 1997, have covered resonator response to liquid loading [7], the modeling of resonator response under loading conditions [12], the modeling of viscoelastic film response [13], and a model for wetting characteristics of roughened surfaces [14]. This collective work has provided a better understanding of loading responses in QCM and has supported more rigorous evaluation of important systems.

Another area of QCM sensor development has been multi-channel QCM. Using arrays of resonators or micro-fluidics to partition individual resonators could provide the simultaneous measurements needed for multi-analyte detection strategies. A significant challenge in multi-channel QCM is the isolation of each electrode. Circuit shorting or resonance overlap from electrode to electrode could cause considerable problems. Two approaches to this problem have made use of electrode miniaturization [15] and the restriction of oscillation to indented areas surrounded by thicker quartz [16]. Both approaches have had success in the construction of a multi-channel QCM instrument.

3.1.2
QCM Wave Penetration Depth

QCM works well with 3D substrates and multilayer adsorptions often encountered in materials applications [1]. QCM acoustic shear waves are evanescent and decay exponentially in the contact liquid, causing a loss of sensitivity at distances from the sensor surface (Fig. 3.1). Mass that is rigidly coupled to the sensor propagates the wave without loss, whereas viscous material causes immediate wave decay [8]. The depth at which these waves penetrate the contact liquid affects instrument sensitivity. QCM, at 5 MHz, has a calculated penetration depth of 250 nm [2], with a total decay length of 1 μm [8], though experiments have shown no loss of sensitivity for layers as large as 400 nm [1, 8, 17, 18]. In comparison, surface plasmon resonance spectroscopy (SPR) has a calculated penetration depth of only 150 nm and has shown significant peak broadening with layers less than 200 nm [19, 20]. Examples of peak broadening in SPR come from the use of metal nanoclusters for signal amplification [21–23]. Antibody-presenting gold nanoclusters binding to immobilized anti-IgG increased the SPR shift as compared to free antibody [21, 22]. This also caused broadening of the SPR curve, which increased in width as the density or diameter of the colloidal gold was increased [23]. QCM has not suffered from this sensitivity loss, even at extensive adsorption [1].

3.1.3
QCM Sensor Specificity

A challenge in QCM, and in many biosensors, is sensor functionalization for improved specificity. While the gold QCM electrode imparts almost no specificity of its own, it provides a substrate for various functionalizations. The gold electrode is amenable to hydrophobic interactions, is well-suited for self-assembled monolayer (SAM) formation, and enables electrochemical surface reactions. Approaches to improve specificity in QCM have used SAMs [24–26], ionic interactions [27–30], electrochemical deposition [31, 32], and protein adsorption (Fig. 3.3) [33–35]. These methods allow for the immobilization of various biological species and nanostructures and permit various detection schemes. SAMs have been studied extensively [26, 36, 37] and feature a defined orientation, high aerial density, and programmability in the exposed head-group. They have recently been used for the covalent immobilization of oligosaccharides [24] and peptide epitopes [38].

Ionic layering, using polyelectrolyte [28–30] or SAMs with ionic head-groups [27], has allowed for effective immobilization and orientation of various nanoparticles. Though the interaction is generally non-specific, relying on electrostatic and van der Waals interactions, layer-by-layer growth is highly tunable according to concentration, ionic strength, and pH. This promotes high-affinity binding, reversibility, and high packing density, and has been used for single layer deposition and multilayer nanorainbow assembly [28, 30].

Electrode functionalization and nanoparticle immobilization has also been pursued with electrochemical deposition, using an electrochemical quartz crystal microbalance (EQCM). Viologens, such as *N*-methyl-*N*′-ethylamine-viologen dinitrate (MEAV), have been previously studied [32] as electroactive species and adhere to electrodes upon reversible reduction of the $[\text{viologen}]^{2+/1+}$ couple. Attachment of MEAV to nanoclusters allows similar adhesion to electrodes, thereby immobilizing the nanocluster on the QCM surface [31].

Another route employs Protein A for immobilization of antibodies in an accom-

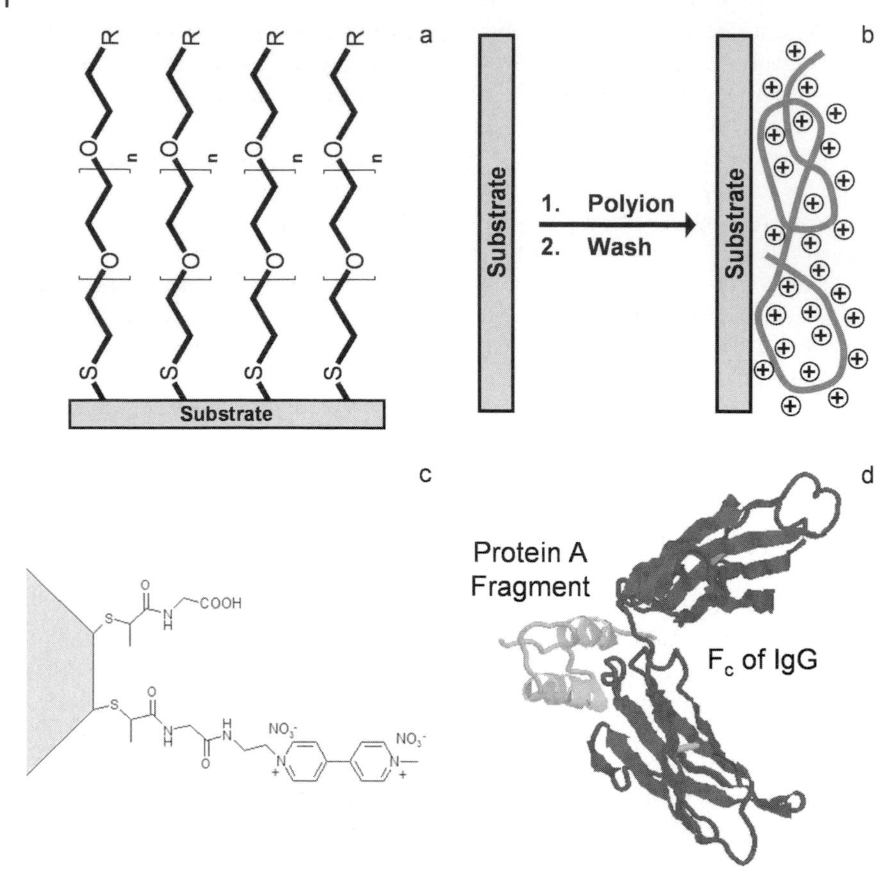

Fig. 3.3. Examples of QCM functionalization techniques to impart specificity: (a) Self-assembled monolayers immobilize ligands in a high density, and often use poly(ethylene glycol) spacers. (b) A polyelectrolyte provides a thin layer of charge (positive or negative) for ionic interactions. (c) Viologen functionalized electroactive nanoclusters can be electrochemically deposited on the QCM surface. (d) X-ray crystal structure of a Protein A fragment commonly used in the immobilization of antibody [35].

modating conformation. Protein A is a cell wall protein that forms stable complexes with gold through van der Waals interactions ($K_a \sim 10^8$ M^{-1}) [33] and contains five tandem domains that bind the Fc region of IgG with high affinity and selectivity. These properties optimize the presentation of antibodies such that both antigen binding regions (Fab) are unobstructed (see also Figs. 3.4 and 3.6) [33–35].

3.1.4
Calculation of Equilibrium and Kinetic Constants

An important advantage of real-time binding measurements is the ability to derive equilibrium binding constants and kinetic rate constants. These constants provide valuable information concerning the affinity or avidity of one material for another, the rate at which they interact, and the interaction order. A molecule's affinity is defined through the equilibrium association constant (K_a) for the interaction and is generally valid for monovalent, one-to-one interactions. Avidity is related to the same equilibrium process (K_v), but describes multivalent processes. For example, at low ligand concentrations a multivalent ligand could bind to several monovalent receptors, increasing the affinity of the ligand for the receptor. There are also cases of multivalent receptors with monovalent ligands as well as multivalent receptors with multivalent ligands [39]. The complexity of these interactions can be daunting, especially for molecules like lectins that are designed for increased affinity through multivalency. The simplest case would be the interaction of a monovalent ligand (e.g., an antigen) with a monovalent receptor (e.g., a ScFv antibody): $\text{Ag} + \text{Ab} \leftrightarrow \text{Ag–Ab}$. The elementary equilibrium expression for the formation of the complex is [Eq. (4)],

$$K_a = [\text{Ag–Ab}]/([\text{Ab}][\text{Ag}]) \qquad (4)$$

where K_a is the equilibrium association constant and is generally in the range of 10^6 to 10^{12} M^{-1} for monoclonal antibody interactions, indicating a large equilibrium shift to product formation when antigen is in the presence of antibody. If this association and equilibrium shift occurs at a surface, where the first component adsorbs to the surface-supported second component, the equilibrium expression can be rearranged to an adsorption isotherm. The adsorption isotherm represents the connection between the amount of substance adsorbed, the concentration of the bulk solution, and the equilibrium constant, K_a. The Langmuir isotherm is commonly used and relies on three important assumptions: (1) there is no lateral interaction between adsorbed species; (2) the surface is homogeneous; (3) the maximum adsorption is saturation to a monolayer. Adsorption isotherms are sometimes written in terms of fractional coverage (Θ) or percent of monolayer formation. A generalized Langmuir isotherm in this form is given by Eq. (5).

$$\Theta/(1 - \Theta) = K_a C \qquad (5)$$

For QCM studies, the fractional coverage is related to the change in mass (Δm) and the initial concentration of surface immobilized antibody is related to the maximum change in mass (Δm_m). Rearranging to solve for Δm provides a commonly used form of the Langmuir isotherm [Eq. (6)].

$$\Delta m = \Delta m_m[K_a C/(1 + K_a C)] \qquad (6)$$

Plotting Δm as a function of bulk concentration (C) shows a steep increase in adsorption that levels off with increasing concentration as a complete monolayer is achieved (Fig. 3.9d below). This line can be fit to nonlinear regressions, but does not provide a simple means for extracting the equilibrium constant. Instead a reciprocal plot can be used to obtain a straight line fit. There is more than one way to obtain a reciprocal plot by rearranging the Langmuir isotherm [Eq. (6)]. Two options are Eqs. (7) and (8).

$$\Delta m = -(K_a)^{-1}(\Delta m/C) + \Delta m_m \quad (7)$$

$$C/\Delta m = (\Delta m_m)^{-1}(C) + (\Delta m_m K_a)^{-1} \quad (8)$$

With Eq. (7), Δm is plotted against $\Delta m/C$ and the inverse of the slope provides K_a. With Eq. (8) $C/\Delta m$ is plotted against C and the K_a must be extracted from a combination of the slope and y-intercept. Both techniques have been used in the literature and provide reliable K_a [1, 24, 40–42]. Once the equilibrium association constant is known, it can be used to calculate both the fractional coverage at an infinite time point [Θ_∞, Eq. (9)] and the Gibbs free energy of adsorption [ΔG_{ads}, Eq. (10)].

$$\Theta_\infty = C/(C + K_a{}^{-1}) \quad (9)$$

$$\Delta G_{ads} = -RT \ln K_a \quad (10)$$

This method offers a simple way to determine equilibrium constants for systems that generally follow the assumptions outlined for a Langmuir isotherm, though variations on the Langmuir isotherm can account for more complex systems.

Equilibrium association constants for antibody/antigen interactions can be determined in different ways, not necessarily involving real-time measurements. Kinetic information, however, requires time-resolved information. The rate at which an interaction takes place can be valuable in evaluating component efficacy. The time course for monolayer formation is given by Eq. (11).

$$\Theta(t) = [C/(C + K_a{}^{-1})]\{1 - \exp[-(k_f C + k_r)t]\} \quad (11)$$

According to Eq. (9), Θ_∞ can be substituted in and the exponent can be simplified by calling $(k_f C + k_r)$ equal to the time constant (τ^{-1}) [Eq. (12)],

$$\Theta(t) = \Theta_\infty[1 - \exp(-\tau^{-1}t)] \quad (12)$$

For a given concentration of the bulk solution (C), $\Theta(t)$ is related to Δm at a particular time and Θ_∞ is related to the maximum change in mass [Δm_m, Eq. (13)],

$$\Delta m_t = \Delta m_m[1 - \exp(-\tau^{-1}t)] \quad (13)$$

Fitting Eq. (13) to each time point in the real-time adsorption binding curve yields the time constant. Knowing the concentration of the bulk solution allows the extraction of forward (k_f) and reverse (k_r) rate constants. The ratio of these kinetic constants is equal to the equilibrium constant $(K_a = k_f/k_r)$ and can be used for comparison with isotherm methods.

3.1.5
QCM Application to Life Sciences

The benefits of QCM and detailed functionalization strategies have allowed the study of many biological and chemical systems [2, 43]. An early and notable example of a chemical study is the direct kinetic measurements of thiolate molecules self-assembling on two-dimensional gold surfaces [37]. This helped provide a foundation for QCM kinetic studies and was quickly followed by a kinetic study of anti-fluorescyl antibody binding to fluorescein lipids in Langmuir–Blodgett films [41]. The fluorescein hapten was coupled to lipids and mixed with unfunctionalized lipids to form a bilayer with ~5% fluorescein lipid. Through QCM measurements, monoclonal antibody was found to have an affinity (K_a) in the range 10^7–10^8 M^{-1} and forward and reverse rate of reaction constants of approximately 2×10^5 $\text{M}^{-1}\ \text{s}^{-1}$ and $2 \times 10^{-3}\ \text{s}^{-1}$, respectively.

More recently, Zeng et al. conducted two immunoassay experiments using QCM [24, 44]. The first was a study of α-Gal carbohydrate antigen as anti-Gal antibodies are of interest for therapeutics in xenotransplantation. Thiolated trisaccharides were immobilized on the surface of the QCM through the formation of a self-assembled monolayer and exposed to polyclonal anti-Gal antibodies, a lectin from *Griffonia simplicifolia*, and a lectin from *Marasmius oreades*. The antibody displayed the strongest binding, with a dissociation constant (K_d) three orders of magnitude greater than either of the lectins. They concluded that this QCM approach is competitive with established label-free techniques [24]. The second immunoassay made use of single-chain fragment variable (ScFv) antibodies to increase the surface density of antigen binding sites (Fig. 3.4). The recombinant antibodies were genetically engineered to contain a linker arm and cysteine residue to ensure self-assembly in a defined orientation. Considering their size (27 kDa) and ease of engineering, they have a considerable advantage over Fab fragments or full-sized IgG for the detection of antigen. The ScFv antibodies were initially expressed and evaluated using SDS-PAGE and Western blot analysis. After immobilization on the QCM gold electrode, the monolayer assembly was verified electrochemically with cyclic voltammetry and electrochemical impedance. This confirmed surface coverage, but provided no information on antibody orientation. The ScFv rabbit anti-IgG antibody was successful in detecting rabbit IgG down to 1.1 nM, which is more than 7 times lower than that detected with an anti-IgG Fab fragment sensor. Another benefit of this system is the reversibility of IgG binding coupled with the

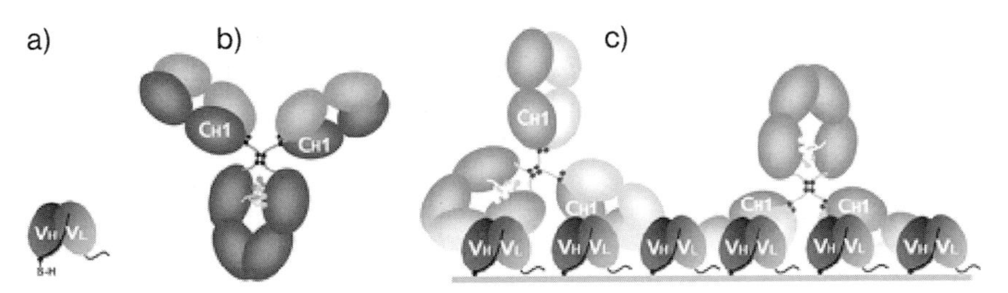

Fig. 3.4. (a) Single-chain fragment variable recombinant antibody with cysteine incorporated into the linker; (b) rabbit IgG; (c) interaction between immobilized ScFv and rabbit IgG. (Reprinted with permission from the American Chemical Society.)

stability of the ScFv monolayer to allow regeneration and reusability of the sensor [44].

A piezoelectric immunosensor has also been developed to detect aerosolized SARS-associated coronavirus [45]. Since the introduction of circuitry able to operate QCM in liquids, gas phase measurements have primarily been used in metal vapor deposition techniques, rather than biological studies. A functionalized QCM crystal could be used as a biological "nose" for the detection of aerosolized agents, such a coronavirus. In this example, polyclonal anti-SARS-CoV was presented on the gold QCM electrode through a specific Protein A intermediate. Antigen powder was then reconstituted in the saliva of a healthy volunteer and aerosolized in the presence of the sensor. The antigen bound quickly to the antibody and was detected at concentrations down to 0.60 mg mL^{-1}.

Many other interesting and pertinent biological systems have been studied, including the use of glycosphingolipids for detection of ricin [46], the use of high resonant frequency quartz crystals (39 to 110 MHz) for the detection of phages [47], and the study of annexin A1 binding to solid-support membranes [48]. A complete review of QCM applications to the life sciences is not the scope of this chapter, though other sources can be consulted for further examples [2, 43].

3.2 Interface Between Biology and Nanomaterials

Fundamental advances in chemistry and biology have allowed biotechnology and materials science to develop over the past decades. Biotechnology has attempted to emulate naturally occurring functional assemblies, from the double helix of DNA to the multi-subunit protein cage, ferritin, while material science has taken advantage of chemical methods for the miniaturization of functional devices. Further evolution of these disciplines into the emerging field of bionanotechnology depends on the successful development and evaluation of structurally well-defined interfaces that bridge biology and inorganic materials chemistry. Interdisciplinary collaboration between these fields will allow for improved biological components to generate new materials while advanced materials will be used to ameliorate biological problems [49]. The design, synthesis, and characterization of readily programmable, structurally well-defined biological interfaces for inorganic materials represent significant challenges for the realization of these goals.

Progress has been made towards the development of such interfaces, though many options have yet to be explored. Some encouraging synthetic attempts have used DNA, semiconductor-binding peptides, genetically-engineered viruses, and silica-precipitating peptides. Specifically, synthetic single-stranded oligonucleotides coordinated to nanoparticles have been shown to self-assemble with the appropriate complement to form higher ordered structures (Fig. 3.5a) [50]. Phage-display libraries have been used in the successful selection of 12-mer peptides that specifically bind to the 100 face of GaAs single crystals (Fig. 3.5b). Bivalent peptides of this nature could be used in the directed assembly of nanoscale components [51].

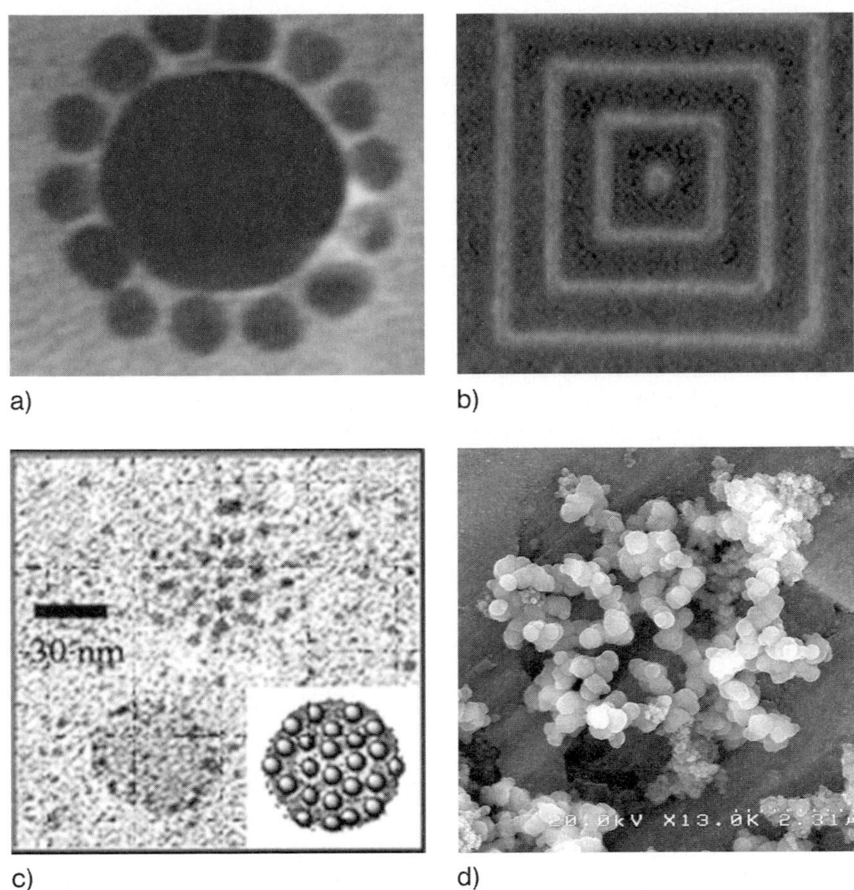

Fig. 3.5. Examples of successful interfaces between biology and inorganic materials: (a) TEM of 8 nm diameter Au nanoparticles connected to a 31 nm diameter Au nanoparticle through complementary strands of DNA [50]. (Reprinted with permission from the American Chemical Society.) (b) Fluorescently labeled antibody bound to peptide presenting phage, which specifically recognized the 100 face of GaAs [51]. (Reprinted with permission from Nature Publishing Group.) (c) TEM of virus templated synthesis of Au^0 nanoparticles [52]. (Reprinted with permission from the Royal Society of Chemistry.) (d) Electron micrograph of silica particles precipitated via a synthetic peptide [53]. (Reprinted with permission from the Royal Society of Chemistry.)

Recently, the iron-storage protein ferritin, which contains its own interface between biology and inorganic iron oxide, has been used as inspiration for the design of a genetically-engineered viral cage that can precipitate nanoclusters. The virus, cowpea chlorotic mottle virus, was engineered with HRE peptide epitopes on its surface and provided a template for the symmetry-directed synthesis of Au^0 nanoparticles (Fig. 3.5c) [52]. Another attempt at interfacing biology and inorganic materials has made use of a synthetic peptide based on silaffin from various eukaryotic algae or diatoms. This peptide was shown to efficiently precipitate silica under mild conditions, mimicking diatom activity (Fig. 3.5d) [53]. These functional materials have enjoyed success in the assembly of nanoscale materials, but have not shown results for sensor applications. Still, they have succeeded at interfacing biology and inorganic materials and have served as inspiration for the subsequent design of sensor interfaces.

One biomolecular recognition method that has not been widely employed in the assembly of nanoscale materials is immunomolecular recognition. Immunoglobulin (IgG) antibodies have affinities (association constant, K_a) on the order of 10^6–10^{12} M^{-1} and have been well studied for sensor applications [54–57]. Applications of such immunoassays have had a tremendous impact on medical diagnosis and the treatment and understanding of disorders and diseases. Interfacing immunology and nanomaterials will expand these applications and provide potential for improvements on existing assays and treatments. There are several cases where immunology and inorganic materials have been successfully brought together to provide a ground work for further development.

Progress began with the non-specific immobilization of antibody on micron-sized, hydrophobic, latex beads [58–60]. Antigen binding to immunoreactive beads causes agglutination, which can be quantified by changes in solution turbidity [60]. Similarly, antibodies have been non-specifically immobilized on 11 nm diameter colloidal gold particles. These particles were used in traditional sandwich immunoassays, where the gold particle provided signal amplification when binding was detected with surface plasmon resonance spectroscopy (SPR) [21]. More recently, Nam et al. created an immunoassay using micron-diameter immuno-magnetic particles in conjunction with 13 nm diameter DNA/antibody derivatized gold clusters. Magnetization allowed for facile separation of agglutinated particle, while double stranded DNA provided an amplification avenue through polymerase chain reaction (PCR) [61]. Other strategies have used nanocluster/antibody interfaces as a means for multiplexing and amplification [62–64]. Another approach made use of a peptide epitope known to bind monoclonal antibody associated with the human malarial parasite, *Plasmodium falciparum*, and was the first example of antigen encapsulated nanoclusters assembling with antibodies through the antibody (paratope)/antigen (epitope) interface [65]. This important advancement appears to provide a robust, functional nanostructure that has the potential of successfully mimicking a biological entity. Together these interface-dependent techniques have suggested new routes for the study and detection of human IgG, prostate specific antigen, hepatitis B surface antigen, and the human malarial parasite.

3.2.1
Antibodies

Extraordinary binding affinities, ease of use, and relevance to medical disorders make antibodies attractive analytical reagents and recognition units for interfacing with nanoscale materials. The classic IgG immunoglobular has a prototypical Y-shape made up of two heavy and light chains (Fig. 3.6). The antigen binding

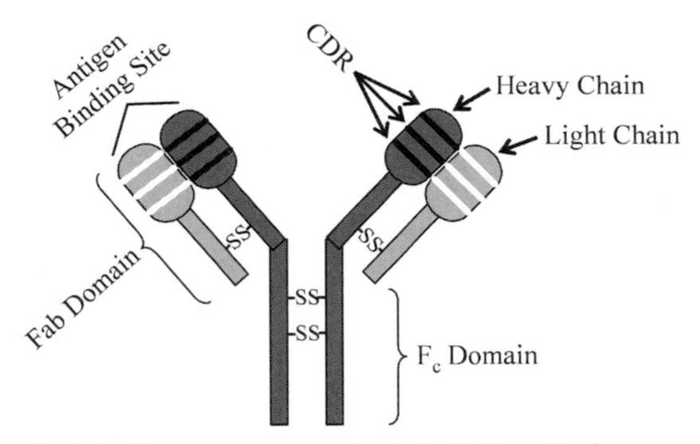

Fig. 3.6. Cartoon structure of a typical IgG antibody, highlighting heavy and light chains joined by disulfides, a constant fragment (Fc), two fragments of antigen binding (Fab), and complementarity determining regions (CDRs).

site (paratope), within the Fab (fragment of antigen binding), is formed by the three-dimensional (3D) arrangement of the complementarity determining regions (CDR). In antibodies, this is the key region of immunorecognition and results in the proteins valency. Three different types of antibody preparations are in use for immunoassays: polyclonal, monoclonal, and single chain fragment variable (ScFv). Polyclonal antibodies (pAb) are generally collected from animals and result as a response to the injection of an antigen. They are produced by many different B lymphocytes and recognize different epitopes within the same antigen, resulting in a mixture of antibodies, with only a subset of total IgG recognizing the epitope of interest. Monoclonal antibodies (mAb), however, are produced in a colony of identical B cells. Each antibody binds to the same epitope and the mAb sample is homogeneous [54]. Single chain fragment variable (ScFv) antibodies represent the smallest functional domain of a traditional monoclonal or polyclonal antibody as they consist of one linked Fab domain [44, 66]. Phage-display recombinant antibody technology has produced high-quality, antigen-specific ScFv antibodies with affinities (binding strengths) comparable or greater than those of traditional antibodies [51]. This technique and these antibodies are particularly useful for antigens that are poorly immunogenic, readily degrade, or for which monoclonal or polyclonal antibodies are difficult to obtain. These three types of antibodies provide a range of options for interfacing immunology with nanoscale materials.

3.2.2 Nanoparticles

The interface between biology and inorganic materials has manifested in many different forms, covering the range of biological systems and the range of inorganic materials. Immunology has interfaced with many substrates, including two-dimensional (2D) surfaces [24, 38, 44], organic polymers [58, 60, 61], large metal particles [21, 65], semiconductor quantum dots [67], and small, ligand-capped metal cores [1]. The interface of antibodies with each of these substrates has produced interesting functional materials, though ligand-capped noble-metal clusters (monolayer-protected clusters, MPC) have received attention for their unique consolidation of self-assembly techniques and metal nanoparticle chemistry [68–70]. Furthermore, their chemical, electronic, and physical properties, a lack of air and water sensitivity, and convenient characterization make them robust materials [65, 71–74]. Programmed specificity through the introduction of biologically relevant molecules is one of their most promising features and lends to the development of an immuno-interface [75, 76].

Synthesis and assembly of MPCs can occur through several routes, using various metals (Au, Ag, Cu, Pt, CdS, ZnS, Ag_2S) [77], reductants (citrate, Na_2S, $NaBH_4$), and capping ligands (thiolate, disulfide, amine, imidazole, carboxylic acid, phosphine, iodine). Surfactants, templates, and physical methods (photochemistry, sonochemistry, radiolysis, thermolysis) have also been employed [78]. Conventional methods have primarily been based on the Brust–Schiffrin method, published in 1994 [79]. This was originally a two-phase synthesis with stabilization by organic soluble thiolates and reduction by $NaBH_4$ (Fig. 3.7a). This method was subsequently adjusted for the synthesis of water-soluble MPCs, which is generally a one-phase synthesis and follows a simple equation [Eq. (14)] [72].

$$HAuCl_4\ xH_2O(aq) + 3RSH(aq) + 10NaBH_4(aq) \rightarrow Au^{\circ}{}_xSR_y + RS{-}SR_z \qquad (14)$$

Templeton et al. have reported the synthesis of water-soluble tiopronin [*N*-(2-mercaptopropionyl)glycine] MPCs 1.8 ± 0.7 nm in diameter by this method. Characterization of these nanoclusters follows from traditional materials techniques, as previously described in depth [69, 72, 80]. Figure 3.7(b) and (c) shows examples of a transmission electron microscopy (TEM) picture and proton nuclear magnetic resonance spectroscopy (^1H NMR) spectrum, respectively. Daniel et al. have recently published a more complete review of gold nanoparticle synthesis, characterization, and applications [78].

One of the most promising features of MPCs is their ability to exchange thiolate ligands with ligands in solution. This allows for the specific introduction of biologically active molecules, programmable specificity, and the synthesis of functional nanoclusters. In this way, nanoscale inorganic materials make potentially interesting solid supports for the immobilization and presentation of antibody or antigen, leading to a successful interface with an immunological system. Place exchange of free thiol onto 3D MPCs is a relatively facile process that is regulated by solvent, reaction temperature, reaction time, and properties and concentration of original and replacement thiol [75]. Investigation of the dynamics, kinetics, and mechanism of place exchange are abundant and generally support an associative, S_N2 mechanism [75, 76, 81, 82]. Place exchange on MPCs has been compared to preceding studies of place exchange on 2D self-assembled monolayers. Exchange is more likely to occur on 3D surfaces (MPCs) than 2D surfaces (SAMs) due to the higher propensity for defect sites in structures with a substantial radius of curvature [36, 75, 83]. Another important feature of place exchange reactions is the

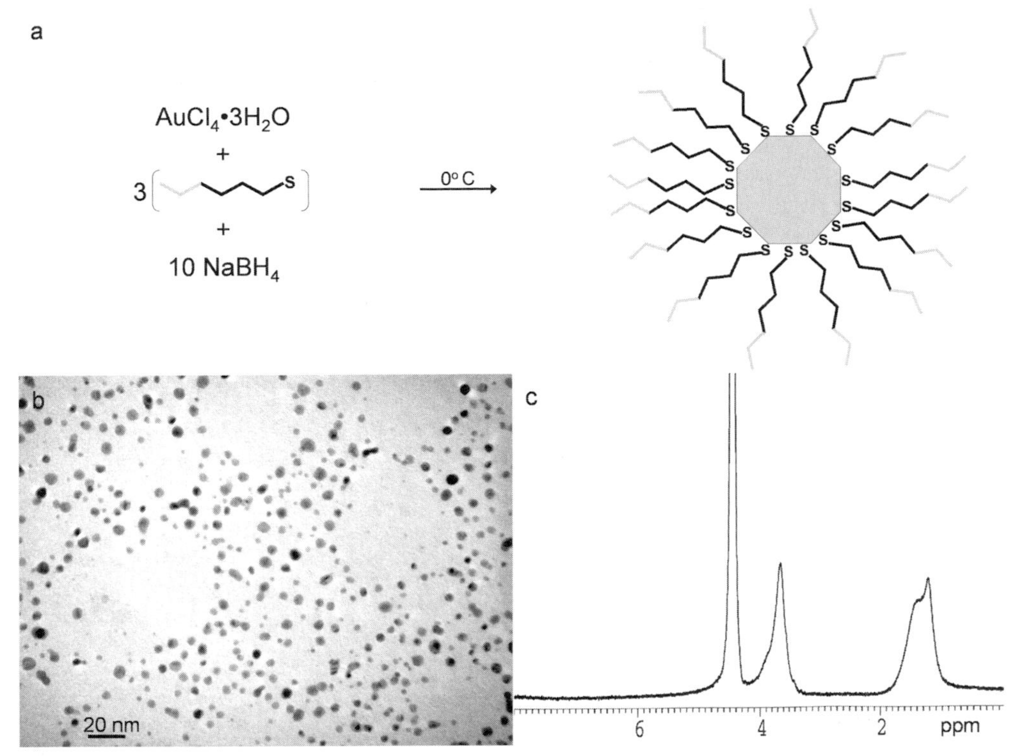

Fig. 3.7. Representative synthesis and characterization for a tiopronin (*N*-2-mercaptopropionylglycine)-protected cluster. (a) Synthesis conditions based on a modified Brust method and cartoon schematic of a monolayer-protected cluster (MPC). MPC ligands are connected through thiolate moieties at one end and generally have a functional group at the opposing end (gray). (b) TEM image of MPC showing spherical shape and size dispersity. (c) ^1H NMR of tiopronin MPC, showing broad $-CH_3$ (~1 ppm) and $-CH_2-/-CH-$ (~3.5 ppm) peaks. The water peak is at 4.8 ppm.

idea that fast-exchange sites are not static. Evidence for this has come from the exchange of several different ligands onto the same MPC [69, 76] and from an inability to completely remove ligand that had been previously exchanged on a cluster [75, 76]. A significant implication of these results is the ability of a thiolate to migrate across the monolayer of an MPC and position itself for an optimal interface [83].

There are many examples of successful place exchange and coupling reactions with organic and water-soluble MPCs, resulting in functional nanostructures. For example, Templeton et al. [76] used traditional amide coupling techniques to attach 12 different functional ligands to nanoclusters, introducing spin labels, amino acids, fluorophores, sugars, and electroactive moieties. Place exchange has also introduced fluorescein and an electroactive viologen species. Electrochemical quartz crystal microbalance of this MPC showed redox activity of the viologen as well as deposition of the MPC upon viologen reduction [31]. Antimicrobial nanoclusters have been prepared by functionalization with vancomycin [84], glutathione-protected nanoclusters have been shown to interact with the enzyme glutathione transferase [85], and biotinylated nanoclusters bind to streptavidin [85]. This type of MPC functionalization produces nanostructures with specific activity and suggests potential for further nanocluster modulation in the design of bioreactive materials.

3.3
QCM Nanoparticle-based Chemical Sensors

The evaluation of interfaces between biology and nanotechnology has been an area of recent research interest, but originally started through the early development of chemical sensors. Originally, QCM was designed for gas-phase measurements and has been used in the control of metal vapor deposition. QCM continues to be used for gas-phase measurements, but has employed polymer and nanoparticle films for organic vapor sensing [86–88]. In 2001, dodecanethiol MPCs of 2 and 5 nm diameter were covalently assembled on a QCM substrate through dithiol linkages or non-covalently through carboxylic acid hydrogen bonding [86]. The nanoparticle thin-layers were then studied for their ability to partition toluene, hexane, methanol, ethanol, and water vapors. Vapor sorption was detected by simultaneous monitoring of frequency and resistance. Results suggested the viability of nanostructure films for vapor sensing and highlighted similarities and differences between the size of MPC used and the assembly technique.

Shortly thereafter, Zamborini et al. used electrochemical QCM to simultaneously measure conductivity and vapor partitioning in MPC films [87]. This approach used small (1.6 nm diameter) nanoclusters with mixed monolayers of alkanethiolate and ω-carboxyalkanethiolate. The free carboxylic acid was used in conjunction with Cu^{2+} ions to assemble a network polymer of carboxylate-Cu^{2+}-carboxylate bridges. These ion bridges allowed for electron transport through the polymer, though film swelling due to vapor partitioning (ethanol or dichloromethane) diminished electron flow.

Another example [88] made use of a greater variety of nanoclusters, protected with dodecanethiol, benzenethiol, 4-chlorobenzenethiol, 4-bromobenzenethiol, 4-(trifluoromethyl)benzenethiol, 4-hydroxybenzenethiol, and 4-aminobenzenethiol. Nanoparticle films were spray coated onto a QCM substrate and monitored by QCM to a final change in frequency of $\sim$10 000 Hz. It was important to deposit a film with a thickness large enough to partition vapor, but small enough to rigidly couple the entire film to the sensor, maintaining an accurate sorption response. Motional resistance was monitored during film deposition to ensure rigidity as resistance is related to film viscoelastic properties. Results implied that different nanocluster films had increased sensitivity for different vapors (hexane, toluene, butanone, and butanol) and compared with vapor partitioning into organic polymers.

3.4 QCM Nanoparticle-based Biosensors

Inspired by the success of QCM bioassays (Section 3.1.5) and bionanotechnology, researchers have developed nanoparticle-based detection schemes using QCM for improved sensitivity. A popular area of analytical biochemistry involves the detection of DNA for a range of applications from gene analysis to forensic applications, where low detection limits are required. Willner et al. have developed three amplification paths for the detection of single-base mismatches in DNA with QCM detection [89–91]. Each path begins with the immobilization of ssDNA (25 bases) through SAM technology. Its complementary strand with an extra 16 bases is then introduced and allowed to base pair. In some cases a complement DNA strand with a single base-pair mismatch is purposely assembled. This allows for a single nucleotide, coupled with biotin, to bind at the mismatch site. At this point, three different approaches can be used: (1) free avidin can bind to the base-pair biotin, followed by biotin-labeled liposome binding; (2) avidin-labeled nanocluster can bind, followed by controlled nanoparticle growth for amplification; (3) avidin-labeled alkaline phosphatase can bind and catalyze the precipitation of an insoluble organic product for amplification. These methods lead to mismatch detection with detection limits ranging from 10^{-12} to 10^{-16} M^{-1}. QCM gravimetry is well-suited for such precipitation/amplification schemes.

A similar approach to DNA hybridization detection and amplification also uses avidin-labeled alkaline phosphatase (as above). In this case, the enzyme catalyzes the production of the reducing agent, *p*-aminophenol, from a *p*-aminophenyl phosphate precursor. The reducing agent reduces Ag^+ ions in solution, which biomineralize into Ag nanoparticles on the DNA strand or on the QCM sensor surface, causing QCM signal amplification. The deposited silver can then be used in anodic stripping voltammetry to further confirm DNA binding, down to 100 aM concentrations.

Another example involves a glucose oxidase-based glucose sensor, which results from a complex nanostructure assembly. In this case, the QCM sensor was not used in the final analyte detection, but is a good example of the ability to assemble and monitor the assembly of complex nanostructures. First, polyethyleneimine (PEI) was immobilized, followed successively by 9–45 nm silica nanoparticles and PEI until several layers were assembled. This provided a roughened surface that allowed the immobilization and increased density of glucose oxide enzyme. QCM was used to monitor these depositions and provided information for subsequent layer-by-layer construction on latex particles.

3.5 QCM Nanoparticle-based Immunosensors

The design, synthesis, and assembly of functional nanostructures are important challenges in the interface with immunology. Finding key recognition units and presenting them in the appropriate environment and conformation are crucial to programming material specificity and affinity. A unique and creative idea for interface assembly may provide a good starting point, but redesign and optimization is difficult without a method for evaluating the proposed interface. The understanding of an interaction places interface development and application within reach. Traditional techniques have supplied qualitative information on antibody recognition for interface design and are widely used. Nanotechnology has expanded immunoassay options for the study of more diverse systems and, combined with QCM, provides a label-free, quantitative alternative. These analytical techniques can determine structural integrity of assembled nanoarchitectures, can provide equilibrium and kinetic binding constants of biological entities, and can detect analytes (antibodies, toxins, etc.) for medical diagnostic applications.

3.5.1 Traditional Immunoassays

Radiolabeled immunoassays were one of the first techniques used in the detection of antibody or antigen in biological systems. In this assay, radioisotopes, commonly ^{125}I, were used to label the antibody or antigen, and scintillation counters measured the gamma or beta emission of the isotope. This provided the low detection limits needed for immunoassays, but regulation of radioactive isotopes made this technique inconvenient [92].

Another standard immunoassay is the enzyme-linked immunosorbant assay (ELISA). Its success comes from the ability to amplify binding through an enzyme reaction, which produces a spectroscopic signal. There are many formats for an ELISA experiment, though the indirect sandwich assay format has been widely accepted. In this format, the primary antibody is immobilized on a solid support (typically a well-plate) and antigen is allowed to bind. A second, polyclonal antibody for the antigen from a different species than that used as the primary antibody is then added and binds to the other side of the immobilized antigen, creating a "sandwich". An antibody that recognizes the second antibody (an anti-antibody) is functionalized with an enzyme (typically horseradish peroxidase) and allowed to bind. Finally, a substrate for the enzyme is introduced, which produces an enzyme product that is chromogenic. The chromophore is detected by conventional spectroscopic methods [57]. While the signal amplification from the enzyme reaction is beneficial, there are certain limitations to ELISA experiments. A lack of simple quantitation and excessive time required to assemble the complex immunomolecular biosensor, as in the case of an indirect sandwich assay, are poignant drawbacks. An example is the recent development of a quantitative ELISA assay for the detection of human IgG, which requires 19 hours from analyte immobilization to chromophore detection [93]. Other limitations include the need for labeling with a bulky enzyme, which could interfere with the antibody/antigen interaction, and the nonspecific adsorption of analyte to a hydrophobic well plate could lead to random orientation of binding sites and possible denaturation of substrate [57].

Surface plasmon resonance spectroscopy (SPR) has recently been used to detect antibody binding. Details of SPR phenomena have been previously outlined [20, 94]. Briefly, SPR is an optical technique that takes advantage of plasmon excitation in bulk metal by wave vector matched photons. The photons induce oscillations of free electrons in the metal, which then propagate an additional field into the contacting dielectric medium. The plasmon excitation requires a transfer of energy from the photons, which can be observed through the sharp minimum of reflectivity during resonance, leading to an SPR signal [94]. Therefore, measuring the change in reflection angle provides real-time, label-free detection of antigen/antibody interaction. SPR has been used to study several different systems, an example being the characterization of FLAG peptide epitope arrays [38]. Important drawbacks of the SPR method are the complicated and expensive optics required for operation, loss of sensitivity at distances from the sensor, and interference from molecules with high molar absorptivity.

3.5.2 Immunoassays using Nanotechnology

Radiolabeling, ELISA, and SPR are important and effective techniques for the evaluation of traditional antibody/antigen interactions. Inorganic materials, such as nanoclusters, offer useful spectroscopic and microscopic properties that can make analysis convenient, but can also introduce added challenges. One of the first interfaces between materials and antibodies [58] was designed because radiolabel immunoassays were cumbersome and enzyme immunoassays require delicate procedures [59]. The potential of materials was harnessed to design a simpler method and successfully used antibody-functionalized latex beads and turbidity measurements to detect agglutination [58, 60]. Improvements to this method quickly followed, using electric pulses to promote antibody/antigen interaction and decrease reaction time [59].

A similar, though more recent, nanoimmunoassay also used antibody-functionalized particles. These 70 nm diameter silica particles had exterior antibody functionalization and interior fluorophore entrapment. Fluorescent-labeling techniques have enjoyed long-lived success, though low fluorescence intensity and photoinstability have been recurring problems. Encapsulation of $Ru(bpy)_3{}^{2+}$ fluorophore in silica nanoparticles yielded high intensity fluorescence and increased photostability due to exclusion of damaging oxygen [95]. These nanoparticles were compared to popular quantum dots (Qdots), which are semiconductor nanoparticles with intense intrinsic fluorescence, and were found to have similar intensity and stability. Antibody-functionalized Qdots have also been used as fluorescent tags for the imaging of live cells [67, 96].

Metal nanoparticles have been used as electrochemical labels for the simultaneous detection of four antigens, β_2-microglobulin, IgG, bovine serum albumin, and C-reactive protein [64]. In this experiment, a large magnetic bead was functionalized with four types of antibodies corresponding to the four different antigens. After binding antigen to the antibody on the magnetic bead, antibody with unique nanoparticle labels bound to the immobilized antigen. Collection of the nanoparticle-labeled antibody and detection with square-wave stripping voltammetry provided four unique signals from the reduction of four unique metal nanoparticles [64].

A microscopic technique often used in nanocluster characterization, but not in immunoassay, is transmission electron microscopy (TEM). Metal nanoclusters absorb electrons, making them visible in TEM, while small carbon-based molecules do not. In a recent study, antibody-functionalized nanoparticles were incubated with pathogens *Staphylococcus saprophyticus* and *S. aureus* and examined with TEM. Functionalized nanoclusters bound to antibody binding sites and were detected. Furthermore, bacteria-bound magnetic particles were collected, thereby concentrating target pathogens [63].

Success with traditional immunoassay formats prompted SPR research in the area of nanoimmunotechnology. In 1998, Natan's group showed SPR signal amplification of antibody-nanocluster complex binding as compared to free antibody [21]. These results were confirmed with the detection of human complement factor 4 (C4) and C4 attached to colloidal Au particles [97].

These techniques, along with others, have had some success in the evaluation of interfaces between immunology and nanoscale materials. They also have their drawbacks. Many suffer from a lack of sensitivity and use large diameter particles (50–1000 nm) with relatively low surface area. Others rely on labeling to provide a detectable signal, which can interfere with recognition events and change the immunoassay dynamics. With SPR, nanoparticle labeling is used to acquire enhanced binding signals. Unfortunately, this also significantly broadens peaks, leading to a loss of sensitivity, complicated time-resolved measurements, and limited structural information [19, 20]. Efforts to develop a label-free, time-efficient, quantitative assay format that allows for 3D substrates and multilayer adsorptions have involved the quartz crystal microbalance (QCM) [1].

3.5.3 QCM Nanoparticle-based Immunosensors

Piezoelectric biosensors developed since 1990 have gained ground on traditional labeling experiments and on competing label-free instrumentation such as SPR [2, 98, 99]. The field of QCM immunosensors has also developed rapidly [43]. Considering the age of the new, though explosive, field of nanotechnology (the Brust nanoparticle synthesis was published in 1994), QCM immunosensors for nanotechnology applications are limited. One example used gold nanoparticle growth to amplify antibody-mediated lung carcinoma cell detection using QCM [100]. In these experiments a monoclonal antibody to cell surface antigen was immobilized on a polystyrene film and captured lung carcinoma cells. The same antibody conjugated to 10 nm diameter citrate-reduced gold nanoparticles bound to the immobilized cell in a typical sandwich scheme. Auric acid and NH_2OH were then introduced, reacted with the pre-existing gold nanoparticle, and caused the nanoparticle to grow. This growth created an increased QCM signal and allowed for the detection of cells at levels as low as 100 cells mL^{-1}. This method provided results similar to ELISA, but was less time consuming.

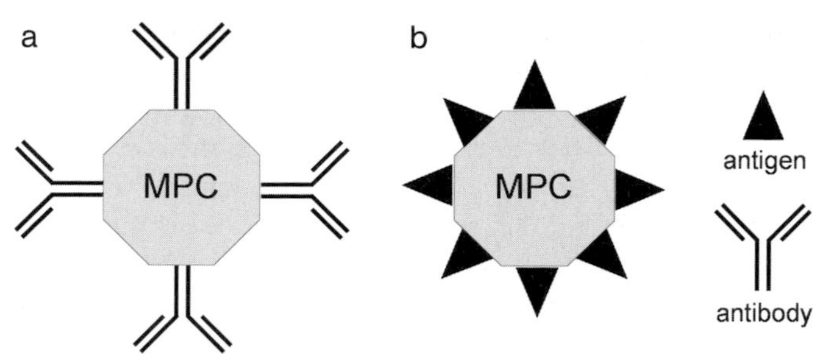

Fig. 3.8. Cartoon schematic of (a) nanocluster presenting antibody (antibody mimic) and (b) nanocluster presenting antigen (antigen mimic). Epitope antigens are generally smaller than antibodies and can be presented at a higher aerial density than antibodies.

Another example used polymer immobilized colloidal gold particles (~15 nm diameter) as an alternative approach to antibody immobilization [101]. Thiol-terminated Fab′ antibody fragments will often denature on an unprotected hydrophobic gold QCM electrode, but retained activity when bound to the polymer/nanogold mixture. This nanoparticle-based immunosensor supplied a means for phenotyping leukemia cells for medical diagnostics with detection limits of 6000 cells mL^{-1}. An important technical detail of this work was the ability to detect four different phenotypes simultaneously using an array of QCM sensors. Other nanoparticle-based QCM immunosensors have made use of peptide epitope functionalized nanoclusters.

3.5.3.1 Antigen Mimic Design

There are two ways that functionalized nanoclusters can interface with immunology through paratope/epitope recognition: an antibody can be covalently attached to an MPC or an MPC can play the role of the antigen (Fig. 3.8). Antibody/nanocluster complexes have been previously studied as mentioned above. Briefly, immunoassays for human IgG have been developed using antibody bound latex beads [59, 60], signal amplification was observed with antibody bound gold nanoclusters in SPR immunoassays [21], and antibody bound magnetic microparticles have been used as capture agents [61]. There is further interest in using nanoparticle labeled antibodies as nanoprobes [102].

An alternative approach would be to attach an antigen to an MPC or functionalize an MPC with a peptide epitope (Fig. 3.8b). An epitope is the antigenic determinant of a protein and corresponds to the region of the protein that is specifically recognized by the antibody [103]. Accordingly, epitopes are only defined in an operational or functional sense: anything that is bound by an antibody is, by definition, an epitope. Therefore, epitope regions from proteins have been synthesized and expressed as smaller functional units and have been shown to bind antibodies. Examples of these would be the FLAG or E tags commonly used in expressed proteins [38, 44]. The current approach to this type of antigen mimetics is to synthesize the linear epitope and rely on primary structure (amino acid sequence) to provide specificity. Combining this type of mimetics with nanoscale materials produced the first example of antigen encapsulated nanoclusters assembling with antibodies through the paratope/epitope interface. A linear peptide epitope known to bind monoclonal antibody associated with the human malarial parasite, *P. falciparum*, was used in the assembly of a robust, functional nanostructure that successfully mimicked a biological entity [65]. A more recent example used glutathione-protected nanoclusters and polyclonal anti-glutathione antibodies. This approach confirmed the ability to assemble epitope-protected nanoclusters with antibodies and provided quantitative binding information through improved analytical techniques [1].

These two examples used nanoclusters with peptide epitopes completely covering their surface and were synthesized with large amounts of the ligand, rather than specifically functionalized with small amounts of peptide. The next step in antigen mimic design was to functionalize pre-existing nanoclusters through place exchange reactions described above. This type of assembly provided a more complex nanostructure with three or more components and allowed for the use of more complicated linear and conformational epitopes. In a very recent study a linear peptide epitope, from the hemagglutanin (HA) protein related to influenza, has been synthesized and specifically presented at a controlled density on the surface of a pre-existing tiopronin-protected cluster [104]. This mimic of the HA protein was shown to interface with monoclonal anti-HA antibody and was compared to a self-assembled monolayer of the same peptide epitope.

Linear peptide epitopes used in the malaria, glutathione, and HA examples rely on primary amino acid sequence to provide antigenicity and specificity. This does not take into account the complex secondary structure (local conformation) exhibited by native proteins, which is essential to antibody recognition. Cyclization of peptides to approximate a loop structure has been used to introduce epitope conformation [105], but the development and presentation of a peptide epitope that reconstitutes a physiological conformation is an interesting alternative. This was recently achieved through the bidentate presentation of a peptide epitope from a loop region in the protective antigen (PA) of *B. anthracis* [83]. The conformational antigen mimic was able to interface with monoclonal anti-PA antibodies and showed a greater than two-fold increase in affinity over a linear antigen mimic.

This variety of antigen mimics using small molecule, linear, and conformational epitopes suggests multiple routes to interfacing inorganic nanoclusters with biological antibodies. Many other options are also available and, together, supply a "toolbox" of interfaces that can be used and studied in a multitude of systems. The three examples of GSH-MPC, HA-MPC, and PA-MPC are reviewed in further detail below.

3.5.3.2 Glutathione-protected Nanocluster

This research aimed to develop an immunoassay to study the test case of anti-glutathione antibody recognition of glutathione-protected nanocluster. The immunoassay was designed to be label-free, time-efficient, and quantitative and allow 3D substrates, multilayer adsorptions, and non-rigid biological recognition. QCM technology was used for many of the reasons previously mentioned, including high sensitivity, real-time detection, low cost, and a large wave penetration depth. Glutathione (GSH) monolayer-protected clusters (MPC) were used for their ease of synthesis, water solubility, high surface area and ligand valency, and the commercial availability of polyclonal anti-GSH antibody. Both GSH and tiopronin (Tiop), a glycine derivative and truncate of the GSH tripeptide, nanoclusters were synthesized according to a modified Brust method and characterized. The GSH-MPC had an average diameter of 3.7 ± 1.2 nm, average composition of $Au_{953}GS_{199}$, and average molecular weight of 220 kDa.

The immunosensor consisted of the gold QCM electrode, Protein A, bovine serum albumin (BSA), and polyclonal anti-GSH antibody (Fig. 3.9a). Protein A was used to conveniently immobilize the antibody on the sensor surface in a defined orientation and BSA was used to block any exposed gold that might contribute to non-specific binding. Protein A bound to the gold electrode as a multilayer, BSA binding was minimal, and IgG bound specifically to Protein A, completing the immunosensor assembly (Fig. 3.9b). One drawback of this immunosensor design is the lack of reversibility and the need to reassemble the immunosensor structure before each experiment.

Analyte, GSH-MPC, was detected in a dose-dependent manner that was designed to limit non-specific adsorption and aggregation. Samples of different concentration were introduced to the immunosensor in short (5 min) doses and provided changes in mass that fit a logarithmic curve, revealing saturation of the immunosensor (Fig. 3.9d). This showed that the antibody was able to recognize the glutathione tripeptide epitope presented on the surface of a nanocluster. Furthermore, control experiments showed a lack of antibody binding to Tiop-MPC, suggesting antibody recognition of the γ-glutamic acid portion of GSH as opposed to the glycine portion (Fig. 3.9c). Equilibrium and kinetic constants for polyclonal anti-GSH antibody binding to GSH-MPC were calculated from Langmuir isotherm fits (Fig. 3.9e) and from individual binding curves (Fig. 3.9c and f). The equilibrium association constant, K_a, was found to be $3.6 \pm 0.2 \times 10^5$ M^{-1}, the rate of forward reaction, k_f, was $5.4 \pm 0.7 \times 10^1$ M^{-1} s^{-1}, and the rate of reverse reaction, k_r, was $1.5 \pm 0.4 \times 10^{-4}$ s^{-1}. These values are reasonable, considering the use of polyclonal antibody. These results confirm the usefulness of a QCM immunosensor and suggest that epitope-presenting nanoclusters can be immunoreactive materials.

3.5.3.3 **Hemagglutanin Mimic Nanocluster**

Hemagglutanin (HA) and neuraminidase are two virus glycoproteins that are responsible for influenza infection and are targets for antibody neutralization. A peptide epitope-presenting nanocluster was designed to mimic one aspect of HA (Fig. 3.10a) and a QCM immunosensor was developed to examine the interface between monoclonal anti-HA and the HA-MPC. The peptide spanning amino acids 98–106 in the protein amino acid sequence (sequence: YPYDVPDYA) have been involved in influenza vaccine studies [106, 107]. The HA mimic was assembled by specific presentation of the synthetic peptide epitope on the surface of a pre-formed Tiop-

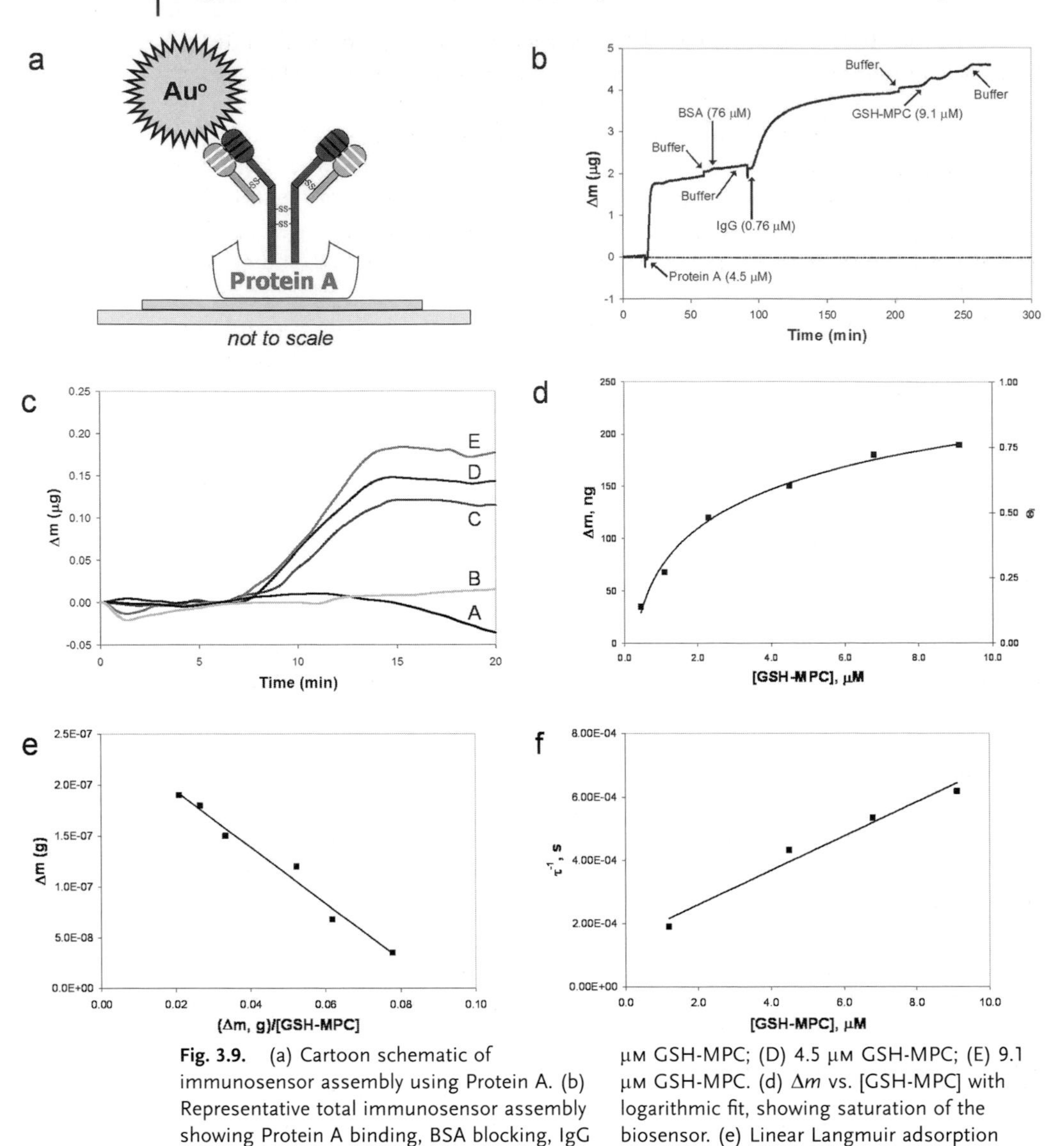

Fig. 3.9. (a) Cartoon schematic of immunosensor assembly using Protein A. (b) Representative total immunosensor assembly showing Protein A binding, BSA blocking, IgG capture, and MPC detection. (c) Binding curves (A) 1 mM free GSH, showing mass stripping; (B) 4.1 μM tiopronin-MPC; (C) 2.3 μM GSH-MPC; (D) 4.5 μM GSH-MPC; (E) 9.1 μM GSH-MPC. (d) Δm vs. [GSH-MPC] with logarithmic fit, showing saturation of the biosensor. (e) Linear Langmuir adsorption isotherm. (f) Adsorption kinetics. (Reprinted with permission from the American Chemical Society.)

MPC through place exchange reaction. This produced a three-component system consisting of 3.5 ± 1.0 nm colloidal gold, tiopronin ligand, and HA peptide, with a final composition of $Au_{807}T_{242}HA_4$. A QCM immunosensor was then used in the evaluation of the interface between HA-MPC and anti-HA antibody.

In this immunoassay, a different assembly technique was used than for the preceding GSH-MPC experiments. This approach made use of ionic interactions to immobilize the nanocluster and detect antibody binding. One reason for using this approach was the available option of analyzing complex samples containing one more antibody and various other components. The immunosensor consisted of the gold QCM electrode, a positively charged polyelectrolyte [poly(diallyldimethylammonium chloride)], nanocluster, and BSA (Fig. 3.10b). Polyelectrolyte had been previously used in the multi-layering of quantum dots and provided a thin layer of positive charge on the surface of the QCM. Negatively charged HA-MPC bound non-specifically and was immobilized for antibody binding. Before analyte was introduced, BSA was used to block non-specific polyelectrolyte binding sites (Fig. 3.10b). Again, a limitation of this immunoassay approach is the lack of reversibility and the time required to reassemble the sensor before each experiment.

This immunosensor was used for two applications, the first being an examination of antibody binding as related to the density of peptide presented on the Tiop-MPC. Three nanoclusters with 4, 11, and 33 peptides per cluster were examined and suggested that the lower density of peptide allowed the highest ratio of antibody to peptide interaction. The second application was a thorough study of the antibody interface with the HA-MPC displaying four peptides per cluster. Samples with different concentrations of antibody were introduced to the sensor in dose amounts and produced changes in mass that fit a logarithmic curve and showed saturation of the biosensor. This suggested that the antibody did recognize the peptide epitope specifically presented on the surface of a nanocluster. Control experiments using Tiop-MPC presenting no peptide showed a lack of antibody binding, signifying the specificity of the antibody for the epitope. Preliminary calculations have suggested an equilibrium binding constant ($K_a = 1.0 \pm 0.3 \times 10^7$ M^{-1}) and kinetic rate constants ($k_f = 5.1 \pm 0.6 \times 10^5$ M^{-1} s^{-1} and $k_r = 9.4 \pm 0.7 \times 10^{-2}$ s^{-1}) for the association of anti-HA and HA-MPC.

These results were compared to traditionally used peptide epitope arrays. Self-assembled monolayers (SAMs) consisting of tiopronin ligand and mixed-monolayers of both peptide and tiopronin were formed on the QCM gold electrode. Antibody was introduced to the mixed-monolayer peptide array in dose amounts and bound in a manner similar to that seen with the functionalized nanocluster (Fig. 3.10c). Preliminary equilibrium association constant ($K_a = 4.1 \pm 0.7 \times 10^6$ M^{-1}) and kinetic constants ($k_f = 1.4 \pm 0.2 \times 10^5$ M^{-1} s^{-1}, $k_r = 3.5 \pm 0.2 \times 10^{-2}$ s^{-1}) were calculated and compared to those for the antigen mimic HA-MPC. The K_a for the nanocluster is more than double that for the SAM, indicating a more accurate reconstitution of the peptide epitope on the surface of the MPC.

3.5.3.4 Protective Antigen of *B. anthracis* Mimic Nanocluster

This study [83] made use of a QCM immunosensor, four independently functionalized nanoclusters, and monoclonal antibodies to effectively map one antibody clone to one peptide epitope. Not only did the identified antibody distinguish between three different peptide epitopes from the protective antigen (PA) of *B. anthracis*, it had an increased affinity for a conformational epitope as compared to a linear version with the same amino acid sequence. Epitopes from PA were used for

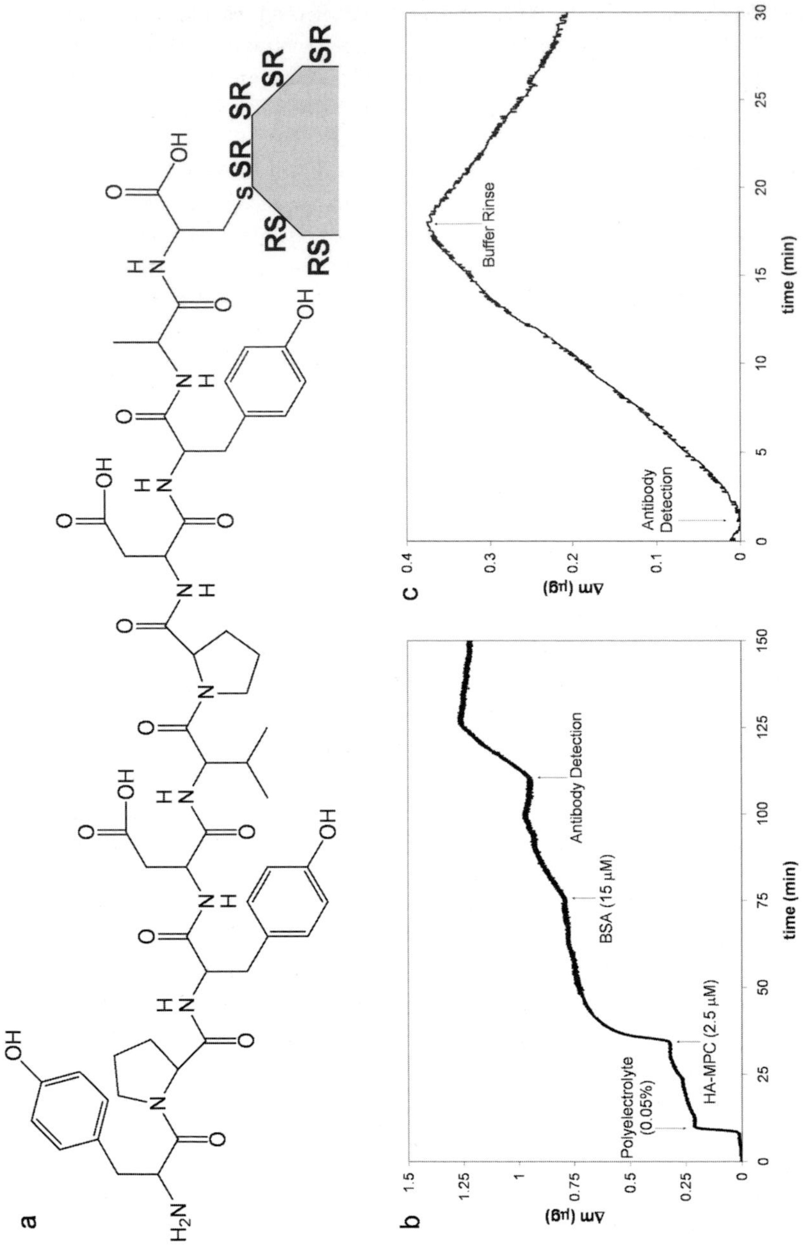

Fig. 3.10. (a) Schematic of a HA peptide epitope-functionalized MPC (not to scale). (b) Representative total immunosensor assembly, showing polyelectrolyte and MPC binding, BSA blocking, and antibody detection. (c) Representative antibody detection using HA/tiopronin SAM.

four reasons: (1) epitopes had been previously identified through various mapping techniques; (2) PA has been a target for immunological studies, as anti-PA antibodies confer immunity to anthrax; (3) PA has both conformational and linear epitopes; (4) anthrax has been identified as a potential bioterrorism agent. Identified PA peptide epitopes [108–111] span amino acids 680–692 (PA680), 703–722 (PA703), and 730–735 (PA730) in the intact protein, and were each synthesized for use in antigen mimic design. The functional nanostructures were assembled by specifically presenting conformational or linear epitopes on the surface of pre-existing Tiop-MPCs through place exchange reactions. Four antigen mimics had final peptide densities of 4 (PA680B, conformational), 4 (PA680M, linear), 8 (PA703B, conformational), and 10 (PA730M, linear) peptides per MPC. A cartoon schematic of PA680B-MPC and the crystal structure of PA [109] are given in Fig. 3.11(a). Labeling experiments suggested the bidentate attachment of loop epitopes to MPCs, implying the formation of conformational antigen mimics. Increased antibody affinity for the conformational over the linear epitope provided a strong case for faithful reconstitution of the loop structure.

The QCM immunoassay used ionic interactions for the immobilization of MPC, similar to that used with HA-MPC (Section 3.5.3.3). Polyelectrolyte acted as an intermediate between the gold electrode and the MPC for successful immobilization (Fig. 3.11b). Again, BSA was used to block non-specific binding sites. This immunosensor assembly was used for two different studies, the first being the screening of seven monoclonal anti-PA antibodies against the four antigen mimics. The antibodies were known to bind the full-sized 83 kDa PA protein, but had not been tested for neutralization ability [112]. Screening provided qualitative binding information and pointed to one antibody as being specific for the conformational antigen mimic, PA680B-MPC. The magnitude of antibody binding as a function of antibody concentration was the second system studied. Antibody did not bind to PA703B-MPC or PA730M-MPC, but PA680B-MPC ($K_a = 5.9 \pm 0.7 \times 10^6$ M^{-1}) and PA680M-MPC ($K_a = 2.6 \pm 0.9 \times 10^6$ M^{-1}) were evaluated to obtain equilibrium association constants. Both were also evaluated in buffer of high ionic strength. In this system, PA680B-MPC had a K_a similar to that previously reported, though PA680M-MPC showed no binding. Kinetic studies of antibody binding for PA680B-MPC yielded forward and reverse reaction rate constants of $9 \pm 2 \times 10^3$ M^{-1} s^{-1} and $2.3 \pm 0.5 \times 10^{-3}$ s^{-1}, respectively. Free peptide (amino acids 680–692), unencumbered or unconstrained by nanocluster or protein scaffold, showed only non-specific binding to the antibody. In this immunoassay, anti-PA antibody was immobilized using Protein A, as previously described (Section 3.5.3.2), and free peptide was introduced. This implies that the peptide epitope was presented in a more accommodating conformation on the surface of the MPC, as opposed to being free in solution.

These results point out that peptide epitopes and monolayer protected nanoclusters can be brought together in the successful assembly of an immunoreactive material and be interfaced with biological antibodies. They also show the screening and effective mapping of an antibody to a peptide epitope, as well as differentiation between conformational and linear presentation.

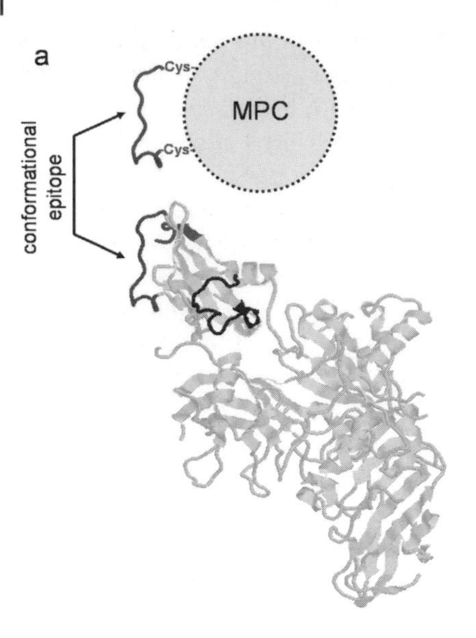

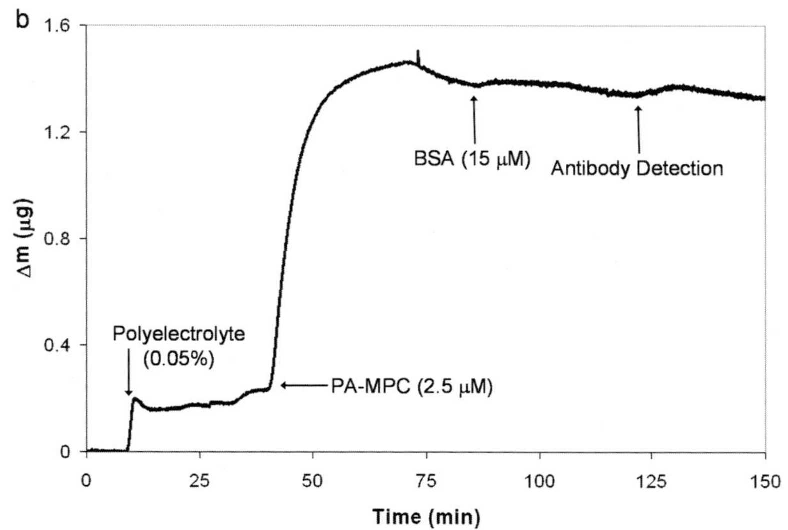

Fig. 3.11. (a) X-ray crystal structure of the protective antigen (PA) of *B. anthracis* with epitope regions highlighted (dark). Schematic of PA680B-MPC antigen mimic, highlighting conformational peptide epitope. (b) Representative total immunosensor assembly, showing polyelectrolyte and MPC binding, BSA blocking, and antibody detection.

3.6
Conclusions and Future Directions

The future of bionanotechnology lies at the interface between biology and inorganic nanomaterials and relies on the ability to probe and evaluate that interface. Many analytical techniques outlined in this chapter, including AFM, TEM, NMR, and molecular imaging, have been developed for this very purpose, for the analysis of all types of interfaces. The quartz crystal microbalance is another technique that has evolved to become a central tool in the study of both biological and nanomaterials applications. QCM has principal advantages for the investigation of bionanotechnology over existing immunological techniques: low cost and ease of operation, quantitation, real-time measurements, and large wave penetration depth, to name a few. For these reasons QCM technology has been applied to the exploration of various life science applications. QCM nanoparticle-based immunoassay is a relatively new field, but has been designed for the evaluation of several antigen mimics. Specifically, QCM has shown that antibody can recognize and interface with immunoreactive nanomaterials designed with linear and conformational peptide epitopes presented on the surface of monolayer-protected clusters. This is only one interface that QCM has successfully characterized. Given the virtues of this technique and the vast number of possible combinations between biology and nanotechnology, more systems will doubtlessly be studied.

The prospects of biotechnology and nanomaterials science are most promising at their interface, where they overlap in the emerging field of bionanotechnology. This interfacial discipline further relies on analytical techniques that can probe the boundary where they meet. The quartz crystal microbalance is certain to be an integral tool in the exploration of this frontier. Specifically, QCM can expand its applicability through the design of new immobilization strategies for repeated, reversible interactions. An improved, commercially available, multichannel detection apparatus will improve throughput and multiplexing. In the realm of medical diagnostics and defense-based toxin detection, a blood sample, water-way, or gaseous area could be largely characterized through multi-analyte detection. This is possible because of the portability, low cost, and ease of use of existing QCM instruments and potential devices. Furthermore, advances in nanotechnology and improved understanding of biology will produce smart sensors and functional devices capable of carrying out an environmental analysis or capable or modulating an environment. Continued research on advanced nanomaterials in a biological paradigm using an effective QCM transducer can make this a reality.

Acknowledgments

A.E.G. would like to thank the Chemical Biology Interface Training Grant (T32 GM065086) and the Vanderbilt Institute of Nanoscale Science and Engineering (VINSE) for support. D.W.W. acknowledges support from the SouthEast Regional Center of Excellence for Biodefense (NIH U54 AI57157-03).

Symbols

A area of the QCM electrode
$[C]$ concentration of a bulk solution
$[C]_0$ initial concentration, initial concentration of surface immobilized analyte

C_1	motional capacitance of the unperturbed crystal
c_{66}	piezoelectrically stiffened quartz elastic constant
C_f	Sauerbrey sensitivity factor, $2f_o{}^2/A(c_{66}\rho_q)^{1/2}$
f_o	frequency of crystal in air (unperturbed) (Hz)
k_f	forward kinetic rate constant
k_r	reverse kinetic rate constant
N	overtone number
R	gas constant (8.3145 J K^{-1} mol^{-1})
R_L	loading resistance
T	temperature (K)
Δf	change in frequency (Hz)
ΔG_{ads}	Gibbs free energy of adsorption
Δm	change in mass (g)
$\Delta\eta$	change in viscosity
$\Delta\rho$	change in density
η_q	effective quartz viscosity
Θ	fractional coverage of a surface by a species
Θ_∞	fractional coverage at infinite time
ρ_q	quartz mass density
τ^{-1}	kinetic time constant, equal to $k_f C + k_r$

References

1 Gerdon, A.E., Wright, D.W., Cliffel, D.E. Quartz crystal microbalance detection of glutathione-protected nanoclusters using antibody recognition, *Anal. Chem.* **2005**, *77*, 304–310.

2 Janshoff, A., Galla, H.J., Steinem, C. Piezoelectric mass-sensing devices as biosensors – an alternative to optical biosensors?, *Angew. Chem. Int. Ed.* **2000**, *39*, 4004–4032.

3 Cady, W.G., *Piezoelectricity*. **1946**, New York: McGraw-Hill.

4 Nomura, T., Okuhara, M. Frequency shifts of piezoelectric quartz crystals immersed in organic liquids, *Anal. Chim. Acta* **1982**, *142*, 281–284.

5 Kanazawa, K.K., Gordon, J.G. Frequency of a quartz microbalance in contact with liquid, *Anal. Chem.* **1985**, *57*, 1770–1771.

6 Martin, S.J., Granstaff, V.E., Frye, G.C. Characterization of a quartz crystal microbalance with simultaneous mass and liquid loading, *Anal. Chem.* **1991**, *63*, 2272–2281.

7 Martin, S.J., Spates, J.J., Wessendorf, K.O., Schneider, T.W. Resonator/oscillator response to liquid loading, *Anal. Chem.* **1997**, *69*, 2050–2054.

8 Rickert, J., Brecht, A., Gopel, W. QCM operation in liquids: constant sensitivity during formation of extended protein multilayers by affinity, *Anal. Chem.* **1997**, *69*, 1441–1448.

9 Chagnard, C., Gilbert, P., Watkins, A.N., Beeler, T., Paul, D.W. An electronic oscillator with automatic gain control: EQCM applications, *Sens. Actuators B* **1996**, *32*, 129–136.

10 Maxtek, Inc. Home Page, http://www.maxtekinc.com, **2003**.

11 Hillier, A.C., Ward, M.D. Scanning electrochemical mass sensitivity

mapping of the quartz crystal microbalance in liquid media, *Anal. Chem.* **1992**, *64*, 2539–2554.

12 Bandey, H.L., Martin, S.J., Cernosek, R.W., Hillman, A.R. Modeling the responses of thickness-shear mode resonators under various loading conditions, *Anal. Chem.* **1999**, *71*, 2205–2214.

13 Martin, S.J., Bandey, H.L., Cernosek, R.W. Equivalent-circuit model for the thickness-shear mode resonator with a viscoelastic film near film resonance, *Anal. Chem.* **2000**, *72*, 141–149.

14 Theisen, L.A., Martin, S.J., Hillman, A.R. A model for the quartz crystal microbalance frequency response to wetting characteristics of corrugated surfaces, *Anal. Chem.* **2004**, *76*, 796–804.

15 Hung, V.N., Abe, T., Minh, P.N., Esashi, M. High-frequency one-chip multichannel quartz crystal microbalance fabricated by deep RIE, *Sens. Actuators A* **2003**, *108*, 91–96.

16 Tatsuma, T., Watanabe, Y., Oyama, N. Multichannel quartz crystal microbalance, *Anal. Chem.* **1999**, *71*, 3632–3636.

17 Fawcett, N.C., Craven, R.D., Zhang, P., Evans, J.A. QCM response to solvated, tethered macromolecules, *Anal. Chem.* **1998**, *70*, 2876–2880.

18 Thompson, M., Kipling, A.L., Duncan-Hewitt, W.C. Thickness-shear-mode acoustic wave sensors in the liquid phase, *Analyst* **1991**, *116*, 881–890.

19 Wijekoon, W.M.K.P., Asgharian, B., Casstevens, M., Samoc, M., Talapatra, G.B., Prasad, P.N., Geisler, T., Rosenkilde, S. Electrooptic effect in Langmuir–Blodgett films of 2-(docosylamino)-5-nitropyridine probed by surface plasmon waves, *Langmuir* **1992**, *8*, 135–139.

20 Salamon, Z., Macleod, H.A., Tollin, G. Surface plasmon resonance spectroscopy as a tool for investigating the biochemical and biophysical properties of membrane protein systems. II: Applications to biological systems, *Biochim. Biophys. Acta* **1997**, *1331*, 131–152.

21 Lyon, L.A., Musick, M.D., Natan, M.J. Colloidal Au-enhanced surface plasmon resonance immunosensing, *Anal. Chem.* **1998**, *70*, 5177–5183.

22 Lyon, L.A., Musick, M.D., Smith, P.C., Reiss, R.D., Pena, D.J., Natan, M.J. Surface plasmon resonance of collodial Au-modified gold films, *Sens. Actuators B* **1999**, *54*, 118–124.

23 Lyon, L.A., Pena, D.J., Natan, M.J. Surface plasmon resonance of Au colloid-modified Au films: particle size dependence, *J. Phys. Chem. B* **1999**, *103*, 5826–5831.

24 Zhang, Y., Telyatnikov, V., Sathe, M., Zeng, X., Wang, P.G. Studying the interactions of α-Gal carbohydrate antigen and proteins by quartz-crystal microbalance, *J. Am. Chem. Soc.* **2003**, *125*, 9292–9293.

25 Bain, C.D., Whitesides, G.M. Formation of monolayers by the coadsorption of thiols on gold: variation in the length of the alkyl chain, *J. Am. Chem. Soc.* **1989**, *111*, 7164–7175.

26 Pale-Grosdemange, C., Simon, E.S., Prime, K.L., Whitesides, G.M. Formation of self-assembled monolayers by chemisorption of derivatives of oligo(ethylene glycol) of structure $HS(CH_2)_{11}(OCH_2CH_2)_mOH$ on gold, *J. Am. Chem. Soc.* **1991**, *113*, 12–20.

27 Templeton, A.C., Zamborini, F.P., Wuelfing, W.P., Murray, R.W. Controlled and reversible formation of nanoparticle aggregates and films using Cu^{+2}-carboxylate chemistry, *Langmuir* **2000**, *16*, 6682–6688.

28 Mamedov, A.A., Belov, A., Giersig, M., Mamedova, N.N., Kotov, N.A. Nanorainbows: graded semiconductor films from quantum dots, *J. Am. Chem. Soc.* **2001**, *123*, 7738–7739.

29 Decher, G. Fuzzy Nanoassemblies: toward layered polymeric multicomposites, *Science* **1997**, *277*, 1232–1237.

30 Hicks, J.F., Seok-Shon, Y., Murray, R.W. Layer-by-layer growth of polymer/nanoparticle films containing

monolayer-protected gold clusters, *Langmuir* **2002**, *18*, 2288–2294.

31 Templeton, A.C., Cliffel, D.E., Murray, R.W. Redox and fluorophore functionalization of water-soluble, tiopronin-protected gold clusters, *J. Am. Chem. Soc.* **1999**, *120*, 4845–4849.

32 Katz, E., de Lacy, A.L., Fierro, J.L.G., Palacios, J.M., Fernandez, V.M. Covalent binding of viologen to electrode surfaces coated with poly(acrylic acid) formed by electropolymerization of acrylate ions, *J. Electroanal. Chem.* **1993**, *358*, 247–259.

33 Davis, K.A., Leary, T.R. Continuous liquid-phase piezoelectric biosensor for kinetic immunoassays, *Anal. Chem.* **1989**, *61*, 1227–1230.

34 Bohinski, R.C. Immunoprecipitation of serum albumin with protein A-Sepharose, *J. Chem. Educ.* **2000**, *77*, 1460–1462.

35 Deisenhofer, J. Crystallographic refinement and atomic models of a human fc fragment and its complex with fragment B of protein A from *Staphylococcus aureus* at 2.9- and 2.8-A resolution, *Biochemistry* **1981**, *20*, 2361–2370.

36 Bain, C.D., Troughton, E.B., Tao, Y.T., Evall, J., Whitesides, G.M., Nuzzo, R.G. Formation of monolayer films by the spontaneous assembly of organic thiols from solution onto gold, *J. Am. Chem. Soc.* **1989**, *111*, 321–335.

37 Karpovich, D.S., Blanchard, G.J. Direct measurement of the adsorption kinetics of alkanethiolate self-assembled monolayers on a microcrystalline gold surface, *Langmuir* **1994**, *10*, 3315–3322.

38 Wegner, G.J., Lee, H.J., Corn, R.M. Characterization and optimization of peptide arrays for the study of epitope-antibody interactions using surface plasmon resonance imaging, *Anal. Chem.* **2002**, *74*, 5161–5168.

39 Lauffenburger, D.A., Linderman, J.J., *Receptors: Models for Binding, Trafficking, and Signaling.* **1996**, New York: Oxford University Press.

40 Ebara, Y., Itakura, K., Okahata, Y. Kinetic studies of molecular recognition based on hydrogen bonding at the air-water interface by using a highly sensitive quartz-crystal microbalance, *Langmuir* **1996**, *12*, 5165–5170.

41 Ebato, H., Gentry, C.A., Herron, J.N., Muller, W., Okahata, Y., Ringsdorf, H., Suci, P.A. Investigation of specific binding of antifluorescyl antibody and Fab to fluorescein lipids in Langmuir–Blodgett deposited films using quartz crystal microblance methodology, *Anal. Chem.* **1994**, *66*, 1683–1689.

42 Smith, E.A., Thomas, W.D., Kiessling, L.L., Corn, R.M. Surface plasmon resonance imaging studies of protein-carbohydrate interactions, *J. Am. Chem. Soc.* **2003**, *125*, 6140–6148.

43 Su, X., Chew, F.T., Li, S.F.Y. Design and application of piezoelectric quartz crystal-based immunoassay, *Anal. Sci.* **2000**, *16*, 107–114.

44 Shen, Z., Stryker, G.A., Mernaugh, R.L., Yu, L., Yan, H., Zeng, X. Single-chain fragment variable antibody piezoimmunosensors, *Anal. Chem.* **2005**, *77*, 797–805.

45 Zuo, B., Li, S., Guo, Z., Zhang, J., Chen, C. Piezoelectric immunosensor for SARS-associated coronavirus in sputum, *Anal. Chem.* **2004**, *76*, 3536–3540.

46 Stine, R., Pishko, M.V., Schengrund, C.-L. Comparison of glycosphingolipids and antibodies as receptor molecules for ricin detection, *Anal. Chem.* **2005**, *77*, 2882–2888.

47 Uttenthaler, E., Schraml, M., Mandel, J., Drost, S. Ultrasensitive quartz crystal microbalance sensors for detection of M13-phages in liquids, *Biosens. Bioelectron.* **2001**, *16*, 735–743.

48 Kastl, K., Ross, M., Gerke, V., Steinem, C. Kinetics and thermodynamics of annexin A1 binding to solid-supported membranes: A QCM study, *Biochemistry* **2002**, *41*, 10087–10094.

49 Niemeyer, C.M. Nanoparticles, Proteins, and nucleic acids: biotechnology meets materials science, *Angew. Chem. Int. Ed.* **2001**, *40*, 4128–4158.

50 Storhoff, J.J., Mirkin, C.A.

Programmed materials synthesis with DNA, *Chem. Rev.* **1999**, *99*, 1849–1862.

51 Whaley, S.R., English, D.S., Hu, E.L., Barbara, P.F., Belcher, A.M. Selection of peptides with semiconductor binding specificity for directed nanocrystal assembly, *Nature* **2000**, *405*, 665–668.

52 Slocik, J.M., Naik, R.R., Stone, M.O., Wright, D.W. Viral templates for gold nanoparticle synthesis, *J. Mater. Chem.* **2005**, *15*, 749–753.

53 Knecht, M.R., Wright, D.W. Functional analysis of the biomimietic silica precipitating activity of the R5 peptide from *Cylindrotheca fusiformis*, *Chem. Commun.* **2003**, *24*, 3038–3039.

54 Nelson, D.L., Cox, M.M., *Lehninger Principles of Biochemistry*. Third edn., **2000**, New York: Worth Publishers.

55 Morris, G.E., Choosing a method for epitope mapping, in *Epitope Mapping Protocols*. **1996**, Humana Press: Totowa, NJ. pp. 1–9.

56 Mernaugh, R., Mernaugh, G., *Molecular Methods in Plant Pathology*, ed. Singh, R.P., Singh, U.S. **1995**, Boca Raton: CRC Lewis Publishers.

57 Crowther, J.R., *The ELISA Guidebook*. Methods in Molecular Biology. vol. 149. **2001**, Totowa, New Jersey: Humana Press.

58 Bernard, A.M., Lauwerys, R.R. Comparison of turbidimetry with particle counting for the determination of human B2-microglobulin by latex immunoassay (LIA), *Clin. Chem. Acta* **1982**, *119*, 335–339.

59 Tamiya, E., Wantanbe, N., Matsuoka, H., Karube, I. Pulse immunoassay for human immunoglobulin G using antibody bound latex beads, *Biosensors* **1988**, 139–146.

60 Ortega-Vinuesa, J.L., Molina-Bolivar, J.A., Hidalgo-Alvarez, R. Particle enchanced immunoaggregation of F(ab′)2 molecules, *J. Immunol. Methods* **1996**, *190*, 29–38.

61 Nam, J.-M., Thaxton, C.S., Mirkin, C.A. Nanoparticle-based bio-bar codes for the ultrasensitive detection of protein, *Science* **2003**, *301*, 1884–1886.

62 Tang, D.P., Yuan, R., Chai, Y.Q., Zhong, X., Liu, Y., Dai, J.Y., Zhang, L.Y. Novel potentiometric immunosensor for hepatitis B surface antigen using a gold nanoparticle-based biomolecular immobilization method, *Anal. Biochem.* **2004**, *333*, 345–350.

63 Ho, K.-C., Tsai, P.-J., Lin, Y.-S., Chen, Y.-C. Using biofunctionalized nanoparticles to probe pathogenic bacteria, *Anal. Chem.* **2004**, *76*, 7162–7168.

64 Liu, G., Wang, J., Kim, J., Jan, M.R. Electrochemical coding for multiplexed immunoassays of proteins, *Anal. Chem.* **2004**, *76*, 7126–7130.

65 Slocik, J.M., Moore, J.T., Wright, D.W. Monoclonal antibody recognition of histidine-rich peptide encapsulated nanoclusters, *Nano Lett.* **2002**, *2*, 169–173.

66 Bird, R.E., Hardman, K.D., Jacobson, J.W., Johnson, S., Kaufman, B.M., Lee, S.-M., Lee, T., Pope, S.H., Riordan, G.S., Whitlow, M. Single-chain antigen-binding proteins, *Science* **1988**, *242*, 423–426.

67 Bentzen, E.L., House, F., Utley, T.J., Crowe, J.E.J., Wright, D.W. Progression of respiratory syncytial virus infection monitored by fluorescent quantum dot probes, *Nano Lett.* **2005**, *5*, 591–595.

68 Hostetler, M.J., Green, S.J., Stokes, J.J., Murray, R.W. Monolayers in three dimensions: synthesis and electrochemistry of ω-functionalized alkanethiolate-stabilized gold cluster compounds, *J. Am. Chem. Soc.* **1996**, *118*, 4212–4213.

69 Ingram, R.S., Hostetler, M.J., Murray, R.W. Poly-hetero-ω-functionalized alkanethiolate-stabilized gold cluster compounds, *J. Am. Chem. Soc.* **1997**, *119*, 9175–9178.

70 Wuelfing, W.P., Zamborini, F.P., Templeton, A.C., Wen, X., Yoon, H., Murray, R.W. Monolayer-protected clusters: molecular precursors to metal films, *Chem. Mater.* **2001**, *13*, 87–95.

71 Schaaf, T.G., Knight, G., Shafigullin, M.N., Borkman, R.F., Whetten, R.L. Isolation and selected properties of a 10.4 kDa gold:

glutathione cluster compound, *J. Phys. Chem. B* **1998**, *102*, 10643–10646.

72 Templeton, A.C., Chen, S., Gross, S.M., Murray, R.W. Water-soluble, isolable gold clusters protected by tiopronin and coenzyme A monolayers, *Langmuir* **1999**, *15*, 66–76.

73 Alvarez, M.M., Khoury, J.T., Schaaf, T.G., Shafigullin, M.N., Vezmar, I., Whetten, R.L. Optical absorption spectra of nanocrystal gold molecules, *J. Phys. Chem. B* **1997**, *101*, 3706–3712.

74 Lee, D., Donkers, R.L., Wang, G., Harper, A.S., Murray, R.W. Electrochemistry and optical absorbance and luminescence of molecule-like Au_{38} nanoparticles, *J. Am. Chem. Soc.* **2004**, *126*, 6193–6199.

75 Hostetler, M.J., Templeton, A.C., Murray, R.W. Dynamics of place-exchange reactions on monolayer-protected gold cluster molecules, *Langmuir* **1999**, *15*, 3782–3789.

76 Templeton, A.C., Hostetler, M.J., Warmoth, E.K., Chen, S., Hartshorn, C.M., Krishnamurthy, V.M., Forbes, M.D.E., Murray, R.W. Gateway reactions to diverse, polyfunctional monolayer-protected gold clusters, *J. Am. Chem. Soc.* **1998**, *120*, 4845–4849.

77 Slocik, J.M., Knecht, M.R., Wright, D.W., *Biogenic Nanoparticles*. Encyclopedia of Nanoscience and Nanotechnology, ed. Nalwa, H.S. vol. 1. **2004**, Stevenson Ranch: American Scientific Publishers. pp. 293–308.

78 Daniel, M.-C., Astruc, D. Gold nanoparticles: Assembly, supramolecular chemistry, quantum-size-related properties, and applications towards biology, catalysis, and nanotechnology, *Chem. Rev.* **2004**, *104*, 293–346.

79 Brust, M., Walker, M., Bethell, D., Schiffrin, D.J., Whyman, R. Synthesis of thiol-derivatized gold nanoparticles in a two-phase liquid-liquid system, *Chem. Commun.* **1994**, *7*, 801–802.

80 Hostetler, M.J., Wingate, J.E., Zhong, C.-J., Harris, J.E., Vachet, R.W., Clark, M.R., Londono, J.D., Green, S.T., Stokes, J.J., Wignall, G.D., Glish, G.L., Porter, M.D., Evans, N.D., Murray, R.W. Alkanethiolate gold cluster molecules with core diameters from 1.5 to 5.2 nm: core, monolayer properties as a function of core size, *Langmuir* **1998**, *14*, 17–30.

81 Montalti, M., Prodi, L., Zaccheroni, N., Baxter, R., Teobaldi, G., Zerbetto, F. Kinetics of place-exchange reactions of thiols on gold nanoparticles, *Langmuir* **2003**, *19*, 5172–5174.

82 Song, Y., Murray, R.W. Dynamics and extent of ligand exchange depend on electronic charge of metal nanoparticles, *J. Am. Chem. Soc.* **2002**, *124*, 7096–7102.

83 Gerdon, A.E., Wright, D.W., Cliffel, D.E. Epitope mapping of the protective antigen of *B. anthracis* using nanoclusters presenting conformational peptide epitopes, **2005**, *J. Am. Chem. Soc.*, submitted.

84 Gu, H., Ho, P.L., Tong, E., Wang, L., Xu, B. Presenting vancomycin on nanoparticles to enhance antimicrobial activities, *Nano Lett.* **2003**, *3*, 1261–1263.

85 Zheng, M., Huang, X. Nanoparticles comprising a mixed monolayer for specific bindings with biomolecules, *J. Am. Chem. Soc.* **2004**, *126*, 12047–12054.

86 Han, L., Daniel, D.R., Maye, M.M., Zhong, C.-J. Core-shell nanostructured nanoparticle as chemically sensitive interfaces, *Anal. Chem.* **2001**, *73*, 4441–4449.

87 Zamborini, F.P., Leopold, M.C., Hicks, J.F., Kulesza, P.J., Malik, M.A., Murray, R.W. Electron hopping conductivity and vapor sensing properties of flexible network polymer films of metal nanoparticles, *J. Am. Chem. Soc.* **2002**, *124*, 8958–8964.

88 Grate, J.W., Nelson, D.A., Skaggs, R. Sorptive behavior of monolayer-protected gold nanoparticle films: implications of chemical vapor sensing, *Anal. Chem.* **2003**, *75*, 1868–1879.

89 Willner, I., Patolsky, F., Weizmann, Y., Willner, B. Amplification

detection of single-base mismatches in DNA using microgravimetric quartz-crystal-microbalance transduction, *Talanta* **2002**, *56*, 847–856.

90 Weizmann, Y., Patolsky, F., Willner, I. Amplified detection of DNA and analysis of single-base mismatches by the catalyzed deposition of gold on Au-nanoparticles, *Analyst* **2001**, *126*, 1502–1504.

91 Patolsky, F., Lichtenstein, A., Willner, I. Amplified microgravimetric quartz-crystal-microbalance assay of DNA using oligonucleotide-functionalized liposomes or biotinylated-liposomes, *J. Am. Chem. Soc.* **2000**, *122*, 418–419.

92 Price, C.P., Newman, D.J., *Principles and Practice of Immunoassay*, ed. Price, C.P., Newman, D.J. **1991**, New York: Stockton Press. pp. 265–295.

93 Quinn, C.P., Semenova, V.A., Elie, C.M., Romero-Steiner, S., Greene, C., Li, H., Stamey, K., Steward-Clark, E., Schmidt, D.S., Mothershed, E., Pruckler, J. Specific, sensitive, and quantitative enzyme-linked immunosorbent assay for human immunoglobulin G antibodies to anthrax toxin protective antigen, *Emerging Infectious Dis.* **2002**, *8*, 1103–1110.

94 Salamon, Z., Macleod, H.A., Tollin, G. Surface plasmon resonance spectroscopy as a tool for investigating the biochemical and biophysical properties of membrane protein systems. I: Theoretical principles, *Biochim. Biophys. Acta* **1997**, *1331*, 117–129.

95 Lian, W., Litherland, S.A., Badrane, H., Tan, W., Wu, D., Baker, H.V., Gulig, P.A., Lim, D.V., Jin, S. Ultrasensitive detection of biomolecules with fluorescent dye-doped nanoparticles, *Anal. Biochem.* **2000**, *334*, 135–144.

96 Wu, X., Liu, H., Liu, J., Haley, K.N., Treadway, J.N., Larson, J.P., Ge, N., Peale, F., Bruchez, M.P. Immunofluorescent labeling of cancer marker Her2 and other cellular targets with semiconductor quantum dots, *Nat. Biotechnol.* **2003**, *21*, 41–46.

97 Liu, X., Sun, Y., Song, D., Zhang, Q., Tian, Y., Bi, S., Zhang, H. Sensitivity-enhancement of wavelength-modulation surface plasmon resonance biosensor for human complement factor 4, *Anal. Biochem.* **2004**, *222*, 99–104.

98 Kosslinger, C., Uttenthaler, E., Drost, S., Aberl, F., Wolf, H., Brink, G., Stanglmaier, A., Sackmann, E. Comparison of the QCM and the SPR method for surface studies and immunological applications, *Sens. Actuators B* **1995**, *24–25*, 107–112.

99 Spangler, B.D., Wilkinson, E.A., Murphy, J.T., Tyler, B.J. Comparison of the Spreeta surface plasmon resonance sensor and a quartz crystal microbalance for detection of *Escherichia coli* heat-labile enterotoxin, *Anal. Chim. Acta* **2001**, *444*, 149–161.

100 Ma, Z., Wu, J., Zhou, T., Chen, Z., Dong, Y., Tang, J., Sui, S.-F. Detection of human lung carcinoma cell using quartz crystal microbalance amplified by enlarging Au nanoparticles, *New J. Chem.* **2002**, *26*, 1795–1798.

101 Wang, H., Zeng, H., Liu, Z., Yang, Y., Deng, T., Shen, G., Yu, R. Immunophenotyping of acute leukemia using an integrated piezoelectric immunosensor array, *Anal. Chem.* **2004**, *76*, 2203–2209.

102 Hainfeld, J.F., Powell, R.D. New frontiers in gold labeling, *J. Histochem. Cytochem.* **2000**, *484*, 471–480.

103 Van Regenmortel, M.H.V. The concept and operational definition of protein epitopes, *Phil. Trans. R. Soc. London B* **1989**, *323*, 451–466.

104 Gerdon, A.E., Wright, D.W., Cliffel, D.E. Epitope presentation on monolayer-protected clusters for multi-component functional nanostructures, *Biomacromolecules* **2005**, accepted.

105 Misumi, S., Endo, M., Mukai, R., Tachibana, K., Umeda, M., Honda, T., Takamune, N., Shoji, S. A novel cyclic peptide immunization strategy for preventing HIV-1/AIDS infection and progression, *J. Biol. Chem.* **2003**, *278*, 32335–32343.

106 Muller, G.M., Shapira, M., Arnon, R. Anti-influenza response achieved by immunization with a synthetic conjugate, *Proc. Natl. Acad. Sci. U.S.A.* **1982**, *79*, 569–573.

107 Lu, Y., Ding, J., Liu, W., Chen, Y.-H. A candidate vaccine against influenza virus intensively improved the immunogenicity of a neutralizing epitope, *Int. Arch. Allergy Immunol.* **2002**, *127*, 245–250.

108 Leppla, S., Robbins, J., Schneerson, R., Shiloach, J. Development of an improved vaccine for anthrax, *J. Clin. Invest.* **2002**, *109*, 141–144.

109 Petosa, C., Collier, R.J., Klimpel, K.R., Leppla, S.H., Liddington, R.C. Crystal structure of the anthrax toxin protective antigen, *Nature* **1997**, *385*, 833–838.

110 Santelli, E., Bankston, L.A., Leppla, S.H., Liddington, R.C. Crystal structure of a complex between anthrax toxin and its host cell receptor, *Nature* **2004**, *430*, 905–908.

111 Wild, M.A., Xin, H., Maruyama, T., Nolan, M.J., Calveley, P.M., Malone, J.D., Wallace, M.R., Bowdish, K.S. Human antibodies from immunized donors are protective against anthrax toxin in vivo, *Nat. Biotechnol.* **2003**, *21*, 1305–1306.

112 Biodesign International, http://www.biodesign.com, **2004**.

4
NMR Characterization Techniques – Application to Nanoscaled Pharmaceutical Carriers

Christian Mayer

4.1
Introduction

The use of nanoparticles for biomedical applications has been widely discussed in recent years. Among their wide variety, organic nanoparticles in stable aqueous dispersions are most promising candidates for pharmaceutical carrier systems. Controlled release and drug targeting are the main issues motivating their development [1–6]. In many cases, the preparation of these dispersions is quite straightforward, even on a technical scale. However, the analysis and detailed characterization of nanoparticle systems still represent a considerable challenge for common analytical methods. Most problems are connected to the fact that the particle structures are evidently nanoscopic, sensitive to many types of sample preparation techniques and often represent inhomogeneous and quite complex systems. Different approaches of transmission and electron microscopy can be applied successfully, but are often complicated by the necessity of tedious sample preparation. Atomic force microscopy is an important tool for studies on the particle assembly, but, as for electron microscopy, this method asks for isolated particles on a surface, a condition that may easily induce the formation of artifacts and destroy sensitive structures.

With the desire to analyze the particles in their "natural habitat", the liquid dispersion, researchers are left to choose from a selection of spectroscopic methods. While most spectroscopic measurements do not differentiate between solid and liquid components, i.e., the particles and the surrounding or encapsulated liquid, nuclear magnetic resonance spectroscopy is a notable exception. Its unique ability to simultaneously detect chemical structure together with the lateral and rotational molecular mobility of every single system component makes it perfectly suitable for the study of particle dispersions [7–15]. This particular advantage of NMR spectroscopy is especially helpful in studying pharmaceutical carrier systems [16–22]. As most NMR methods are run under extremely mild physical conditions, measurements on nanoparticle dispersions may be extended over long periods under a wide variety of chemical and thermal influences. This allows for long-term time-resolved observations and opens up the possibility for continuous monitoring of

Nanotechnologies for the Life Sciences Vol. 3
Nanosystem Characterization Tools in the Life Sciences. Edited by Challa S. S. R. Kumar

ISBN: 3-527-31383-4

slow release and degradation processes. In most cases, the influence of the NMR measurement on the particle system can be completely neglected. The main disadvantage of NMR spectroscopy, its significant lack of sensitivity, is compensated by increasing the particle concentrations or sample volumes.

Clearly, NMR is not a routine technique for process or quality control and will always require special equipment and experimental skills. However, it may play a key role for the development and design of nanoscaled pharmaceutical carriers. Unlike any other single method, it yields comprehensive data on the structure and the function of these systems under a large variety of conditions. The combination between high resolution and solid-state NMR together with the application of pulsed field gradient (PFG) turns out to be especially promising. The experimental techniques are quite straightforward and can be performed on standard commercial NMR spectrometers. Analysis of the resulting data is often complicated by the fact that slow rearrangements on the nanometer scale, e.g. rotational diffusion of the particles, affect line shapes and require numerical simulation of the spectra. However, these phenomena offer the chance to study motions on the microsecond scale such as particle tumbling. Thus, NMR becomes a valuable tool for the time-dependent observation of processes like particle degradation or agglomeration.

In the following, examples are presented that demonstrate the power and versatility of NMR spectroscopy in its application to nanoparticle dispersions. They include studies on basic particle characteristics such as the nanoparticle structure, phase transitions of the particle matrix, exchange processes on the particle surface, permeability of nanocapsule walls, release and particle degradation. Analysis of the NMR data is performed using adapted simulation procedures based on rotational and lateral diffusion of nanoparticles and their individual system components.

4.2 Structural Analysis of Nanoparticles

The unique power of NMR spectroscopy lies in its ability to simultaneously detect the chemical nature and molecular mobility of individual chemical components in a complex system. Identification of the chemical composition of a substance in the liquid or solid phase is a well-known feature of NMR and is practiced on a routine basis. With solid components, it requires solid-state techniques, with magic-angle spinning (MAS) as the most prominent representative. In addition, broad line NMR techniques and relaxation measurements have been applied for the analysis of rotational molecular mobility. Recently, commercial NMR instruments have also acquired the capability of detecting lateral motion, a method that generally combines classical echo sequences with pulsed magnetic field gradients. All three approaches may be successfully combined for studies on dispersed nanoparticles.

In a spectrum of a static sample, all solid constituents (like the particle matrix and capsule walls) are easily identified by their wide spectral lines, whereas all liquid and dissolved components yield relatively narrow signals due to their

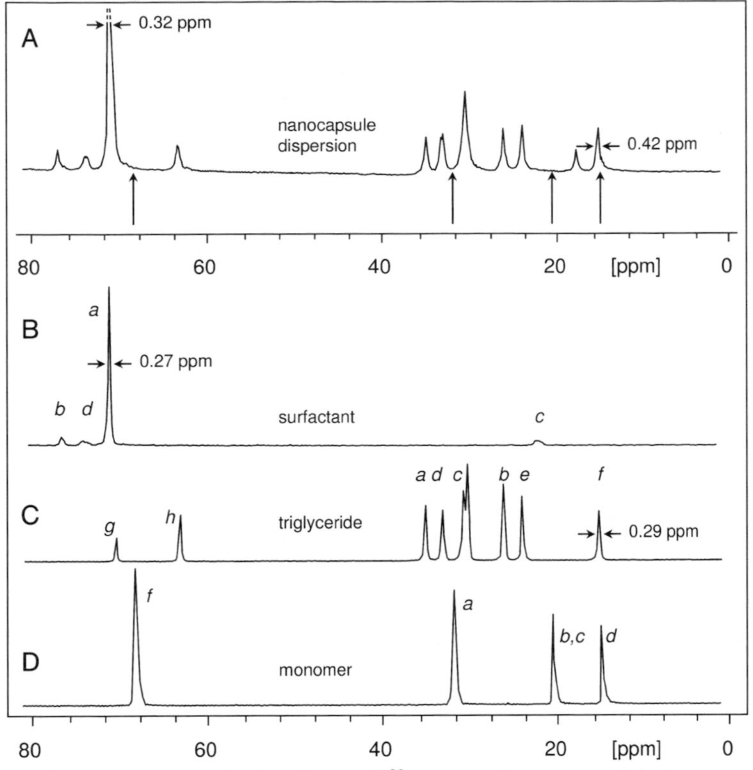

Fig. 4.1. Comparison of a directly excited ^{13}C spectrum of dispersed poly-n-butylcyano-acrylate nanocapsules (A) with the corresponding spectra of dissolved surfactant (B), of liquid triglyceride (C) and of a solution of the monomer used for capsule formation (D) [8]. Capsule spectrum A shows no traces of the resonances of spectrum D (arrows), proving the absence of any residual monomer after capsule formation. For assignment of the signals see Fig. 4.2.

rapid isotropic rotational diffusion. Due to their spectral widths, the signals of the solid components usually appear at very small amplitude and often remain undetectable.

Figure 4.1(A) shows an example of a directly excited ^{13}C spectrum of dispersed nanoparticles. It was obtained on an aqueous dispersion of poly-n-butylcyanoacrylate nanocapsules, a capsule system where the liquid core is formed by a commercial triglyceride (Miglyol ®) and which is stabilized by a block-copolymer surfactant (Synperonic F68®) [9, 23–25]. The spectrum lacks all traces of the polymer forming the solid capsule walls, but clearly shows all liquid and dissolved components [8]. A comparison with reference spectra of the individual system components (Fig. 4.1B and C) allows for the identification of all capsule constituents in the liquid phase. The complete absence of the monomer signals (spectrum D in Fig. 4.1) proves the completion of the polymerization process during capsule formation. The increased line widths in the presence of the capsules are assigned to partial immobilization of molecules on the capsule surface and the local inhomogeneity of the magnetic field.

To enhance the spectral contribution of the solid material, the (^{1}H)–^{13}C cross-polarization (cp) technique may be applied, which strongly amplifies signals from molecules with slow tumbling mobility [8, 17, 26] and therefore allows for the simultaneous observation of all constituents. The resulting (^{1}H)–^{13}C cp spectrum of the dispersion clearly shows relatively narrow signals superimposed on the characteristic wide line of the solid polymer (Fig. 4.2A). The broad contribution to the capsule spectrum is reproduced by a cp solid-state spectrum of a bulk sample of poly-n-butylcyanoacrylate (Fig. 4.2B). The positions of the narrow lines of the cap-

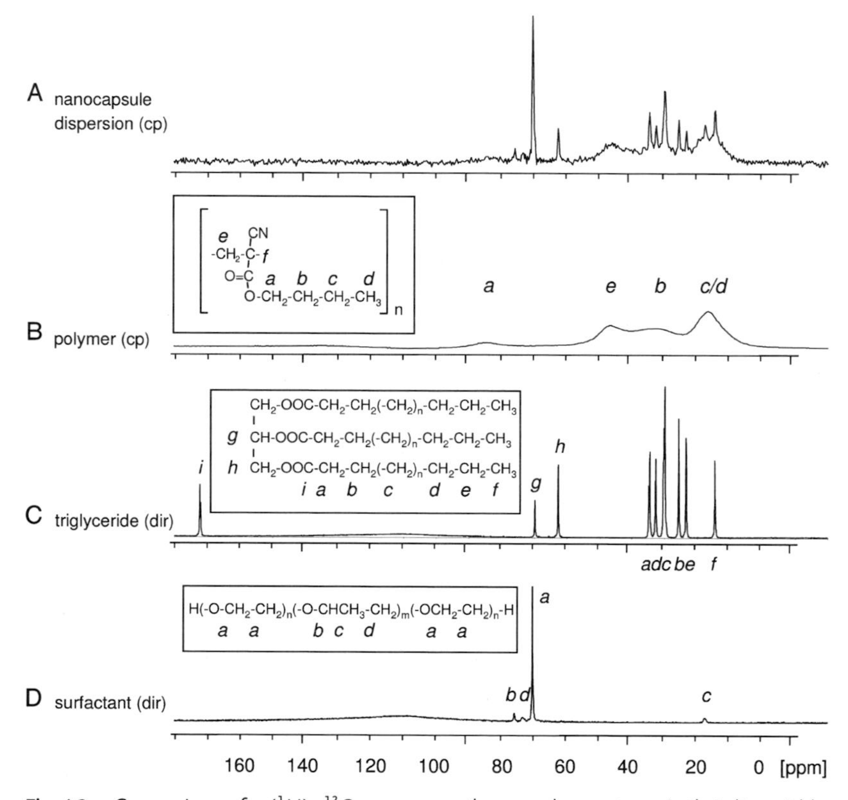

Fig. 4.2. Comparison of a (^{1}H)–^{13}C cross-polarization spectrum of dispersed poly-n-butylcyanoacrylate nanocapsules (A) with a corresponding spectrum of bulk solid poly-n-butylcyanoacrylate (B) and directly excited ^{13}C-spectra of liquid triglyceride (C) and of an aqueous solution of the surfactant (D). All system components (B–D) are detectable in the capsule spectrum A; their line widths depend strongly on their phase state (see text). The carbonyl signal of the triglyceride (signal *i* in spectrum C) and two carbon signals of the polymer remain invisible in cp spectra A and B due to lack of hydrogen at these positions. Signal assignments are given in the inserts.

sule spectrum are in good agreement with those of reference spectra of the liquid oil phase and the dissolved surfactant (Fig. 4.2C and D). Again, an increase in line width is detected in the presence of the capsules.

Clearly, identification of the solid components is difficult given the low resolution of the corresponding wide signals. Spectral amplitude and resolution can be improved significantly using sample spinning conditions such as magic-angle spinning (MAS), which has been widely used for bulk samples of solid polymers [27–29]. For particle dispersions, one generally faces the problem that a typical sample spinning process induces a gravitational field similar to the conditions of a centrifugation experiment [7]. However, if the dispersion is sufficiently stable, or the density of the particles is matched to the density of the liquid phase, the method can be applied to particle dispersions [7, 18].

Figure 4.3 shows an example for a sample-spinning experiment on a nanosized carrier system [18, 30]. The spectrum was obtained on an aqueous dispersion of

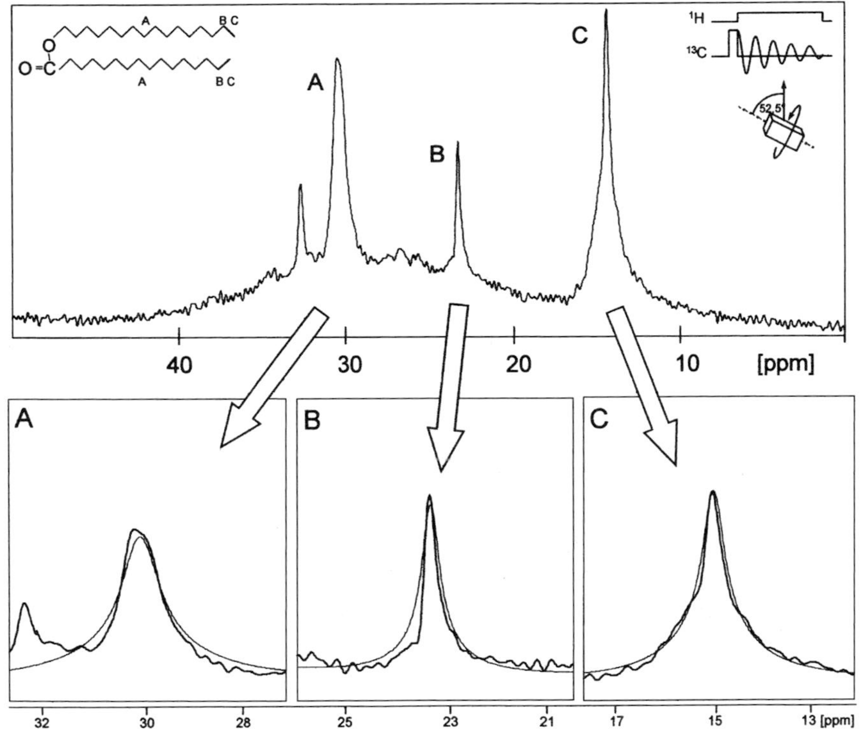

Fig. 4.3. Top: Result of a sample-spinning experiment of an aqueous dispersion of cetyl palmitate nanoparticles (SLN) obtained under direct excitation on ^{13}C nuclei. The sample was spun at a rate of $\omega_r/2\pi = 1660$ Hz and at an angle of 52.5°, which deviates from the magic angle. Signals A–C can be assigned to the central methylene, end methylene and end methyl groups, respectively. Due to incomplete relaxation, the integrals of the signals do not reflect the corresponding number of nuclei. Bottom: comparison between signal line shapes A–C and their simulated fits based on the expected rotational diffusion of the particles.

solid lipid nanoparticles (SLNs) that consist of cetyl palmitate nanospheres with diameters between 100 and 200 nm [31]. Sample spinning was performed at a spinning rate of $\omega_r/2\pi = 1660$ Hz and at an angle of 52.5°, which differs slightly from the magic angle (54.74°). Under these conditions, one obtains sufficient resolution for the assignment of the signals to the central methylene, end methylene and end methyl groups (signal A, B, and C, respectively, in Fig. 4.3). At the same time, the deviation from the magic angle leads to a characteristic variation of line shape that depends on the rate of rotational diffusion of the particles [18, 30]. Thus, the chemical composition of the particles can be studied together with their rotational diffusion. However, the influence of isotropic tumbling on signal line shapes is not at all straightforward. Analysis of the effect on the NMR spectrum requires a numeric simulation procedure that has been developed based on a finite element scheme: the time axis as well as the orientation in space is segmented into a set of discrete numbers and positions, such that the NMR time signal can be numerically calculated based on given sets of system parameters and experimental data [26, 32, 33]. The influence of rotational diffusion of the molecules is approximated by a corresponding spin exchange between adjacent "sites", with rate constants depending on the equilibrium population, orientation and rotational diffusion constant. After Fourier transformation, the simulation procedure yields sets of NMR spectra that can be compared with experimental ones. A typical example for the end methyl group of cetyl palmitate under the experimental parameters given above is shown in Fig. 4.4. It refers to correlation times of isotropic rotational diffusion from 15 ms (front) to 15 μs (back), showing the strong dependence of line shape on the rate of the motion [7, 18].

In some cases, the particle matrix undergoes local phase separations, leading to an inhomogeneous solid phase on the nanometer scale. On these occasions, analy-

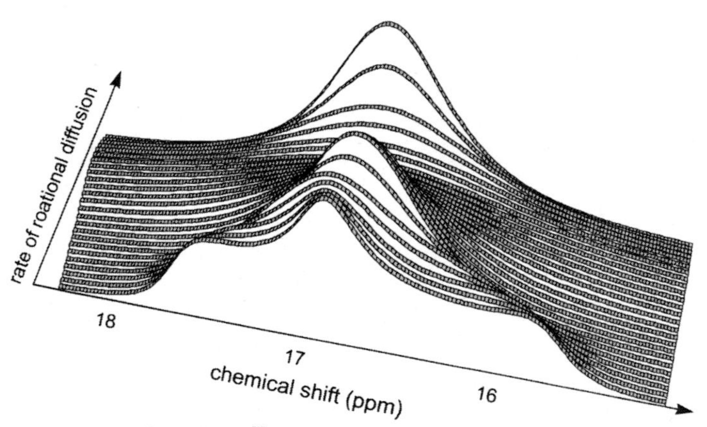

Fig. 4.4. Set of simulated ^{13}C spectra for the end methyl group of cetyl palmitate (signal C in Fig. 4.3) for different rates of isotropic rotational diffusion. The correlation time varies from 15 ms (front) to 15 μs (back). Rotational sidebands are weak and not included in the given frequency window [7, 18]. For all spectra, the simulated sample spinning has been adjusted to a frequency of $\omega_r/2\pi = 1660$ Hz. The spinning axis is tilted at 52.5° with respect to the direction of the external magnetic field.

sis of spin diffusion between different components of the matrix material allows for the detection of inhomogeneous distribution and the determination of domain sizes [21]. Due to spin diffusion between all protons in the solid phase, only a single common relaxation curve is observed for all proton signals of a homogeneous matrix domain. As soon as different domains with sizes exceeding approximately 5 nm are formed, it is possible to identify different decay curves for the proton spin–lattice relaxation ($T_{1\rho}$) in the rotating frame. Hereby, phase separations between a solid matrix and a solid active ingredient may be detected and quantified [21].

The methods introduced so far allow for the identification of chemical components together with their assignment to the liquid and to the solid phase. However, with nanocapsules, we often deal with two separate liquid compartments: the continuous (solvent) phase and the encapsulated liquid domain. The widths of the spectral signals or their development in a cross polarization spectrum does not differentiate between the two, as their rotational diffusion is within the same order of magnitude. The only physical property that differs between the two liquid compartments is represented by the characteristics of lateral self-diffusion. While molecules of the continuous phase undergo almost free self-diffusion, lateral motion of the encapsulated counterparts is strongly hindered by the capsule walls. Let us assume that we can observe the average lateral dislocation of the molecules in the liquid phase. If the observation period is very short, the difference between free and encapsulated becomes almost negligible as the number of molecules that collide with the capsule boundaries is very small. On extending the observation period, the number of wall collisions increases and the difference between the inner and the outer phase becomes increasingly evident. Finally, the residual mobility of the molecules in the capsule content appears to be identical to the mobility of the capsules as a whole.

In NMR, this theoretical consideration is experimentally reproduced by application of the pulsed field gradient (PFG) technique. In combination with echo sequences, like the Hahn or the stimulated echo, it allows the determination of mean-square displacements over given time intervals Δ, which can be varied over a wide range [34–37]. This method has served for the characterization of heterogeneous systems such as emulsions [36, 38, 39]. Figure 4.5 shows an example for the application to poly-n-butylcyanoacrylate nanocapsules [9]. The basic pulse sequence (insert in Fig. 4.5) consists of a so-called stimulated echo sequence that produces a proton echo signal after three pulses of equal duration. In combination with two gradient pulses (FG) of a given duration (δ), the amplitude of the signal (and of the corresponding line spectrum) depends on the gradient strength and on the average lateral dislocation during the gradient pulse spacing Δ. While the dependence on the gradient strength G_{max} allows one to determine the average dislocation pattern, the variable Δ determines the time window for the observation. The resonance signals shown in Fig. 4.5 correspond to different liquid or dissolved components in the capsule system: water, ethanol, triglyceride and the block-copolymer surfactant. With increasing gradient strength G_{max}, the signals of water and ethanol show rapid decay, indicating free, efficient lateral self-diffusion. The

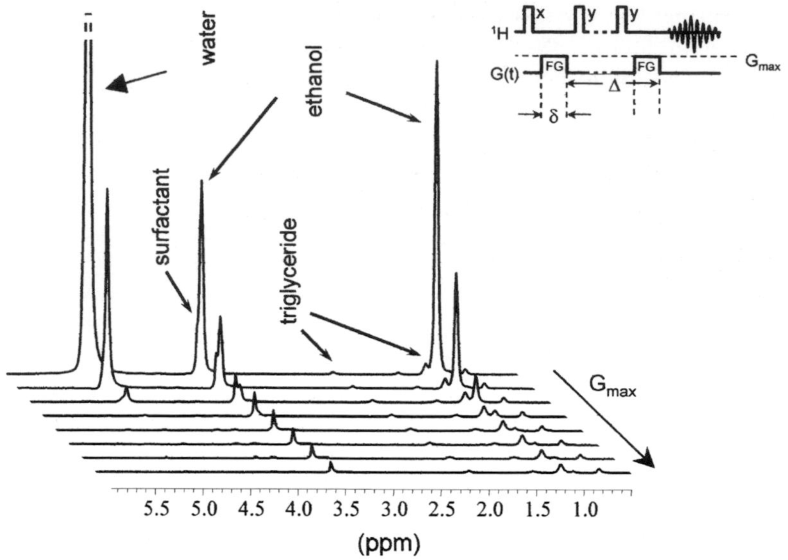

Fig. 4.5. Set of proton spectra obtained in a pulsed field gradient experiment on a dispersion of poly-n-butylcyanoacrylate nanocapsules. The pulse sequence (insert) consists of a stimulated echo sequence on protons combined with two gradient pulses FG. All signals in the displayed region can be assigned to liquid or dissolved system components: water, ethanol, triglyceride and surfactant. The dependence of signal intensity on gradient strength (G_{max}) indicates lateral dislocation of the molecules within the given time window of $\Delta = 30$ ms. Efficient self-diffusion leads to strong decays of the signals for water and ethanol, while those of the large surfactant molecules and the encapsulated oil show much smaller dependencies [9].

signal for the surfactant exhibits a much shallower decay, caused by slower self-diffusion connected to its much larger molecular mass. An even smaller dependence is observed for the triglyceride signal, which almost stays constant on the given scale. As the molecular mass of the triglyceride is much smaller than that of the surfactant, this observation strongly hints that the triglyceride oil is enclosed inside the capsule walls.

This becomes more evident if the echo decay of the triglyceride signal at 2.1 ppm is studied under variation of the diffusion time Δ. Figure 4.6 shows a plot of the negative logarithmic echo intensity ($-\ln I_{rel}$) vs. the parameter $\gamma^2 G_{max}^2 \delta^2 (\Delta - \delta/3)$, with γ standing for the gyromagnetic ratio of the observed nucleus [9]. This type of plot yields linear dependencies with slopes identical to the apparent self-diffusion constant D_{app}. Evidently, the resulting apparent diffusion constant strongly depends on the diffusion time Δ (Fig. 4.7). For very short observation periods ($\Delta < 10$ ms), it comes close to the value of 2.3×10^{-11} m^2 s^{-1} found for a bulk triglyceride sample [9]. On this time scale, we observe local dislocations where the presence of the capsule walls affects only a small fraction of the oil molecules. Consequently, there is little difference between the behavior of free and encapsulated oil. On increasing

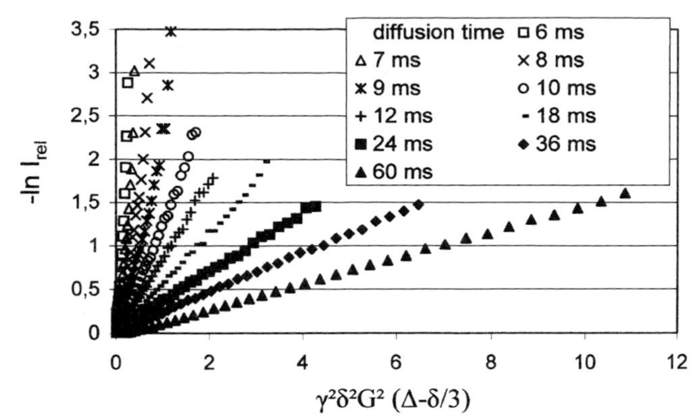

Fig. 4.6. Reciprocal echo decay plots for the triglyceride signal from an aqueous dispersion of poly-n-butylcyanoacrylate nanocapsules. Negative logarithmic echo intensities are plotted vs. the parameter $\gamma^2 G_{max}^2 \delta^2(\Delta - \delta/3)$ as derived from the triglyceride ^{1}H signal at 2.1 ppm (Fig. 4.5). As typical for restricted diffusion, the observed slopes depend strongly on the diffusion time Δ. Apparent diffusion constants $D_{app}(\Delta)$ are determined from the slopes of the individual plots (Fig. 4.7).

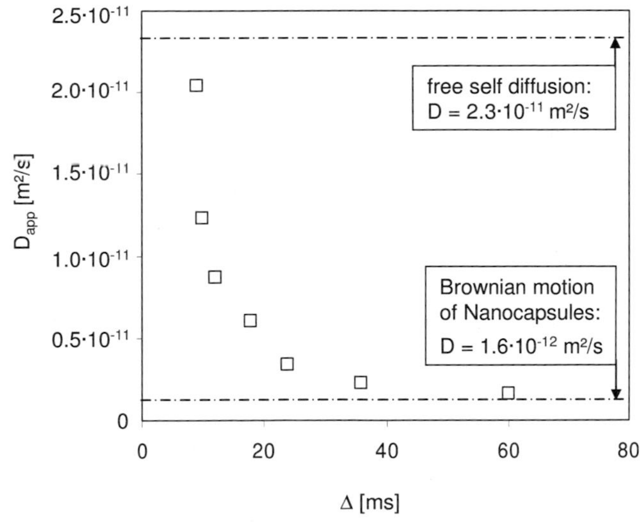

Fig. 4.7. Dependence of the apparent self-diffusion constant (D_{app}) of the triglyceride oil component on the observation time interval Δ (diffusion time). For short Δ, D_{app} comes close to the self-diffusion coefficient of the bulk oil ($D = 2.3 \times 10^{-11}$ m^2 s^{-1}, top line). For long diffusion times, D_{app} approaches the diffusion constant that describes the Brownian motion of the capsules in aqueous medium ($D = 1.6 \times 10^{-12}$ m^2 s^{-1}, bottom line).

diffusion time, the apparent diffusion constant decreases significantly and finally approaches the value for the Brownian motion of the capsules in the aqueous dispersion (Fig. 4.7) [9]. In this case, the lateral shift of the encapsulated molecules becomes seriously restricted by the capsule walls. Finally, for $\Delta > 50$ ms, the position of all oil molecules average out in the capsule center and we observe the mobility of the complete capsules.

The similarity between the apparent oil diffusion constant for short time intervals with that observed for the bulk oil can serve as proof that we actually deal with "hollow" capsules with obviously no diffusion barriers in the internal volume. This important structural information could hardly have been obtained with other techniques. In summary, the NMR approach allows chemical identification of the solid matrix and the external liquid as well as the liquid-encapsulated components. In addition, it yields data about the molecular dynamics of the system, from which further conclusions on the particle structure can be derived.

4.3 Phase Transitions of the Particle Matrix

Even if we restrict ourselves to solid particle matrices, we still have to consider certain varieties of the solid phase. A prominent example is the gel phase formed by extensively swollen polymers, which is often encountered when particles are formed by polymer precipitation [40]. This gel phase may be reversibly or irreversibly transformed into a "real" solid state by removal of the solvent. An example of an irreversible phase transition is represented by poly-ε-caprolactone nanoparticles formed by the emulsion-diffusion process [20, 41] (Fig. 4.8). Preparation starts with the polymer being dissolved in the organic phase of an o/w emulsion, the continuous water phase being saturated with the organic solvent ethyl acetate. In the next step, the aqueous phase is diluted, causing the ethyl acetate to diffuse from the droplets into the surrounding water. Hereby, the polymer, which is dissolved in ethyl acetate at the maximum concentration, precipitates and forms nanospheres (Fig. 4.8, left-hand column). In the presence of a hydrophobic oil component, the process leads to nanocapsules with the oil forming the liquid core (Fig. 4.8, right-hand column). In both cases, the organic solvent is subsequently removed by evaporation.

Initially, the solid matrix of the particles consists of a physically crosslinked network that is swollen by water and some residual ethyl acetate. A solid-state ^{13}C NMR spectrum of the freshly prepared nanospheres (Fig. 4.9, top) reveals a very high degree of local mobility for the polymer chains. Compared with a solid-state spectrum of the bulk polymer (Fig. 4.9, bottom), the spectrum of the particles exhibits relatively narrow lines that are consistent with rotational diffusion on the microsecond time scale [7]. This means that, on a local scale, most of the polymer behaves as though in a dissolved state while it still forms the framework of a solid particle on the sub-micrometer size range. This apparent contradiction is explained

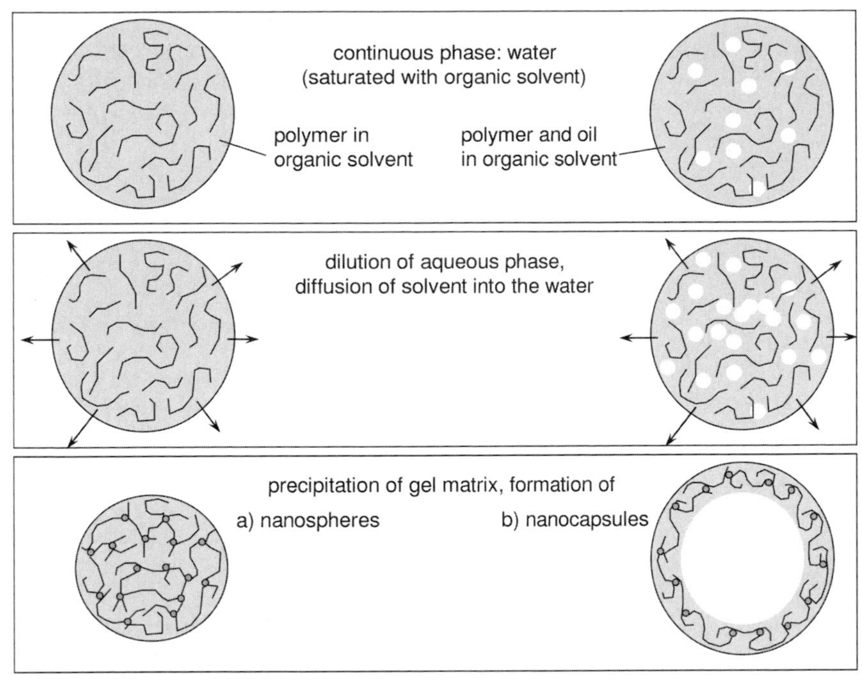

Fig. 4.8. Schematic representation of processes leading to the formation of gel phase nanospheres (left) and nanocapsules (right) by the emulsion-diffusion process. Top: a polymer dissolved in an organic solvent (optionally containing an oil component, right-hand column) is dispersed in an aqueous phase. Center: water is added to the dispersion, causing the organic solvent to diffuse into the aqueous phase in connection with reestablishment of the distribution equilibrium. Bottom: polymer precipitation is induced by loss of solvent in the organic phase, forming gel nanospheres (a) or gel nanocapsules (b). In both cases, the polymer becomes physically crosslinked and is swollen beyond the point that accords with the thermodynamic swelling equilibrium [7, 20].

by assuming a gel state of the polymer induced by a high water content that is much beyond the value expected from the swelling equilibrium.

On removing the water, an irreversible phase transformation is observed (Fig. 4.10). Careful evaporation under lowered pressure induces the appearance of an increasing contribution of polymer in the classical solid state. At the same time, the process is accompanied by a slight decrease in particle diameter [20]. The transformation is irreversible, as addition of water does not lead back to the gel state. Alternatively, removal of water from the particle matrix can be induced by simple freezing of the sample [7, 20]. At all stages of water removal, the amount of polymer in both phase states can be quantitatively determined by solid-state NMR [7].

This phenomenon opens up the possibility for an interesting application of gel particles: the matrix of nanospheres or nanocapsules in the gel state may be re-

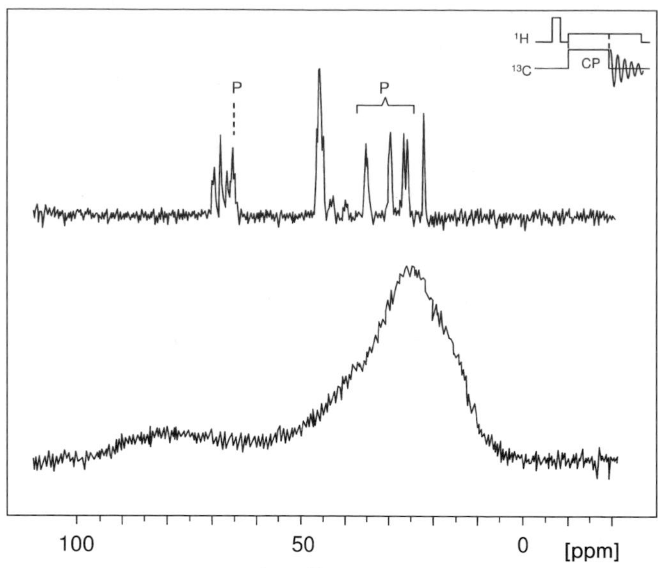

Fig. 4.9. Comparison of a (^{1}H)–^{13}C cross-polarization spectrum of freshly prepared poly-ε-caprolactone nanospheres (top) with a corresponding spectrum of the bulk polymer (bottom). Obviously, the dispersed particles do not contain organic material in the classical solid state [7, 20]. A high degree of local chain mobility leads to narrow signals that can be assigned to the polymer (P). All other peaks derive from the surfactant [poly(vinyl alcohol)] and small amounts of poly(vinyl acetate). The insert indicates the pulse sequence used.

garded as an "open" carrier system, allowing ingredients to diffuse into or through the matrix. After the particles are loaded, they may be irreversibly "sealed" by a simple freezing step that turns the gel matrix into a relatively dense, impermeable solid polymer, a state that is preserved after melting the dispersion. The whole process of phase transition of the polymer and encapsulation of an active ingredient may be conveniently monitored by NMR spectroscopy.

4.4
Adsorption to the Particle Surface

For medical applications of nanoparticles, it is of special interest to elucidate adsorption phenomena on the external particle surface. The most important adsorbent is the surfactant used to stabilize the particle dispersion. Additional components, such as encapsulated ingredients, may also adsorb to the particle surface. Surface adsorption usually means an equilibrium state, where adsorption and desorption processes take place simultaneously at equal rates. NMR offers an experimental approach for the study of adsorption phenomena and the molecular exchange connected to a continuous adsorption–desorption process [7, 17].

The use of cross-polarization in solid-state NMR generally favors the spectral

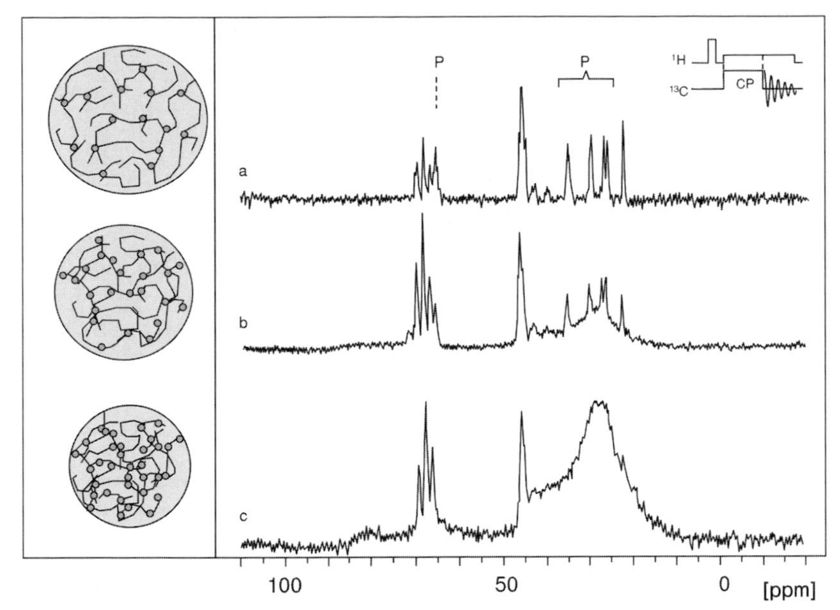

Fig. 4.10. (^{1}H)–^{13}C cross-polarization spectra of dispersed poly-ε-caprolactone nanospheres in different stages of irreversible dehydration. With the original water content after the emulsion-diffusion process (a), all polymer signals (P) indicate a high degree of local mobility. After removal of ca. 50% of the swelling water (c), the complete polymer content has transformed into a rigid state. Spectrum (b) represents an intermediate situation [7, 20]. Pictograms on the left symbolize the corresponding stages of network formation.

contribution of solid constituents. It requires reduced mobility of the molecular environment and is therefore quite inefficient for dissolved molecules. However, there is a notable exception: if dissolved molecules become temporarily adsorbed to a solid surface, they remain relatively immobilized for a time interval in the millisecond range, which is sufficient to efficiently allow for the polarization transfer, e.g., between protons and carbon nuclei. The small adsorbed fraction is usually not directly detectable by NMR. However, after the molecules desorb, they again undergo rapid rotational diffusion and their cross-polarization induced carbon signal becomes a dominant contribution to the spectrum (Fig. 4.11).

In an aqueous dispersion of poly-n-butylcyanoacrylate nanocapsules, this phenomenon is observed for the block-copolymer surfactant as well as, to a somewhat smaller degree, for the encapsulated oil [7, 17]. Indications for an unusual mechanism of cross-polarization are obvious from the characteristics of cross-polarization development. While the solid-state signal of the nanocapsule wall material is already fully developed after a cross-polarization period of 0.1 ms, the narrow signals of the triglyceride and the surfactant still gain intensity after 1 ms and reach their maximum only after 5 ms [7]. After 10 ms of cross-polarization, the polymer signal is almost fully relaxed, while the signals for the triglyceride and the surfactant re-

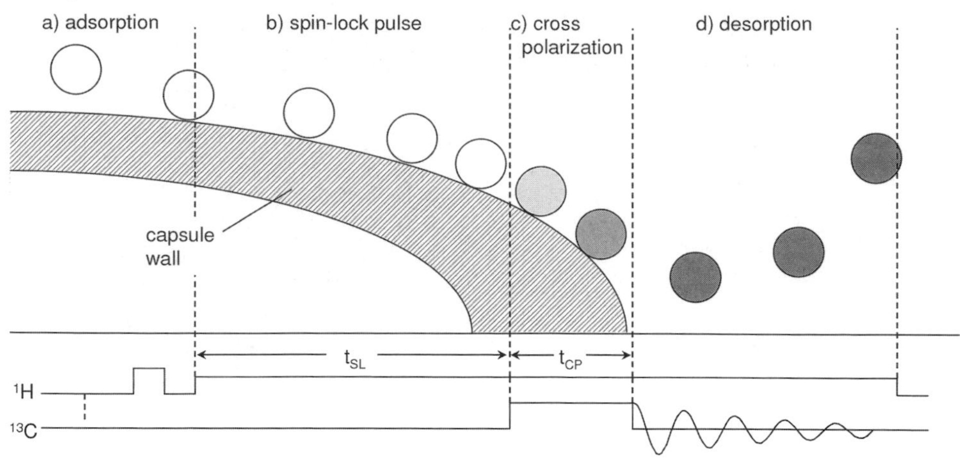

Fig. 4.11. Schematic representation of the process of (^{1}H)–^{13}C cross-polarization induced by temporary adsorption of dissolved molecules to a nanocapsule wall. The degree of polarization is indicated by the intensity of the grey shading of the symbol. First, we focus on a molecule absorbed shortly after an initial pulse is set to the proton nuclei (a). A spin-lock pulse may be included at this time to study spin–lattice relaxation [7, 17] (b). Later, during the period of adsorption, simultaneous pulses on ^{1}H and ^{13}C nuclei are adjusted to fulfill the Hartmann–Hahn condition, which allows for an efficient (^{1}H)–^{13}C polarization transfer within the immobilized molecules (c). Finally, the molecule is desorbed, taking the strong polarization of the ^{13}C nuclei into the fluid state, where it can be easily detected [7, 17] (d).

main constant for more than 30 ms. Such a cross-polarization and relaxation pattern is quite uncommon and indicates a special mechanism of polarization transfer.

Studies under variation of the spin-lock time have been performed to elucidate this polarization phenomenon, assuming an adsorption–desorption cycle on the capsule walls (Fig. 4.12). For comparison, corresponding relaxation curves obtained under direct excitation have been measured that focus on the triglyceride and surfactant molecules in the liquid phase. In this case, all data could be fitted assuming a mono-exponential decay and long relaxation times between 34 and 79 ms for the

Fig. 4.12. Relaxation curves showing the dependence of two aliphatic signals of triglyceride oil (top and center) and a single aliphatic signal of the surfactant (bottom) on the duration of a spin-lock period. With direct excitation experiments (left-hand column), which focus on molecules in the liquid phase, the spin-lock period follows the initial pulse on the protons. All resulting relaxation curves can be fitted mono-exponentially. For cross-polarization experiments (right-hand column) the spin-lock pulse is set before the cp period (see phase b in Fig. 4.11) and the signal intensity is observed via the corresponding carbon signals. The resulting relaxation curves are bi-exponential and consist of an initial steep decay, which corresponds to immobilized molecules (dotted line), and a subsequent shallow decay that can be assigned to molecules in the liquid phase (dashed line). Pulse sequences are indicated schematically at the top of each column [17].

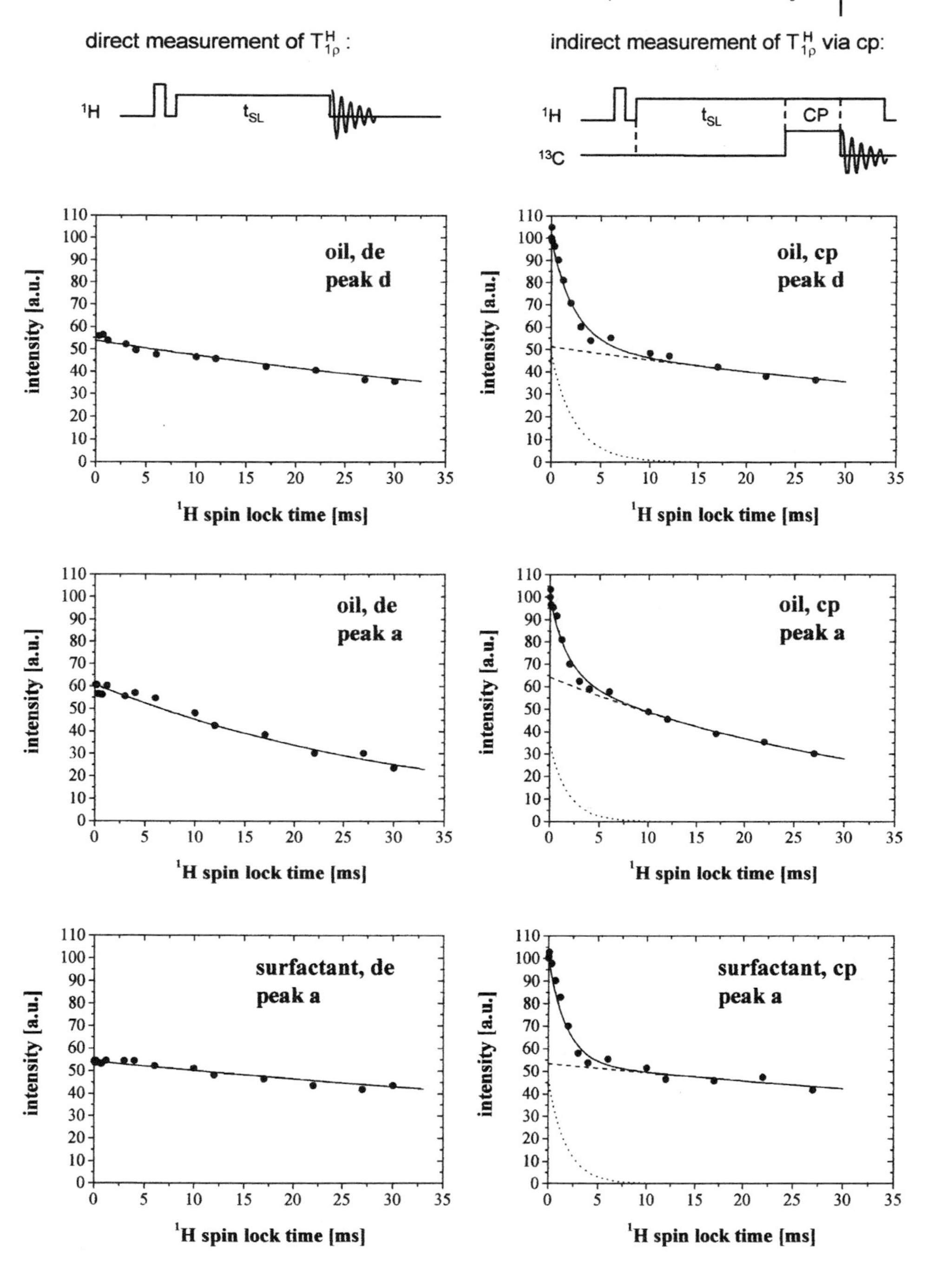

direct measurement of $T_{1\rho}^{H}$:
indirect measurement of $T_{1\rho}^{H}$ via cp:
^{1}H
t_{SL}
CP
^{13}C
oil, de
peak d
oil, cp
peak d
oil, de
peak a
oil, cp
peak a
surfactant, de
peak a
surfactant, cp
peak a
intensity [a.u.]
^{1}H spin lock time [ms]

triglyceride oil and 131 ms for the strongest signal of the surfactant [17]. With cross-polarization experiments (right-hand column in Fig. 4.12), the spin-lock pulse is set prior to the cp period (see Fig. 4.11b) and the signal intensity is observed via the corresponding carbon signals. No significant change in the resonance line shapes is observed during the full course of the experiment [17]. However, all resulting relaxation curves are distinctly bi-exponential and consist of an initial steep decay and a following shallow section of the relaxation curve. The latter is very similar to the results obtained by direct excitation with relaxation times that differ only slightly from those mentioned above [17]. This part clearly corresponds to the fraction of molecules in the desorbed state. The initial steep decay is connected to relaxation times of 1–2 ms, which resemble those of the solid polymer. This contribution obviously reflects the fraction of molecules that are in the adsorbed state. The complete signal decay curves from the cp experiment can be analyzed for adsorption–desorption characteristics using a relatively simple expression for relaxation under molecular exchange [42].

Figure 4.13 summarizes the results of the best fit to the complete set of relaxation data. They reveal that the average duration (τ_{ads}) of the adsorption step of an individual molecule is around 1 ms for the surfactant and 3 ms for the oil component. The overall signal intensities for both components consist of the signal

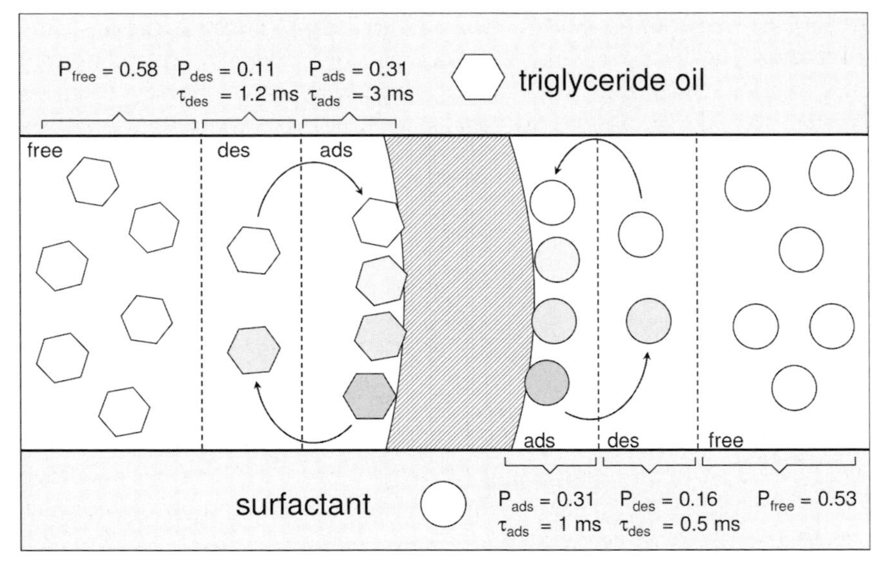

Fig. 4.13. System parameters obtained from a least-squares fit on the data shown in Fig. 4.12, based on an analytical expression for relaxation under molecular exchange [42]. τ_{ads} and τ_{des} for the triglyceride oil and the surfactant refer to the average duration of adsorbed and desorbed states, respectively, of molecules taking part in the adsorption–desorption cycle. P_{ads}, P_{des}, and P_{free} stand for signal contributions of the adsorbed, desorbed and "free" molecules, respectively. The latter fraction derives from those molecules in the liquid phase that do not take part in the adsorption–desorption cycle during the observation period.

fraction P_{ads} of the adsorbed molecules, the signal fraction P_{des} of the desorbed molecules, and finally the fraction P_{free} of the molecules that take no part in the adsorption–desorption cycle during the observation period. Notably, these fractions are relative signal contributions and do not reflect the actual number of molecules involved in these processes. The resulting average residence times of molecules on the particle surface compare well with results on similar systems obtained with pulsed field gradient NMR [43, 44]. Using comparable non-ionic surfactants, residence times between 7.0 and 13.4 ms were detected. These values were obtained with much lower surfactant concentrations than for the results illustrated in Fig. 4.13; it was also shown that the residence time decreases with increasing surfactant concentration. Therefore, the value of 1 ms found for the highly concentrated block-copolymer on the poly-n-butylcyanoacrylate nanocapsules is within the expected range.

Principally, cross-polarization NMR (as well as pulsed field gradient experiments) offers another promising experimental approach for the study of adsorption phenomena on surfaces. In contrast to field gradient techniques where lateral diffusion is the key parameter, it distinguishes adsorbed molecules from molecules in the fluid phase by their degree of rotational diffusion. As the immobilized fraction is only observed indirectly after desorption into the fluid phase, it is generally more sensitive in its detection. However, there is also a significant disadvantage. As cross polarization is involved, there is no possibility of quantitative analysis of the adsorbed and desorbed molecules. Conversely, the cp experiment gives access to the time scale of the adsorption–desorption cycle of all system components. Such adsorption–desorption phenomena on the nanoparticle surface are of key importance for the primary physiological response to nanoparticles when applied to living organisms.

4.5 Molecular Exchange through Nanocapsule Membranes

According to the nomenclature of Kreuter [40], nanocapsules are nanoparticles that form a shell-like wall, encapsulating a generally liquid content. Nanocapsule walls can be very thin, with thicknesses down to few nanometers. For polymer capsule walls, this corresponds to only a few molecular layers. Obviously, this can not be considered as an impermeable barrier for compounds of small molecular weight. At the same time, the main purpose of nanocapsules is based on their ability to separate two fluid domains and to block the diffusion of encapsulated components into the continuous phase. Therefore, the capsule wall permeability represents a key parameter for the characterization of the capsule system.

A very straightforward approach for a physico-chemical determination of the capsule wall permeability involves fluorescent labels and the use of a dialysis cell. In many cases, however, transfer through the capsule membranes is too rapid to be followed in this manner. On these occasions, NMR offers an extremely versatile and reliable technique for the study of trans-membrane exchange in capsule dis-

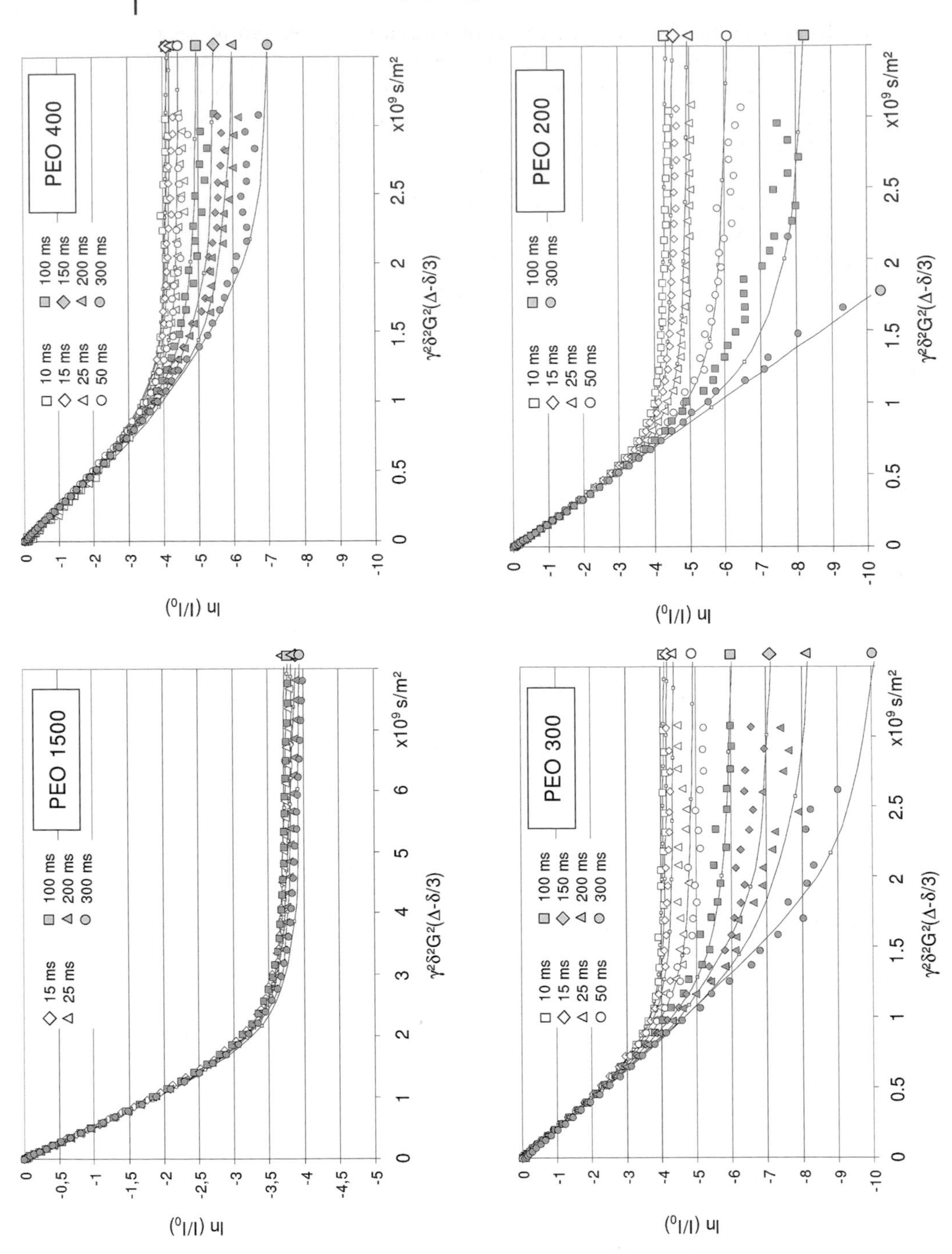
PEO 1500
15 ms
25 ms
100 ms
200 ms
300 ms
ln (I/I0)
γ²δ²G²(Δ-δ/3)
x10⁹ s/m²
PEO 400
10 ms
15 ms
25 ms
50 ms
100 ms
150 ms
200 ms
300 ms
ln (I/I0)
γ²δ²G²(Δ-δ/3)
x10⁹ s/m²
PEO 300
10 ms
15 ms
25 ms
50 ms
100 ms
150 ms
200 ms
300 ms
ln (I/I0)
γ²δ²G²(Δ-δ/3)
x10⁹ s/m²
PEO 200
10 ms
15 ms
25 ms
50 ms
100 ms
300 ms
ln (I/I0)
γ²δ²G²(Δ-δ/3)
x10⁹ s/m²

persions. It relies on the fact that free and encapsulated molecules can be separated by the characteristics of their lateral diffusion (Section 4.2). This criterion can be used to follow release processes in a time-resolved manner, similar to experiments using fluorescent labels. In addition, for rapid exchange through capsule membranes, this method offers an especially promising alternative. It allows for the observation of trans-membrane exchange of a tracer molecule in the equilibrium state, a method that has been successfully applied to nanocapsules [9], micellar carriers [45, 46] and vesicles [47].

The basic idea of this experimental approach may be shown by the example of block-copolymer vesicles [48, 49]. The result of a pulsed field gradient (PFG) measurement on vesicles from poly(2-vinylpyridine-*block*-ethylene oxide) [47] is represented in Fig. 4.14. The external as well as the internal domain are completely identical in their chemical composition and mainly made up from water. Polyethylene oxide (PEO) molecules of variable molecular mass are used as water-soluble tracers to sample the permeability of the vesicle membrane. Figure 4.14 displays four of the resulting echo decay curves (cf. Figs. 4.5 and 4.6), showing dramatic changes of the exchange pattern depending on molecular mass of the PEO tracer. All echo intensities were obtained from the proton signal at 3.2 ppm and are plotted logarithmically in relation to the original echo intensity I_0. In the absence of trans-membrane exchange, interpretation of the plots is quite straightforward. In general, the steep initial decay corresponds to the free molecules, while the encapsulated fraction of the tracer is represented by the subsequent plateau. The level of the plateau indicates the fraction of encapsulated molecules, e.g., a $\ln(I/I_0)$ of -4 corresponds to a relative fraction of $\exp(-4) = 0.0183$, which means that 1.83% of the tracer is in the encapsulated state. The slope of the initial decay is identical to the negative self-diffusion coefficient of the tracer in the continuous phase, while the slope of the final plateau is dominated by the (relatively slow) Brownian motion of the vesicles themselves.

For tracers with high molecular mass (e.g., PEO-1500 with $M_w = 1500$ g mol^{-1}), the result of the PFG measurement basically follows this simple interpretation. As expected from impermeable vesicle membranes, the decay curves show no significant dependence on the diffusion time: for $\Delta = 15$ to 300 ms, the plots are more or less identical. With both fluid domains strictly separated, the experimental result is not affected by a change in observation period. This result changes dramatically for a PEO tracer with $M_w = 400$ g mol^{-1}. Here, the plateau value decreases with Δ. This is due to the relatively high permeability of the vesicle membrane for PEO-400. With increasing time period (Δ), more and more of the initially encapsulated

Fig. 4.14. Experimental and simulated echo decay curves for a block-copolymer vesicle dispersion containing poly(ethylene oxide) tracer molecules of different molecular mass: PEO-1500 ($M_w = 1500$ g mol^{-1}), PEO-400 ($M_w = 400$ g mol^{-1}), PEO-300 ($M_w = 300$ g mol^{-1}), and PEO-200 ($M_w = 200$ g mol^{-1}). Echo intensities are plotted as $\ln(I/I_0)$ vs. $\gamma^2\delta^2G^2(\Delta - \delta/3)$ such that the resulting slope is equivalent to the negative self-diffusion coefficient for free diffusion. Calculated curves (solid lines) are assigned by the symbols on the right-hand axis. Table 4.1 lists the parameters for the simulated curves [47].

PEO molecules will be exchanged with those from the external domain, which already have suffered a large lateral dislocation. Therefore, the resulting slope is determined by a certain mix of the external self-diffusion constant, the apparent constant for the restricted diffusion in the confined space and the Brownian motion of the vesicles. The influence of the exchange is increasingly relevant with increasing observation period Δ. As expected, this dependence is even more dominant for PEO-300 and PEO-200, with $M_w = 300$ and 200 g mol^{-1}, respectively. Finally, with PEO-200 and $\Delta = 300$ ms, the rapid exchange does not allow for an experimental distinction between external and encapsulated tracer molecules and the encapsulated fraction is not detectable.

All in all, detailed analysis of the PFG data yields extensive information on the exchange characteristics of the different tracer molecules. However, such an analysis is not always straightforward. For very small nanocapsules where, to a first approximation, the internal diffusion may be neglected it is possible to use an analytical approach for molecular exchange between two domains with different signal decay properties [9, 45, 46], similar to that developed for magnetic relaxation [42]. If internal self-diffusion of the encapsulated tracer is significant, restricted diffusion within the nanocapsules has to be accounted for. Analysis of PFG NMR measurements for restricted diffusion in a spherical confinement has been treated analytically [50–52]. However, in combination with molecular exchange through the capsule walls, there is no simple solution to this problem [53]. This again asks for a numerical analysis approach that is once more based on a finite element approximation [7, 47]. In this case, the external as well as the internal volume is segmented into small finite elements in space along the axis of the field gradient. Lateral diffusion is then approximated by an exchange of spins between adjacent space elements with rates defined by the diffusion constant. An additional spin exchange between space elements for the internal and those for the external domain accounts for the simultaneous transfer of molecules through the capsule walls [47]. Except for the exchange of spin contributions, all processes are assumed to be in steady state equilibrium, therefore the net flow during the simulated tracer diffusion is zero.

The result of a simulated best fit to the experimental data is shown by solid lines in Fig. 4.14 [47]. Basically, all plots for a given PEO tracer are fitted by a single parameter that is the average exchange rate constant $\bar{k}_{ex}$. The encapsulated fraction x_{in} of the tracer, which is identical to the partial volume of the vesicles, can be determined from the echo decay plots as well as by independent measurements. The complete set of simulation data (including one for PEO-600, which is not shown in Fig. 4.14) is listed in Table 4.1. As expected, the trans-membrane exchange rate decreases with increasing molecular mass of the tracer. For the given mass range, the rates vary between 0.05 and 1.6 s^{-1}, which is equivalent to half-life times of the encapsulated state between 13.8 s and 430 ms. Of course, all molecules that leave the encapsulated domain are simultaneously replaced by molecules from the external domain in an equilibrium exchange. With exchange rates in the given range, the equilibrium state between the internal and the external domain is sufficiently established within minutes after addition of the tracer. So the actual experiment is

Tab. 4.1. Simulation parameters used for the numerical reproduction of the echo decay curves shown in Fig. 4.14 (including data for PEO-600) [47].

Tracer molecule	*Approx. molar mass ($g\ mol^{-1}$)*	*Encapsulated fraction (x_{in})*	*Exchange rate ($\bar{k}_{ex}$) (s^{-1})*
PEO-200	200	0.022 ± 0.001	1.6 ± 0.2
PEO-300	300	0.023 ± 0.001	0.85 ± 0.05
PEO-400	400	0.0205 ± 0.001	0.38 ± 0.02
PEO-600	600	0.015 ± 0.001	0.075 ± 0.01
PEO-1500	1500	0.028 ± 0.001	0.05 ± 0.01

kept quite simple: the tracer is added to a given vesicle dispersion, the PFG-NMR experiment is started after a waiting period of approximately one hour. After completion of the measurement, the tracer A can be removed by successive washing/dialysis cycles and be replaced by a tracer B.

A systematic study of wall permeability using a series of tracers plays a significant role in a general characterization of capsule membranes. For block-copolymer vesicles, it is used to understand the mechanism of the molecular transfer through the vesicle walls. Figure 4.15 shows a logarithmic plot of the average exchange constant $\bar{k}_{ex}$ vs. hydrodynamic radius of the PEO molecules in aqueous solution [47]. For molecular masses between 200 and 600 g mol^{-1}, the plot shows a roughly linear dependence between log $\bar{k}_{ex}$ and the hydrodynamic radius. Based on the Arrhenius law, this indicates that the activation energy of the molecular transfer through the vesicle membrane is proportional to the radius of the molecule. Possibly, the energy barrier for the transfer is connected to a local disruption of the membrane

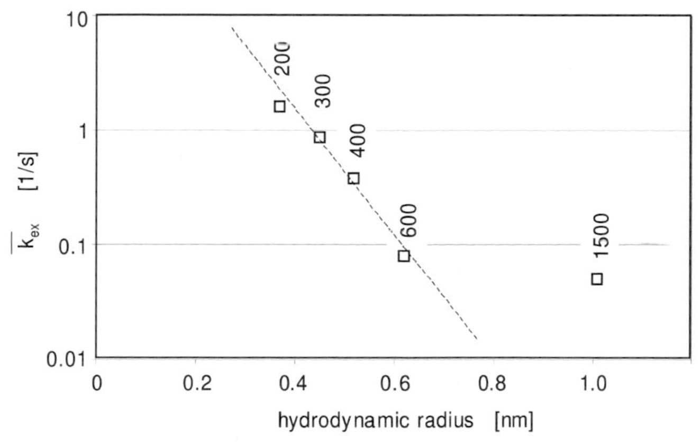

Fig. 4.15. Logarithmic plot of rate constants ($\bar{k}_{ex}$) for the trans-membrane exchange (Table 4.1) vs. hydrodynamic radii of PEO tracers in aqueous solution. For molecular masses between 200 and 600 g mol^{-1}, the dependence of log $\bar{k}_{ex}$ on hydrodynamic radius is nearly linear, indicating that the activation energy for the membrane transfer is proportional to the size of the molecule [47].

– one that grows linearly with increasing size of the tracer molecule. Obviously, the tracer PEO-1500 does not obey this rule. This could be linked to a specific unfolding mechanism necessary for the transfer of larger molecules. In general, the use of tracers under systematic variation of size, polarity, conformational flexibility etc. leads to specific information on the membrane characteristics. Such information is crucial for the capsule design in order to optimize the stability, permeability and release properties.

4.6 Particle Degradation and Release

The final steps during the life cycle of a particulate carrier system usually consist in the release of the active ingredients and, more or less simultaneously, degradation of the particles. Given the ability of NMR to distinguish between chemical components that are embedded in nanospheres, confined in nanocapsules or part of the continuous phase, it is not surprising that it gives easy access to release profiles. In general, all techniques described in Section 4.2 can be applied in a time-resolved manner; therefore, NMR represents an efficient method to follow structural changes such as the release of an active ingredient over time.

Figure 4.16 shows an example of a simple time-resolved measurement on polyelectrolyte nanospheres containing poly(ethylene oxide) as a model ingredient [54]. After dispersion in an aqueous environment, the directly-excited ^1H signal of the PEO is followed over time in occasional measurements. While the solid PEO inside the nanospheres is practically invisible under these conditions, it is possible to determine the amount of dissolved PEO by detecting the integral of the single

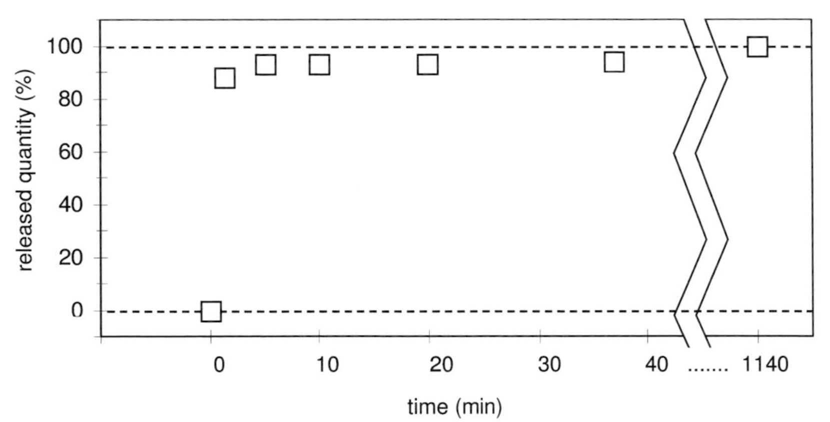

Fig. 4.16. Time-resolved observation of the ^1H signal of the model compound PEO incorporated in polyelectrolyte nanospheres. The signal is obtained under direct excitation, therefore only the released fraction of PEO is detected as a function of time after the addition of water at $t = 0$.

^{1}H peak at 3.2 ppm. In direct excitation, the signal integral is strictly proportional to the quantity of the observed component. In the given case, the variation in peak intensity over time clearly shows a phenomenon that was quite undesired: on addition of water, the system undergoes a burst release of ~90% of the model active ingredient within the first 3 min. The residual 10% of the PEO then slowly escapes into the fluid phase within the following 20 h. This observation yields the release profile, but also gives a certain clue about the particle structure. Most probably, the largest part (90%) of the PEO is loosely adsorbed onto the particle surface; only about 10% seems to be integrated into the particle matrix.

The solid matrix of nanospheres and nanocapsules is subject to various degradation processes. They make take place during storage as well as during their *in vivo* application and may be connected to chemical as well as physical phenomena. They all affect the solid contribution to the magnetic resonance spectrum, so cross polarization is a promising technique to study particle degradation. However, some mechanisms of matrix decay also lead to the formation of low molecular weight components in the liquid phase. In this case, the direct excitation approach is especially suitable. Of course, both experimental methods can be combined to yield a complete analysis of the degradation process.

Figure 4.17 shows an example for such an analysis on an aqueous dispersion of poly-n-butylcyanoacrylate nanocapsules. It refers to thermal decomposition of the dispersion induced by storage for 3 h at different temperatures (50, 100, 130 °C) [55]. As mentioned before, the cp spectra (left-hand column of Fig. 4.17) focus on the solid matrix of the capsules. While the broad solid-state spectrum is virtually unchanged after treatment at 50 °C (a), it shows increasing loss in intensity after storage at 100 °C (b) and completely disappears after 3 h storage at 130 °C (c). The narrow peaks assigned to the triglyceride and the surfactant remain visible but suffer significant losses at higher temperatures. Their residual cp signal intensity corresponds to the fraction P_{free} (Fig. 4.13), which is independent of polarization transfer on the particle surface. Clearly, the solid-state components have vanished after 3 h treatment at 130 °C. Since obviously no new solid phase has been generated, the material that originally formed the solid matrix has to show up in the liquid phase. Traces of it can be seen in the directly excited spectrum (right-hand column in Fig. 4.17) after heat treatment at 130 °C (c, arrows). Figure 4.18 shows a scaled-up comparison between the "new" signals (top) and the counterparts from a spectrum of the monomer n-butylcyanoacrylate (bottom). Obviously, these "new" signals show up near corresponding peaks expected for the product of depolymerization. The slight deviations for the chemical shifts of 1 to 3 ppm may derive from differences in the chemical environment: the reference spectrum was taken in $CDCl_3$ as a solvent, while the depolymerization product was dissolved in an aqueous environment containing some residual ethanol. Hence, the loss in solid polymer seems to be accompanied by the appearance of monomer in the fluid phase. This indicates that the matrix has been subject to a thermally induced depolymerization that finally leads to complete disintegration of the particles.

At this point, two varieties of particle defragmentation are possible: (a) depolymerization of the matrix could cause the capsule to separate into smaller solid

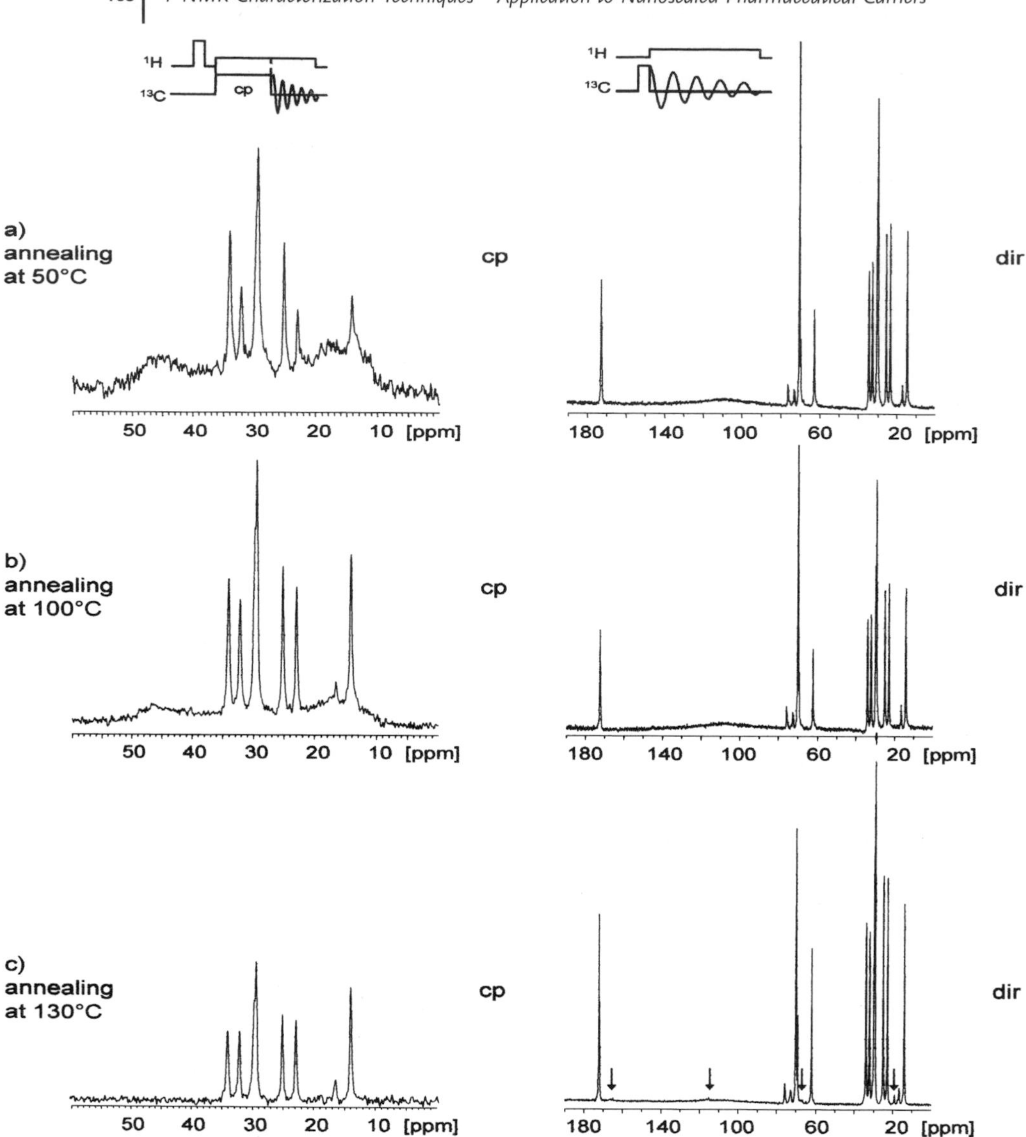

Fig. 4.17. Proton-decoupled ^{13}C spectra of aqueous dispersions of poly-n-butylcyanoacrylate nanocapsules after heat treatment for 3 h at (a) 50 (b) 100 and (c) 130 °C. Wide lines in the (^{1}H)–^{13}C cross-polarization spectra (left-hand column) indicate the loss of solid matrix material at higher temperatures. The superimposed narrow signals derive partially from adsorption of liquid components to the capsule walls, and partially from the residual cp in the liquid phase. Direct excitation spectra (right-hand column) show the liquid and dissolved components and, for $T = 130$ °C (c, arrows), indicate the formation of a depolymerization product (Fig. 4.18). Inserts: pulse sequences used for cross-polarization and direct excitation experiments.

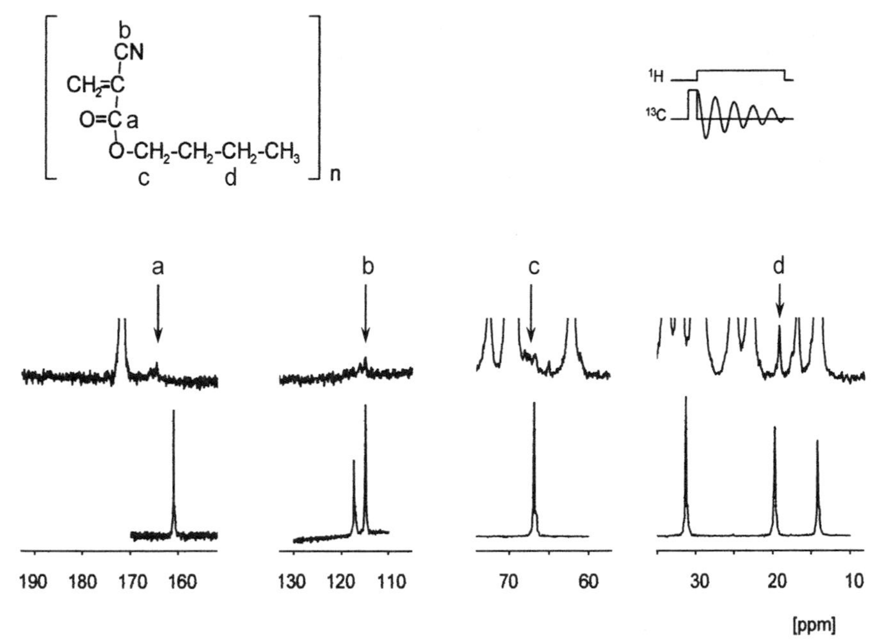

Fig. 4.18. Comparison between enlarged sections of direct excitation ^{13}C spectra obtained on poly-n-butylcyanoacrylate nanocapsules after heat treatment for 3 h at 130 °C (top line) and on a solution of n-butylcyanoacrylate in $CHCl_3$ (bottom line). The "new" signals a–d can be assigned to traces of monomer resulting from depolarization of the capsule matrix. The obvious variation of the chemical shifts could result from the different solvent environments. Inserts: signal assignments and pulse sequence used for direct excitation.

fragments or (b) the capsule could essentially preserve its spherical geometry while the capsule wall corrodes and finally disappears. In the first case, the smaller solid fragments should lead to an increase in the tumbling rate, leading to a corresponding narrowing of the cp resonance lines of the solid-state spectrum. Figure 4.19 shows that this is clearly not the case: direct comparison of the spectra after annealing at 100 and 130 °C shows that the solid line shapes are virtually identical – no indications for a motional narrowing of the cp signal are found. This leaves the second alternative: while suffering increasing material loss, the capsule sphere preserves its overall geometrical shape and therefore keeps its original low tumbling rate.

One of the principal advantages of decomposition studies based on NMR detection is its variability for chemical and physical conditions. The NMR experiment can be performed under a wide variety of thermal, chemical or other influences on the sample during the time-resolved observation. At the same time, the impact caused by the static NMR measurement itself is insignificant, which allows for undisturbed and extended measurements over long periods.

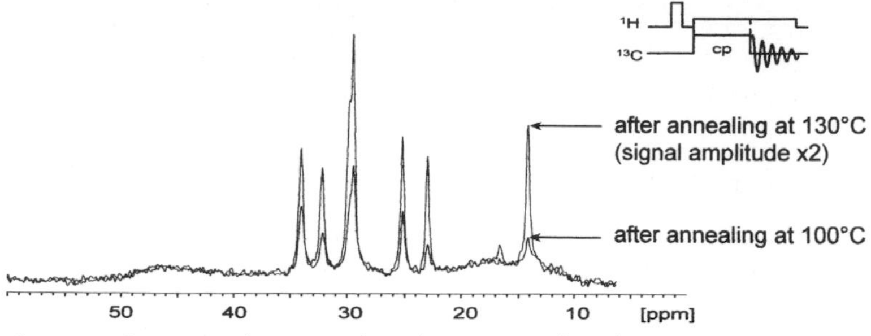

Fig. 4.19. Comparison between enlarged sections of cross-polarization spectra of dispersed poly-n-butylcyanoacrylate nanocapsules after heat treatment for 3 h at 100 and 130 °C. The spectrum for 130 °C has been vertically expanded to match that for 100 °C in its solid-state contribution. Both solid-state line shapes (superimposed by narrow signals for the triglyceride and surfactant) are virtually identical. For the solid matrix, no motional narrowing is detected that would indicate the formation of smaller fragments with increased tumbling rate. Insert: pulse sequence used for the cp measurement.

4.7 Summary and Outlook

This selection of NMR experiments on nanoparticles is meant to show the power of NMR in this field. Unlike any other analytical technique, it combines a distinctly non-invasive character with the ability to analyze for a chemical composition as well as for local mobility of individual system components. Its main disadvantages, most prominently represented by its relatively low sensitivity and the time consuming acquisition of experimental data, can be overcome by a suitable choice of pulse sequences and experimental conditions. Applied to nanoscaled pharmaceutical carrier systems, it gives access to several important structural details and system parameters. Table 4.2 gives an overview of various experimental NMR techniques that are suitable for studies focusing on different particle characteristics. Evidently, this scheme is by no means complete, but it may serve as a starting point to elucidate further approaches in NMR in this field.

One new technique that is presently under development may serve as an example: Studies are in progress that focus on the characterization of magnetic nanoparticles in liquid dispersion (magnetofluids) [56–61] used for medical applications, e.g., for the magnetically controlled local application of drug-loaded organic carriers. Here, the local perturbation of the magnetic field caused by the nano-sized magnets is detected by the observation of spins in the solid or fluid environment of the particles. It is used to characterize system parameters such as the magnetism of the particles and their distribution in space. The particle-induced magnetic perturbation may also be used in the opposite sense: by creating a very strong local field gradient it could allow for the observation of lateral molecular mobility on a very small scale. As for many other approaches of this kind, analysis of such experimental data requires a numerical simulation procedure that accounts for the local structures on the nanometer scale. Such algorithms are being developed based on the idea of small-scale susceptibility effects [61].

Tab. 4.2. Overview of the application of different NMR techniques for the analysis of different particle system characteristics.

	Direct excitation (incl. multi-dimensional NMR)	Cross polarization	Magic-angle spinning	Pulsed field gradient NMR	Relaxation curves (T_1,$T_{1\rho}$,T_2,...)
Identification of structural elements	Identification of fluid components	Observation of solid matrix	Identification of solid matrix	Identification of capsule contents	
Heterogeneity of the solid matrix					Detection of solid domain sizes
Phase transitions	Detection of new fluid components	Detection of new solid components	Identification of new solid matrix		
Adsorption to particle surface		Identification of adsorbed components		Time scale of adsorption–desorption cycle	Time scale of adsorption–desorption cycle
Permeability of capsule membranes				Detection of exchange rates	
Release	Release from nanospheres			Release from nano-capsules	
Degradation	Detection of fluid degradation products	Observation of disappearing solid components			

References

1 W. Mehnert, K. Mäder, Solid lipid nanoparticles – production, characterization and applications, *Adv. Drug Deliv. Rev.* 47, 165–196 (**2001**).

2 P. Couvreur, G. Barrat, E. Fattal, P. Legrand, C. Vauthier, Nanocapsule technology: a review, *Critical Rev. Therap. Drug Carrier Systems* 19, 99–134 (**2002**).

3 G. Barratt, Colloidal drug carriers: achievements and perspectives, *Cell. Mol. Life Sci.* 60, 21–37 (**2003**).

4 J. K. Vasir, K. Tambwekar, S. Garg, Bioadhesive microspheres as a

controlled drug delivery system, *Int. J. Pharm.* 255, 13–32 (**2003**).

5 F. Marcucci, F. Lefoulon, Active targeting with particulate drug carriers in tumor therapy: fundamentals and recent progress, *Drug Discovery Today* 9, 219–228 (**2004**).

6 M. J. Alonso, Nanomedicines for overcoming biological barriers, *Biomed. Pharmacother.* 58, 168–172 (**2004**).

7 C. Mayer, NMR on dispersed nano-particles, *Prog. NMR Spectrosc.* 40, 307–366 (**2002**).

8 C. Mayer, D. Hoffmann, M. Wohlgemuth, Structural analysis of nanocapsules by nuclear magnetic resonance, *Int. J. Pharm.* 242, 37–46 (**2002**).

9 M. Wohlgemuth, C. Mayer, Pulsed field gradient NMR on polybutyl-cyanoacrylate nanocapsules, *J. Colloid Interface Sci.* 260, 324–331 (**2003**).

10 K. Jores, W. Mehnert, K. Mäder, Physicochemical investigations on solid lipid nanoparticles and on oil-loaded solid lipid nanoparticles: a nuclear magnetic resonance and electronic spin resonance study, *Pharm. Res.* 20, 1274–1283 (**2003**).

11 M. Garcia-Fuentes, D. Torres, M. Martin-Pastor, M. J. Alonso, Application of NMR spectroscopy to the characterization of PEG-stabilized lipid nanoparticles, *Langmuir* 20, 8839–8845 (**2004**).

12 M.-H. Chen, R. Kumar, V. S. Parmar, J. Kumar, L. A. Samuelson, A. C. Watterson, Self-organization of amphiphilic copolymers into nanoparticles: study by ^{1}H NMR longitudinal relaxation time, *J. Macromol. Sci., Part A: Pure Appl. Chem.* A41(12), 1489–1496 (**2004**).

13 M. Garcia-Fuentes, D. Torres, M. Martin-Pastor, M. J. Alonso, Application of NMR spectroscopy to the characterization of PEG-stabilized lipid nanoparticles, *Langmuir* 20(20), 8839–8845 (**2004**).

14 J. Kriz, J. Plestil, H. Pospisil, P. Kadlec, C. Konak, L. Almasy, A. I. Kuklin, ^{1}H NMR and small-angle neutron scattering investigation of the structure and solubilization behavior of three-layer nanoparticles, *Langmuir* 20(25), 11255–11263 (**2004**).

15 Y. Zhang, Q. Zhang, L. Zha, W. Yang, C. Wang, X. Jiang, S. Fu, Preparation, characterization and application of pyrene-loaded methoxy poly(ethylene glycol)-poly(lactic acid) copolymer nanoparticles, *Colloid Polym. Sci.* 282(12), 1323–1328 (**2004**).

16 K. Westesen, A. Gerke, M. H. J. Koch, Characterization of native and drug-loaded human density lipoproteins, *J. Pharm. Sci.* 84, 139–147 (**1995**).

17 D. Hoffmann, C. Mayer, Cross polarization induced by temporary adsorption: NMR investigation on nanocapsule dispersions, *J. Chem. Phys.* 112, 4242–4250 (**2000**).

18 C. Mayer, G. Lukowski, Solid state NMR investigations on nanoparticular carrier systems, *Pharm. Res.* 17, 486–489 (**2000**).

19 H. Y. Huang, K. L. Wooley, J. Schaefer, REDOR determination of the cross-linked amphiphilic core-shell nanoparticles and the partitioning of sequestered fluorinated guests, *Macromolecules* 34, 547–551 (**2001**).

20 S. Guinebretiere, S. Briancon, J. Lieto, C. Mayer, H. Fessi, Study on the emulsion-diffusion process: formation and characterization of nanoparticles, *Drug Develop. Res.* 57, 18–33 (**2002**).

21 S. Wissing, R. Müller, L. Manthei, C. Mayer, Structural characterization of Q10-loaded SLN by nuclear magnetic resonance, *Pharm. Res.* 21, 400–405 (**2004**).

22 P. A. Bertin, K. J. Watson, S. T. Nguyen, Indomethacin-containing nanoparticles derived from amphi-philic polynorbornene: a model ROMP-based drug encapsulation system, *Macromolecules* 37(22), 8364–8372 (**2004**).

23 P. Couvreur, B. Kante, M. Rd, P. Guiot, P. Baudhuin, P. Speiser, *J. Pharm. Pharmacol.* 31, 331 (**1979**).

24 A. T. Florence, T. L. Whateley, D. A. Wood, *J. Pharm. Pharmacol.* 31, 422 (**1979**).

25 M. Wohlgemuth, W. Mächtle, C. Mayer, Improved preparation and physical studies of polybutylcyano-acrylate nanocapsules, *J. Microencapsulation* 17, 437–448 (**2000**).

26 C. Mayer, Calculation of cross-polarization spectra influenced by slow molecular tumbling, *J. Magn. Reson.* 145, 216–229 (**2000**).

27 M. Mehring, High resolution NMR spectroscopy in solids, in *NMR Basic Principles and Progress*, P. Diel, E. Fluck, R. Kosfeld (eds.), Springer Verlag, Berlin (**1976**).

28 V. J. McBrierty, K. J. Packer, *Nuclear Magnetic Resonance in Solid Polymers*, Cambridge University Press, Cambridge (**1993**).

29 K. Schmidt-Rohr, W. Spiess, *Multidimensional Solid-State NMR and Polymers*, Academic Press, London (**1994**).

30 G. Lukowski, D. Hoffmann, P. Pflegel, C. Mayer, *Solid State NMR Investigations On Nanosized Carrier Systems*, Proc. 3rd World Meeting APV/APGI Berlin (**2000**).

31 R. H. Müller, K. Mäder, S. Gohla, Solid lipid nanoparticles (SLN) for controlled drug delivery – a review of the state of the art, *Eur. J. Pharm. Biopharm.* 50, 161–178 (**2000**).

32 C. Mayer, Lineshape calculations on spreadsheet software, *J. Magn. Reson.* 138, 1–11 (**1999**).

33 C. Mayer, Calculation of MAS spectra influenced by slow molecular tumbling, *J. Magn. Reson.* 139, 132–138 (**1999**).

34 E. O. Stejskal, J. E. Tanner, *J. Chem. Phys.* 42, 288 (**1965**).

35 E. O. Stejskal, J. E. Tanner, *J. Chem. Phys.* 43, 3597 (**1965**).

36 J. E. Tanner, E. O. Stejskal, *J. Chem. Phys.* 49, 1768 (**1968**).

37 R. Kimmich, *NMR Tomography Diffusometry Relaxometry*, Springer Verlag, Berlin (**1997**).

38 I. Lönnqvist, A. Khan, O. Södermann, *J. Colloid Interface Sci.* 144, 401 (**1990**).

39 O. Södermann, *Progr. Colloid Polym. Sci.* 106, 34 (**1997**).

40 J. Kreuter, Nanoparticles, in *Colloidal Drug Delivery Systems*, J. Kreuter (ed.), Marcel Dekker, New York (**1994**).

41 D. Quintanar-Guerrero, Etude du mecanisme de formation de nanoparticules polymériques d'émulsification-diffusion, *Colloid Polym. Sci.* 275, 640–647 (**1997**).

42 D. E. Woessner, Relaxation effects of chemical exchange, in *Encyplopedia of Magnetic Resonance*, Wiley, New York (**1996**), Vol. 6, pp. 4018–4028.

43 M. Schönhoff, O. Södermann, PFG-NMR diffusion as a method to investigate the equilibrium adsorption dynamics of surfactants at the solid/liquid interface, *J. Phys. Chem. B* 101, 8237–8242 (**1997**).

44 M. Schönhoff, O. Södermann, Exchange dynamics of surfactants in adsorption layers investigated by PFG NMR diffusion, *Magn. Reson. Imag.* 16, 683–685 (**1998**).

45 K. I. Momot, P. W. Kuchel, B. E. Chapman, D. Whittaker, NMR study of the association of propofol with nonionic surfactants, *Langmuir* 19, 2088–2095 (**2003**).

46 K. I. Momot, P. W. Kuchel, Pulsed field gradient nuclear magnetic resonance as a tool for studying drug delivery systems, *Concepts Magn. Reson. Part A* 19, 51–64 (**2003**).

47 A. Rumplecker, S. Förster, M. Zähres, C. Mayer, Molecular exchange through vesicle membranes: a pulsed field gradient NMR study, *J. Chem. Phys.* 120, 8740–8747 (**2004**).

48 D. E. Discher, A. Eisenberg, Polymer vesicles, *Science* 297, 967–973 (**2002**).

49 E. Krämer, S. Förster, C. Göltner, M. Antonietti, Synthesis of nanoporous silica with new pore morphologies by templating the assemblies of ionic block copolymers, *Langmuir* 14, 2027–2031 (**1998**).

50 B. Balinov, B. Jönsson, P. Linse, O. Södermann, The NMR self-diffusion method applied to restricted diffusion. Simulation of echo attenuation from molecules in spheres and between planes, *J. Magn. Reson. A* 104, 17–25 (**1993**).

51 P. T. Callaghan, Pulsed-gradient

spin-echo NMR for planar, cylindrical, and spherical pores under conditions of wall relaxation, *J. Magn. Reson. A* 113, 53–59 (**1995**).
52 S. L. Codd, P. T. Callaghan, Spin echo analysis of restricted diffusion under generalized waveforms: planar, cylindrical, and spherical pores with wall relaxivity, *J. Magn. Reson.* 137, 358–372 (**1999**).
53 J. Pfeuffer, U. Flögel, W. Dreher, D. Leibfritz, Restricted diffusion and exchange of intracellular water: theoretical modelling and diffusion time dependence of ^{1}H NMR measurements on perfused glial cells, *NMR Biomed.* 11, 19–31 (**1998**).
54 D. Hoffmann, M. Zähres, unpublished result.
55 D. Hoffmann, PhD thesis, University of Duisburg (**2000**).
56 E. K. Jang, I. Yu, ^{1}H NMR of water influenced by suspended magnetic particles, *Colloids Surf. A* 72, 229–235 (**1993**).
57 C. E. Gonzalez, D. J. Pusiol, A. M. F. Neto, M. Ramia, A. Bee, Nuclear magnetic resonance study of the internal magnetic field distribution in water base ionic and surfacted ferrofluids, *J. Chem. Phys.* 109, 4670–4674 (**1998**).
58 S. Saito, M. Ohaba, Proton nuclear magnetic resonance and optical microscopic studies of magnetic fluids, *J. Phys. Soc. Jpn.* 68, 1357–1363 (**1999**).
59 A. Terheiden, M. Michaelsen, G. Dyker, C. Mayer, Nuclear magnetic resonance as a tool for the characterization of magnetic fluids, *Magnetohydrodynamics* 39, 17–22 (**2003**).
60 A. Terheiden, C. Mayer, NMR study on colloidal magnetic fluids of various viscosities, *Phase Transitions* 77, 81–87 (**2004**).
61 C. Mayer, A. Terheiden, Numerical simulation of magnetic susceptibility effects in nuclear magnetic resonance spectroscopy, *J. Chem. Phys.* 118, 2775–2782 (**2003**).

5
Characterization of Nano Features in Biopolymers using Small-angle X-ray Scattering, Electron Microscopy and Modeling

Angelika Krebs and Bettina Böttcher

5.1
Introduction

Biological molecules and systems have several attributes that make them highly suitable for nanotechnology applications. For example, proteins fold into precisely defined three-dimensional (3D) shapes, and nucleic acids assemble according to structural rules. Antibodies are highly specific in recognizing and binding their ligands, and biological assemblies such as molecular motors can perform transport operations. Because of these, and other favorable properties, they are ideal for applications in nanotechnology. Gaining structural information of such complexes is crucial for understanding their functionality and for their successful incorporation in nano technologies.

Various techniques are available that provide information at various degrees of reliability and detail. Protein crystallography and multidimensional NMR are well-established methods that provide models of biological macromolecules on an atomic scale. With small-angle X-ray scattering the structure of macromolecules in solution can be investigated. Typical samples range from small proteins (from 5–10 kDa) to multi-domain proteins and protein complexes (up to MDa). The scattering data are sensitive to domain orientations and conformational changes and/or flexibility as well as to molecular associations; atomic resolution, however, will not be achieved. The technique is most powerful when used as a complementary tool with other structural techniques, such as three-dimensional electron microscopy (3D-EM). Three-dimensional image reconstructions from electron micrographs have become important in recent years, because near atomic resolution has been achieved with cryo-EM for several important systems. At the moment, low-resolution EM structures are obtained almost routinely for complex biopolymers and this information can be used advantageously in combination with other methods. One example is the merging of complex structures determined by EM at medium resolution with high-resolution crystal structures of their components. Another possibility, which is the scope of this chapter, is the merging of structural information obtained by EM with information obtained from SAXS.

After a general introduction to SAXS and EM, this chapter explains how both

Nanotechnologies for the Life Sciences Vol. 3
Nanosystem Characterization Tools in the Life Sciences. Edited by Challa S. S. R. Kumar

ISBN: 3-527-31383-4

methods together can be used successfully to gain structural information on the nanoscale.

5.2
Small-angle X-ray Scattering

Small-angle X-ray scattering is a well-established technique for structural investigations of proteins on the nanometer length scale (1–300 nm), having its origins more than 50 years ago [1–4]. Recently, biological SAXS has gained additional importance, mainly due to outstanding hardware and software development. Very important in this context are new SAXS instruments and advanced methods to analyze SAXS data from macromolecular solutions (e.g., *ab initio* low-resolution structure analysis and rigid body refinement). Common applications in nanotechnology are the investigation of micro-emulsions or liquid crystals, the particle sizing of suspended nano-powders or investigations concerning structural information of polymer films and fibers. Applications in life science and biotechnology mainly deal with proteins, viruses and DNA complexes, whereby the border between nanotechnology and life science constantly diminishes.

5.2.1
Scattering Technique

5.2.1.1 Scattering Phenomenon

X-rays are photons with wavelengths in the range 0.1–100 Å and, depending on the frequency of electromagnetic waves, different interactions with matter are observable. They are usually generated by bombarding a target with electrons of energies of 10 000 eV or more. Upon collision, these high-energy electrons can knock electrons out of the target atoms, leaving vacancies in atomic shells. If, for example, a vacancy is produced in the innermost (K) shell of an atom, it rapidly will be filled by an electron descending from the next (L) shell, or one from the one after that (M). The photons emitted as a result of these transitions are the X-rays; they vary in wavelength according to the path of the electron. Copper is commonly used as a target, where CuK_α radiation is emitted by electrons propagating from the L to the K shell and CuK_β is the radiation emitted by an electron changing from the M to the K shell.

X-rays interact with matter in different ways, i.e., absorption, elastic and inelastic scattering (Fig. 5.1). Elastic scattering is the basic event for X-ray scattering techniques. There, the incident X-ray photons undergo perfectly elastic collisions with the electron, leaving their energy unchanged and leading to radiation propagating away from the sample in all directions [1, 3–5].

The event behind that is the following: When a beam of electromagnetic radiation strikes an electron, some of the energy is momentarily absorbed and the electron becomes displaced from its unperturbed position due to the force exerted on it by the electric field. As a result, the electron is set into periodic motion with a fre-

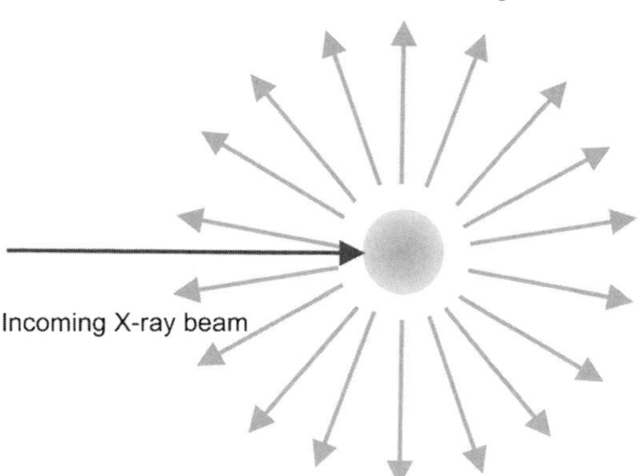

Fig. 5.1. Scattering phenomenon. When a beam of electromagnetic radiation strikes an atom it is mainly the electrons that react. Some of the energy is momentarily absorbed and the electron becomes displaced from its unperturbed position due to the force exerted on it by the electric field. As a result, the electron is set into periodic motion with a frequency equal to that of the exciting radiation. Since, according to the Maxwell equations, any accelerating or decelerating charge must radiate an electromagnetic wave in all directions, and since the radiation reemitted by the electron has the same frequency as the exciting radiation, experimental observation gives the impression that the incident radiation is scattered in all directions by the electron.

quency equal to that of the exciting radiation. Since, according to the Maxwell equations, any accelerating or decelerating charge must radiate an electromagnetic wave in all directions and since the radiations reemitted by the electron has the same frequency as the exciting radiation, the experimental observation gives the impression that the incident radiation is scattered in all directions by the electron. Thus the electrons can be considered as centers of secondary waves and the wavelength of the incident and the scattered radiation is the same (elastic scattering).

The amount of scattering is described by the Thomson equation [Eq. (1)], which states that scattered intensity from a single independent electron $I(\theta)$ is given by:

$$I(\theta) = I_0 \frac{e^4}{m_0^2 c^4 a^2} \left\{ \frac{1 + \cos^2(2\theta)}{2} \right\} \tag{1}$$

where e is the electronic charge, m is the mass of the electron, c is velocity of light, a is the sample–detector distance and 2θ is the scattering angle. The product $e^4/m_0^2 c^4$ is known as the electronic scattering factor (also termed electron-scattering cross section and Thomson's constant), and the quantity in parentheses is known as polarization factor. The angle θ is defined as one-half the angle of deflection of the incident beam relative to the scattered beam. In so-called reciprocal space the scattering vector h (nm^{-1}) is usually used [Eq. (2)], where with λ is the wavelength.

$$h = (4\pi/\lambda) \sin \theta \tag{2}$$

While scattering from a single electron is extremely weak, in real systems, such as a protein in aqueous solution, we measure the total scattering from all electrons in the irradiated volume (which is of the order of 0.1 mL). Since the dimensions of macromolecules are always large relative to the wavelength of the incident X-ray radiation, interference occurs between the radiation scattered from individual electrons. Thus the spatial distribution of electrons in one molecule leads to several secondary waves that are able to interfere. The phase difference of the scattered waves depends on the path distance of the electrons in the molecule. This path difference increases with larger scattering angles and is the cause of diminishing scattering at larger angles. At very large scattering angles the scattered intensity is 0. This is why we observe small-angle X-ray scattering.

5.2.1.2 Scattering Curve and Pair Distance Distribution Function

In general, because electrons are not localized, it is better to describe an electron density $\rho(\mathbf{r})$ in a volume element $\mathrm{d}V$; the scattering then is proportional to $\rho(\mathbf{r})\,\mathrm{d}V$. Therefore, for a continuous electron distribution, the sum of all secondary (scattered) waves e^{-ihr} is replaced by an integral [Eq. (3)].

$$F(h) = \int_V \rho(\mathbf{r})\mathrm{e}^{-ihr}\,\mathrm{d}V \tag{3}$$

The integral is over the entire sample. This equation is the single fundamental equation that governs all X-ray scattering and diffraction. If the electron density distribution $\rho(\mathbf{r})$ of a sample is known, one can compute the structure factor, and from this one can compute the expected X-ray scattering for all scattering geometries.

In solution small-angle X-ray scattering, however, the information contained in the 3D electron-density distribution $\rho(\mathbf{r})$, which describes the whole structure of the particle, is reduced to a one-dimensional distance distribution function $p(r)$, because the particles we measure are non-oriented in solution. [For solution scattering $\rho(\mathbf{r})$ stands for the difference in electron density between sample and solvent.] The $p(r)$ function is proportional to the number of lines with length r that connect any volume element i with any volume element k of the same particle. The spatial orientation of these connection lines is not important and the connection lines r are weighted by the product of the number of electrons situated in the volume elements i and k respectively. Each distance between two electrons of the sample as part of the function $p(r)$ leads to an angular dependent scattering intensity. This physical process of scattering can be mathematically expressed by a Fourier transformation that defines how the information in real space (distance distribution function) is transferred into reciprocal space (scattering function) [Eq. (4)], with $\gamma(r)$ as a measure of the probability of finding two points with the distance r within the particle of interest (correlation function).

$$I(h) = 4\pi \int_0^\infty \gamma(r) r^2 \frac{\sin hr}{hr}\,\mathrm{d}r \tag{4}$$

The correlation function can be calculated from the scattering intensity using a Fourier transform [Eq. (5)].

$$\gamma(r) = \frac{1}{2\pi^2}\int_0^\infty I(h)h^2 \frac{\sin hr}{hr}\,\mathrm{d}h \tag{5}$$

With $p(r) = \gamma(r)r^2$ one can calculate the pair distance distribution function from the scattering intensity according to Eq. (6).

$$p(r) = \frac{r^2}{2\pi^2}\int_0^\infty I(h)h^2 \frac{\sin hr}{hr}\,\mathrm{d}h \tag{6}$$

5.2.1.3 **Determination of Scattering Parameters**

SAXS parameters, such as the radius of gyration R_G, maximum particle size D_{max}, volume V and molecular weight MW can be determined directly from the scattering behavior of the sample. In the following the procedures are outlined briefly:

(a) The radius of gyration, R_G, i.e., the root-mean-square of the distances of all the electrons of the particle from its center of electronic mass, is a characteristic geometric parameter of any particle. It can be determined according to a Guinier plot [ln $I(h)$ vs. h^2] [Eq. (7), with $I(0)$ as scattering intensity at zero angle].

$$I(h) = I(0)\mathrm{e}^{-(h^2 R_G{}^2/3)} \tag{7}$$

Another way is to use the information given by the distance distribution function [Eq. (8)].

$$R_G = \sqrt{\frac{\int_V \rho(\mathbf{r})r^2\,\mathrm{d}V}{\int_V \rho(\mathbf{r})\,\mathrm{d}V}} = \sqrt{\frac{1}{2}\frac{\int_V \rho(r)r^2\,\mathrm{d}r}{\int_V p(r)\,\mathrm{d}r}} \tag{8}$$

(b) The maximum diameter D_{max} of a particle can be determined by a linear diagram of $p(r)$ vs. distance r. D_{max} is then given by the crossover of the $p(r)$ function with the x-axis.

(c) The hydrated volume can be assessed according to Eq. (9), with $Q = \int I(h)h^2\,\mathrm{d}h$ being the invariant according to Porod.

$$V = 2\pi^2 \frac{I(0)}{Q} \tag{9}$$

(d) Finally, the relation of the scattered intensity to the primary intensity is a measure of the molecular weight of the biological sample [Eq. (10)].

$$\mathrm{MW} = \frac{I(0)a^2}{N_A T_e P_0 (z_2 - v_2'\rho_1)^2\, dc} \tag{10}$$

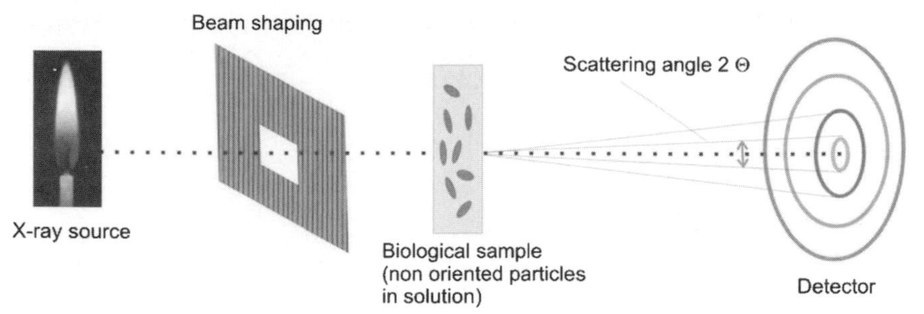

Fig. 5.2. Typical SAXS experiment.

where P_0 is the primary intensity of the X-ray beam, N_A is the Loschmidt number, T_e is the Thomson factor, (7.9×10^{-26} cm^2), z_2 is the number of electrons per gram of the soluble sample in (mol g^{-1}), ρ_1 is the electron density of the solvent in (mol mL^{-1}), v_2' is the isopotential specific volume in (mL g^{-1}), d is the thickness of the sample in cm, and c is the concentration (g mL^{-1}).

5.2.1.4 **Experimental Setup**

With solution SAXS it is possible to study monodisperse as well as polydisperse systems. For monodisperse systems one can determine size, shape, interparticle interactions and internal structure of the particles. For polydisperse systems a size distribution can be calculated under the assumption that all particles have the same shape. In the SAXS experiments, the sample is irradiated by a well-defined, monochromatic X-ray beam (Fig. 5.2).

There exist laboratory instruments based on more conventional sources and synchrotron-based instrumentation. Synchrotron radiation sources with their intense brightness and natural collimation are ideal because biological materials are very poor scatterers. There is always some form of beam shaping required to maintain the small cross-section in going from the source to the sample and to reduce distortions from parasitic scattering from whatever obstacles, including air, are encountered. This is where much of the experimental effort is required and additional mathematical corrections take care of the non-ideal circumstances of a normal scattering experiment. A sample stage that may or may not involve heating/cooling elements then follows, ideally all within an in-vacuum path. No preparation such as staining of the material is required (in contrast to EM), and thicknesses of the biological sample between 1 and 3 mm are usually sufficient. Most commonly, small vials are used. As shown here the SAXS technique is performed in transmission mode. An extended sample–detector distance is usually required to give the barely scattered photons room to spread out from the main beam and also to reduce the detected X-ray background. Finally, a proportional counter, a scintillation counter, or a position sensitive detector, ideally two-dimensional, is required to measure the scattered intensity. As can be seen here, the black spot would be the beamstop that is absolutely essential to block the main beam. To reduce errors, measurements are performed repeatedly, because the statistical error is inversely proportional to the square root of the counts.

5.2.2 Interpretation of Data

5.2.2.1 Direct Methods

Interpreting SAXS data can be very difficult, unless one is very lucky and the sample fits one of the many idealized models developed over the years. Regular wide-angle X-ray scattering tends to focus on the location, width, shifts, etc. of Bragg peaks that arise from crystalline lattice structures. One can still observe Bragg peaks in SAXS but these will result from regular spacings that are of the order of hundreds of angstroms. Most of the time, however, the observed curves tend to be apparently featureless.

Direct methods of analysis give us information based on interpretation of the clean (background corrected) data with no further manipulation [6]. All of these parameters are based on well-defined assumptions, such as the existence of uniform density within our particle, uniform density in the background, sharp interfaces between the two, etc.

At very small angles, the slope of the scattering in the so-called Guinier region (Fig. 5.3) can be used to get an idea of the radius of gyration of any distinct structure [see Eq. (7)]. For unisometric particles the radius of gyration of the cross-section and the radius of gyration of the thickness can be determined with equations similar to Eq. (7) (for further details see Ref. [4], p. 155). At higher angles, if we have a system of relatively identical particles, dilute enough for there to be no interactions, we may be able to see broad peaks that give us information on

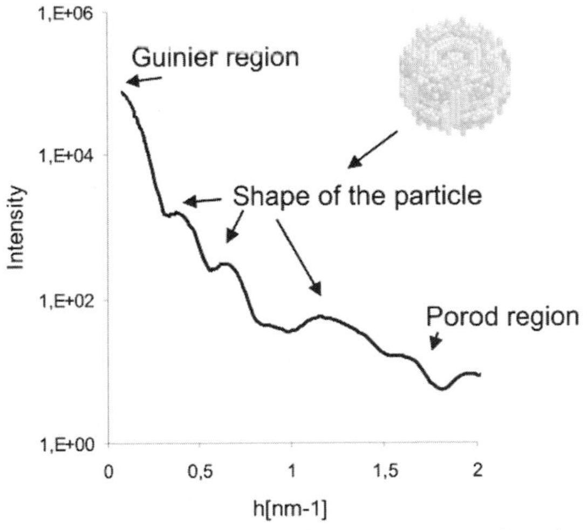

Fig. 5.3. Scattering curve and SAXS model from a biopolymer on the nanometer scale (hemoglobin from the earthworm *Lumbricus terrestris*).

Tab. 5.1. Experimental SAXS data and model data of *Lumbricus terrestris* hemoglobin.

	Lumbricus terrestris hemoglobin			
	Experimental SAXS data.[a]	***SAXS "consensus model"***[b]	***EM model "expected volume"***[c]	***EM model "hydrated volume"***[d]
R_G (nm)	10.71 ± 0.02	10.68 ± 0.04	11.30	10.73
D_{max} (nm)	29.37 ± 0.21	29.61 ± 0.05	29.39	29.03
V (nm^3)	6200 ± 200	6200 ± 400	4500	7200
MW (MDa)	3.5 ± 0.2			
Number of spheres		6844	11671	23021
Radius of spheres (nm)		0.66	0.45	0.426

[a] Experimental data are taken from Ref. 90.
[b] Mean values and standard deviations result from averaging the 22 models included in the consensus model shown in Fig. 5.3.
[c] Threshold = 171, voxel-size = $0.726 \times 0.726 \times 0.726$.
[d] Threshold = 115, voxel-size = $0.687 \times 0.687 \times 0.687$.

the shape of the particles. At still higher angles, the so-called Porod region, the shape of the curve is useful in obtaining information on the surface-to-volume ratio of the scattering objects. This can also be used to gain information on the dimensions of our scattering particles. The area under the curve gives us the invariant, which is a measure of how much scattering material is seen by the beam and allows us to estimate the hydrated volume according to Eq. (9). The volume is a very important parameter, which we can determine with SAXS [the scattering amplitude is well defined according to Eq. (1)], whereas this is not easy in EM. Changes in the invariant can be used, for instance, to monitor the crystallization process in polymer materials. Finally, the shape of the $p(r)$ function gives additional information on the size and overall structure of the particle as well as the maximum diameter. Thus we gain useful information rather quickly by careful analysis of the scattering behavior of the particle (Table 5.1).

5.2.2.2 **Indirect Methods**

For further interpretation purposes, it may be helpful if we propose specific structures and simulate their scattering behavior. The model-scattering data are then compared with the experimental data and the fit is a measure of the quality of the proposed structure. This approach, however, tends to imply that we already know the answer. Another possibility, therefore, if we don't really have any clear order to base our interpretation on, would be to assume a strongly disordered structure and to change this structure as long as it takes to gain a similar scattering behavior of the model and the experiment. This then is the basis of any indirect interpretation. In recent decades, various indirect modeling techniques have been described, including simple whole-body approaches, multi-body procedures and advanced *ab initio* modeling techniques (Fig. 5.4) (for summaries see Refs. [7, 8]).

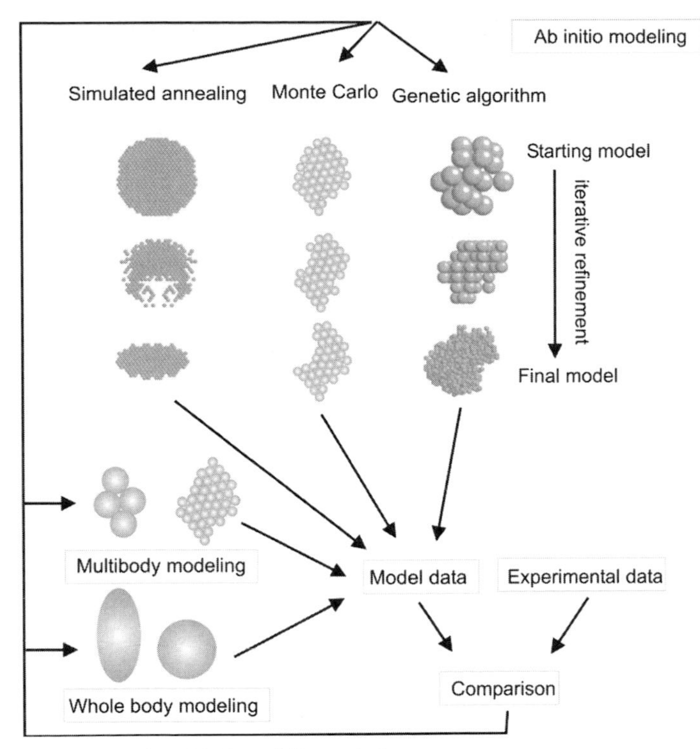

Fig. 5.4. Modern SAXS modeling techniques.

Conventional Whole-body and Multi-body Modeling Approaches As a first approximation, biopolymers can be modeled as spheres, prolate or oblate ellipsoids or triaxial bodies with unequal axes. Approaches along this line are called whole-body modeling [9–11]. They consider hydration contributions by using experimental molecular properties such as the hydrated volume, V, which can be directly obtained by SAXS (see above). The surface of whole-body models is entirely smooth and therefore whole-body approaches are not used to determine structural details on the nanoscale.

For nano-applications the use of an assembly of spheres (beads) instead of a single geometrical body seems appropriate, e.g., Refs. [7, 12–18], because proteins and other types of biopolymers are often composed of various constituents. In multi-body modeling approaches the target experimental profile is simulated by a finite element method ("bead modeling") and we can either use a small number of objects (each object would then simulate a subunit of the whole protein complex under investigation) or a large number of small spheres (in that case the spheres might be chosen to represent single amino acids or slightly larger details of the nano particle). Model scattering curves are usually calculated by means of Debye's formula [4].

An example of a two-step multi-body approach is the consensus model of a giant multisubunit protein complex with a diameter of about 30 nm; this complex with a molecular weight of about 3600 kDa and an overall *D*6 symmetry [19, 20] is responsible for oxygen transport in the earthworm *Lumbricus terrestris* (Fig. 5.3). Due to considerable differences in mass estimations, the exact stoichiometry of the components was unclear for a long time [21, 22]. Two models have been proposed to explain the architecture of *Lumbricus* Hb: The first consists of 24 octameric subassemblies of globin chains and 24 linker chains [23]. In the second, the so-called "bracelet model" [21, 24–26] 12 dodecameric 200 kDa subunits, each composed of 3 monomeric and 3 trimeric heme-containing globin chains, and 36 heme-deficient linker chains form the HBL complex with a total mass of about 3.6 MDa. Several 3D reconstructions from cryo-EM have been presented [27–29] and the crystal structure at 5.5 Å resolution has been published [30], revealing an organization of 144 oxygen-binding hemoglobin subunits and 36 non-hemoglobin linker subunits, similar to the bracelet model.

Solution SAXS studies and subsequent multi-body modeling led to the consensus model, which consists of 6844 spheres of equal size (radius 0.66 nm) but unequal weight. In the first step, about 600 different models, all biased to represent the biopolymer structure in eclipsed arrangement (top and bottom half exactly on top of each other), were generated by trial and error and tested for equivalence in scattering with the protein complex. In the second step, the 22 best-fitting models were superimposed and averaged, a procedure that resulted in spheres of different weights according to the different occupation densities of positions [18]. No structural bias other than hexagonal symmetry (known from early EM studies) was imposed in the model calculations. The model shown here is an early example of generation of a SAXS model that allows us to determine better and worse defined areas of the model. This is an important step to produce more reliable SAXS models. The scattering behavior of this model is compared with the experimental data in Table 5.1.

Structure Reconstruction by *Ab Initio* Modeling Approaches A fascinating aspect of modern evaluation procedures for SAXS data is the possibility to establish *ab initio* reconstructions of low-resolution biopolymer shapes [7, 31–35], even without resorting to any kind of spatial information. Among these highly advanced methods, in particular the approaches based on simulated annealing [36–38], a genetic algorithm [39], Monte-Carlo approaches [40, 41] or a molecular-dynamics algorithm [42] need to be addressed.

In the simulated annealing approach [37] models from densely packed dummy atoms (beads) are established. Each dummy atom is ascribed either to the particle or the solvent. Starting from a random initial configuration in a chosen search space (in general a sphere of diameter exceeding slightly the particle diameter, or a chain-like ensemble of dummy residues to mimic the backbone of the protein structure and dummy water molecules to simulate hydration), simulated annealing is used as a global minimization algorithm to find a configuration matching the SAXS data. In the beginning symmetry information or other information on the shape of the particle (from other methods, such as EM) may be introduced to increase the reliability of the generated models.

Using the genetic algorithm the scattering curves may also be iteratively fitted [39, 43]. There, a population of genes codifying a given mass distribution on a hexagonal lattice in a confined volume (e.g., an ellipsoidal search space of selected dimensions) is randomly generated. From each genotype the obtained model structure is compared to the observed SAXS data (by calculating the scattering profile of the model body by the Debye approach) and a fitness criterion is used to generate the next population by genetic operators (crossover and mutation), until, as a consequence of the selection pressure, the system converges, i.e., the best-fitting model is found. The fitting starts with a low number of large spheres and uses, incipiently, only the innermost portion of the scattering curve; during the run, the size of the spheres is gradually scaled down and the limit of resolution is increased by addition of further portions of the scattering curve. Monte Carlo approaches add and remove beads on a lattice until an optimum fit to the experimental SAXS profile is reached and the value of the score is at its minimum [40, 41].

In general, a thorough analysis to develop a structural model from SAXS data requires the execution of multiple runs for each condition chosen (e.g., bead radius, i.e., resolution) to avoid misinterpretations owing to unfavorable or ill-posed calculation conditions. A comparison of the results requires visualization, alignment and superimposing of the obtained models, followed by some kind of averaging and filtering. In any case, the resultant models cannot guarantee absolute uniqueness. In particular, for very complex structures or if considerable particle inhomogeneities exist, a note of caution is advised. In such cases, structural knowledge from other methods (such as EM) is extremely helpful for the creation of SAXS models and should therefore be introduced at the beginning of the model calculations.

Generation of Averaged Models Averaging of SAXS models is a legitimate procedure to point out structural tendencies by emphasizing the most persistent features [17, 44]. In this way, important, recurring features in the protein structure can be spotted and interpreted accordingly. Structural details at high resolution, however, may be lost during the averaging procedure and the averaged model may also provide a worse fit of the experimental data. Despite these reservations, averaging can be a valuable tool to reduce the problem of uniqueness of shape reconstructions and, nowadays, best-matching alignment and superposition of the bead models and subsequent averaging may be done automatically [44, 45].

5.3 Electron Microscopy

As we have seen in preceding paragraphs, interpretation of SAXS data greatly benefits from the incorporation of a priori knowledge of the structural organization of the nano-particles. Here EM can provide useful information on size and shape.

In addition, merging of different views allows reconstruction of the 3D image information. Such image reconstructions provide quite accurate phase information of the object but often show a modulation of the amplitude profile, which, in turn, can be corrected by combination with SAXS measurements.

5.3.1 Image Formation

Microscopy, as other methods, relies on the interference of electromagnetic radiation with the object. In contrast to diffraction methods, images contain the complete image information, consisting of amplitude and phase information. In EM the source of electromagnetic radiation is electrons accelerated by high voltage in a vacuum. The velocity of these electrons is close to the speed of light. With a typical accelerating voltage of 200 kV the electrons already reach a speed of 210 000 km s^{-1}, which is ~70% of the velocity of light. At these high relativistic speeds, electrons can be considered as waves with wavelengths of the order of a few picometer (2.5 pm at 200 kV).

5.3.1.1 Interference of Electrons with Matter

Electron waves interfere with an object in several modes (Fig. 5.5). Most of the incoming electrons pass a thin object without interference. A certain fraction of electrons is elastically scattered by small angles and interferes with the unscattered

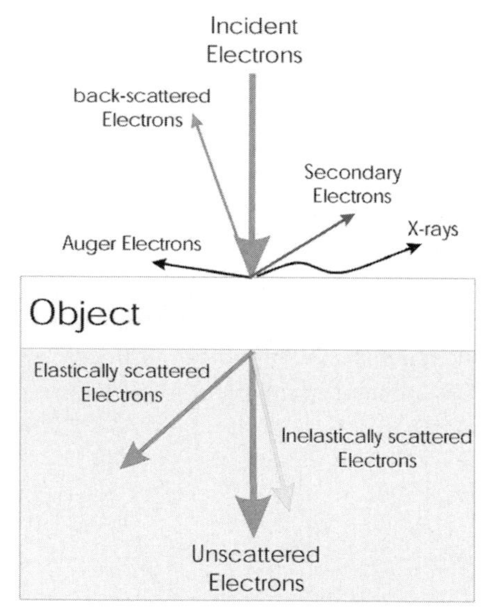

Fig. 5.5. Interaction of electrons with the object in EM. Phase contrast, which is the main source of contrast for thin unstained specimens composed of elements with low atomic numbers, is generated by interaction of the elastically scattered electrons with the unscattered electrons.

beam. These electrons contribute to the phase contrast, which is the main source of contrast in transmission EM of thin objects. Another fraction of electrons is scattered by higher angles and does not reach the back-focal plane. These electrons contribute to amplitude contrast. Some of the incoming electrons are scattered inelastically and deposit energy on the object, which leads to beam-damage and consequently to loss of high-resolution information. The event of inelastic scattering is not highly localized. Therefore, the inelastically scattered electrons do not contribute to the high-resolution information but contain image information at low spatial frequencies [46].

The cross-section of atoms in the sample for scattering electrons increases with the atomic number. Heavy metal atoms have much larger cross-sections than carbon, oxygen or nitrogen atoms, which are the main building blocks of biopolymers and organic polymeric materials. Therefore, heavy metal compounds play a crucial role as stains in EM. As a consequence of their larger cross-sections more electrons are scattered and contribute to the phase contrast. The stronger interaction also increases the number of electrons scattered by larger angles, leading to an increase in amplitude contrast. For a thin stained catalase crystal the amplitude contrast is of the order of 37% [47] whereas for unstained biological samples the amplitude contrast is only about 7% [48].

The cross-section for the inelastic scattering also depends on the atomic number. Up to an atomic number of about 10 the inelastic cross section is larger than the elastic cross section [49], meaning that for every elastic scattering event, which contributes to phase or amplitude contrast, at least one or more electrons are scattered inelastically and contribute to beam damage. Especially for biological objects and organic polymers, which consist mainly of elements with low atomic numbers, this leads to a serious limitation in resolution by beam damage. For example, in carbon, for each electron that is elastically scattered, three electrons are inelastically scattered (for an accelerating voltage of 100 kV), depositing an average energy between 10 and 25 eV on the sample. Therefore, to recover high-resolution information, electron doses are limited to about 5–10 e Å^{-2}. However, at these small doses images are noisy. At large spatial frequencies the signal is much lower than the noise level. To recover the high-resolution information the image information of many low-dose images has to be merged. In theory the combination of only 3000–10000 individual images [50, 51] is sufficient to calculate a 3D image reconstruction to 3 Å resolution.

5.3.1.2 Contrast Transfer Function

These considerations assume perfect transfer of contrast in the electron microscope. However, this is far from the real situation, where image contrast is modulated, depending on different parameters. In theory a thin object can be considered as a weak phase, weak amplitude object. Interference of the electron wave with the phase object introduces a phase shift χ. This phase shift depends on the defocus Δ, the spatial frequency R and the wavelength of the electrons λ. With a perfect lens a pure phase object in focus has no contrast. However, electron lenses are far from perfect. They deflect electrons much more strongly further from the optical axis than in their centre. This causes an aberration, which introduces an additional phase shift that depends on the "hardware" of the electron microscope, such as the spherical aberration of the lens C_S, the wavelength of the electrons (λ) and the spatial frequency R. The phase-shift χ is given by Eq. (11).

$$\chi = \frac{\pi}{2}(c_S\lambda^3 R^4 - 2\Delta\lambda R^2) \tag{11}$$

In phase contrast the contrast is proportional to the sine of the phase shift, whereas amplitude contrast is proportional to the cosine. For an accurate description of the image information the exact ratio of amplitude and total contrast (a) has to be known. For most conditions it is a sufficient approximation to assume that the ratio between amplitude and phase contrast is constant over the spatial frequencies. For the contrast transfer function (CTF) it follows that

$$\mathrm{CTF} = a\cos(\chi) + \sqrt{1 - a^2}\sin(\chi) \tag{12}$$

According to Eq. (12), a pure phase object has almost no contrast at low spatial frequencies, whereas a pure amplitude object shows optimal contrast.

At high spatial frequencies, additionally to the modulation of the contrast-transfer due to the phase-shift, a decrease in contrast transfer occurs that is caused by imperfections of the electron optics such as spherical and chromatic aberrations. This dampening of contrast transfer can be described by exponential decay functions [52]. For spherical aberration, the function depends on the defocus and the illumination half-angle. This has serious implications for high-resolution imaging of low contrast phase objects. Usually, micrographs are taken with a defocus of 2–5 μm to increase contrast at low spatial frequencies (visibility of the particles). At this defocus and with moderate exposure times of 1–2 s, the dampening of contrast transfer is already evident at spatial frequencies of about 1/(2 nm) in electron microscopes with thermionic guns. For better contrast-transfer at higher spatial frequencies either electron microscopes with brighter electron sources such as field emission guns or longer exposure times, which require extremely stable sample holders, have to be used.

5.3.2 Sample Preparation

In contrast to SAXS, sample preparation in EM is an important issue. Sample preparation has three major aims: (1) The object has to be reduced to a suitable size (40–500 nm thick, less than 3 mm in diameter). (2) The object has to be stabilized to resist the high vacuum inside the electron microscope. (3) The structure has to be preserved to withstand beam-damage over a high electron dose.

5.3.2.1 Vitrification of Biological Specimens

Nano-particles such as biological complexes, viruses or artificial polymeric materials are in general small enough to be imaged as a whole. However, the high vacuum inside the electron microscope makes it impossible to image these particles in an aqueous environment, where the vacuum leads to instant evaporation of the solution. This evaporation can be avoided by lowering the vapor pressure of the water by cooling the sample [53]. However, slow cooling leads to the formation of ice crystals, which can destroy the fine-structure of the object. Therefore, the aim of preservation is to maintain a state of water that is as close to liquid as possible. Such a state is the vitrified form of water, where water solidifies in an amorphous modification [54–57]. Vitrification is achieved by high cooling rates, which are of the order of 3×10^6 K s^{-1} [58]. These cooling rates are so rapid that water does not form sizeable ice crystals. Electron diffraction patterns of vitrified solutions show smooth rings, demonstrating the amorphous nature, with some variability in the average spacing between molecules [54]. The average nearest neighbor distance between two oxygen atoms in the vitrified state is 2.76 Å. The structural model, which had been proposed for the vitrified form of water, is based on cubic ice but with a greater variation of the second nearest neighbor distance [59].

In practice, vitrified samples are prepared by forming a thin film of particle suspension and plunge freezing it in liquid ethane. To form the thin film, 2–5 μL of particle suspension are applied to a carbon-coated copper grid. The carbon film is used as support for the sample. Because strong interactions between the object and the carbon film can lead to distortions and the carbon adds an undesirable background, instead of a continuous film, often a film with holes (diameter 1–5 μm) is used, where the particle suspension spreads over the holes. For a good spread of particles and an even thickness of the vitrified water, a hydrophilic surface of the support film is required, which is achieved by mild plasma etching.

For formation of the thin film most of the sample is removed with filter paper. The thickness of the remaining film is ideally between 20 and 100 nm. Thicker films are possible but usually lead to micrographs of poor quality (low contrast, multiple scattering, charging). This is a major limitation for the investigation of large nano-particles, which are thicker than 100 nm. If the thickness of the film is below the diameter of the object, the particle experiences a considerable pressure, forcing it out of the aqueous environment into the air. This pressure can be so large that deformation (e.g., liposomes) or even bursting of the object (e.g., whole cells) occurs.

Freezing Apparatus Immediately after film formation the sample is rapidly frozen by plunging it into liquid ethane. The cooling rates in ethane are sufficient to prevent formation of ice crystals. The vitrified modification is stable and can be stored below the devitrification temperature of about 150–160 K for a prolonged time. To date, different types of apparatus have been used for vitrification. The simplest one is a guillotine in which the forceps holding the grid are mounted [57]. The sample is applied to the grid and then removed by pressing a filter paper against the liquid. Exact observance of the area where filter paper and sample touch allows determination of the best moment when the film has the right thickness. Then the guillotine is released and the forceps with the grid are plunged into a pot with liquid ethane, which is cooled by a surrounding bath of liquid nitrogen.

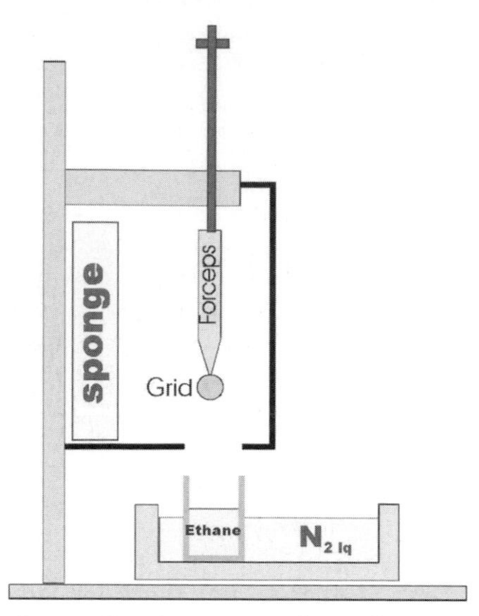

Fig. 5.6. Freezing apparatus for vitrifying samples. Forceps holding the grid are mounted in a humidified chamber (water soaked sponges). After release the grid is plunged by a spring driven mechanism into liquid ethane, which is cooled by a bath of liquid nitrogen.

The timing for blotting depends on ambient humidity, viscosity of the sample and concentration of solutes and requires some experience before reproducible results are achieved. Furthermore, evaporation of the sample leads to an increase of solute concentrations and a measurable drop in temperature. Evaporation can be reduced by surrounding the sample with a humidified chamber (Fig. 5.6), which can be temperature controlled if necessary [60]. Here blotting times are constant and no longer depend on ambient parameters. Alternatively, evaporation can be minimized by blotting the sample from both sides (e.g., Ref. [61]) where the surrounding filter paper wetted with the sample provides a defined microenvironment.

The ethane pot also requires special attention in the setup. At liquid nitrogen temperature ethane is solid. After condensation of ethane gas, ethane freezes within a relatively short time when cooled by a bath of liquid nitrogen. This provides only a short time window in which the thin film has to be formed before plunge freezing. This period can be increased if cooling is slowed by surrounding the ethane pot with an insulating layer. Alternatively, ethane can be heated with a thermo-foil to just above its freezing point. This provides constant conditions over long periods.

After freezing, the sample has to be transferred from the ethane tank to liquid nitrogen, where it is stored. During transfer, the grid is vulnerable to accumulating surface contaminations such as small hexagonal ice crystals formed by condensation of ambient humidity. Therefore, a dry environment for the transfer is desirable, which can easily be achieved by increasing the height of the walls of the surrounding nitrogen bath above the top of the ethane pot. Evaporation of liquid nitrogen forms a dry nitrogen atmosphere, which is usually sufficient to minimize surface contamination.

Vitrification preserves the sample in an environment that is similar to the aqueous phase. However, there are also important differences, such as the density of the vitrified phase, which is more similar to cubic ice than to liquid water [56, 59]. Consequently, the packing density of the water molecules must have decreased during the short period of freezing. Whether this sudden change in the structure of the water induces structural alterations in the nano-particles is still unclear. For the many biological objects investigated so far, there is no evidence that vitrification by itself induces damage. The change in density from the liquid to vitrified state increases the contrast between the nano-particle and the solvent. This has to be taken into account when combining data from EM and SAXS.

During imaging of vitrified samples constant cooling below the devitrification temperature is required. This cooling has the advantageous effect that beam damage is significantly reduced compared to that at room temperature [62–65]. One reason for this reduction is the decreased mobility of breakdown products at lower temperatures. Consequently, at even lower temperatures (4–10 K) this protective effect is further increased, permitting higher doses [66] without losing high-resolution information. In the past, this has been exploited in acquiring high-resolution information (2.8–4 Å) of two-dimensional crystals of membrane proteins (e.g., Refs. [67–71]). However, at low resolution (>20 Å) the protective effect is not evident; on the contrary, with increasing dose a massive loss in contrast occurs. This is most likely caused by an increase in density of the solute that surrounds the particles. The cause for such a change can be either a change in the modification of the water by a transition to a high density vitrified state [56] or/and the beam-induced enrichment of high density break down fragments, which usually evaporate at higher temperatures and therefore do not alter the contrast. These changes in density have to be taken into account when merging data from EM and SAXS.

5.3.3
Two-dimensional Merging of Electron Microscopic Data

To preserve a certain resolution of an organic or biological material the permissible dose is limited to about 5–20 e Å^{-2}. At higher doses, in vitrified samples bubbles form at the surface of protein or carbon [54]. This event marks severe damage to the gross structure and defines the highest tolerable dose for structure determination. As a consequence, micrographs have to be taken with low electron doses, which results in poor counting statistics. To improve the signal-to-noise ratio, and thus recover the high-resolution information, many images of different particles have to be merged. As a prerequisite for a coherent merging of the data, particles must exist in multiple identical copies. For many biological complexes, which have defined stoichiometries and assembly pathway, this is fulfilled. For artificially created nano-particles made of organic polymeric material the exact stoichiometry and shape is often not precisely determined. In this case, the assumption that the particles exist in multiple identical copies is no longer valid. For such particles, which are usually similarly sensitive to radiation as biological complexes, high-resolution structure determination becomes impossible.

5.3.3.1 Cross Correlation Function

Another prerequisite for coherent merging of image data is that the particle images show the same view of the object. If this is the case, the images are only variable by in-plane operations such as image-shift and image-rotation (x and y for the origin of the particle and ϕ for the in-plane-rotation). These parameters can be calculated by cross-correlation functions between the noisy particle images and a common reference (Fig. 5.7). The relative image shift between the reference and the raw image is determined from the position of the maximum of the cross-

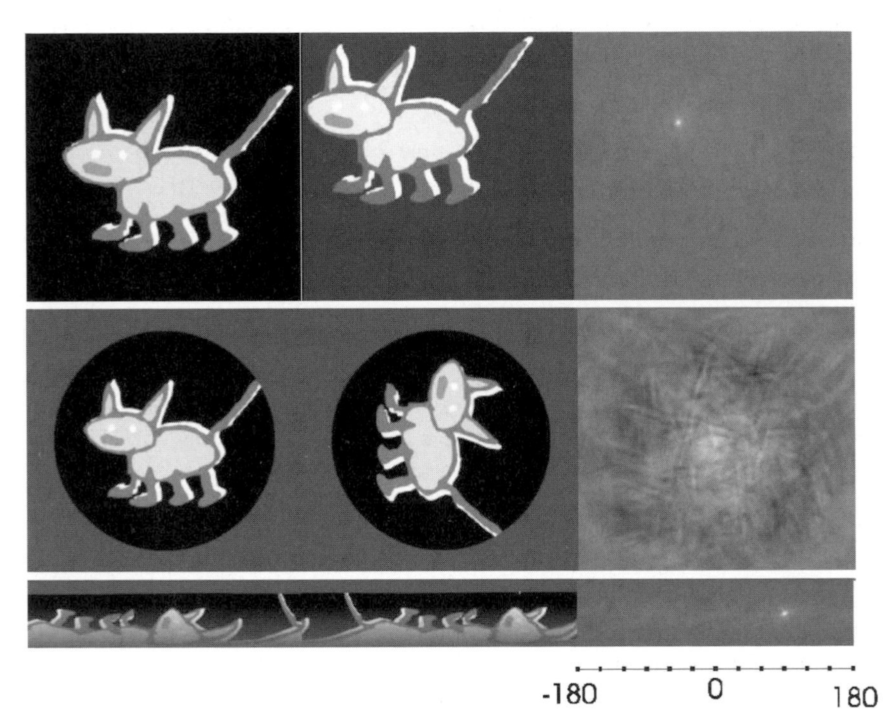

Fig. 5.7. Cross correlation function: Upper row: The cat on the left is correlated to the cat in the central panel. The cross-correlation function between the two cats shows a sharp peak (right-hand side). The position of this peak relative to the centre gives the translational shift between the two cats. Central row: The two cats vary by their in-plane rotation. The correlation function between the two cats (right-hand panel) shows a broad peak. Lower row: Cats from the central row are transformed into cylindrical coordinates. The cross-correlation function (right-hand panel) shows a defined peak. The position of the peak relative to the centre gives the relative in-plane rotation.

correlation function relative to the image-centre. The height of the peak above background is a direct measure for the correlation and thus for the signal-to-noise ratio. If the reference and raw image do not match, the peak is broad without a well-defined shape. In this case, determination of the common origin from the cross correlation function becomes unreliable. A direct measure of how well the reference and raw image agree is the cross correlation coefficient, which ranges between -1 and $+1$, where 0 means no correlation and 1 indicates that the compared images are identical. If the two images have opposite contrast, the cross-correlation coefficient becomes negative.

The rotational angle between reference and raw image is also determined by a cross correlation function but this time a polar-coordinate grid is used (Fig. 5.7). The position of the maximum in the cross-correlation function in this coordinate system gives the relative rotational angle. Exact determination of the rotational angle requires that reference image and raw image have the same origin. However, precise determination of the origin also requires the same rotational angle between the two compared images. To solve this dilemma, both origin and rotational angle are often searched alternately in an iterative process.

If all particle images in the data set are brought to a common origin and have the same in-plane rotation, the images can be averaged pixel-wise. Noise has random distribution whereas image information has the same position in all aligned images. By averaging many images the signal from the localized image information grows much faster than the one from the randomly distributed noise, increasing the signal-to-noise ratio.

For precise alignment of very noisy images the choice of reference plays a crucial role. Alignment of pure noise to a well-defined reference reproduces the basic features of the reference in the average [72, 73]. Averaging large numbers of pure noise images preserves the features of the reference over many iterative steps. This makes it difficult to spot whether a certain feature in an average is a genuine feature of the particle or arises from noise correlation. Recently, several strategies have been suggested to test data for such artifacts. One possibility is the exclusion of certain bands of spatial frequencies from the reference image. If the correlation in this resolution band is only due to noise correlation, after alignment to a reference with the missing spatial frequencies, the correlation will drop to levels below significance in this band. Another possibility is to mask off small areas in the reference image. If, after alignment, the feature in this "blind" spot of the reference is reproduced faithfully in the average, it represents a real feature of the particle and is not solely caused by noise correlation.

5.3.3.2 Identification of the Different Views

Particles prepared for EM only rarely assume a single unique orientation – an exception is two-dimensional crystals. For single particle preparations a unique orientation of particles is uncommon. Instead particles usually have different orientations and often show a random distribution of views. This creates the problem of identifying the different views and merging only those particle images that represent the same view. Identification of the different views requires statistical analysis of the image data, in which differences caused by noise are separated from those related to overall changes in shape related to a difference in particle orientation. The most common approach is a statistical method, which is similar to the principal component analysis used in the analysis of correlated data. The basic idea is that certain details of an image that are related to positional variations change in a concerted way, whereas noise varies independently across the image.

For the analysis of similarity between images, each particle image is represented as a single point in a multidimensional space. The space is defined by linear independent axes that point into the directions of the major differences between all images in the data set [74, 75]. The advantage of this representation is that comparable particle images will be in close proximity whereas images of different views will be further apart. Therefore, the distance between particle images can be directly used to group particle images according to similarity. The axes of the coordinate system can also be regarded as images (Eigenimage), representing the major changes between different views, and are therefore quite informative to identify common properties of the images. For example, in data of centered particles with random in plane rotation, rotational symmetry can be easily spotted in the Eigenimages [76]. In this case two of the major Eigenimages will show a ring of alternating bright and dark areas, which have the same rotational symmetry. Another property of the particles, which can be easily identified by analysis of the Eigenimages, is heterogeneity in size distribution [77]. Representation of particle images in Eigenimage space is the bases for the subsequent grouping of particles. Classification algorithms sort particle images in close proximity into the same class. These images represent approximately the same view and therefore fulfill the requirements for coherent merging, which leads to improvement of the signal-to-noise ratio.

For particle images representing different views (or different types of conformations or architectures), a single reference is no longer the best choice for alignment. Instead, multiple references, which represent all characteristic views, are more suitable and allow a more precise alignment. Appropriate sets of references can be generated from class averages to which each noisy particle image is aligned. As origin and rotational angle for the raw image the parameters from the alignment to the reference image with the highest cross correlation coefficient are chosen.

To reduce the bias introduced by the choice of an unrepresentative first reference a combination of classification and multi-reference alignment is favorable. This strategy was coined alignment by classification [76]. The basic idea is to classify a data-set of pre-centered particle images. A first set of references is generated from selected class-averages. By doing so only representative views are used as references and biasing by choosing an unsuitable single reference image is avoided. The limitations of this approach are connected to the accuracy of pre-centering. For approximately spherical particles quite accurate pre-centering is achieved by alignment to a rotationally averaged mean of all particle images. However, for elongated particles, where the different views vary significantly in their dimensions or for spherical particles belonging to different size classes this gives unsatisfactory results.

Alternatively to the alignment by classification, neuronal networks are used to generate self-organizing maps to avoid biasing by the first reference [78]. Although this is a quite promising stable approach, it also relies on accurately pre-centered particles.

5.3.4
Merging of EM-data in Three Dimensions

Due to the high depth of field caused by the small permissible apertures, an electron micrograph can be considered a good approximation of a projection of particle density. Consequently, no 3D information is obtained if the object is imaged from only one side. This is in contrast to light microscopy where the depth of field is much smaller than the thickness of the object. Here 3D image information is reconstructed by stacking consecutive focus layers.

5.3.4.1 Sinogram Correlation

To reconstruct a three-dimensional volume from electron micrographs different projections have to be combined. With single particles, where the object has random orientations, the class averages can be considered as random projections of the object. A prerequisite for reconstructing the 3D volume is that the relative orientations of the projected particles are known. These orientations can be determined in real space by sinogram correlation [79] or in Fourier space by the equivalent method of cross common lines [80].

In 3D-Fourier space the 2D-Fourier-transform of a projection is a single plane. Fourier transforms of different projections have the same origin but different orientations, which depend on the direction of projection in real space. The planes in Fourier-space intersect along lines at which the phases and amplitudes in the two intersecting Fourier transforms are equal. These lines are called cross-common lines. The position of the cross-common lines in a pair of intersecting transforms reflects the angular relationship between the two projections in real space. Pair wise comparison of three transforms is sufficient to allow the determination of the complete angular relationship to which further projections can be fitted. The relative orientation of the first three projections can be determined by an exhaustive search. For finding the orientation of further projections, the cross-common lines between the transform of the projection with the unknown orientation and the set of transforms with known spatial relationship are searched for all possible orientations.

The real space approach uses sinograms [79], which are stacks of regularly angular-spaced 1D-line-projections of a 2D projection. Two different 2D-projections will share one 1D-line-projection with the same profile (Fig. 5.8). This line can be identified by sinogram correlation. The position of this 1D-line projection in the two sinograms depends on the angular relationship between the 2D projections.

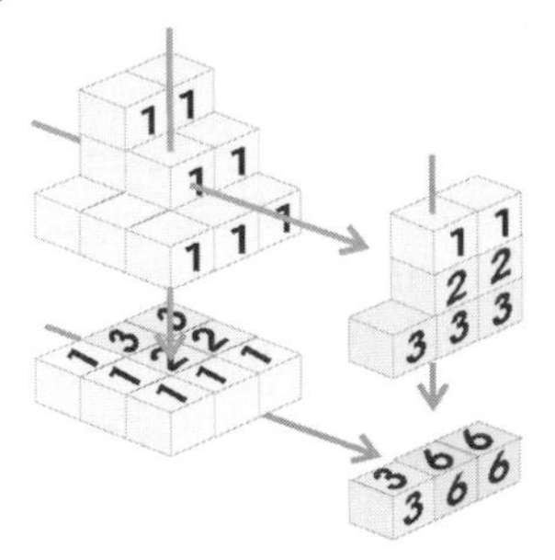

Fig. 5.8. Sinogram-correlation: Two 2D-projections of the same object share a common 1D-line projection.

The common 1D-line-projection in sinograms is equivalent to the cross-common lines between the two Fourier transforms in the cross common line approach.

For asymmetric particles these two approaches are problematic because only a very small part of the image information is used for determining the relative angular relationship between the different projections (three pairs of line projections in real space or three pairs of cross-common lines in Fourier space). Therefore, this method is only reliable for very accurate projections, which have a very high signal-to-noise ratio and are centered precisely. To cover the space optimally three approximately orthogonal views should be chosen for the first three projections.

Other approaches for the determination of orientations work with multiple micrographs of the same object where the relative tilt angle between consecutive images and the position of the tilt axis are known. These methods are much more robust for determining the initial orientations from projections of asymmetric particles. One example of these strategies is the conical tilt reconstruction [81], where two micrographs of the same object are recorded. In the first micrograph the object is highly tilted (preferably 60–70°) and in the second micrograph it is un-tilted. The same particles are selected pair wise from both micrographs. When the orientation of the tilt axis and the tilt angle are known the relative spatial orientation between the two projections of the particle can be calculated. For unambiguous determination of the spatial orientations only those particle images are considered that show the same projection in the un-tilted micrograph.

For a particular view the in-plane rotation of the particles is determined by standard alignment procedures of the images of the un-tilted object. Knowing the tilt axis and tilt angle gives the complete information on the relative spatial orientations for the complementing tilted images.

5.3.4.2 **Reconstruction of the Three-dimensional Model**

The next step is to reconstruct the 3D model from the different projections with known orientations. This can be either done in real space or in Fourier space. In real space the back-projection is used [82], where each projection is projected back to a three-dimensional volume. The direction for the back-projections is the same as the original direction of projections. For a single projection, back-projection

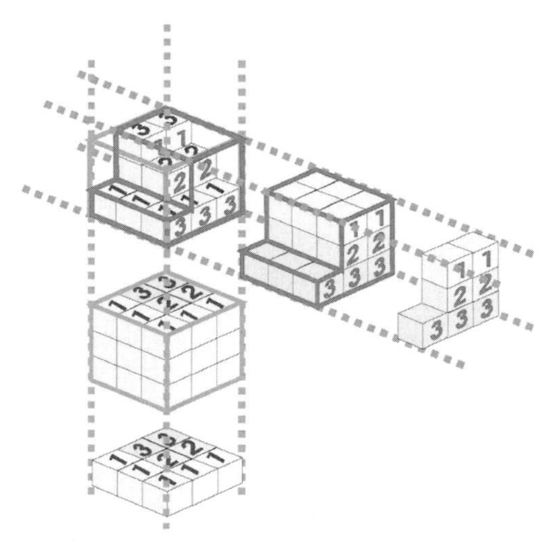

Fig. 5.9. Back-projection: 2D-projections are projected back into 3D space using the same direction for back-projection as that initially used for generating the 2D-projections. Averaging of the back-projected volumes regenerates the original volume.

results in a prism, where the image information is smeared out in the direction of projection (Fig. 5.9). Many of these back-projected volumes are averaged to reconstruct the three-dimensional volume.

This approach leads to an overestimation of the low spatial frequencies, resulting in a blurred appearance. To reduce this problem filter algorithms are employed that take the angular spread of projections and the resulting degree of over-sampling into account. This leads to a faithful representation of the 3D volume.

Alternatively to reconstruction in real space, reconstruction can also be carried out in Fourier space. Here, for the different projections, Fourier-transforms are calculated, which are combined in 3D-Fourier space. The 2D-Fourier-transforms in 3D-Fourier space are given the same orientations as determined for the corresponding projections in real space. Fourier synthesis provides the final reconstructed three-dimensional volume [83]. As in real space, over-sampling of the low-resolution information occurs and has to be properly corrected. For an object that is tilted around a single axis the maximal obtainable resolution d is given by the diameter D of the particle and the number n of evenly spaced views between -90° and $+90^\circ$ [Eq. (13)].

$$d = \pi \frac{D}{n} \tag{13}$$

For a nano-particle of 50 nm diameter, for instance, about 80 regularly spaced projections are needed to reconstruct the volume at 2 nm resolution. At higher resolution the Fourier-space is under-sampled and can therefore not be reconstructed properly. According to the formula thinner objects need less independent projections. Importantly, resolution depends not only on these spatial considerations but also on the signal-to-noise ratio, which is related to the cumulative number of scattered electrons [51] and the general quality of the micrograph.

Often the projections are not equally spaced over the asymmetric unit but cover only a certain area. In this case the resolution is no longer isotropic and is worse in the direction where views are missing. Such image reconstructions look blurred and elongated in one direction.

For further refinement of orientations, reference-based approaches are chosen where the 3D model is taken into account. For example, in sinogram correlation the orientation of a projection is determined by sinogram correlation against a set of sinograms of equally spaced "anchor"-projections of the three-dimensional volume. Alternatively a projection matching algorithm is employed, where the volume is projected in different directions and the particle images are aligned by multi-reference alignment to this set of references. After alignment the particle is assigned the same orientation as the direction of projection of the reference image to which the particle correlated best [84]. A similar approach is followed in Fourier-space. Here Fourier transforms of the particle images are compared to the 3D-Fourier-transform of the reference-volume, testing different orientations for the best match [85].

For high-resolution image reconstructions, the transfer of contrast by the electron microscope plays an important role. This modulation is described by the contrast transfer function. Depending on the defocus and other imaging conditions this function has zeros at certain spatial frequencies, where no information is transferred and pass-bands at other spatial frequencies, where the information is transferred alternately with positive or negative contrast. In the pass-bands the strength of contrast transfer varies between no transfer and complete transfer of information. The simplest approach to correction is the contrast-inversion in every other pass-band, which leads to transfer of image information with the same contrast. Doing that is sufficient for obtaining image reconstructions where the phases are accurately preserved but the amplitudes show a clear deviation from their true values. Especially at low spatial frequencies the amplitudes are severely underestimated (e.g., no phase contrast transfer at 0 spatial frequencies). More sophisticated approaches also attempt to correct for the amplitude information. However, at spatial frequencies, where the overall contrast transfer is close to zero or zero, the information is often too noisy and should not be amplified by huge factors. This problem is solved by a Wiener filtering algorithm in which a maximal permissible amplification factor is set (e.g., Ref. [86]). The choice of maximal amplification factor depends on the signal-to-noise ratio and is chosen close to 1 when correction for the contrast transfer function is done on raw images and can be considerably higher (e.g., 10; Ref. [86]) if the correction is done on averaged images. Even with these more sophisticated approaches to contrast transfer correction, low spatial frequencies are usually underestimated. Therefore, the amplitude profile is not correctly maintained. However, the relative amplitudes in a certain resolution band are preserved accurately.

5.4 Merging of Methods

5.4.1 Comparison of EM and SAXS Data

SAXS data provides information on the accurate amplitude profile. For a combination with electron microscopic data a relative scaling of the amplitude profiles is required. For such scaling it is necessary to understand the underlying mechanisms that influence the amplitude profile in the electron microscopic data. We have already discussed in detail the effect of the contrast transfer function. An exact knowledge of if and how it was corrected is a prerequisite for the interpretation of the amplitude profile. In addition, at low spatial frequencies the contrast between particles and solvent determines the amplitude profile. Here differences in density between water in the SAXS measurements and vitrified water in the electron microscopic measurements have to be taken into account. Furthermore, in many image reconstructions of electron microscopic images band-pass filtering of the data has been used to suppress information at low and high spatial frequencies. Again the exact filter parameters and the shape of the filter have to be known to reproduce the true amplitude profile. Another serious effect on the amplitude profile is decay in contrast-transfer at higher spatial frequencies, which is best described by a temperature factor, similar to the one known from X-ray crystallography. Many parameters contribute to this overall temperature factor, such as spherical and chromatic aberration of the microscope, stability of the holder, beam damage, charging, inaccuracy in the determination of particle orientations, effects of interpolation in image processing, and the modulation transfer function of the CCD or the photographic film and scanner. Typical cumulated temperature factors have been estimated to range between ca. 500 and 1000 Å^2 [51, 87], which is considerably larger than those observed in X-ray crystallography. Accurate estimates of the temperature factors in the raw data come from comparison of the amplitude profile from electron micrographs and SAXS measurements on Herpes Simplex Virus [88]. These temperature factors ranged between 50 and 200 Å^2, depending on the defocus.

To illustrate that, we discuss here a direct comparison of SAXS and EM data from *Lumbricus terrestris* hemoglobin. SAXS data include the scattering curve, $I(h)$, pair distance distribution function, $p(r)$, radius of gyration, R_G, hydrated volume, V and the maximum diameter of the particle, D_{max}. EM data provide information on the 3D electron density distribution (Table 5.1). For direct comparison the EM data need to be converted into data suitable for SAXS model calculations. As outlined in previous sections, in SAXS models, the protein mass is simulated with a large number of small spheres. Accordingly, to compare EM data with SAXS profiles, EM data are used to calculate models with a large number of small spheres, whereby the position of the spheres is given by the EM model coordinates. Then model scattering functions, pair distance distribution functions and molecular parameters (radius of gyration, maximum particle diameter, volume) of the EM model can be calculated in a way similar to the calculation of the SAXS models (Table 5.1). To convert the EM electron density in a meaningful weighting scheme for the model calculations, the weight of spheres is chosen to represent density values between 0.3345 (electron density of water) and 0.4395 (averaged electron density of a protein). Here the value for the electron density of water was chosen, because vitrified water, as used in the EM experiment, forms no ice crystals (see above). Additionally, appropriate rescaling is necessary to fit the positions of the maxima of the experimental scattering curve and the model parameters (scaling due to magnification differences).

A common problem in EM is the exact determination of the volume of the biological sample and here knowledge from SAXS studies may help. Figure 5.10

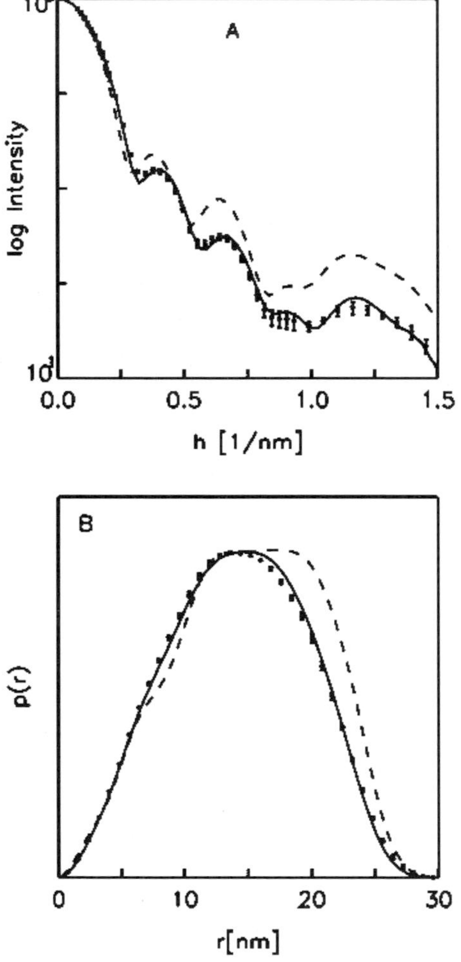

Fig. 5.10. Comparison EM (dashed line) and SAXS data (experimental data dotted with error bars; model data solid line) for *Lumbricus terrestris* hemoglobin.

shows that the threshold of the EM model has to be chosen in a way that the model represents the hydrated particle volume rather than the "expected" volume of the protein to give good agreement of data. The "expected" particle volume is calculated by the molecular weight of the protein, and the partial specific volume. This is usually much smaller than the hydrated volume [18].

We see that SAXS and EM data agree very well up to a scattering angle of $h = 1.2\ \text{nm}^{-1}$. Careful analyses of the scattering curves reveal that the 1st minimum of the experimental scattering curve is not fitted within the limits of error by the scattering curve calculated from the EM data. The minimum of the model scattering curve seems to be more pronounced than the minimum of the experimental curve. This, however, must not be taken too seriously. In scattering experiments the randomly oriented particles in solution together account for the observed scattering and the structural information from all the particles in solution is summarized in one scattering curve. This summarizing of structural information may lead to an additional "smearing" of the minima of the scattering curve. If, on the other hand, the scattering curve of a model (e.g., an EM model) is calculated, the information of only one particle is present, giving rise to a very "clear" scattering curve. Therefore, model scattering curves sometimes show more pronounced minima than the experimental curves and consequently, as long as the positions of the minima and the heights and positions of the maxima are fitted correctly by the model data, this does not decrease the quality of a model.

The observed differences in the outer part of the scattering curve, however, may be due to slightly different structures of the protein observed in vitreous ice and in solution. Another explanation might be the contrast transfer of the magnetic lens of the electron microscope or the application of an additional filtering procedure during the EM reconstruction (as mentioned above).

In general, it is important to state that the value of the hydrated volume can only be determined with sufficient certainty from the SAXS data, whereas information on the 3D electron density distribution is contained in the EM data. Thus, both methods together give maximal information.

5.4.2 SAXS Modeling Approaches using EM Information

We have seen that SAXS modeling is not unique and therefore SAXS data are usually interpreted by comparing the experimental data with model profiles. The creation of model profiles is not trivial and, as stated above, it is very useful to implement structural knowledge from other methods to increase the reliability of any SAXS modeling attempt. As an example we want to discuss here an indirect incorporation of information gained by EM into automated SAXS modeling: In this approach, first templates are created using the structural information gained by EM. Subsequently, those templates are used as starting models for the automated SAXS model creation.

In Fig. 5.11 different models of the hemoglobin from the leech *Macrobdella decora* are shown. Overall the protein complex is related to the hemoglobin from

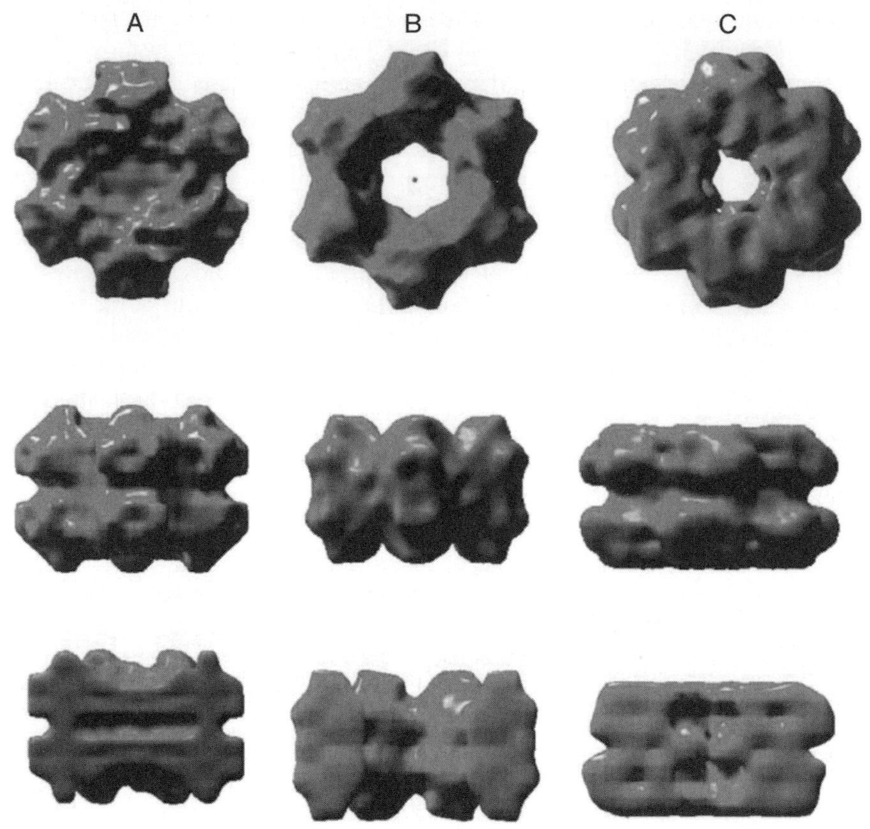

Fig. 5.11. Models of *Macrobdella decora* hemoglobin: (A) two-step trial-and-error SAXS model, (B) automatic generation without information other than *D*6 symmetry, (C) automatic generation employing a template created with information from EM data.

Lumbricus terrestris and it has a similar appearance (hexagonal bilayer architecture). In Fig. 5.11 the trial-and-error approach from SAXS multi-body modeling is shown in part (A) (similar to the approach shown earlier for *L. terrestris* hemoglobin). Fig. 5.11(B) shows the automatically generated SAXS model calculated without structural information other than *D*6 symmetry (*ab initio* modeling). For Fig. 5.11(C) additional information concerning shape and internal structure was incorporated in the calculations by the use of a template derived from 3D EM reconstructions [89]. Clearly, this final example shows that SAXS modeling procedures can be increased in accuracy and reliability through the use of suitable EM templates. SAXS and EM model values are given in Table 5.2.

In conclusion, if characterization of a nano-particle is attempted by either SAXS or EM, a combination of knowledge gained by different methods is advisable. Although such approaches are not yet well established, the results shown here illustrate that this is the way forward for nano applications.

Tab. 5.2. Experimental SAXS parameters and data of SAXS models of *Macrobdella decora* hemoglobin.

	Exp.[a,b]	*Two-step model*[c]	*Automatic generation with D6*	*Automatic generation with EM template*
R_G (nm)	10.75 ± 0.01	10.73 ± 0.02	10.79 ± 0.01	10.78 ± 0.01
D_{max} (nm)	29.54 ± 0.02	29.6 ± 0.1		
V (nm^3)	5800 ± 300	6000 ± 300	5895 ± 155	5746 ± 111
r_b (nm)		0.66	0.8	0.8
d_z (nm)[d]		19.4 ± 1.7	22.5 ± 2.2	17.6 ± 0.0
Consensus model:				
Number of models included (N_m)		4	11	8
Total number of spheres (N_b)		5970	1662 ± 44	1620 ± 31

[a] Experimental data taken from Ref. 90.
[b] Mean values and standard deviations result from averaging experimental data.
[c] Mean values and standard deviations result from averaging the four models included in the consensus model shown in Fig. 5.11.
[d] Values for d_z refer to the largest dimension in the z direction, obtained from the difference of sphere coordinates by adding the diameter of one sphere.

References

1 Guinier, A., Fournet, J., *Small-angle Scattering of X-rays*, **1995**, Wiley, New York.

2 Pessen, H., Kumosinski, F., Timasheff, S.N., Small-angle X-ray scattering, *Methods Enzymol.* **1973**, *27*, part D, 151–209.

3 Pilz, I., Glatter, O., Kratky, O. Small-angle X-ray scattering, *Methods Enzymol.* **1979**, *61*, 148.

4 Glatter, O., Kratky, O. (eds), *Small-angle X-ray Scattering*, **1982**, Academic Press, New York, London.

5 Kratky, O., Pilz, I. Recent advances and applications of diffuse X-ray small-angle scattering on biopolymers in dilute solutions. *Quart. Rev. Biophys.* **1972**, *5*, 481–537.

6 Feigin, L.A., Svergun, D.I. *Structure Analysis by Small-Angle X-ray and Neutron Scattering*, **1987**, Plenum Press, New York.

7 Zipper, P., Durchschlag, H. Modeling of protein solution structures. *J. Appl. Crystallogr* **2003**, *36*, 509.

8 Durchschlag, H., Zipper, P. Modeling the hydration of proteins: prediction of structural and hydrodynamic parameters from X-ray diffraction and scattering data. *Eur. Biophys. J.* **2003**, *32*, 487.

9 Kumosinski, T.F., Pessen, H. Estimation of sedimentation coefficients of globular proteins: an application of small-angle X-ray scattering. *Arch. Biochem. Biophys.* **1982**, *219*, 89.

10 Durchschlag, H., Zipper, P. Calculation of structural parameters from hydrodynamic data. *Prog. Colloid Polym. Sci.* **1999**, *113*, 87.

11 Durchschlag, H., Zipper, P. Prediction of hydrodynamic

parameters of biopolymers from small-angle scattering data. *J. Appl. Crystallogr.* **1997**, *30*, 1112.

12 Byron, O. Hydrodynamic bead modeling of biological macro-molecules. *Methods Enzymol.* **2000**, *321*, 278.

13 Svergun, D., Barberato, C., Koch, M.H.J. *CRYSOL* – a program to evaluate X-ray solution scattering of biological macromolecules from atomic coordinates. *J. Appl. Crystallogr.* **1995**, *28*, 768.

14 Perkins, S.J., Ashton, A.W., Boehm, M.K., Chamberlain, D. Molecular structures from low angle X-ray and neutron scattering studies. *Int. J. Biol. Macromol.* **1998**, *22*, 1.

15 Zipper, P., Krebs, A., Durchschlag, H. Comparative modeling of giant annelid hemoglobins. *Prog. Colloid Polym. Sci.* **2004**, *127*, 126.

16 Zipper, P., Krebs, A., Durchschlag, H. Prediction of hydrodynamic parameters of *Lumbricus terrestris* hemoglobin from small-angle X-ray and electron microscopic structures. *Prog. Colloid Polym. Sci.* **2002**, *119*, 141.

17 Krebs, A., Durchschlag, H., Zipper, P. Small angle X-ray scattering studies and modeling of *Eudistylia vancouverii* chlorocruorin and *Macrobdella decora* hemoglobin. *Biophys. J.* **2004**, *87*, 1173.

18 Krebs, A., Lamy, J., Vinogradov, S.N., Zipper, P. *Lumbricus terrestris* hemoglobin: a comparison of small-angle x-ray scattering and cryoelectron microscopy data. *Biopolymers* **1998**, *45*, 289.

19 Royer, W.E. Jr, Hendrickson, W.A. Molecular symmetry of *Lumbricus* erythrocruorin. *J. Biol. Chem.* **1998**, *263*, 13762.

20 Boekema, E.J., van Heel, M. Molecular shape of *Lumbricus terrestris* erythrocruorin studied by electron microscopy and image analysis. *Biochim. Biophys. Acta* **1988**, *957*, 370.

21 Martin, P.D., Kuchumov, A.R., Green, B.N., Oliver, R.W.A., Braswell, E.H., Wall, J.S., Vinogradov, S.N. Mass spectrometric composition and molecular mass of *Lumbricus terrestris* hemoglobin: a refined model of its quaternary structure. *J. Mol. Biol.* **1996**, *255*, 154.

22 Zhu, H., Ownby, D.W., Riggs, C.K., Nolasco, N.J., Stoops, J.K., Riggs, A.F. Assembly of the gigantic hemoglobin of the earthworm *Lumbricus terrestris. J. Biol. Chem.* **1996**, *271*, 30007.

23 Ownby, D.W., Zhu, H., Schneider, K., Beavis, R.C., Chait, B.T., Riggs, A.F. The extracellular hemoglobin of the earthworm, *Lumbricus terrestris.* Determination of subunit stoichio-metry. *J. Biol. Chem.* **1993**, *268*, 13539.

24 Vinogradov, S.N., Lugo, S.D., Mainwaring, M.G., Kapp, O.H., Crewe, A.V. Bracelet protein: a quaternary structure proposed for the giant extracellular hemoglobin of *Lumbricus terrestris. Proc. Natl. Acad. Sci. U.S.A.* **1986**, *83*, 8034.

25 Vinogradov, S.N., Sharma, P.K., Qabar, A.N., Wall, J.S., Westrick, J.A., Simmons, J.H., Gill, S.J. A dodecamer of globin chains is the principal functional subunit of the extracellular hemoglobin of *Lumbricus terrestris. J. Biol. Chem.* **1991**, *266*, 13091.

26 Martin, P.D., Eisele, K.L., Doyle, M.A., Kuchumov, A.R., Walz, D.A., Arutyunyan, E.G., Vinogradov, S.N., Edwards, B.F.P. Molecular symmetry of the dodecamer subunit of *Lumbricus terrestris* hemoglobin. *J. Mol. Biol.* **1996**, *255*, 170.

27 Schatz, M., Orlova, E.V., Dube, P., Jäger, J., van Heel, M. Structure of *Lumbricus terrestris* hemoglobin at 30 Å resolution determined using angular reconstitution. *J. Struct. Biol.* **1995**, *114*, 28.

28 Taveau, J.-C., Boisset, N., Vino-gradov, S.N., Lamy, J.N. Three-dimensional reconstruction of *Lumbricus terrestris* hemoglobin at 22 Å resolution: intramolecular localization of the globin and linker chains. *J. Mol. Biol.* **1999**, *289*, 1343.

29 Mouche, F., Boisset, N., Penczek, P.A. *Lumbricus terrestris* hemoglobin – the architecture of linker chains and

structural variation of the central toroid. *J. Struct. Biol.* **2001**, *133*, 176.

30 Royer, W.E. Jr, Strand, K., van Heel, M., Hendrickson, W.A. Structural hierarchy in erythrocruorin, the giant respiratory assemblage of annelids. *Proc. Natl. Acad. Sci. U.S.A.* **2000**, *97*, 7107.

31 Svergun, D.I. Advanced solution scattering data analysis methods and their applications. *J. Appl. Crystallogr.* **2000**, *33*, 530.

32 Svergun, D.I., Koch, M.H.J. Advances in structure analysis using small-angle scattering in solution. *Curr. Opin. Struct. Biol.* **2002**, *12*, 654.

33 Koch, M.H.J., Vachette, P., Svergun, D.I. Small-angle scattering: a view on the properties, structures and structural changes of biological macromolecules in solution. *Quart. Rev. Biophys.* **2003**, *36*, 147.

34 Hammel, M., Kriechbaum, M., Gries, A., Kostner, G.M., Laggner, P., Prassl, R. Solution structure of human and bovine beta(2)-glycoprotein I revealed by small-angle X-ray scattering. *J. Mol. Biol.* **2002**, *321*, 85.

35 Takahashi, Y., Nishikawa, Y., Fujisawa, T. Evaluation of three algorithms for *ab initio* determination of three-dimensional shape from one-dimensional solution scattering profiles. *J. Appl. Crystallogr.* **2003**, *36*, 549.

36 Svergun, D.I. Restoring low resolution structure of biological macromolecules from solution scattering using simulated annealing. *Biophys. J.* **1999**, *76*, 2879.

37 Svergun, D.I., Petoukhov, M.V., Koch, M.H.J. Determination of domain structure of proteins from X-ray solution scattering. *Biophys. J.* **2001**, *80*, 2946.

38 Petoukhov, M.V., Svergun, D.I. New methods for domain structure determination of proteins from solution scattering data. *J. Appl. Crystallogr.* **2003**, *36*, 540.

39 Chacón, P., Morán, F., Díaz, J.F., Pantos, E., Andreu, J.M. Low-resolution structures of proteins in solution retrieved from X-ray scattering with a genetic algorithm. *Biophys. J.* **1998**, *74*, 2760.

40 Walther, D., Cohen, F.E., Doniach, S. Reconstruction of low-resolution three-dimensional density maps from one-dimensional small-angle X-ray solution scattering data for biomolecules. *J. Appl. Crystallogr.* **2000**, *33*, 350.

41 Vigil, D., Gallagher, S.C., Trewhella, J., García, A.E. Functional dynamics of the hydrophobic cleft in the N-domain of calmodulin. *Biophys. J.* **2001**, *80*, 2082.

42 Kojima, M., Timchenko, A.A., Higo, J., Ito, K., Kihara, H., Takahashi, K. Structural refinement by restrained molecular-dynamics algorithm with small-angle X-ray scattering constraints for a biomolecule. *J. Appl. Crystallogr.* **2004**, *37*, 103.

43 Chacón, P., Díaz, J.F., Morán, F., Andreu, J.M. Reconstruction of protein form with X-ray solution scattering and a genetic algorithm. *J. Mol. Biol.* **2000**, *299*, 1289.

44 Zipper, P., Durchschlag, H., Krebs, A. Modeling of biopolymers, **2005**, in Scott, D.J., Harding, S.E., Rowe, A.J. (eds.) Modern analytical ultrazentrifugation: techniques and methods. Royal Society of Chemistry, Cambridge, U.K.

45 Volkov, V.V., Svergun, D.I. Uniqueness of *ab initio* shape determination in small-angle scattering. *J. Appl. Crystallogr.* **2003**, *36*, 860.

46 Langmore, J.P., Smith, M.F. Quantitative energy-filtered electron microscopy of biological molecules in ice. *Ultramicroscopy*, **1992**, *46*, 349–373.

47 Erickson, H.P., Klug, A. Measurement and compensation of defocusing and aberrations by Fourier processing of electron micrographs. *Phil. Trans. Roy. Soc. Lond. B*, **1971**, *261*, 105–118.

48 Toyoshima, C., Unwin, N. Contrast transfer for frozen-hydrated specimens: determination from pairs of defocused images. *Ultramicroscopy*, **1988**, *25*, 279–291.

49 Colliex, C., Jeanguillaume, C., Mory, C. Unconventional modes for STEM imaging of biological structures. *J. Ultrastruct. Res.*, **1984**, *88*, 177–206.

50 Henderson, R. The potential and limitations of neutrons, electrons and X-rays for atomic resolution microscopy of unstained biological molecules. *Quart. Rev. Biophys.*, **1995**, *28*, 171–193.

51 Rosenthal, P.B., Henderson, R. Optimal determination of particle orientation, absolute hand, and contrast loss in single-particle electron cryomicroscopy. *J. Mol. Biol.*, **2003**, *333*, 721–745.

52 Wade, R.H. A brief look at imaging and contrast transfer. *Ultramicroscopy*, **1992**, *46*, 145–156.

53 Taylor, K.A., Glaeser, R.M. Electron diffraction of frozen, hydrated protein crystals. *Science*, **1974**, *186*, 1036–1037.

54 Dubochet, J., Lepault, J., Freeman, R., Berriman, J., Homo, J.C. Electron microscopy of frozen water and aqueous solutions. *J. Microscop.*, **1982**, *128*, 219–237.

55 Dubochet, J., Adrian, M., Chang, J.J., Homo, J.C., Lepault, J., McDowall, A.W., Schultz, P. Cryo-electron microscopy of vitrified specimens. *Q. Rev. Biophys.*, **1988**, *21*, 129–228.

56 Heide, H.G. Observation on ice layers. *Ultramicroscopy*, **1984**, *14*, 271–278.

57 Heide, H.G., Zeitler, E. The Physical Behaviour of solid water at low temperatures and the embedding of electron microscopical specimens. *Ultramicroscopy*, **1985**, *16*, 151–160.

58 Bald, W.B. On crystal size and cooling rate. *J. Microsc.*, **1986**, *143* (Pt 1), 89–102.

59 Narten, A.H. Diffraction pattern and structure of amorphous solid water at 10 and 77 K. *J. Chem. Phys.*, **1976**, *64*, 1106–1120.

60 Bellare, J.R., Davis, H.T., Scriven, L.E., Talmon, Y. Controlled environment vitrification system: An improved sample preparation technique. *J. Electron. Microsc. Techn.*, **1988**, *10*, 87–111.

61 White, H.D., Thirumurugan, K., Walker, M.L., Trinick, J. A second generation apparatus for time-resolved electron cryo-microscopy using stepper motors and electrospray. *J. Struct. Biol.*, **2003**, *144*, 246–252.

62 Egerton, R.F. Chemical measurement of radiation damage in organic samples at and below room temperature. *Ultramicroscopy*, **1980**, *5*, 521–523.

63 Knapek, E., Dubochet, J. Beam damage to organic material is considerably reduced in cryo-electron microscopy. *J. Mol. Biol.*, **1980**, *141*, 147–161.

64 Lamvik, M.K. Radiation damage in dry and frozen hydrated organic material. *J. Microscop.*, **1991**, *161*, 171–181.

65 Siegel, G. Der Einfluß tiefer temperature auf die Strahlenschädigung von organischen kristallen durch 100 keV-elektronen. *Z. Naturforsch.*, **1971**, *27*, 325–332.

66 Stark, H., Zemlin, F., Boettcher, C. Electron radiation damage to protein crystals of bacteriorhodopsin at different temperatures. *Ultramicroscopy*, **1996**, *63*, 75–79.

67 Baldwin, J.M., Henderson, R., Beckman, E., Zemlin, F. Images of purple membrane at 2.8 A resolution obtained by cryo-electron microscopy. *J. Mol. Biol.*, **1988**, *202*, 585–591.

68 Henderson, R., Baldwin, J.M., Ceska, T.A., Zemlin, F., Beckmann, E., Downing, K.H. Model for the structure of bacteriorhodopsin based on high-resolution electron cryo-microscopy. *J. Mol. Biol.*, **1990**, *213*, 899–929.

69 Kühlbrandt, W., Wang, D.N., Fujiyoshi, Y. Atomic model of plant light-harvesting complex by electron crystallography. *Nature*, **1994**, *367*, 614–621.

70 Miyazawa, A., Fujiyoshi, Y., Unwin, N. Structure and gating mechanism of the acetylcholine receptor pore. *Nature*, **2003**, *423*, 949–955.

71 Murata, K., Mitsuoka, K., Hirai, T.,

Walz, T., Agre, P., Heymann, J.B., Engel, A., Fujiyoshi, Y. Structural determinants of water permeation through aquaporin-1. *Nature*, **2000**, *407*, 599–605.

72 Shaikh, T.R., Hegerl, R., Frank, J. An approach to examining model dependence in EM reconstructions using cross-validation. *J. Struct. Biol.*, **2003**, *142*, 301–310.

73 Stewart, A., Grigorieff, N. Noise bias in the refinement of structures derived from single particles. *Ultramicroscopy*, **2004**, *102*, 67–84.

74 van Heel, M. Multivariate statistical classification of noisy images (randomly oriented biological macromolecules). *Ultramicroscopy*, **1984**, *13*, 165–183.

75 van Heel, M., Frank, J. Use of multivariate statistics in analysing the images of biological macromolecules. *Ultramicroscopy*, **1981**, *6*, 187–194.

76 Dube, P., Tavares, P., Lurz, R., van Heel, M. The portal protein of bacteriophage SPP1: a DNA pump with 13-fold symmetry. *Embo J.*, **1993**, *12*, 1303–1309.

77 White, H.E., Saibil, H.R., Ignatiou, A., Orlova, E.V. Recognition and separation of single particles with size variation by statistical analysis of their images. *J. Mol. Biol.*, **2004**, *336*, 453–460.

78 Pascual-Montano, A., Donate, L.E., Valle, M., Barcena, M., Pascual-Marqui, R.D., Carazo, J.M. A novel neural network technique for analysis and classification of EM single-particle images. *J. Struct. Biol.*, **2001**, *133*, 233–245.

79 van Heel, M. Angular reconstitution: A posteriori determination of projection directions for 3D reconstructions. *Ultramicroscopy*, **1987**, *21*, 110–113.

80 DeRosier, D.J., Klug, A. Reconstruction of three dimensional structures from electron micrographs. *Nature*, **1968**, *217*, 130–134.

81 Radermacher, M. Three-dimensional reconstruction of single particles from random and nonrandom tilt series. *J. Electron. Microsc. Tech.*, **1988**, *9*, 359–394.

82 Hoppe, W., Schramm, H., Sturm, M., Hunsmann, N., Gassman, J. 3-dimensional electron-microscopy of individual biological objects. 1. Methods. *Z. Naturforsch*, **1976**, *A31*, 645–655.

83 Crowther, R.A., Klug, A. Structural analysis of macromolecular assemblies by image reconstruction from electron micrographs. *Annu. Rev. Biochem.*, **1975**, *44*, 161–182.

84 Sander, B., Golas, M.M., Stark, H. Corrim-based alignment for improved speed in single-particle image processing. *J. Struct. Biol.*, **2003**, *143*, 219–228.

85 Grigorieff, N. Three-dimensional structure of bovine NADH:ubiquinone oxidoreductase (complex I) at 22 A in ice. *J. Mol. Biol.*, **1998**, *277*, 1033–1046.

86 Böttcher, B., Crowther, R.A. Difference imaging reveals ordered regions of RNA in turnip yellow mosaic virus. *Structure*, **1996**, *4*, 387–394.

87 Böttcher, B., Wynne, S.A., Crowther, R.A. Determination of the fold of the core protein of hepatitis B virus by electron cryomicroscopy. *Nature*, **1997**, *386*, 88–91.

88 Saad, A., Ludtke, S.J., Jakana, J., Rixon, F.J., Tsuruta, H., Chiu, W. Fourier amplitude decay of electron cryomicroscopic images of single particles and effects on structure determination. *J. Struct. Biol.*, **2001**, *133*, 32–42.

89 de Haas, F., Boisset, N., Taveau, J.-C., Lambert, O., Vinogradov, S.N., Lamy, J. Three-dimensional reconstruction of *Macrobdella decora* (Leech) hemoglobin by cryoelectron microscopy. *Biophys. J.* **1996**, *70*, 1973–1984.

90 Krebs, A., Zipper, P., Vinogradov, S.N. Lack of size and shape alteration of oxygenated and deoxygenated Lumbricus terrestris hemoglobin? *Biochim. Biophys. Acta* **1996**, *1297*, 115.

6
In Situ Characterization of Drug Nanoparticles by FTIR Spectroscopy

Michael Türk and Ruth Signorell

6.1
Introduction

Infrared (IR) spectroscopy is an important tool to characterize nanomaterials in life sciences. Refs. [1–10] highlight only a few recent examples in this large field, including catalysis using nanoparticles, the targeted synthesis of nanoparticles, the use of nanoparticles as biosensors, and the characterization of drug nanoparticles. This chapter focuses on particulate matter built from molecules and in particular on drug nanoparticles. For the characterization of these molecular nanoparticles IR spectroscopy is particularly well suited. It is relatively sensitive and non-invasive, which is decisive for instance for the investigation of sensitive molecular particles. Furthermore, IR spectroscopy can be used for different kinds of particulate samples (aerosols, particles on a holder, or particles in a matrix) and under a broad range of different experimental conditions (temperature, pressure, etc.). For molecular particles, it opens a direct window to the characteristic intermolecular and intramolecular vibrational dynamics. The vibrational spectra of such large aggregates contain a wealth of information not only about the chemical composition or the phase behavior of the particles, but also about intrinsic particle properties such as the particle size, the particle shape, or structural changes in the particles' surface.

Approximately 80% of all pharmaceutical products are in solid dosage form. Thus, in biomedical applications both the size and shape of the solid particles are important quantities to know, for instance because they can affect the bioavailability of drug particles. The size of the particles manifests itself in the IR extinction spectra by scattering phenomena. They lead to slanted baselines and to a dispersion shape of the absorption bands. To determine the size distribution for a particle ensemble directly from the infrared spectrum it is crucial to know the frequency-dependent optical data (indices of refraction) of the particles. Nowadays, few databases provide refractive index data for particulate systems and most of these data bases focus more on molecular ice particles than on biomolecular substances. Since the optical properties of particles can differ markedly from those of the solid bulk, corresponding bulk data are often not suitable either. This clearly illustrates the need for optical data of particles, especially for substances of biomolecular in-

Nanotechnologies for the Life Sciences Vol. 3
Nanosystem Characterization Tools in the Life Sciences. Edited by Challa S. S. R. Kumar

ISBN: 3-527-31383-4

terest. The particles' shape can lead to characteristic band structures in IR extinction spectra. These phenomena, however, are pronounced only for very strong vibrational transitions. The structure and position of weak absorption bands are dominated by rather local effects, i.e., by interactions with neighboring molecules, such as the formation of specific hydrogen-bonded networks in the particles.

The main focus of this chapter lies in the *in situ* characterization of nanoparticles by Fourier-transform infrared (FTIR) spectroscopy. Most often, off-line techniques are used to investigate particles, which have, consequently, first to be collected on a holder. This approach is not only very slow, but the process of collecting can also affect the particles' properties. *In situ* characterization by contrast is quick, thus allowing us to investigate and control the particle properties during their formation. We compare and complement the results from direct absorption FTIR spectroscopy *in situ* with particle sizing using a scanning mobility particle sizer, with 3-wavelengths-extinction measurements, with scanning electron microscopy, with differential scanning calorimetry, and with X-ray diffraction.

The infrared investigations are combined here with two different particle generation methods, *viz.* Rapid Expansion of Supercritical Solutions (RESS) and electro-spraying. Rapid expansion of supercritical CO_2 solutions is a particularly useful method to micronize thermally labile drugs, which often consist of lipophilic compounds. The main advantage of the solvent CO_2 lies in its low critical data and in the fact that it can easily be separated from the product after particle formation. As an example, the mixing/coating of phytosterol with polymers (Eudragit©, L-PLA) is discussed. Encapsulation of drugs with polymers is especially attractive to avoid agglomeration and to control drug release in the body. Electro-spraying of aqueous or alcoholic solutions allows us to generate nanoscale particles of hydrophilic compounds. Here, the particle formation of sugar-like substances is discussed. This class of substances is attractive as carrier for administering drugs in solid dosage form.

6.2 Particle Generation Methods

6.2.1 Rapid Expansion of Supercritical Solutions (RESS)

The potential of RESS for the micronization of molecular substances has been demonstrated previously [11–15]. The key idea behind RESS is to dissolve the solute of interest in a supercritical fluid followed by a rapid expansion of the supercritical solution. This leads to an extremely fast phase change from the supercritical to the gas-like state, resulting in high supersaturation in the supersonic free jet and the formation of submicron particles. Since the solvent is a dilute gas after expansion, the RESS process offers a highly pure final product. Dissolution experiments demonstrate that the RESS processing of Griseofulvin leads to a signifi-

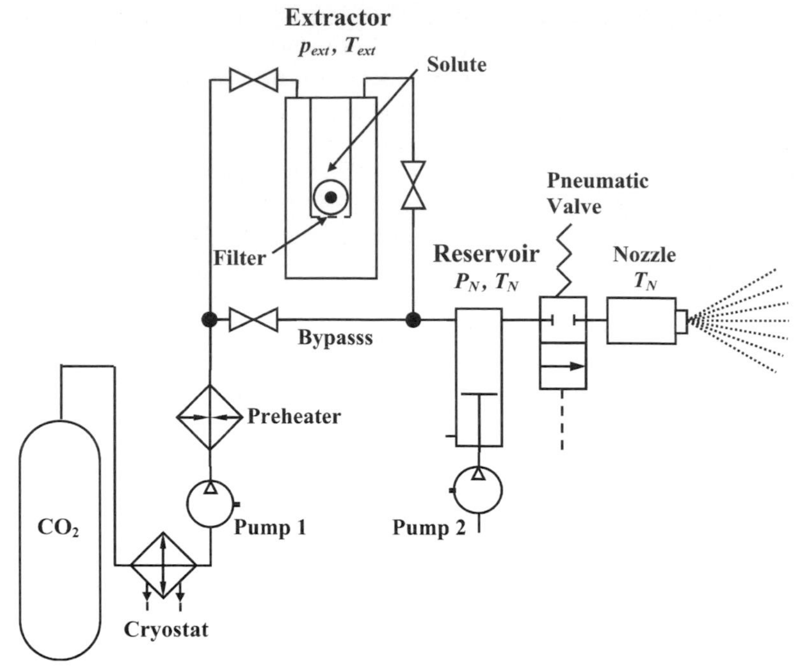

Fig. 6.1. Scheme of the RESS apparatus.

cantly better dissolution rate of the drug, resulting in an improved bioavailability [14].

We have built up several types of RESS apparatuses [14–18], which differ in size (continuous and batch RESS apparatuses) and in the mode of expansion, i.e. continuous expansion versus pulsed expansion. A general scheme of a RESS apparatus is depicted in Fig. 6.1. Our setup is designed for experiments in the temperature range 300–600 K and for pressures up to 60 MPa. The gaseous solvent (CO_2, N_2O, CHF_3) is taken from a reservoir, liquefied by a cryostat, and pressurized (pump 1: diaphragm pump or pneumatic pump) to the desired pressure in the extractor (p_{ext}). For the formation of nanoscale particles of thermally labile drugs, CO_2 is a particularly suitable solvent since it has low critical data ($T_{crit} = 304$ K, $p_{crit} = 7.38$ MPa), is non-toxic, and can be easily separated from the particulate phase after particle formation. The extractor is a heated (T_{ext}) high-pressure autoclave that is packed with the solute. For continuous operation, the supercritical solution is expanded directly through the pinhole nozzle. For pulsed operation, a heatable high-pressure reservoir (T_N, p_N) is filled with the solution first. Expansion then takes place from this reservoir, the pressure of which is kept constant during the pulses (duration $t_{puls} > 100$ ms) by a movable piston connected to pump 2. The length of the pulses is controlled by the pneumatic valve in front of the nozzle. The pinhole nozzles used have variable inner diameters of 20–200 μm and lengths of

50–300 μm. The nozzle is positioned in the expansion chamber, as further explained in Section 6.3.1.1 and depicted in Fig. 6.3 below.

Particle generation by RESS can be used, in principle, for a wide range of different solutes [11–14]. This statement, however, has to be qualified to some extent for thermally labile substances such as drugs for which the solvent has to meet special requirements. Here, carbon dioxide is often the only suitable solvent so that the application of RESS is restricted to compounds with reasonable solubility in supercritical CO_2. Micronization with RESS, in general, produces particles with rather broad size distributions [geometric standard deviation $\sigma > 1.5$, see Eq. (2)]. In this context, an important factor is the agglomeration of the primary particles, which leads to a broadening of the size distribution. Different methods (RESSAS, CORESS, CPD) for stabilizing the primary particles against agglomeration have recently been realized [19–26]. They are discussed briefly in Section 6.5.

6.2.2 Electro-Spraying

Electro-spraying is a method of generating particles of molecular compounds that are preferentially soluble in polar solvents such as water or different alcohols. A scheme of the Electrospray Aerosol Generator (EAG, TSI 3480) is depicted in Fig. 6.2 (see also Refs. [9, 27–30] for further information). The compound of interest is

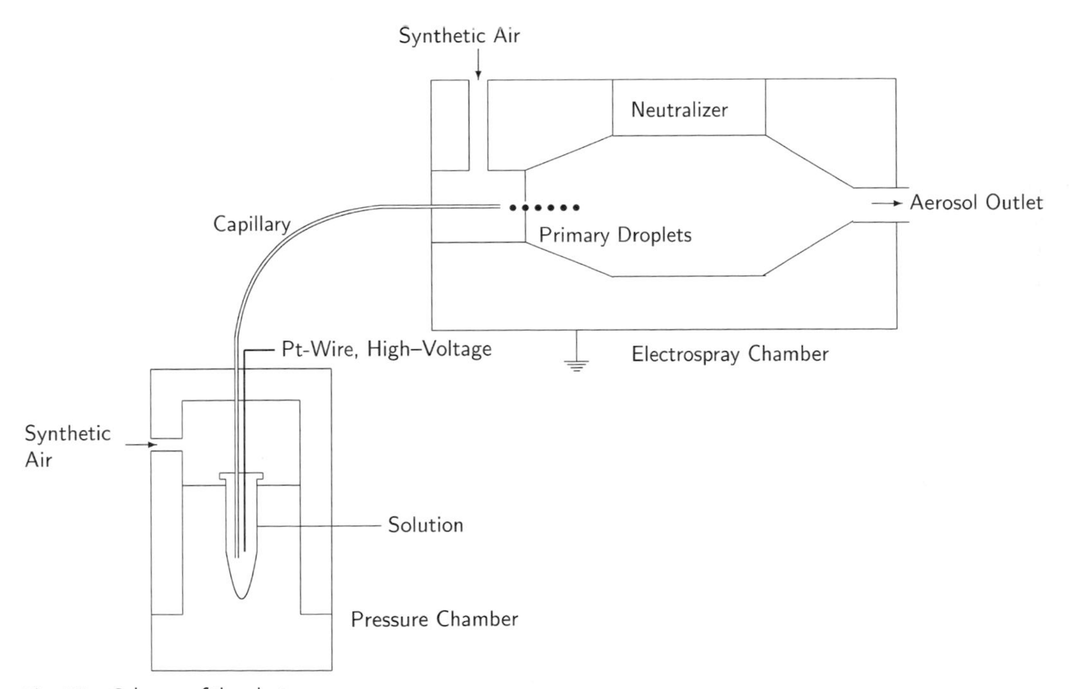

Fig. 6.2. Scheme of the electrospray.

dissolved in an aqueous buffer solution (ammonium acetate/ammonium hydroxide in H_2O). The buffer solution serves to adjust the pH and to increase the electric conductivity of the solution (usually, 500–2000 $\mu S\ cm^{-1}$). Typical concentrations of the solute lie between 2 and 10 vol%. The solute concentration determines the diameters of the final particles, which lie below ~100 nm. The setup consists of a capillary that is immersed in the solution. The solution is driven through the capillary (~70 nL min^{-1}) by a pressure gradient. Between the capillary tip and an electrode with a pinhole, a high voltage is applied so that highly charged primary droplets of the solution (diameters ~ 150 nm) are formed. The droplets are neutralized by a radioactive source and the solvent evaporates from these primary droplets in a sheath-flow of clean dry synthetic air (N_2/O_2; ~2 L min^{-1}). The neutralized aerosol flow then enters the multireflection sheath-flow cell described in Section 6.3.1.1.

The electrospray generates small particles with relatively narrow size distributions. Typical geometric standard deviations amount to $\sigma = 1.3$. This is a great advantage over many other particle generation methods, which for molecular substances often lead to particles with fairly broad distributions ($\sigma > 2$). As monodisperse products exhibit much more homogeneous properties they are usually highly preferred over polydisperse particle ensembles. However, particle generation by electro-spraying has two major disadvantages. One concerns the separation of the solvent. The commonly used solvents are liquid at atmospheric conditions and thus can easily condense on the particles collected from the gas phase. The second more severe problem is the clogging of the capillary, especially of the capillary tip. This reduces the general applicability of electro-spraying for particle generation to substances that are slow to crystallize at the capillary tip (i.e., with a tendency to form supersaturated solutions).

6.3 Particle Characterization Methods

6.3.1 *In Situ* Characterization with FTIR Spectroscopy

6.3.1.1 Experimental Setup

Figure 6.3 shows the experimental setup for the *in situ* characterization with Fourier transform infrared spectroscopy (FTIR). The nozzle of the RESS apparatus (see Section 6.2.1 and Fig. 6.1) is positioned in the expansion chamber and is connected to the reservoir of the RESS apparatus with a high-pressure hose. The expansion chamber and the buffer volume are connected vacuum chambers with a total volume of 0.8 m^3. The buffer volume helps to limit the increase of pressure, p_c, in the expansion chamber. The whole vacuum part is evacuated to $p_c = 0.03$ mbar by two rotary piston vacuum pumps (Leybold DK 100 and E 250) connected in parallel. With the 50 μm nozzle and a reservoir pressure of $p_R = 400$ bar, the pressure in the chamber typically increases to 0.16 mbar after a pulse of 0.5 s duration. A pulse duration of 2 s leads to $p_c = 0.55$ mbar. The IR spectra of nanoparticles were

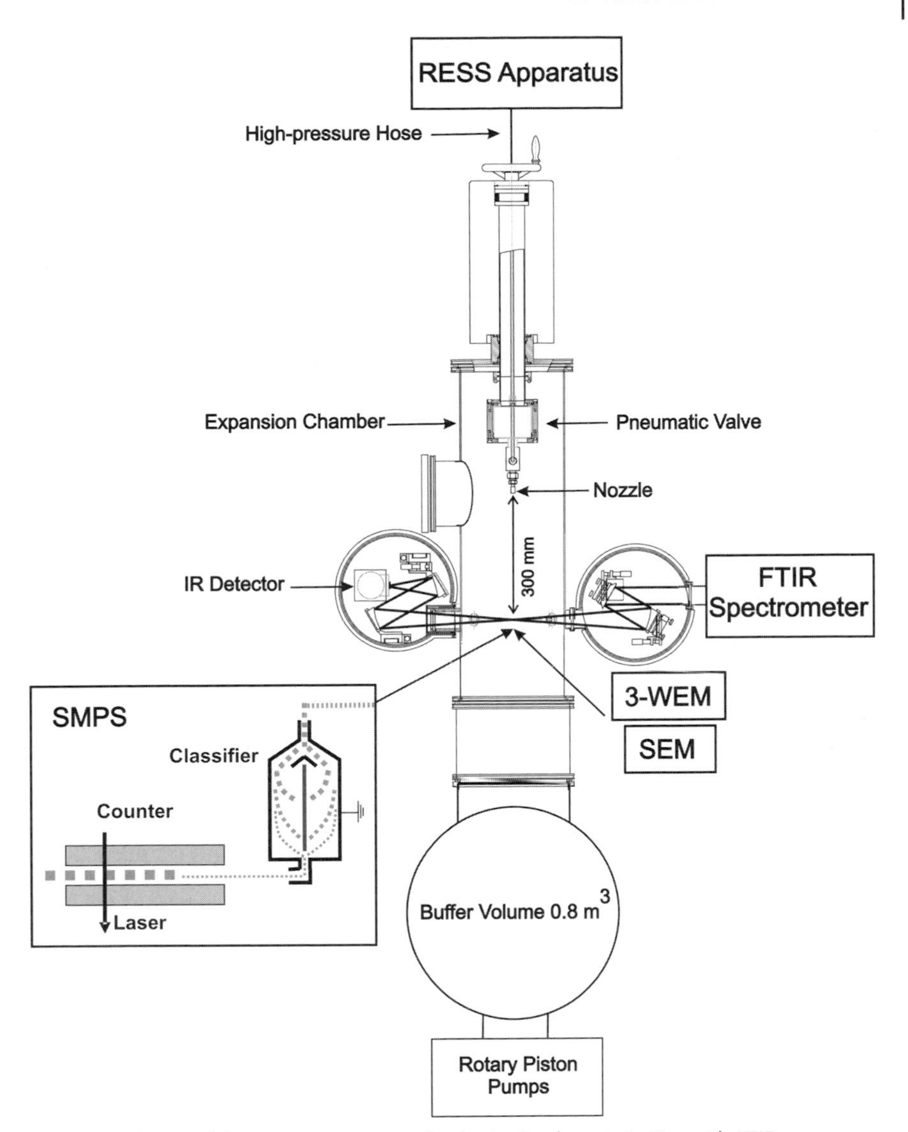

Fig. 6.3. Scheme of the vacuum apparatus for the *in situ* characterization with FTIR.

recorded *in situ* using a Bruker IFS 66v/S FTIR spectrometer. As depicted in Fig. 6.3, the supersonic flow is probed radially by the infrared beam. The infrared beam from the Globar light source is focused into the center of the expansion by a first off-axis parabolic mirror (327 mm focal length). After passing the expansion, the light is collected by a second off-axis parabolic mirror (327 mm focal length) and a third parabolic mirror (109 mm focal length) focusses the light onto a liquid nitrogen cooled MCT detector. The spectrometer and detector chamber are separated from the expansion chamber by two KBr windows. The rapid-scan mode of the spectrometer allows us to synchronize the acquisition of IR spectra with the pulsed expansion. With a scanner velocity of 280 kHz, one scan is recorded in 30 ms at a resolution of 2 cm^{-1}, which means that several scans can be made during one pulse (pulse duration $>$ 100 ms). To get a good signal-to-noise ratio, we typically accumulate more than 500 scans. A similar combination of direct absorption FTIR spectroscopy with intense fluid pulses through a slit-nozzle into the vacuum has been realized for the first time by Suhm and coworkers [31, 32] to study the vibrational dynamics of small clusters.

The investigation of particles during their generation in the aerosol phase by *in situ* FTIR spectroscopy has many advantages. First of all, IR spectra contain information about various particle properties such as the size distribution, chemical composition, and structural aspects. Therefore, the combination with rapid characterization *in situ* represents a very useful method for process control. With the setup described here, the particles do not have to be collected first but can be continuously analyzed during their generation. This allows us, for instance, to control their size by changing the process parameters such as the temperature, pressure, or concentration of the supercritical solution. It also enables us to control the portion of different components in the final product if supercritical solutions with several solutes are used. This is of interest for the coating of drug particles or for the generation of mixed drug/matrix particles (see also Section 6.5.3). In addition, with the vacuum option we can change the generation conditions of the particles by varying the pressure p_c in the expansion chamber. With phenanthrene particles, particles generated under vacuum conditions have smaller mean sizes but broader distributions than those generated at ambient pressure (see Ref. [10] and Section 6.5.1). Section 6.3.1.2 demonstrates that our setup can also be used to study the region of the expansion before the Mach disc, which is essential for a better understanding of the RESS and is crucial to verify corresponding theoretical predictions [12, 33–38].

Particles generated in the electrospray (Fig. 6.2) are investigated spectroscopically in a multireflection sheath-flow cell [30, 39]. The cell consists of two concentric cylinders. The carrier gas is introduced into the gap between the two, enters the inner cylinder through a large number of small holes, and is finally pumped off at the bottom. This creates a smooth sheath-flow for samples injected coaxially at the top of the inner cylinder. The cell is operated at room temperature with a sheath-flow of 1.5 L He per minute. The sheath-flow serves to guide the aerosol through the flow tube and to minimize contaminations of the mirrors. The cell is equipped with White optics with an optical path length of about $h = 16$ m.

6.3.1.2 Characterization of the RESS Process

The mole fraction of the solute in supercritical CO_2 typically lies below 0.1–0.01 [40]. In other words, the supercritical solution expanded in the RESS process mainly consists of CO_2. Evidently, the influence of CO_2 on particle formation is not negligible. Due to the high supersaturation, one expects that not only the solute with its low volatility but also the CO_2 itself condenses into small particles within

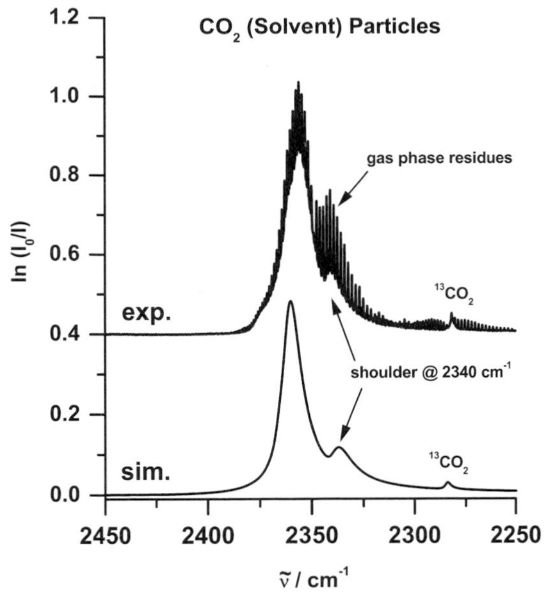

Fig. 6.4. Upper trace: Experimental IR spectrum of CO_2 particles recorded in the collision-free region before the Mach disc. Experimental conditions: nozzle diameter $d = 50$ µm, $p_R = 400$ bar, $T_N = 298$ K. The sharp peaks arise from residual gas phase CO_2 in the expansion. Lower trace: Calculated CO_2 particles spectrum. For the fit we have used the refractive index data from Ref. [44], a Mie-model [42], and a log–normal distribution [Eq. (2)]. $\tilde{\nu}$ is the transition wavenumber.

the nozzle and some nozzle diameters downstream [12, 33–38]. Figure 6.4 shows that these solvent particles really can be observed in the collision-free region before the Mach disc. The experimental IR spectrum in the region of the antisymmetric stretching vibration ν_3 of CO_2 is depicted in the upper trace. The sharp peaks superimposed on the broad band have nothing to do with the solvent particles, they arise from residual gas-phase CO_2 in the expansion. The appearance of the broad band is characteristic of CO_2 particles [41]. It allows us to estimate the size of the solvent particles [10] since both the width and the shoulder at about 2340 cm^{-1} strongly depend on the particle size. As a rough rule one can say that for larger particles the band broadens and the intensity of the shoulder increases. The size is an important quantity to know since the condensation/evaporation behavior of the solvent influences the particle formation processes. It is also of interest to verify corresponding theoretical predictions [12, 33–38]. The sizes are estimated by fitting calculated extinction spectra to the experimental IR spectrum. For that, we have assumed Mie theory [42] and a log–normal size distribution that is characterized by the mean particle radius (r_0), the geometric standard deviation (σ) and the particle number concentration (N) [see Eq. (2)]. The refractive index data have been taken from Refs. [43, 44]. Typical parameters for expansions at temperatures between 298–398 K and pressures between 100–400 bar are listed in Table 6.1. As

Tab. 6.1. Geometric standard deviation σ, mean particle radius r_0, and particle number concentration N [see Eq. (2)]. Data from IR spectra have been obtained from a fit to the experimental data using Mie theory and two different sets of refractive index data [43, 44]. Since r_0 and σ are strongly correlated, the value for σ has been fixed [10].

p_N/(bar)	T_N/(K)	σ	r_0/(nm)	$N^{[a]}/(10^8\ cm^{-3})$
400	298	1.4	300	0.2
100	298	1.4	180	0.1
400	360	1.4	210	0.2
400	298	2.1	70	3.9

[a] Values for N for a path length of $h = 1$ cm.

expected, we find a systematic decrease of r_0 with decreasing reservoir pressure (p_N) and with increasing temperature. The values of the radii of the solvent particles in Table 6.1 lie well above 100 nm. However, because the parameters σ and r_0 turned out to be highly correlated we have fixed the value of σ at 1.4 during the refinement procedure. For comparison, the last row of Table 6.1 gives one example with the geometric standard deviation fixed at $\sigma = 2.1$. Comparison with the first row shows that the mean size decreases if a broader distribution is assumed. Independent of the widths of the distribution, however, we can finally state that the mean particle radius of the solvent particles lies clearly above $r_0 = 50$ nm for temperatures between 298 and 398 K and pressures between 100 and 400 bar [10].

In the collision-free region of the expansion, CO_2 and the solute both exist as small particles due to the supersaturation and subsequent condensation upon expansion. This result is in good agreement with the modeling results published by Helfgen et al. [38]. From these calculations it follows that the decrease of pressure and temperature in the supersonic free-jet can lead to solvent condensation. Depending on the pre-expansion conditions, the condensate mass fraction goes up to around 30% for CO_2 and 25% for CHF_3. As mentioned above, the mole fraction of the solute in supercritical CO_2 lies below 0.1–0.01 so that the amount of condensed CO_2 is much higher than that of condensed solute. But it is not clear whether the two components form (statistically) mixed particles or whether the solute condenses first to small particles that then act as condensation nuclei for the more volatile CO_2 (coated particles). We have investigated this question for an expansion of a n-nonadecane/CO_2 solution. Figure 6.5 shows the corresponding IR spectrum in the region of the CH-stretching vibrations of n-nonadecane ($C_{19}H_{40}$) particles. The region of the antisymmetric stretching vibration of the CO_2 particles is not shown again. It looks the same as already depicted in Fig. 6.4. The shape and the band positions in the spectrum of the nonadecane particles in Fig. 6.5 are the same as those for pure nonadecane particles [10]. We consider this fact a hint that coated particles rather than mixed particles exist in the collision-free region before the Mach disc. The preference of coated particles, however, is much more plausible since nonadecane is much less volatile than CO_2 and thus condenses first, probably already within the nozzle [12, 33–38].

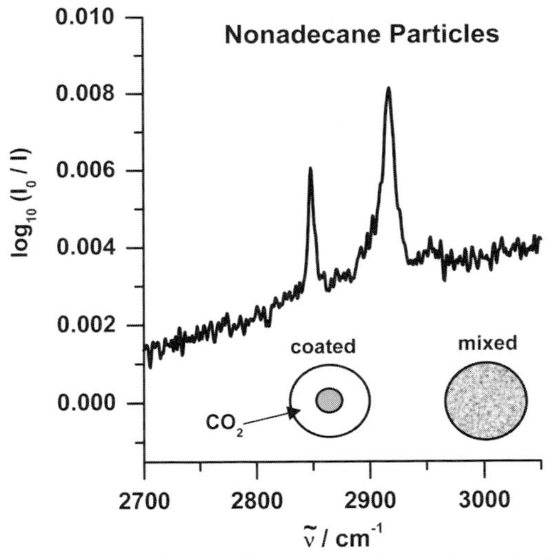

Fig. 6.5. Experimental IR spectrum of nonadecane in the region of the CH-stretching vibrations. The spectrum was recorded in the collision-free region before the Mach disc. In this region both nonadecane and the solvent CO_2 condense to small particles. $\tilde{\nu}$ is the transition wavenumber.

The conditions and processes before the Mach disc described in this section certainly influence the properties of the final particles. But the processes taking place at and after the Mach disc and the manner in which the particles are collected are much more important factors, for instance for the size of the final particles. All these factors can lead to coagulation and agglomeration of primary particles. The sudden temperature and pressure increase taking place at the Mach disc leads to evaporation of the solvent particles so that after the Mach disc only pure solute particles are present. However, we have shown previously [10] that complete evaporation of the solvent particles takes some seconds. Since agglomeration and coagulation seem to be important it is not astonishing that the electron microscopy images of the final products, which are further discussed in Section 6.5, often show agglomerates consisting of relatively small (50–100 nm) primary particles. One possibility to avoid this subsequent agglomeration is by coating the particles as discussed in Section 6.5.3.

6.3.2
In Situ Characterization with 3-WEM

In continuous particle formation processes, fast *in situ* measurement techniques for on-line particle size determination are desirable. In contrast to the usual off-line examination techniques, the particles are therefore measured in the expansion chamber on-line and *in situ* with the Three-Wavelengths-Extinction (3-WEM) measurement technique (see Fig. 6.3). The 3-WEM measuring devices used are equipped with three lasers emitting at wavelengths of 405, 633, and 1064 nm (home-built device) and of 674, 814, and 1311 nm (WIZARD-DQ, Wizard Zahoransky KG, Todtnau, Germany), respectively. The fiber-coupled transmitter of the latter device is placed at the optical access of the test volume. The remaining light is detected with the opposite receiver. The relative transmissions of the three different wavelengths I/I_0 are used for the evaluation on a PC. According to Mie theory [42], the system allows the measurement of particle diameters in a range from 0.2 to 4 μm and particle concentrations ranging from 10^5 to 10^8 particles per cm^3 [10, 45]. The advantages of the 3-WEM measurement technique are that it is a suitable method for measuring particle size and number concentration under sub-, near-, and supercritical conditions. In addition, the 3-WEM enables on-line process control because particle characteristics can be determined within 10^{-6} s. However, for non-spherical and/or agglomerated particles the 3-WEM measurement technique can lead to incorrect results for the particle size distribution. To overcome this, a combination of different measurement methods has to be used to accumulate sufficient information on the size and structure of the particles investigated. Therefore, samples for additional SEM examination (see Section 6.3.3) were taken 300 mm from the nozzle exit directly behind the 3-WEM-probe. A more detailed description of the apparatus and the experimental procedure is given elsewhere [16].

6.3.3
Characterization with SMPS and SEM

The number size distribution of the aerosol is also measured on-line with a Scanning Mobility Particle Sizer (SMPS), which can be connected to the RESS expansion chamber (Fig. 6.3) or to the electro-spray (Fig. 6.2). The SMPS consists of a differential mobility analyzer (TSI 3080L/N), which classifies the particles according to their mobility in an electric field, and of a condensation nuclei counter (TSI 3022A), where the particles are counted optically. The SMPS only works at atmospheric pressure and at room temperature and, therefore, can only be used for a limited range of experiments. Compared with 3-WEM and IR spectroscopy it leads to more detailed size distributions since it measures the number of particles for each single size and does not rely on any assumption about the type of the size distribution (e.g., log–normal distribution). However, the SMPS is much slower than 3-WEM or IR spectroscopy. Typical acquisition times lie around 2 min. Examples for size distributions measured with the SMPS are given in Section 6.5.

To visualize the shape of nanoparticles and also to determine their approximate size, we use Scanning Electron Microscopy (SEM). This off-line characterization requires collection of the particles first, which often leads to agglomeration. We collect them either on a polycarbonate membrane or on a silicon holder previously covered with a thin gold layer (11 nm). After collecting, the sample is covered with a second gold layer (7 nm) to avoid evaporation of substances inside the SEM apparatus. Examples of SEM images are depicted in Section 6.5.

6.4
Determination of Refractive Index Data in the Mid-infrared Region

Knowledge of the frequency-dependent complex refractive index $n + ik$ of the solute particles is a prerequisite for *in situ* characterization with FTIR spectroscopy. Therefore, we have started to derive corresponding optical data from the measured infrared extinction spectra of organic aerosols and from size distributions determined experimentally with the SMPS and with 3-WEM (Section 6.3). The formalism is given below by Eqs. (1–5). The refractive index data thus derived are collected in Ref. [46] together with a short description and an estimate of their accuracy. There exist other databases for refractive indices in the mid-infrared region (see for instance Refs. [47–49]) which are, however, focussed more on molecular ices rather than on particles of biomolecular interest.

Our data are obtained from particle spectra and not from thin films. It is more difficult to derive accurate data from particle spectra than from thin films since it is often very difficult to get accurate information about the correct size distribution or the shape of the particles. On the other hand, the structure and even the chemical composition of the particles (crystalline/amorphous, modified structure of the particle surface [9, 30, 41, 50–52]) can strongly depend on the generation method and conditions. The latter aspect clearly illustrates the need for optical data obtained directly from particle spectra. Ref. [46] contains an accurate description of the methods used, but due to the limited information about size and shape of the particles the tabulated data only provide a first starting point towards the determination of more accurate data. Although we use three different methods (SEM, SMPS, and 3-WEM) for the characterization of the size and shape it is nearly impossible to get very accurate experimental data for these properties. The main disadvantage of SEM is that, upon collecting, the particles often agglomerate. Analysis with SMPS and 3-WEM relies on assumptions for the particle shape or the size distribution. Moreover, the derivation of the infrared optical data is based on models [see Eqs. (1–5)]. The Mie theory used for that purpose gives the solutions for the scattering and absorption of light by spherical particles. The assumption of a spherical shape is often regarded as an adequate approximation even for non-spherical particles. But this is more of an assertion than a fact and would have to be tested for each case [42, 53]. For instance, Mie theory fails for non-spherical particles in the case of strong vibrational transitions when transition dipole coupling is dominant [41, 54, 55]. We also neglect multiple scattering, which will not lead to major errors as long as the particle density is not too high.

The frequency-dependent complex refractive index $n + ik$ of solute particles has been determined directly from the experimental extinction spectrum of the corresponding aerosols in the mid-infrared region by following two alternative approaches: (a) a Lorentz model fit and (b) a Kramers–Kronig inversion. Both proceed via the complex refractive index to characterize the optical properties of the particulate phase. We employed Mie theory for spherical particles [42] to calculate the observed extinction spectrum. The refractive index of the surrounding medium (different gases) was set to unity. The frequency dependent extinction cross section

(C_j) is calculated separately for each value of the particle radius (r_j). The calculated total absorbance (A_{calc}) is obtained by summing over all different radii using the experimentally determined particle size distribution:

$$A_{\text{calc}} = h \sum_j N_j C_j \tag{1}$$

where h is the optical pathlength and N_j is the number density of aerosol particles with size r_j. The experimental size distribution is determined either by the SMPS (Section 6.3.3), by 3-WEM (Section 6.3.2), or by a combination of both [9, 10, 15, 29, 30, 41]. The classifier of the SMPS determines the particle radii (r_j) for different size intervals j. Each decade of particle size is divided into 64 equidistant logarithmic intervals. The particle counter of the SMPS then measures the particle number concentration (N_j) in each interval j. The measurement range of the SMPS is limited to the region 7–1000 nm. Size distributions that exceed this range are extrapolated by using a log–normal distribution. A log–normal distribution for the particle radii (r) is also assumed for the interpretation of the 3-WEM experiments. Its form is given by [56]:

$$df = \frac{1}{\sqrt{2\pi}\ln\sigma} \exp\left\{-\frac{[\ln(r/u) - \ln(r_0/u)]^2}{2(\ln\sigma)^2}\right\} d[\ln(r/u)] \tag{2}$$

with the mean particle radius (r_0) and the geometric standard deviation (σ) as parameters; u is the unit radius. The product $N \cdot h$ of the particle number concentration (N) and the pathlength (h) is determined by the absolute values measured for the extinction [Eq. (1)].

In the first approach to determine the complex refractive index of the nanoparticles we assume a model function for the (complex, frequency dependent) dielectric function (ε) of the solute particles. The empirical parameters of this function are determined from a nonlinear least-squares fit of the calculated extinction, A_{calc} [Eq. (1)], to the experimental aerosol spectra. In this study we have used the so-called Lorentz model, which describes the optical properties of the solid by a collection of isotropic damped harmonic oscillators coupled to the electromagnetic field of the light by their effective charge. Solution of the classical equations of motion leads to the Kramers–Heisenberg dielectric function [42, 57]:

$$\varepsilon = \varepsilon_1 + i\varepsilon_2 = \varepsilon_e + \sum_s \frac{\tilde{\nu}_s^2 f_s}{\tilde{\nu}_s^2 - \tilde{\nu}^2 - i\gamma_s \tilde{\nu}} \tag{3}$$

The dielectric function is directly related to the complex index of refraction (n, k):

$$n = \sqrt{\frac{\sqrt{\varepsilon_1^2 + \varepsilon_2^2} + \varepsilon_1}{2}} \quad k = \sqrt{\frac{\sqrt{\varepsilon_1^2 + \varepsilon_2^2} - \varepsilon_1}{2}} \tag{4}$$

The parameter ε_e is the dielectric function at frequencies that are high compared with the vibrational excitations considered (but low compared with electronic excitations, which are neglected here). Eq. (3) sums over all oscillators in the spectral region of interest. They are characterized by their resonance wavenumber ($\tilde{\nu}_s$), their reduced oscillator strength (f_s) and a width (γ_s) resulting from the damping of the oscillator.

To the extent that a Lorentz oscillator can be compared to a molecular oscillator (e.g., a molecular normal mode), $\tilde{\nu}_s^2 f_s$ is proportional to the absorbance $\int A(\tilde{\nu})\,d\nu$ integrated over the corresponding absorption band (also termed integrated intensity). Eq. (3), however, neglects any local field correction, i.e., it assumes that each oscillator experiences the external electric field, which is not necessarily a good approximation in the condensed phase. An approximate local field correction for isotropic media leads to a formally identical expression [58], but with shifted resonance wavenumbers and modified effective reduced oscillator strengths. For the relatively small values of the reduced oscillator strengths found in the vibrational spectrum of many of the solutes, the wavenumber shifts can be neglected. The modification of the reduced oscillator strengths significantly changes their absolute values, thus affecting their interpretation in terms of molecular transition moments [58]. The relative values, however, remain approximately unchanged by the local field correction.

In the second approach, we exploit the Kramers–Kronig relation between the real (n) and the imaginary (k) part of the complex refractive index [42, 57]. The optical constants can thus be derived from the experimental extinction spectrum without assuming any particular model. In its so-called subtractive form the Kramers–Kronig relation [59–61] is given by Eq. (5).

$$n(\tilde{\nu}') = n(\tilde{\nu}_r) + \frac{1}{\pi}\int_{\tilde{\nu}_{min}}^{\tilde{\nu}_{max}} \left(\frac{k(\tilde{\nu})}{\tilde{\nu}+\tilde{\nu}'} - \frac{k(\tilde{\nu})}{\tilde{\nu}+\tilde{\nu}_r} + \frac{k(\tilde{\nu})}{\tilde{\nu}-\tilde{\nu}'} - \frac{k(\tilde{\nu})}{\tilde{\nu}-\tilde{\nu}_r} \right) d\tilde{\nu} \tag{5}$$

Eq. (5) requires knowledge of the real index of refraction at some reference point, $\tilde{\nu}_r$. Although in principle arbitrary, it is best (for numerical reasons) chosen to lie within the spectral region of interest. Starting with some guess for the real part n of the refractive index its imaginary part k is adjusted (by bisection) so that the Mie scattering calculation reproduces the experimentally measured extinction separately for each value of the wavenumber $\tilde{\nu}$. Eq. (5) is then applied to $k(\tilde{\nu})$ to produce an improved guess for $n(\tilde{\nu})$. The Cauchy principal value of the integral in Eq. (5) was calculated on a grid as described in Ref. [61] with the difference, however, that $k(\tilde{\nu})$ was expanded to fourth order around the singularity at $\tilde{\nu} = \tilde{\nu}'$. The whole procedure is repeated until convergence is reached. Since the reference value $n(\tilde{\nu}_r)$ of the real index of refraction cannot be determined through this procedure it enters as the only free parameter. Here it was set to the value obtained from the Lorentz model fit at 2000 cm^{-1}, where the experimentally measured extinction vanishes. Otherwise this second approach to the determination of optical constants constitutes an inversion of the experimental spectrum rather than a fit.

6.5 Examples

6.5.1 Phenanthrene Particles: Size, Shape, Optical Data

Phenanthrene ($C_{14}H_{10}$) serves here as a model system to demonstrate the different aspects of the particle generation and characterization methods described in Sections 6.2–6.4. Phenanthrene has been used as a representative compound in the modeling of particle formation processes with RESS [33, 36, 37]. Polycyclic aromatic hydrocarbons such as phenanthrene are of importance as pollutants in the atmosphere [62, 63] and have also been discussed in astrophysics as carriers of the "unidentified" infrared bands [64, 65].

We have generated phenanthrene particles with the RESS apparatus (Section 6.2.1) under various experimental conditions. In summary, we find that the phenanthrene particles are nearly spherical, as depicted in the SEM image in Fig. 6.6(a). Various SEM images also show that the tendency to form agglomerates is only minor for this substance, which is in contrast to the behavior of phytosterol particles discussed in Section 6.5.3. For phenanthrene, we found no major differences in the properties of the particles by changing the pre-expansion conditions (temperatures between 298 and 350 K, pressures between 20 and 40 MPa, and for different nozzle diameters). In this context we also mention the partly contradictory discussions in Refs. [15, 66–70]. One parameter that affects the size of the particles is the pressure p_c in the expansion chamber: Particles generated under vacuum conditions ($p_c = 0.03$ mbar) have smaller mean diameters by about 30% but larger geometric standard deviation by about 30% than those formed at $p_c = 960$ mbar. Whether this is a general trend for other substances as well remains to be seen.

Typical size distributions for the expansion against $p_c = 1000$ mbar ($T_{ext} = T_N = 298$ K, $p_R = 400$ bar) are depicted in Fig. 6.6. From different SEM images recorded under equivalent conditions, we have determined a rough size distribution, which is shown by the bars in Fig. 6.6(a). Each bar represents the relative number of particles in the corresponding size interval. This size distribution is obviously very coarse and, in particular, it does not contain any information about the particle number concentration (N) of the corresponding aerosol from which the SEM samples were collected. This information is important for the derivation of optical data from corresponding IR spectra as discussed below. Detailed information about the particle number concentration and the size can be obtained with the SMPS. The corresponding experimental number size distribution is shown by the bars in Fig. 6.6(b). The circles represent a fit to the measured distribution assuming a log–normal distribution [Eq. (2)]. This fit allows us to extrapolate the distribution beyond the measurement range (see Section 6.4) and leads to the following typical data for the phenanthrene particles:

$$r_0 \sim 120 \text{ nm}, \quad \sigma \sim 2.2, \quad \text{and} \quad N \sim 5 \times 10^6 \text{ cm}^{-3} \tag{6}$$

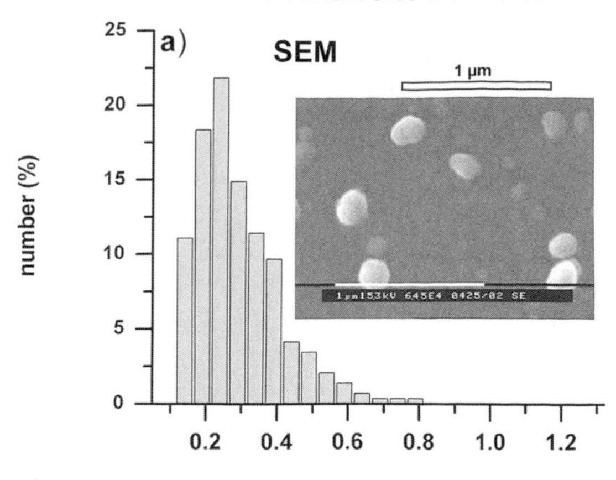

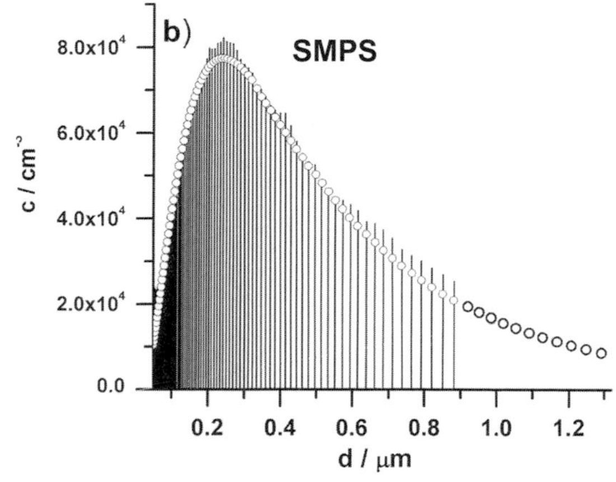

Fig. 6.6. (a) Bars: Size distribution of phenanthrene particles obtained from SEM measurements. Ordinate: Percentage of particles with sizes in size intervals of 50 nm. Inset: Typical SEM image. (b) Bars: Number size distribution of phenanthrene particles measured on-line with the SMPS. c is the number concentration per logarithmic sampling interval. Bars indicate the center of the intervals. Circles: Fit to the measured distribution assuming a log–normal distribution [Eq. (2)]. Conditions for particle formation: $T_{ext} = T_N = 298$ K, $p_{ext} = p_N = 400$ bar.

Comparison of Fig. 6.6(a) and (b) shows that the size distribution from SEM and from SMPS are in good agreement. Since the phenanthrene particles are nearly spherical the 3-WEM should also lead to similar results as SEM and SMPS. In contrast to this expectation, the 3-WEM leads to larger mean radii (40%) and smaller geometric standard deviations (20%) than the two other methods. Similar behavior has already been observed in previous studies that compared 3-WEM with other sizing methods [16, 45, 71]. At the moment we cannot finally explain the origin of this deviation but it is, presumably, due to assumptions implicit in the analysis of the 3-WEM data (Section 6.3.2).

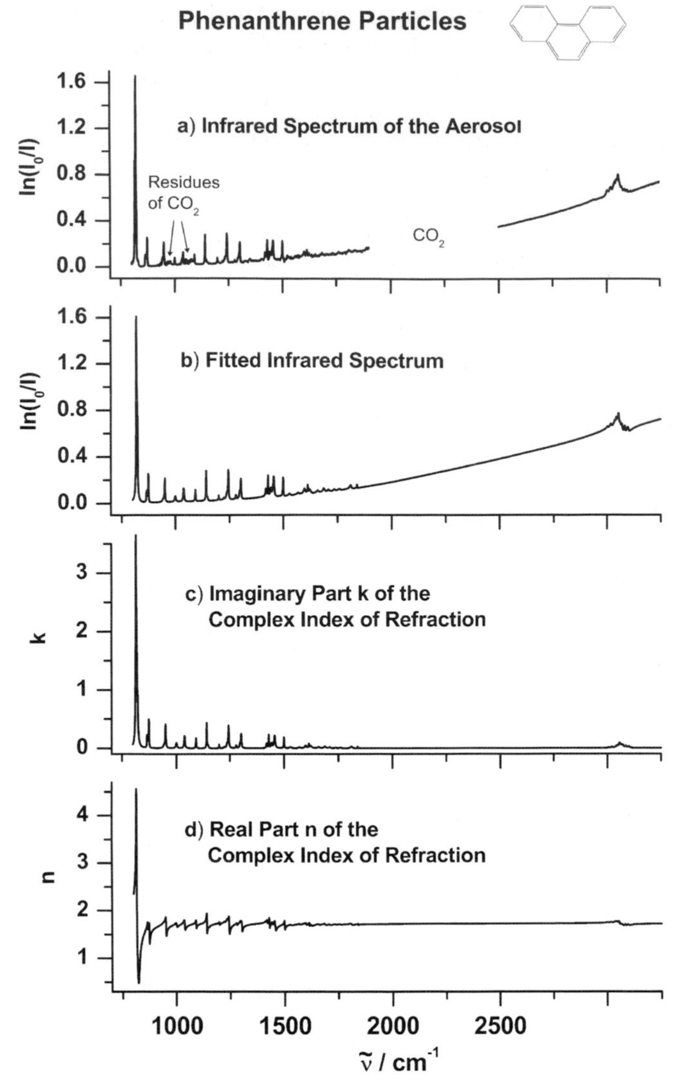

Fig. 6.7. (a) Experimental IR spectrum of phenanthrene aerosol particles. (b) Fitted infrared spectrum (see Section 6.4). (c) Imaginary part k of the complex index of refraction. (d) Real part n of the complex index of refraction. $\tilde{\nu}$ is the transition wavenumber.

The size information is also contained in the IR spectrum of the aerosol depicted in Fig. 6.7(a). Thus, the scattering by the particles leads to the slanted baseline, which is characteristic for the size distribution of the particle ensemble. Note that the strong bands from gas-phase CO_2 (residues of the solvent) below 800 cm^{-1} and between 1900 and 2500 cm^{-1} have been cut from this spectrum. The determination of the particle size distribution from the experimental mid-infrared spectrum requires knowledge of the corresponding refractive index data $n + ik$ for the solute in this region (Section 6.4). Since these data were not known for phenanthrene particles, we have derived them from a fit to the experimental spectrum using the experimentally measured size distributions from the SMPS (Fig. 6.6b). We assumed Mie theory for spherical particles and a Lorentz model for the optical properties [Eqs. (3, 4), and [42]]. More details of this procedure can be found in Refs. [15, 30, 42]. According to Eq. (3) each Lorentz oscillator is characterized by three parameters: The resonance wavenumber, the oscillator strength, and the damping width. The number of oscillators is not determined *a priori*. It depends on the number of intra- and intermolecular vibrations and on interactions between them. The phenanthrene spectrum in Fig. 6.7(a) consists of sharp resolved absorptions. Therefore, we have assigned one Lorentz oscillator to each of these bands. The results from our Lorentz fit are depicted in Fig. 6.7(b–d). Trace (b) shows the fitted IR spectrum, which reproduces the experimental spectrum quite well. The region below 2000 cm^{-1} is dominated by the strong absorption of the CH out-of-plane bending at 819 cm^{-1}. All other vibrations in the mid-infrared region are rather weak. The region below 2000 cm^{-1} consists of well-resolved individual transitions whereas the CH-stretching transitions around 3100 cm^{-1} strongly overlap. Trace (c) shows the imaginary part k and trace (d) the real part n of the complex index of refraction. The corresponding fitted parameters and refractive index data together with a description of the estimated uncertainties (~20%) are available in Ref. [46]. With these data the determination of the size distribution now becomes possible directly from IR spectra for phenanthrene aerosols.

Apart from the size, the IR spectrum also contains information about the chemical composition, about the structure (crystalline, amorphous), and sometimes about the particle shape. Comparison of the experimental spectrum in Fig. 6.7(a) with corresponding spectra for crystalline phenanthrene reveals the crystalline character of the phenanthrene particles [72]. Fig. 6.7(c) and (d) show that there exists a spectral range around 819 cm^{-1} where n is small and k varies widely. In classical scattering theory, this is the condition for the occurrence of shape effects in particle spectra [42]. We have pointed out previously [44] that on a molecular level this can be explained by exciton coupling (transition dipole coupling), which elucidates that shape effects only occur for bands with relatively large molecular transition dipoles ($\delta\mu > 0.1$–0.2 D). For phenanthrene this is fulfilled only for the band at 819 cm^{-1}, which is shown in an expanded view in Fig. 6.8(a). By different examples [52, 54, 55, 73] we have illustrated that different particle shapes lead to different band structures for such strong absorption bands in the IR spectra. Figure 6.8(a) shows the band at 819 cm^{-1} for spherical phenanthrene particles as depicted in the SEM image of Fig. 6.6(a). To modify the shape of the phenanthrene particles we have produced a layer of particles on a NaCl window and recorded its IR spectrum, which is depicted in Fig. 6.8(b). This layer no longer consists of single spherical particles but of strongly agglomerated/coagulated primary particles, as illustrated by the SEM image in trace (b). The deviation from the spherical shape obviously leads to a completely different band shape in the IR spectrum of the par-

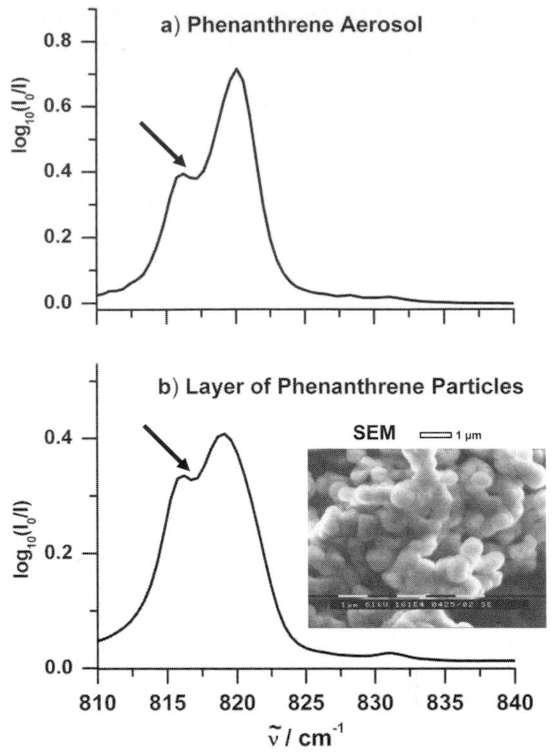

Fig. 6.8. (a) Expanded view of the IR spectrum of phenanthrene aerosol shown in Fig. 6.7(a) in the region of the band at 819 cm^{-1}. (b) IR spectrum of a layer of particles on a NaCl window together with the corresponding SEM image of this layer. The different band shapes in traces a and b most likely arise from the different particle shapes of the two samples. $\tilde{\nu}$ is the transition wavenumber.

ticle layer (trace in Fig. 6.8b) compared with the aerosol (trace a). In accordance with the explanations above, we do not observe such deviations between the two different spectra for the weak absorption bands.

6.5.2
Sugar Nanoparticles

Sugar-like substances such as cyclodextrins are attractive carriers for drugs to administer them in solid dosage form and improve their dissolution. Cyclodextrins are cyclic oligosaccharides of D-glycopyranose units able to form inclusion complexes with many drugs [74, 75]. Controlled particle deposition (CPD, [74, 76]) is one possibility to produce such complexes. In CPD the solute of interest is dissolved in supercritical CO_2. This solution then permeates the pores of the carrier where the drug precipitates after a fast drop of the pressure. Another possibility is to generate mixed particles with an electro-spray or with RESS by using a solution of both the drug and the carrier (CORESS [19–22, 24, 74, 76]). This results in encapsulation of the drug by or mixing of the drug with the carrier. In this section we show some results concerning particle formation of different sugar-like substances with the electro-spray (Section 6.2.2) and with RESS (Section 6.2.1). The complicated network of hydrogen bonds in such sugar particles gives rise to various amorphous and crystalline forms that can affect their use in nanoparticle technology. Infrared spectroscopy again proves to be useful to elucidate the complex inter- and intramolecular hydrogen-bond dynamics in these particles.

We have investigated different sugar particles generated with the electro-spray, including glycolaldehyde, hydroxyacetone, glyceraldehyde dimer, dihydroxyacetone and its dimer, glucose, fructose, saccharose, ribose, 2-deoxyribose, and erythrose [9, 29, 30, 51]. Figure 6.9 shows, from top to bottom, the IR spectra of fructose particles ($C_6H_{12}O_6$), glucose particles ($C_6H_{12}O_6$), ribose particles ($C_5H_{10}O_5$), and 2-deoxyribose particles ($C_5H_{10}O_4$). Common to all of these spectra are the relatively broad absorption bands, which clearly indicate that the aerosol particles are amorphous and not crystalline. This statement has been further corroborated by comparing the particle spectra with spectra of the crystalline substances. In contrast to the particles the latter exhibit narrow bands in the IR spectra, as illustrated for 2-deoxyribose by Fig. 6.10(b), with the corresponding particle spectrum shown in Fig. 6.10(a). Comparison of the four spectra in Fig. 6.9 also shows that the finger-

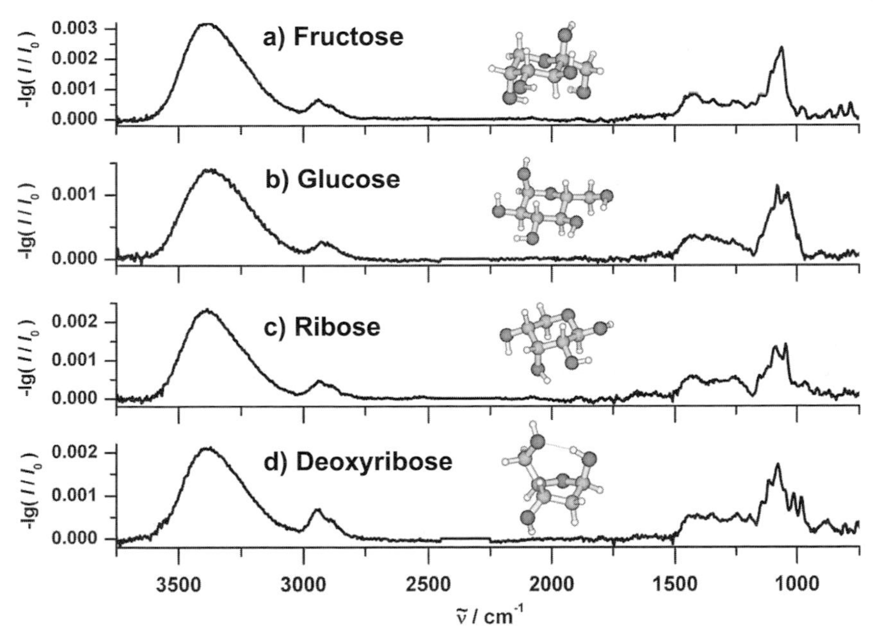

Fig. 6.9. Infrared spectra of aerosol particles of (a) fructose, (b) glucose, (c) ribose, and (d) deoxyribose. Fructose, glucose, and ribose molecules form pyranose rings in the particles, as indicated by the molecular structures. For deoxyribose a furanose ring is shown, but it is not clear in this case whether the particles consist of furanose or pyranose rings. $\tilde{\nu}$ is the transition wavenumber.

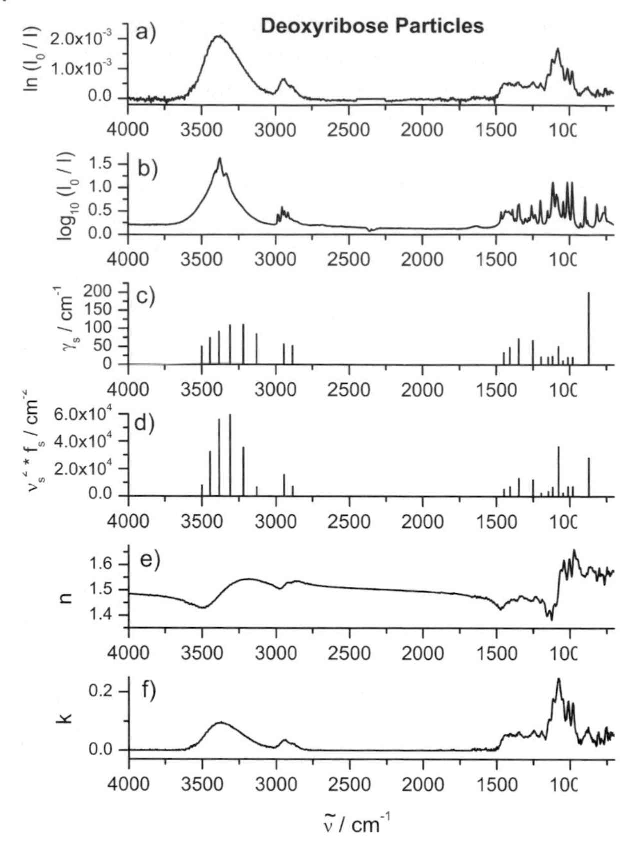

Fig. 6.10. (a) Experimental IR spectrum of deoxyribose aerosol particles. (b) Experimental IR spectrum of crystalline deoxyribose in KBr. (c) Optimized damping constants γ_s and (d) optimized relative transition intensities $\tilde{\nu}_s^2 f_s$ from the Lorentz model [Eq. (3)]. (e) Real part n and (f) imaginary part k of the complex refractive index obtained by a Kramers–Kronig inversion of the experimental particle spectrum [Eq. (5)]. $\tilde{\nu}$ is the transition wavenumber.

print region below 1800 cm^{-1} is characteristic for each compound, even for compounds as similar as the two isomers fructose and glucose. The fingerprint region does not only allow us to distinguish between different sugars, it even provides information about the structure of the molecules in the particles. For instance the missing carbonyl band (expected at 1750 cm^{-1}) in all four spectra proves that the sugar molecules are not present in their open form but that they form ring structures in the particles. In addition, from comparison of the IR spectra with quantum chemical calculations it is possible, for some sugars, to make statements about the kind of ring structure (e.g., furanose or pyranose). For fructose, glucose, and ribose the pyranose rings seem to be predominant, as indicated by the molecular structures (insets) in Fig. 6.9. For deoxyribose, however, it is not clear from the analysis of the IR spectra whether the particles consist of pyranose or furanose rings.

In contrast to the fingerprint region, the OH-stretching region around 3300 cm^{-1} looks almost identical for all four sugars. It is characterized by a single, unstructured, very broad band, which is shaded towards lower frequencies. This spectral region is strongly influenced by the formation of inter- and intramolecular hydrogen bonds. The characteristic band shape seems to reflect a more general behavior that is common to all four carbohydrate particles. To elucidate this phenomenon we have performed a band shape analysis assuming a Lorentz model that leads to the dielectric function given by Eq. (3). The relation to the refractive index data is given by Eq. (4). The same procedure as described in the previous section for phenanthrene particles was used to determine the optical data. As a result, we obtained the oscillator strengths ($\tilde{\nu}_s^2 f_s$) and the damping widths (γ_s) of the different Lorentz oscillators with resonance wavenumbers ($\tilde{\nu}_s$). For deoxyribose, the values for γ_s are depicted in Fig. 6.10(c) and the values for $\tilde{\nu}_s^2 f_s$ in 6.10(d). Here, we are mainly interested in the OH-stretching region. The sequence of oscillator strengths and widths found in this region for the particle spectra reveals a hierarchy of different hydrogen bonds: large redshifts ($\tilde{\nu}_s$) of OH-stretch transitions are accompanied by an increase in transition intensity ($\tilde{\nu}_s^2 f_s$) and a stronger coupling to the environment (γ_s). We interpret this observation as a result of a certain degree of short-range order in the amorphous solid particles caused by intermolecular hydrogen bonds. Since the other three sugars give the same qualitative results as discussed here for deoxyribose this hints at similarities in the kind of short range order for the different sugar-like substances. The band shape analysis with the Lorentz model also yields refractive index data. We have further improved these data by a Kramers–Kronig inversion [Eq. (5)]. The final refractive index data n and k are listed in Ref. [46] and are shown in Fig. 6.10(e) and (f) respectively. With these data, the determination of the size distribution now becomes possible for deoxyribose aerosols.

In addition to the electro-spray generation, a batch RESS apparatus has been used to investigate the formation of submicron sugar particles. This apparatus is designed for experiments in the temperature range from 270 to 370 K and pressures up to 25 MPa. The RESS experiments were performed at a pre-expansion temperature of 338 K and a pre-expansion pressure of 20 MPa. In all experiments the extraction conditions were $T_{ext} = 323$ K and $p_{ext} = 20$ MPa. Figure 6.11 shows, on the left-hand side, maltodextrin particles and, on the right-hand side, lactose particles produced by RESS. These SEM pictures are typical examples of the products obtained and show that the spherical particles are less than 300 nm in diameter. However, notably, due to the very low solubility of both maltodextrin and lactose in supercritical CO_2 a co-solvent should be used in further investigations to enhance the solubility of the sugars in CO_2.

6.5.3
Drug Nanoparticles

One main problem arising from the micronization of drugs with RESS is that in the final products the small primary particles often coagulate or agglomerate to

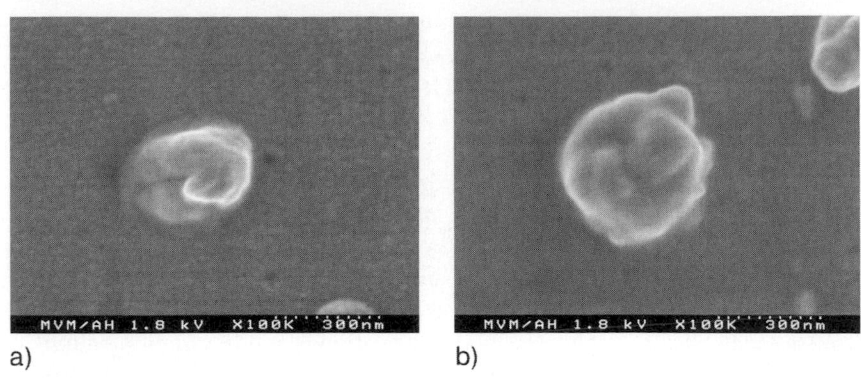

Fig. 6.11. Maltodextrin particles (left) and lactose particles (right) produced by RESS. The particles are 300 nm in diameter.

form much larger aggregates. These processes can take place either upon particle formation, or upon collecting the particles, or during storage of the powders. Coagulation and agglomeration leads to a decrease of the active surface and thus reduces the dissolution rate and the bioavailability of drugs. This can be overcome by mixing with or encapsulating by a second substance that avoids or at least reduces agglomeration. For instance, this can be achieved by rapid expansion of a ternary supercritical mixture which leads to co-precipitation of the two solutes (CORESS). Until now, much work has been done on producing particles of pure solutes by RESS, but only a few studies have been devoted to the simultaneous co-precipitation of two solutes [19–22, 24, 26, 74, 76]. Another way to prevent particle growth through coagulation and agglomeration and to stabilize the particles is to spray the supercritical solution directly into an aqueous surfactant solution (RESSAS [14, 18, 23, 25]). Encapsulation of particles is not only of interest for stabilizing primary particles, it is also a promising way for controlled drug delivery in the body. For instance, the coating can allow for targeted drug release in different intestinal regions (pH-dependent release). We have investigated such aspects for different pharmaceutical agents such as griseofulvin (oral antifungal drug), ibuprofen (anti-inflammatory drug), and phytosterol [10, 14, 18, 24, 74, 76]. In this section, we discuss the agglomeration behavior of phytosterol particles generated with RESS and the influence of co-expanding a polymer in the supercritical solution (CORESS). Phytosterol is a mixture of 85% β-sitosterol, 10% stigmasterol, and 5% campesterol. The main component β-sitosterol can be used as an additive for cosmetic substances or as an agent to reduce the amount of cholesterol in the human blood.

Pure phytosterol particles generated by RESS have a strong tendency to coagulate/agglomerate. This is documented by the two SEM images in Fig. 6.12. The survey in trace (a) shows that micronized phytosterol has a spongy structure that consists of coagulated/agglomerated primary particles. This can be seen best in trace (b), which represents an expanded view of trace (a). The agglomerated

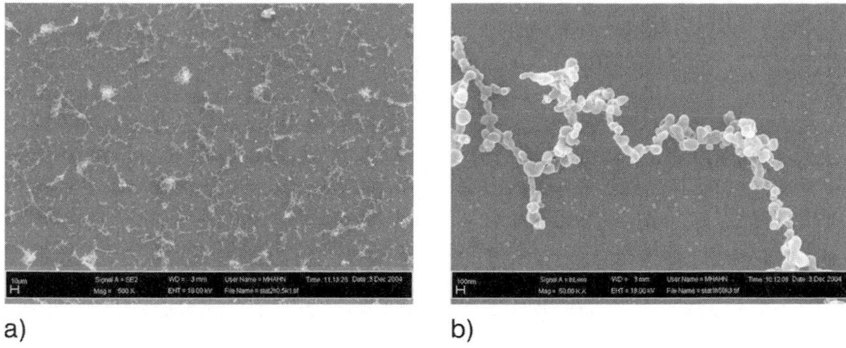

a) b)

Fig. 6.12. SEM images of pure phytosterol particles produced by RESS. (a) Survey and (b) detailed view.

spongy structure consists of primary particles with diameters between 50 and 150 nm. With our *in situ* characterization methods we have tried to find out where the agglomeration of the phytosterol particles takes place. In principle, the particles can agglomerate before and at the Mach disc, after the Mach disc in the gas phase, or during the particle collection on the sample holder for the SEM examination. From our measurements, we cannot make firm statements about agglomeration processes taking place before and at the Mach disc, but the 3-WEM at least indicates that the phytosterol particles are quite small ($r_0 \sim 50$ nm, $\sigma \sim 1.8$) immediately after the pulse and after all CO_2 has evaporated. After the pulse has finished, however, we can observe the behavior of the aerosol in the expansion chamber as a function of time. For the very large agglomerated particles depicted in Fig. 6.12(a), we would expect a decrease of the extinction in the IR spectra and a decrease of the extinction signals recorded for the 3-WEM with increasing time. The reason for that is the settling of the agglomerates. For comparison, settling of the phenanthrene particles with mean diameters around 400 nm leads to a decrease of the extinction by 70% already after 10 min. In contrast to these expectations, both the infrared signal and the 3-WEM signal are almost completely constant for more than 15 min after the end of the pulse. This is shown in Fig. 6.13(a) and (b). Both figures show only a minor change of the signal with time. This means that most of the particles in the gas phase are very small and stay small (no agglomeration). Bigger particles or agglomerates would settle down in our chamber and thus lead to a decrease in the extinctions for FTIR and 3-WEM. This is a clear experimental hint that the agglomeration takes place predominantly during the sample collection for the SEM measurements.

Based on a modified RESS-process (CORESS [19–22, 24, 74, 76]), additional experiments were performed with biodegradable polymers [Eudragit© and poly-L-lactic-acid (L-PLA)] and phytosterol as solutes. In these experiments both phytosterol and the polymer are thoroughly mixed and weighed out and then packed into the extraction vessel. Experiments with Eudragit© were performed with the batch apparatus (see Section 6.2.1) at pre-expansion temperatures between 310

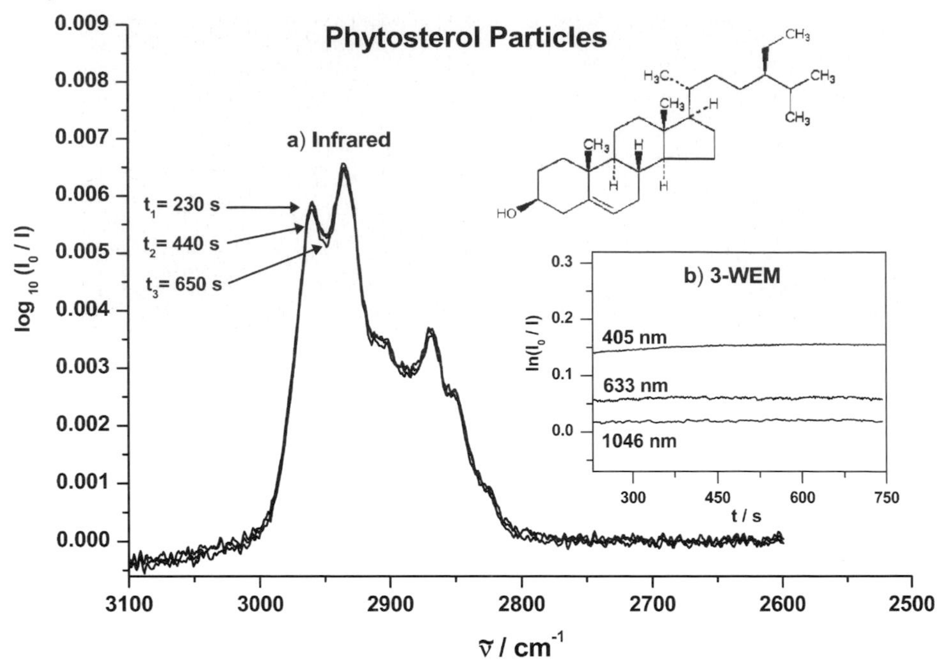

Fig. 6.13. (a) Infrared spectra of pure phytosterol aerosol particles recorded at $t_1 = 230$, $t_2 = 440$, and $t_3 = 650$ s after the end of the expansion pulse ($t = 0$ s). $\tilde{\nu}$ is the transition wavenumber. (b) Extinction signals from 3-WEM for the three lasers with wavelengths $\lambda_1 = 405$, $\lambda_2 = 655$, and $\lambda_3 = 1064$ nm. The signals remain nearly constant as a function of time (t) after the end of the expansion pulse.

and 340 K and a pre-expansion pressure of 20 MPa. With L-PLA the continuous RESS apparatus was used. These experiments were performed at an extraction temperature of 313 K, pre-expansion temperatures between 348 and 388 K and a pre-expansion pressure of 20 MPa. To avoid clogging, the nozzle temperature was 10 K higher than the respective pre-expansion temperature. These pre-expansion conditions were chosen to prevent particle precipitation inside the capillary nozzle. SEM was used to observe the morphology of the particle surface. The size of the particles produced by CORESS was measured online with 3-WEM. Differential Scanning Calorimetry (DSC) was used for physical characterization (melting point, heat of fusion, and crystallinity) of the original and the processed materials. The decrease of the heat of fusion obtained in the first heating-run allowed us to estimate the amount of polymer in the mixture. In addition, the unprocessed and the processed substances were characterized by X-ray diffraction (XRD).

Figure 6.14 shows, on the left, typical results for the co-precipitated phytosterol/Eudragit© particles. As mentioned above, the pure phytosterol particles are usually agglomerated and show a spongy structure. Contrary to these results the CORESS experiments performed with mixtures of phytosterol/Eudragit© (mass ratios of 1:1 and 1:10) lead to finely dispersed particles with diameters in the range of 250 nm.

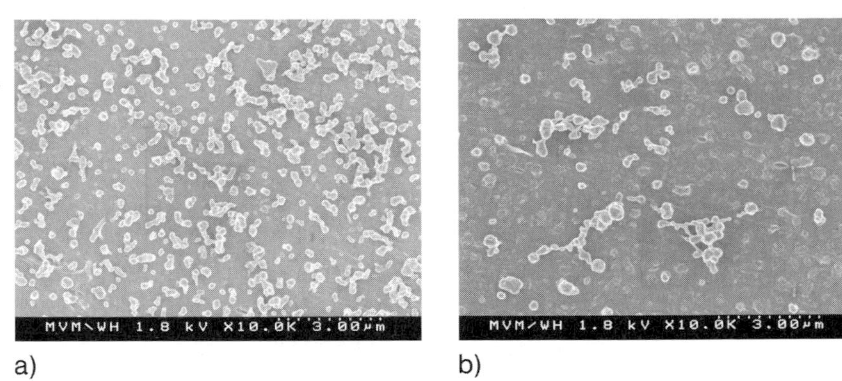

a) b)

Fig. 6.14. Phytosterol/Eudragit© particles (left) and phytosterol/L-PLA particles (right) produced by CORESS. The amount of Eudragit© is ~10 wt% and the amount of L-PLA is ~12 wt%.

The particles thus obtained were analyzed by DSC and compared with the unprocessed materials to determine the amount of polymer in the particles. The melting temperature and the heat of fusion of the original phytosterol were found to be 413 K and 56.9 J g^{-1}, respectively. These values differ markedly from those of the submicron phytosterol/Eudragit© particles (407 K and 37.1 J g^{-1}, respectively, for a mass ratio of 1:10). Depending on the initial mass ratio of the mixture in the extraction vessel, the amount of Eudragit© in the particles varies from ~10 to ~35 wt%. Additional FTIR analysis of the phytosterol/Eudragit© particles obtained confirm these results.

Additional CORESS experiments were performed with mixtures of phytosterol/L-PLA (mass ratios of 1:5 and 1:8). As shown in the right-hand part of Fig. 6.14, the experiments again lead to dispersed particles 250 nm in diameter. The product was characterized by DSC and XRD. Both the melting temperature and the heat of fusion were reduced in comparison to pure phytosterol. Typical DSC curves of pure phytosterol particles, of phytosterol/L-PLA particles, and of pure L-PLA particles are depicted in Fig. 6.15. The curve of pure phytosterol shows a melting peak in the range between 403 and 415 K while the curve of the phytosterol/L-PLA particles shows a broad peak between 388 and 411 K, corresponding to the microencapsulation of phytosterol. Depending on the initial mass ratio of the mixture in the extraction vessel, the amount of L-PLA varied from 12 to 40 wt%. Figure 6.16 displays typical XRD patterns of pure phytosterol particles, of phytosterol/L-PLA particles, and of pure L-PLA particles. For both the pure phytosterol and the phytosterol/L-PLA mixture diffraction peaks characteristic for crystalline phytosterol are obtained. These diffraction peaks disappear for the submicron L-PLA particles. Thus, the L-PLA particles are in the amorphous state. This result is in good agreement with the DSC analyses where no melting peak of crystalline L-PLA was observed (Fig. 6.15).

IR spectroscopy allows us to determine and thus to control the amount of L-PLA

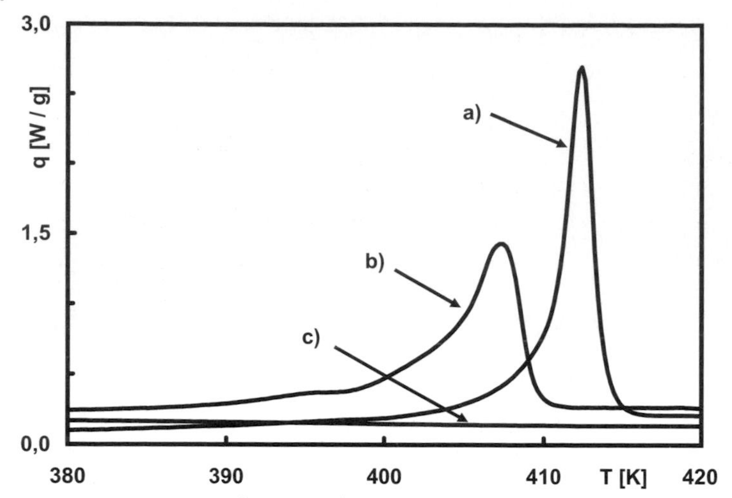

Fig. 6.15. DSC curves of (a) pure phytosterol particles, (b) phytosterol/L-PLA particles, and (c) pure L-PLA particles.

in situ during particle formation. This is much faster and simpler than the analysis with DSC and allows us to optimize the amount of polymer by adjusting the process parameters during particle formation. Figure 6.17 compares the IR spectrum of pure phytosterol particles (trace a) with two spectra of mixed phytosterol/L-PLA particles (traces b and c). The spectrum of pure phytosterol particles has strong absorption bands in the OH/CH-stretching region around 2900 cm^{-1} but only weak absorption bands in the fingerprint region below 1800 cm^{-1}. Although there are many bands in this region, they are barely visible due to their weakness. Traces

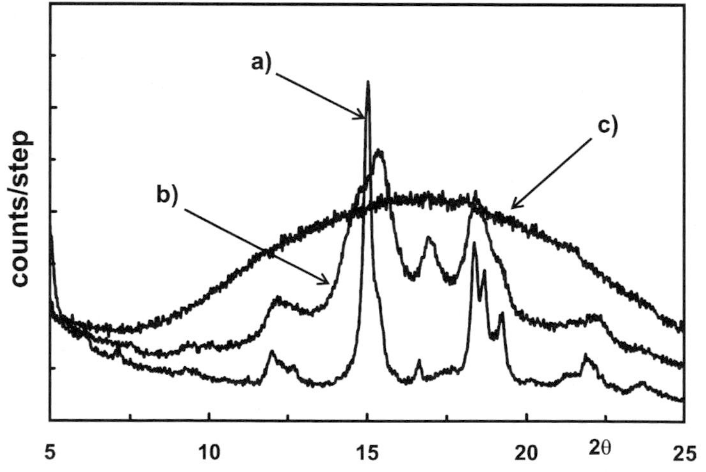

Fig. 6.16. XRD patterns of (a) pure phytosterol particles, (b) phytosterol/L-PLA particles, and (c) pure L-PLA particles.

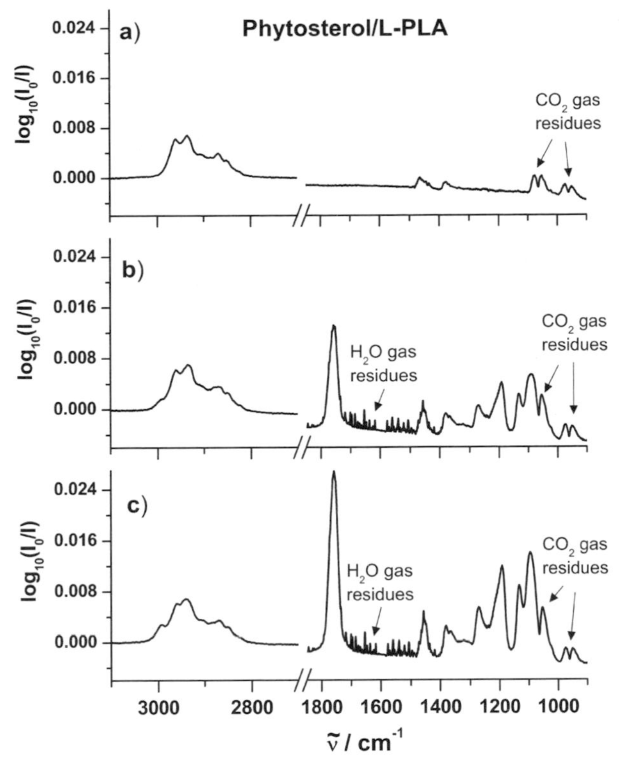

Fig. 6.17. (a) Infrared spectrum of pure phytosterol aerosol particles. (b) and (c) IR spectra of mixed phytosterol/L-PLA aerosol particles. There is ~70% more polymer in the particles of trace (c) than in trace (b). $\tilde{\nu}$ is the transition wavenumber.

(b) and (c) show that the strongest band of L-PLA in the fingerprint region is the carbonyl stretching-band around 1765 cm^{-1}. In contrast to the other absorption bands of L-PLA in the fingerprint region it is well separated and does not overlap with phytosterol bands or CO_2 bands. Moreover, L-PLA shows no significant absorbance above 2800 cm^{-1}. Therefore, the ratio of its carbonyl band intensity to that of the phytosterol OH/CH absorption is well suited to determine the relative content of L-PLA in the particles. From trace (b) to trace (c) the amount of L-PLA in the particles increases by ~70%.

In summary, these results illustrate that the simultaneous co-precipitation of two solutes is a promising method to produce composite particles. The particles obtained from the CORESS process appear as a drug core encapsulated in a polymer coating. This encapsulation helps to prevent agglomeration and to control drug release, which both depend on the amount of polymer in the particles. The easiest way to control the amount of polymer is by *in situ* FTIR spectroscopy during particle formation.

6.6
Summary and Conclusion

FTIR spectroscopy provides a wealth of information about molecular nanoparticles, which makes it an particularly powerful method for the characterization of such systems. Vibrational spectra of particulate systems allow us to draw conclusions about the chemical composition, the size, the shape, and about structural properties of the particles. Especially attractive is the use of *in situ* direct absorption FTIR spectroscopy, which allows us to characterize and control the particle properties during or shortly after their generation. For these investigations it is crucial that *in situ* FTIR spectroscopy is fast, that it can be combined with various particle generation methods, and that it can be used under very different experimental conditions.

The different examples in this chapter document the general applicability of this characterization method and the wide range of information it provides for biomolecular particles. This comprises different aspects of particle characterization. One example is the investigation of the particle formation process itself. In agreement with previous modeling results the RESS experiments show that, depending on the pre-expansion conditions, condensation of the solvent CO_2 can occur in the initial stage of particle formation. Such information is very useful for the evaluation of the obtained particles since the condensation/evaporation behavior of the solvent has a strong influence on particle formation and therewith on the properties of the final solid particles. Another example is the on-line control of particle composition. This is crucial, for instance, for polymer encapsulated drugs when a well defined amount of polymer in the particles has to be maintained.

A prerequisite for a successful analysis of the information contained in the IR spectra is the knowledge of the corresponding refractive index data. These data provide information, for instance, about the size effects and the shape effects to be expected in extinction spectra. We have demonstrated that the determination of accurate optical data from experimental IR spectra is not easy. The main reason lies in the lack of accurate information about the size, shape, and number concentration of the particles and in the model assumptions used in the derivation of the optical data. Therefore, it is not astonishing that, nowadays, only a few databases provide optical data for particulate systems and nearly none for nanoparticles of interest in life sciences. The lack of these data certainly represents one of the major restrictions for the routine application of FTIR spectroscopy for particle characterization. To alleviate that restriction is one of the most important challenges in the field.

Acknowledgment

Our special thanks go to to our coworkers Dr A. Bonnamy, G. Firanescu, D. Hermsdorf, P. Hils, M. Jetzki, Ch. Kornmayer, M. Meier, R. Ueberschaer, and D. Vollmar. We gratefully acknowledge collaboration with Professor Dr M. A. Suhm and PD Dr S. Beuermann. Financial support was provided by the Deutsche Forschungsgemeinschaft (grant Nrs. SI 833/1-1,2 and Tu 94/5-1,2; SFB 602; and GRK 782) and by the Fonds der Chemischen Industrie.

References

1 D. Liu, J. Gao, C. J. Murphy, C. T. Williams. In situ attenuated total reflection infrared spectroscopy of dendrimer-stabilized platinum nanoparticles adsorbed on alumina. *J. Phys. Chem. B*, 108, 12911–12916, **2004**.

2 J. Holman, S. Ye, D. J. Neivandt, P. B. Davies. Studying nanoparticle-induced structural changes within fatty acid multilayer films using sum frequency generation vibrational spectroscopy. *J. Am. Chem. Soc.*, 126, 14322–14323, **2004**.

3 F. Maillard, E. R. Savinova, P. A. Simonov, V. I. Zaikovskii, U. Stimming. Infrared spectroscopic study of CO adsorption and electro-oxidation on carbon-supported Pt nanoparticles: interparticle versus intraparticle heterogeneity. *J. Phys. Chem. B*, 108, 17893–17904, **2004**.

4 V. Stamenvović, M. Arenz, P. N. Ross, N. M. Marković. Temperature-induced deposition method for anchoring metallic nanoparticles onto reflective substrates for in situ electrochemical infrared spectroscopy. *J. Phys. Chem. B*, 108, 17915–17920, **2004**.

5 F. S. Manciu, R. E. Tallman, B. D. McCombe, B. A. Weinstein, D. W. Lucey, Y. Sahoo, P. N. Prasad. Infrared and Raman spectroscopies of InP/II–VI core–shell nanoparticles. *Physica E*, 26, 14–18, **2005**.

6 J. L. Burt, C. Gutiérrez-Wing, M. Miki-Yoshida, M. José-Yacamán. Noble-metal nanoparticles directly conjugated to globular proteins. *Langmuir*, 20, 11778–11783, **2004**.

7 J. M. Merritt, G. E. Douberly, R. E. Miller. Infrared–infrared double resonance spectroscopy of cyano-acetylene in helium nanodroplets. *J. Chem. Phys.*, 121, 1309–1316, **2004**.

8 S. A. Kalele, S. S. Ashtaputre, N. Y. Hebalkar, S. W. Gosavi, D. N. Deobagkar, D. D. Deobagkar, S. K. Kulkarni. Optical detection of antibody using silica–silver core–shell particles. *Chem. Phys. Lett.*, 404, 136–141, **2005**.

9 M. Jetzki, R. Signorell. The competition between hydrogen bonding and chemical change in carbohydrate nanoparticles. *J. Chem. Phys.*, 117, 8063–8073, **2002**.

10 A. Bonnamy, D. Hermsdorf, R. Ueberschaer, R. Signorell. Characterization of the rapid expansion of supercritical solutions by FTIR spectroscopy in situ, accepted. *Rev. Sci. Instrum.*, *76*(5), Art. No 053904 (published online).

11 D. W. Matson, J. L. Fulton, R. C. Petersen, R. D. Smith. Rapid expansion of supercritical fluid solutions: solute formation of powders, thin films, and fibers. *Ind. Eng. Chem. Res.*, 26, 2298–2306, **1987**.

12 J. W. Tom, P. G. Debenedetti. Particle formation with supercritical fluids – a review. *J. Aerosol. Sci.*, 22, 555–584, **1991**.

13 J. Jung, M. Perrut. Particle design using supercritical fluids: Literature and patent survey. *J. Supercrit. Fluids*, 20, 179–219, **2001**.

14 M. Türk, P. Hils, B. Helfgen, K. Schaber, H.-J. Martin, M. A. Wahl. Micronization of pharmaceutical substances by rapid expansion of supercritical solutions (RESS): a promising method to improve bioavailability of poorly soluble pharmaceutical agents. *J. Supercrit. Fluids*, 22, 75–84, **2002**.

15 D. Hermsdorf, A. Bonnamy, M. A. Suhm, R. Signorell. Infrared spectra of phenanthrene particles generated by pulsed rapid expansion of CO_2

solutions. *Phys. Chem. Chem. Phys.*, 6, 4652–4657, **2004**.

16 S. Cihlar. *Mikronisierung organischer Feststoffe durch schnelle Expansion überkritischer Lösungen.* Doctoral Thesis, VDI Verlag GmbH (Düsseldorf), **2000**.

17 M. Türk. Formation of small organic particles by RESS: Experimental and theoretical investigations. *J. Supercrit. Fluids*, 15, 79–89, **1999**.

18 M. Türk, B. Helfgen, P. Hils, R. Lietzow, K. Schaber. Micronization of pharmaceutical substances by Rapid Expansion of Supercritical Solutions (RESS): Experiments and modeling. *Particle Particle Systems Characteriz.*, 19, 327–335, **2002**.

19 J. W. Tom, P. G. Debenedetti. Formation of bioerodible polymeric microspheres and microparticles by rapid expansion of supercritical solutions. *Biotechnol. Prog.*, 7, 403–411, **1991**.

20 J. W. Tom, G.-B. Lim, P. G. Debenedetti, R. K. Prud'homme. Applications of supercritical fluids in the controlled release of drugs. *ACS Symp. Ser.*, 514, 238–257, **1993**.

21 J. W. Tom, P. G. Debenedetti. Precipitation of poly(L-lactic acid) and composite poly(L-lactic acid)–pyrene particles by rapid expansion of supercritical solutions. *J. Supercrit. Fluids*, 7, 9–29, **1994**.

22 J.-H. Kim, T. E. Paxton, D. L. Tomasko. Microencapsulation of naproxen using rapid expansion of supercritical solutions. *Biotechnol. Prog.*, 12, 650–661, **1996**.

23 M. Türk. Herstellung organischer nanopartikel und deren stabilisierung in wässrigen Lösungen (RESSAS). *Chem. Ing. Techn.*, 75, 792–795, **2003**.

24 M. Türk. Untersuchung zum coating von submikronen partikeln mit dem CORESS-Verfahren. *Chem. Ing. Techn.*, 76, 835–838, **2004**.

25 M. Türk, R. Lietzow. Stabilized nanoparticles of phytosterol by rapid expansion from supercritical solutions into aqueous solution. *AAPS PharmSciTech*, 5:Article 56, **2004**.

26 Kh. Hussein, M. Türk, M. A. Wahl. Preparation and evaluation of drug/β-cyclodextrin solid inclusion complexes by supercritical fluid technology. I. Kikic, M. Perrut (eds.), *Proceedings of the 9th Meeting on Supercritical Fluids*, Trieste, Italy, June 13–16, 2004.

27 D.-R. Chen, D. Y. H. Pui, S. L. Kaufman. Electrospraying of conducting liquids for monodisperse aerosol generation in the 4 nm to 1.8 µm diameter range. *J. Aerosol Sci.*, 26, 963–977, **1995**.

28 S. L. Kaufman, R. Caldow, F. D. Dorman, K. D. Irwin, A. Pöcher. Conversion efficiency of the TSI model 3480 electrospray aerosol generator using sucrose. *J. Aerosol Sci.*, 30, S373–S374, **1999**.

29 R. Signorell, M. K. Kunzmann, M. A. Suhm. FTIR investigation of non-volatile molecular nanoparticles. *Chem. Phys. Lett.*, 329, 52–60, **2000**.

30 R. Signorell, D. Luckhaus. Aerosol spectroscopy of dihydroxyacetone: Gas phase and nanoparticles. *J. Phys. Chem. A*, 106, 4855–4867, **2002**.

31 Th. Häber, U. Schmitt, C. Emmeluth, M. A. Suhm. Ragout-jet FTIR spectroscopy of cluster isomerism and cluster dynamics: from carboxylic acid dimers to N_2O nanoparticles. *Faraday Discuss.*, 118, 331–359, **2001**.

32 Th. Häber, U. Schmitt, M. A. Suhm. FTIR-spectroscopy of molecular clusters in pulsed supersonic slit-jet expansions. *Phys. Chem. Chem. Phys.*, 1, 5573–5582, **1999**.

33 P. G. Debenedetti. Homogeneous nucleation in supercritical fluids. *AIChE J.*, 36, 1289–1298, **1990**.

34 M. Türk. Influence of thermodynamic behaviour and solute properties on homogeneous nucleation in supercritical solutions. *J. Supercrit. Fluids*, 18, 169–184, **2000**.

35 B. Helfgen, P. Hils, Ch. Holzknecht, M. Türk, K. Schaber. Simulation of particle formation during the rapid expansion of supercritical solutions. *J. Aerosol Sci.*, 32, 295–319, **2001**.

36 M. Weber, L. M. Russell, P. G. Debenedetti. Mathematical modeling

of nucleation and growth of particles formed by the rapid expansion of supercritical solution under subsonic conditions. *J. Supercrit. Fluids*, 23, 65–80, **2002**.

37 M. Weber, M. C. Thies. Understanding the RESS Process. In Y.-Ping Sun, ed., *Supercritical Fluid Technology in Material Science and Engineering*, pp. 387–437. Marcel Dekker, New York, **2002**.

38 B. Helfgen, M. Türk, K. Schaber. Hydrodynamic and aerosol modelling of the rapid expansion of supercritical solutions (RESS-process). *J. Supercrit. Fluids*, 26, 225–242, **2003**.

39 M. K. Kunzmann, R. Signorell, M. Taraschewski, S. Bauerecker. The formation of N_2O nanoparticles in a collisional cooling cell between 4 and 110 K. *Phys. Chem. Chem. Phys.*, 3, 3742–3749, **2001**.

40 K. D. Bartle, A. A. Clifford, S. A. Jafar, G. F. Shilstone. Solubilities of solids and liquids of low volatility in supercritical carbon dioxide. *J. Phys. Chem. Ref. Data*, 20, 713–756, **1991**.

41 R. Signorell. Infrared spectroscopy of particulate matter: between molecular clusters and bulk. *Mol. Phys.*, 101, 3385–3399, **2003**.

42 C. F. Bohren, D. R. Huffman. *Absorption and Scattering of Light by Small Particles*, Wiley-Interscience, New York, **1998**.

43 J. A. Roux, B. E. Wood, A. M. Smith. *IR Optical Properties of Thin H_2O, NH_3, and CO_2 Cryofilms*. AEDC-TR-79-57 (AD-A074913), September, 1979.

44 A. Bonnamy, M. Jetzki, R. Signorell. Optical properties of molecular ice particles from a microscopic model. *Chem. Phys. Lett.*, 382, 547–552, **2003**.

45 J. Meyer, M. Katzer, E. Schmidt, S. Cihlar, M. Türk. Comparative particle size measurements in lab-scale nanoparticle production processes. J. Bridgewater (ed.), *Proceedings of the World Congress on Particle Technology 3*, July 6–9, 1998, Brighton, UK.

46 http://www.user.gwdg.de/~rsignor/refindex.htm.

47 D. M. Hudgins, S. A. Sandford, L. J. Allamandola, A. G. G. M. Tielens. Mid- and far-infrared spectroscopy of ices: optical constants and integrated absorbances. *Astron. Astrophys. Suppl. Ser.*, 86, 713–722, **1993**.

48 Th. Henning, V. B. Il'in, N. A. Krivova, B. Michel, N. V. Voshchinnikov. WWW database of optical constants for astronomy. *Astron. Astrophys. Suppl. Ser.*, 136, 405–406, **1999**.

49 L. S. Rothman et al. The HITRAN molecular spectroscopic database and HAWKS (HITRAN Atmospheric Workstation): 1996 edn. *J. Quant. Spectrosc. Radiat. Transfer*, 60, 665–710, **1998**.

50 M. K. Kunzmann, S. Bauerecker, M. A. Suhm, R. Signorell. Spectroscopic characterization of N_2O aggregates: from clusters to the particulate state. *Spectrochim. Acta A*, 59, 2855–2865, **2003**.

51 M. Jetzki, D. Luckhaus, R. Signorell. Fermi resonance and conformation in glycolaldehyde particles. *Can. J. Chem.*, 82, 915–924, **2004**.

52 M. Jetzki, A. Bonnamy, R. Signorell. Vibrational delocalization in ammonia aerosol particles. *J. Chem. Phys.*, 120, 11775–11784, **2004**.

53 M. I. Mishchenko, J. W. Hovenier, L. D. Travis. *Light Scattering by Nonspherical Particles*, Academic Press, San Diego, **2000**.

54 R. Signorell. Verification of the vibrational exciton approach for CO_2 and N_2O nanoparticles. *J. Chem. Phys.*, 118, 2707–2715, **2003**.

55 R. Signorell, M. K. Kunzmann. Isotope effects on vibrational excitons in carbon dioxide particles. *Chem. Phys. Lett.*, 371, 260–266, **2003**.

56 W. C. Hinds. *Aerosol Technology*, Wiley Interscience, New York, **1999**.

57 H. Kuzmany. *Solid-State Spectroscopy*, Springer-Verlag, Berlin, **1998**.

58 J. E. Bertie, S. L. Zhang, C. D. Keefe. Infrared intensities of liquids XVI. Accurate determination of molecular band intensities from infrared refractive index and dielectric

constant spectra. *J. Mol. Struct.*, 324, 157–176, **1994**.

59 R. Z. Bachrach, F. C. Brown. Exciton-optical properties of TlBr and TlCl. *Phys. Rev. B*, 1, 818–831, **1970**.

60 R. K. Ahrenkiel. Modified Kramers–Kronig analysis of optical spectra. *J. Opt. Soc. Am.*, 61, 1651–1655, **1971**.

61 J. P. Hawranek, P. Neelakantan, R. P. Young, R. N. Jones. The control of errors in i.r. spectrophotometry – IV. Corrections for dispersion distortion and the evaluation of both optical constants. *Spectrochim. Acta A*, 32, 85–98, **1976**.

62 B. J. Finlayson-Pitts, J. N. Pitts. *Chemistry of the Upper and Lower Atmosphere*, Academic Press, San Diego, **2000**.

63 H. Schönbuchner, G. Guggenberger, K. Peters, H. Bergmann, W. Zech. Particle-size distribution of PAH in the air of a remote Norway spruce forest in northern bavaria. *Water, Air, Soil Pollut.*, 128, 355–367, **2001**.

64 A. Li, B. T. Draine. Infrared emission from interstellar dust. II. The diffuse interstellar medium. *Astrophys. J.*, 554, 778–802, **2001**.

65 M. Schnaiter, H. Mutschke, J. Dorschner, Th. Henning. Matrix-isolated nano-sized carbon grains as an analog for the 217.5 nanometer feature carrier. *Astrophys. J.*, 498, 486–496, **1998**.

66 E. M. Berends, O. S. L. Bruinsma, G. M. van Rosmalen. Nucleation and growth of fine crystals from supercritical carbon dioxide. *J. Cryst. Growth*, 128, 50–56, **1993**.

67 G.-T. Liu, K. Nagahama. Application of rapid expansion of supercritical solutions in the crystallization separation. *Ind. Eng. Chem. Res.*, 35, 4626–4634, **1996**.

68 G.-T. Liu, K. Nagahama. Solubility and RESS experiments of solid solution in supercritical carbon dioxide. *J. Chem. Eng. Jpn.*, 30, 293–301, **1997**.

69 C. Domingo, E. Berends, G. M. van Rosmalen. Precipitation of ultrafine organic crystals from the rapid expansion of supercritical solutions over a capillary and a frit nozzle. *J. Supercrit. Fluids*, 10, 39–55, **1997**.

70 D. M. Ginosar, W. D. Swank, R. D. McMurtrey, W. J. Carmack. Flow-field studies of the RESS process. *Proceedings of the 5th International Symposium on Supercritical Fluids*, Atlanta, USA, April 8–12, 2000.

71 R. Ueberschaer. Charakterisierung der Partikelbildung bei der schnellen Expansion überkritischer Lösungen. In *Diploma Thesis*. University of Göttingen, **2004**.

72 E. T. Peters, A. F. Armington, B. Rubin. Lattice constants of ultrapure phenanthrene. *J. Appl. Phys.*, 37, 226–227, **1966**.

73 A. Bonnamy, R. Georges, E. Hugo, R. Signorell. Infrared signatures of $(CO_2)_N$ clusters: size, shape, and structural effects. *Phys. Chem. Chem. Phys.*, 7, 963–969, **2005**.

74 M. Türk, P. Hils, K. Hussein, M. Wahl. Utilization of supercritical fluid technology for the preparation of innovative carriers loaded with nanoparticular drugs. In U. Teipel, ed., *Produktgestaltung in der Partikeltechnologie*, pp. 337–386, Fraunhofer-IRB-Verlag, Stuttgart, **2004**.

75 J. Szejtli. Introduction and general overview of cyclodextrin chemistry. *Chem. Rev.*, 98, 1753–1753, **1998**.

76 M. Türk, M. Wahl. Utilization of supercritical fluid technology for the preparation of innovative carriers loaded with nanoparticular drugs. In S. E. Pratsinis (ed.). *Proceedings of the International Congress for Particle Technology, PARTEC 2004, Nürnberg, March 16–18 (2004)*.

7
Characterization of Nanoscaled Drug Delivery Systems by Electron Spin Resonance (ESR)

Karsten Mäder

7.1
Introduction

Nanoscaled drug delivery systems (nano-DDS) have been heavily investigated due to their large potential to make drug therapy more efficient. After years of intense research, they have entered the pharmaceutical market and many systems have reached the late clinical state. Nano-DDS are only defined by their submicron size. They are made from different materials (e.g., lipid or polymer based) and include very different kinds of associates and particles. Examples include (mixed) micelles, polymer micelles, liposomes, microemulsions, polymer nanoparticles, solid lipid nanoparticles, nanocapsules. The main applications are in peroral and parenteral drug delivery, although nano-DDS are also considered for other administration routes (dermal, pulmonal). In peroral drug delivery, higher absorption rates and decreased variabilities of poorly soluble drugs are achieved. Nano-DDS make the i.v. injection of poorly soluble drugs feasible and help to push drug targeting strategies from the conception to reality.

The performance of nano-DDS strongly depends on many parameters, including composition, particle size, surface charge and the physical state of the matrix. Appropriate characterization of nanoparticle systems is a serious challenge due to following reasons:

- the submicron size of the objects;
- the sensitivity of many Nano-drug delivery systems to environmental changes, including drying, freezing or dilution;
- the coexistence of several colloidal species;
- the importance of dynamic phenomena;
- material properties might differ significantly from bulk material.

Some commonly used methods require tedious sample preparation (electron microscopy) or fixing of the particles (AFM). Fixation by drying of the samples composed of soft matter such as lipid nanoparticles, however, will change considerably the sample characteristics. Therefore, there is a need for methods that are can char-

Nanotechnologies for the Life Sciences Vol. 3
Nanosystem Characterization Tools in the Life Sciences. Edited by Challa S. S. R. Kumar

ISBN: 3-527-31383-4

acterize Nano-drug delivery systems with no or minimized sample preparation. NMR spectroscopy, fluorescence and other optical techniques, Raman-spectroscopy and ESR are methods that offer possibilities to conduct such studies. The challenge of appropriate characterization is considerable higher for characterization in biological surroundings *in vitro* or *in vivo*. Most techniques are not capable of *in vivo* measurements. Radiolabeling techniques are used to monitor the distribution of Nano-drug delivery systems in the body of animals and man; however, they provide no information about the physicochemical changes of the drug delivery system caused by interaction with the environment or metabolic processes. Consequently, our knowledge of the detailed fate of the drug and drug carrier *in vivo* is still rather limited. This chapter will discuss how ESR can contribute to shedding more light on nanoscaled drug delivery systems. The basic requirements of ESR will be discussed. Examples will show which information can be extracted from the spectra and how this information can be used to characterize drug delivery systems. Finally, the opportunities and limits of ESR for the *in vivo* characterization will be shown.

7.2
ESR Basics and Requirements

Electron Spin Resonance (ESR) or synonymously Electron Paramagnetic Resonance (EPR) can be regarded as a sister method for Nuclear Magnetic Resonance (NMR). Both methods were first described about 60 years ago and both require the presence of a magnetic moment. In NMR, the magnetic moment comes from the nuclear spin (e.g., ^{1}H, ^{13}C, ^{19}F), in ESR it derives from unpaired electrons. Sources of unpaired electrons are free radicals and certain oxidations states of transition metals.

The basic magnetic resonance experiment (NMR and EPR) requires the positioning of the sample in a homogenous magnetic field. The applied magnetic field leads to parallel or anti-parallel alignment of the spins. Irradiation of the sample with an appropriate frequency (= resonance frequency) leads the spins to switch between both states. Under such conditions, part of the energy is absorbed and a signal observed. Higher magnetic fields lead to higher sensitivity and need higher frequencies. Typical NMR frequencies are radiofrequencies (e.g., 20–600 MHz), whereas typical ESR frequencies are in the microwave range (10 GHz). However, the recent development of high-field NMR (800 MHz and higher) and low-frequency ESR (1 GHz and 300 MHz) have led to partial overlapping of the NMR and ESR frequency ranges. Due to the high dielectric loss, the standard ESR frequency of 10 GHz penetrates only 1–2 mm in water-containing samples (e.g., biological tissue). The development of low frequency ESR was necessary to make possible the investigation of larger samples with high water content. It permits ESR experiments on whole living mammals (mice and rats). Refs. [1, 2] give a broad coverage of basic ESR and biological related applications.

NMR is per se less sensitive than ESR due to the smaller magnetic moment of

Tab. 7.1. Comparison of NMR and ESR.

	ESR	*NMR*
Prerequisite	Electron spin	Nuclear spin
Examples	Free radicals, Cu^{2+}, Mn^{2+}, Fe^{3+}	^{1}H, ^{13}C, ^{19}F, ^{35}P
Magnetic moment ($J\ T^{-1}$)	Larger: $\mu_{e-} = 9.274 \times 10^{-24}$	Smaller: $\mu_{H} = 5.051 \times 10^{-27}$
Typical frequency	9–10 GHz	100–800 MHz
Mode of operation	Continuous wave	FT-NMR
Typical line widths	MHz	$<$Hz
"Time window" (s)	10^{-9}–10^{-11}	10^{-5}–10^{-9}

the nuclear spin. However, this disadvantage is compensated by the much higher concentrations of nuclear spins and the development of pulsed NMR. Most ESR spectrometers work in the continuous mode because most radicals relax very rapidly and, therefore, pulsed methods are difficult or impossible to apply. Table 7.1 summarizes the main features of NMR and ESR.

The basic instrumentation of ESR requires a magnet, a microwave bridge, and a resonator with phase sensitive detection. Of course, modern spectrometers are operated via a computer interface. Most spectrometers operate at 9 to 10 GHz (X-band). Due to the high dielectric loss, the measurement of samples with high water content is limited to a size < 1–2 mm. Recent progress in the design of instruments has resulted in small, sensitive benchtop spectrometers at reasonable prices (Fig. 7.1). Larger objects with high water content (tablets or small animals) can be assessed by spectrometers that work at lower frequencies, e.g., 0.3 or 1 GHz (Fig. 7.2).

We will focus only on drug delivery and will not consider other important applications such as ESR oximetry [3], the measurement of nitric oxide [4] or the direct detection [5] of radical metabolites. Spin trapping is a method that transfers very reactive radical intermediates into more stable spin adducts [6]. However, very careful preparation of the experiments and interpretation of the results is necessary to avoid artifacts [7, 8]. Artifacts might be caused by impurities of the com-

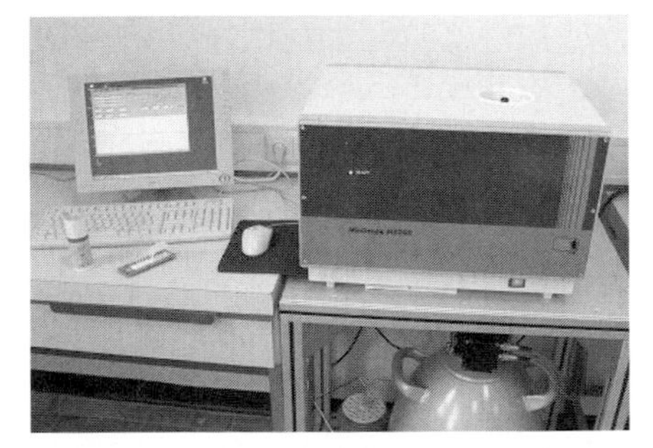

Fig. 7.1. Benchtop 9 GHz ESR-spectrometer (X-band) with temperature unit (below the spectrometer).

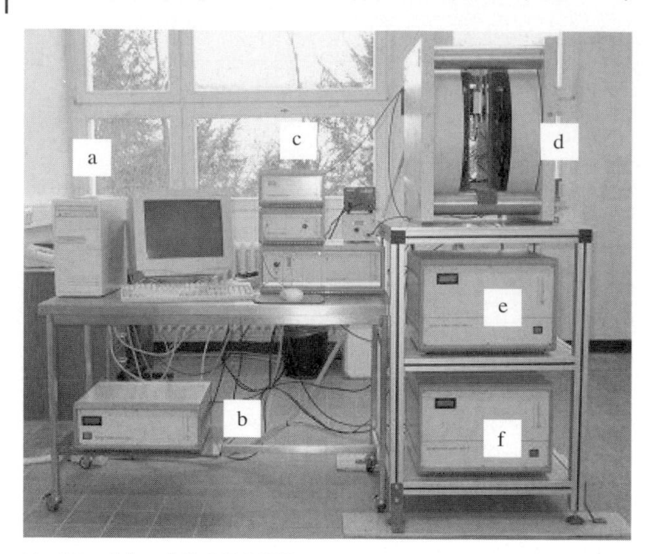

Fig. 7.2. L-band (1 GHz) ESR spectrometer with imaging possibility. The lower frequency of the L-band permits noninvasive measurement of larger samples with high water content, including mice and rats. The spectrometer consists of a computer (a), a magnet power supply (b), a microwave bridge (c), the magnet with resonator (d) and power supply for the imaging gradients (e).

mercial spin traps, non-radical reactions or instability of the spin adduct. Inexperienced researchers often misinterpret the experimental results due to their poor knowledge of the reactivity of the radical, the trapping efficiency, the instability of the trap under the experimental conditions and the adduct stability. ESR spin trapping has been used recently to detect the radical formation from TiO_2 nanoparticles [9] and the antioxidant potential of selenium nanoparticles [10].

Paramagnetic species that can be detected by ESR are free radicals and certain oxidation states of metal ions. ESR spectra of metal ions (e.g., Fe^{3+}, Cu^{2+}, Mn^{2+}) are in most cases very broad due to short relaxation times and magnetic interactions. Sometimes, their detection requires study at very low temperatures (77 K and below). ESR spectroscopy of metal ions permits the characterization of enzymes and organs, but it is of low importance in drug delivery. Some materials like bentonite and kaolin contain Fe^{3+} as impurity. Nearly all ESR studies in the field of drug delivery are based on the detection of free radicals, because most samples are ESR silent. Endogenous radicals could be present in the case of gamma sterilized or mechanical treated samples, if the delivery matrix or the drug is in the crystalline state and has a high melting point [11, 12]. Gamma irradiation induced radicals will rapidly decay in liquid and semisolid matrices and in polymers with a low glass transition temperature, but they can be detected and quantified by spin-trapping techniques or at low temperatures [13]. Gamma irradiation induced radicals can be used to follow water penetration in biodegradable polymers *in vitro* and *in vivo*, if they have sufficient thermostability (e.g., slow decay rates at 37 °C) [14].

Magnetite is a very important material for both diagnostic and therapy, and exhibits endogenous ESR signals. Magnetite is used as an MRI contrast agent as well as for local hyperthermia. Magnetite has very broad ESR lines. These can be used to quantify and characterize the material, including the uptake in cells and tissue [15, 16, 17].

However, most samples are ESR silent and require the addition of radicals. For this reason, the addition of radical probes is a necessity. In most cases, nitroxide radicals are used. A large variety of nitroxides with different hydrophobicity, charge and functional groups is commercially available. Most nitroxides derive from piperidine, pyrrolidine or imidazolidine structures. Examples of nitroxide radicals are illustrated in Fig. 7.3.

The numerous different structures (hydrophilic, hydrophobic, charged, uncharged, aliphatic or aromatic) permit one to choose a molecule with the desired

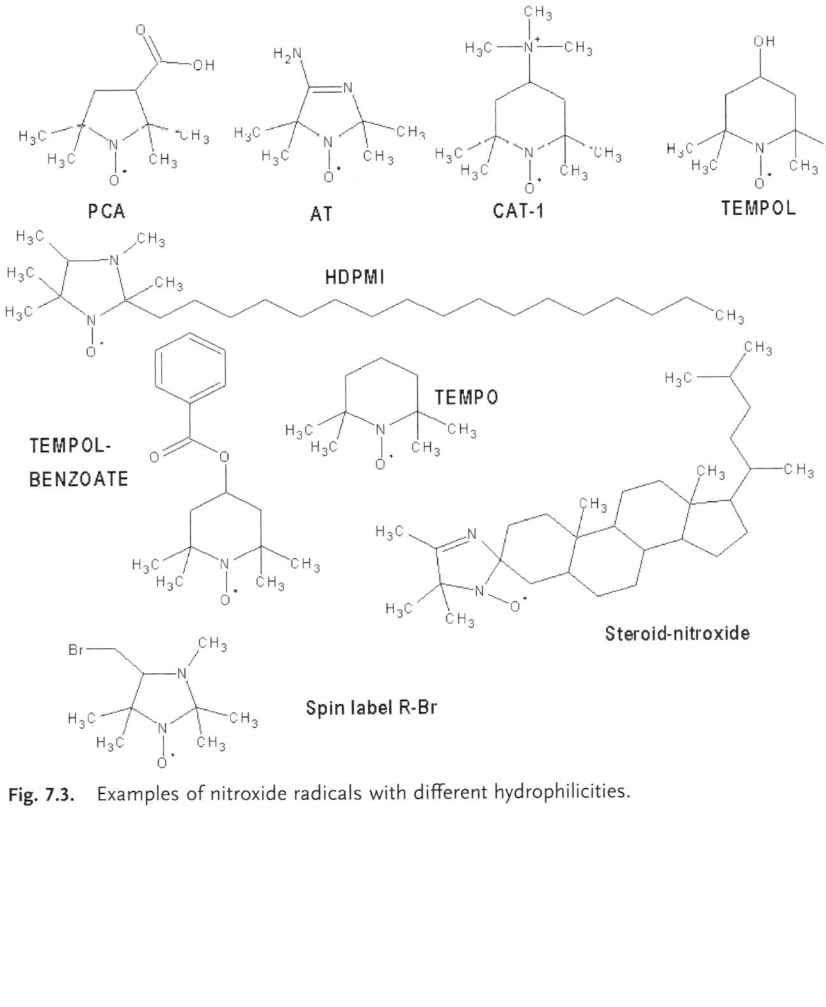

PCA AT CAT-1 TEMPOL HDPMI TEMPOL-BENZOATE TEMPO Steroid-nitroxide Spin label R-Br

Fig. 7.3. Examples of nitroxide radicals with different hydrophilicities.

physicochemical properties. Nitroxides can be used as spin probes (= model drugs) to monitor processes of drug delivery *in vitro* and *in vivo*. Larger molecules (e.g., proteins, PVA, chitosan) can be spin labeled by the covalent linking of a chemically activated nitroxide to a suitable group of the macromolecule (e.g., amino, hydroxyl, carboxyl groups). For most macromolecules, the spin labeling procedure will lead to small and insignificant changes in the properties the molecules; however, this has to be verified. Spin labeling of low molecular weight compounds will lead to significant changes in molecular properties because the nitroxide moiety contributes significantly to the whole molecule.

7.3 Information from ESR Spectroscopy and Imaging

The following information can be obtained by EPR spectroscopy and imaging:

1. nitroxide concentration
2. micropolarity and microviscosity
3. microacidity
4. distribution of radicals and spatial resolution of parameters 1–3 (by imaging techniques or using different nitroxide isotopes).

Let us consider the information content step by step.

7.3.1 Nitroxide Concentration

Nitroxide concentrations can be used to follow the release and distribution kinetics of drug carriers. ESR spectra are recorded in most cases in the form of the first derivative. Therefore, the nitroxide concentration can be obtained by double integration of the spectra. The signal intensity depends on the radical concentration, the parameters of measurement (microwave power, modulation amplitude, resonator type, filling factor, sample positioning, temperature etc.) and the dielectric loss of the sample. Samples with higher dielectric loss will give lower signal intensities for equal radical concentrations. It is highly recommended to use internal standards for quantitative measurements and to take great care about the positioning of the sample, reproducible filling factors, and sufficient scan width.

The hydrophilic ascorbic acid reduces piperidine nitroxides rapidly to the EPR silent hydroxylamines, and the loss of the EPR signal intensity indicates the accessibility of the nitroxide to the aqueous outer phase (Fig. 7.4). Solid lipid nanoparticles (SLN) and nanostructured lipid carriers (NLC) are less protective, respectively, than a nanoemulsion. This finding can be explained by the platelet structure of the SLN and the expulsion of foreign molecules during lipid crystallization.

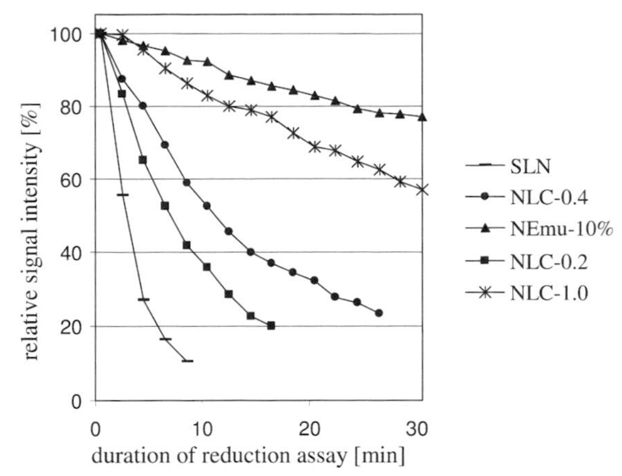

Fig. 7.4. Kinetics of nitroxide reduction of different colloidal lipid carriers. The lipophilic nitroxide TEMPOLBENZOATE is reduced more rapidly from solid lipid nanoparticles (SLN) and nanostructured lipid carriers (NLC) than with a standard nanoemulsion.

7.3.2
Micropolarity and Microviscosity

The unpaired electron of the nitroxide group is distributed between the nitrogen and oxygen of the nitroxyl group. Mesomeric forms of the distribution are illustrated in Fig. 7.5.

The nitrogen nucleus has a nuclear spin of 1. Magnetic interaction between the electron and nuclear spins of the nitrogen leads to a splitting of the ESR signal into three lines, which is also called hyperfine splitting (Hfs). Hfs is given by the Hfs constant $\mathbf{a_N}$ (Fig. 7.6).

The oxygen atom has no nuclear spin and therefore does not contribute to the splitting. Polar solvents (e.g., water) will favor the left-hand form in Fig. 7.5 and increase the spin density (= density of the unpaired electron) at the nitrogen. Non-polar solvents (e.g., oil) will increase localization of the spin at the oxygen and, therefore, decrease the spin density at the nitrogen. As a result, a smaller hyperfine coupling constant $\mathbf{a_N}$ will be observed (Fig. 7.6). Figure 7.7 shows quantitative values of hyperfine splitting in different solvents of pharmaceutical relevance.

For nitroxides, hyperfine coupling between the electron and nuclear spins is anisotropic. The a_Ns are smaller for the x and y direction (around 0.6 mT) than

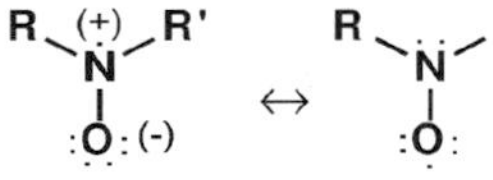

R–N(+)(–R')–O:(-) ↔ R–N(–R')–O:

Fig. 7.5. Mesomeric forms of the nitroxyl group. The free electron is distributed between the nitrogen and oxygen atom.

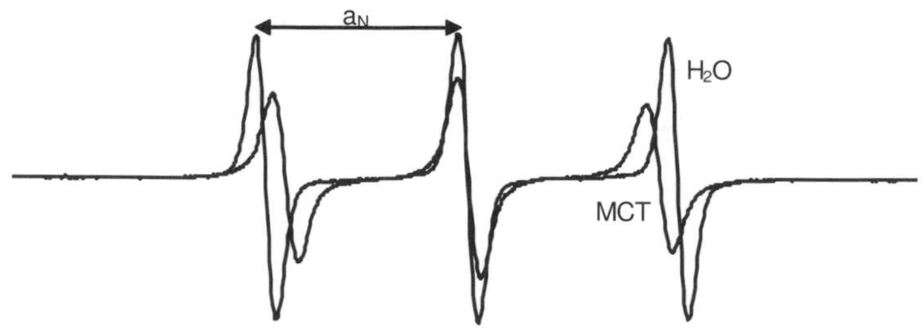

Fig. 7.6. ESR spectra of the spin probe TEMPOL in middle chain triglycerides (MCT) (oil) and water. Note the larger hyperfine splitting (a_N) in the solvent with higher polarity.

for the *z* direction (around 3.3 mT). The anisotropy is averaged by rapid movement of the nitroxide in low viscous environments and an isotropic Hfs of around 1.5 mT results: $a_{N\text{-iso}} = (0.6 + 0.6 + 3.3\ \text{mT})/3 = 1.5\ \text{mT}$.

Viscous or even solid environments permit only partial or no averaging of the anisotropy. Incomplete averaging of the Hfs results in changes of the spectral shape and increased linewidths (Fig. 7.8). Initially, increased line widths, but no changes in line positions, are observed. The third line is most sensitive for mobility changes and shows the largest broadening, which leads to the smallest amplitude (see spectrum at 15 °C). If mobility of the nitroxides is further restricted, the line shape changes and anisotropic spectra are observed (Fig. 7.8). If the nitroxide tumbling is very slow (e.g., microseconds and below), no averaging occurs and the ESR spectrum depends on the orientation of the nitroxide axis to the magnetic field. Typical "powder" ESR spectra are recorded if the nitroxide is randomly orientated (Fig. 7.8, −30 °C).

The following examples demonstrate how the sensitivity of ESR spectra to micropolarity and microviscosity might be used to characterize nanosized drug delivery systems.

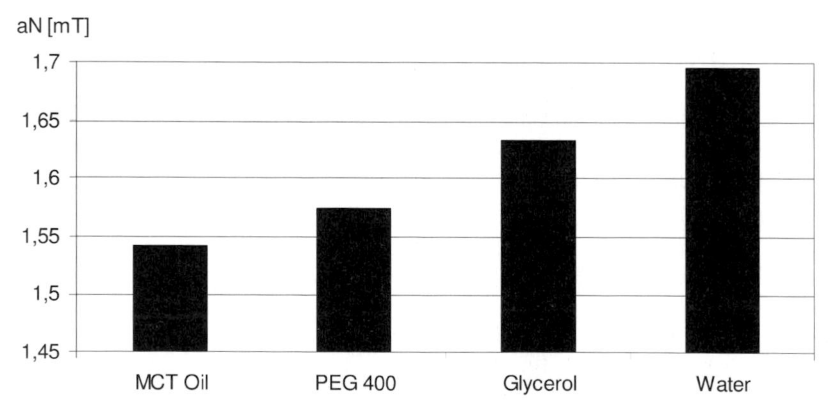

Fig. 7.7. Hyperfine coupling constants of TEMPOL in different solvents of pharmaceutical relevance.

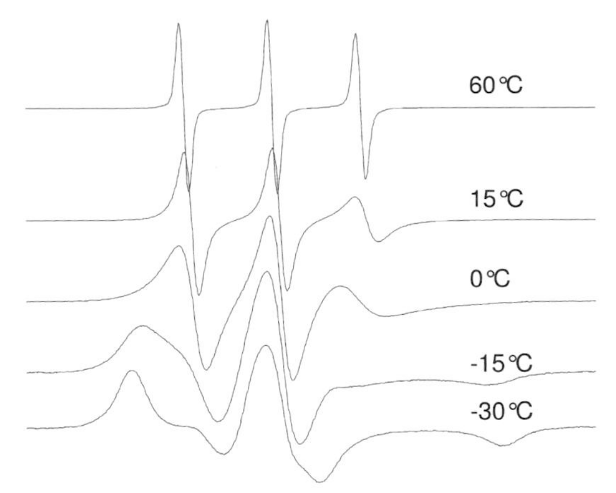

Fig. 7.8. Influence of temperature on ESR spectra of TEMPOL in poly(oxyethylene-pentaerythrol): Lower temperatures increase the viscosity and lead to a decreased averaging of the anisotropy of the hyperfine coupling. The high mobility of TEMPOL at 60 °C leads to a nearly isotropic shape; higher viscosities at lower temperatures lead to imperfect averaging (15 to −15 °C) of the anisotropy. At −30 °C, no averaging occurs on the ESR timescale and fully anisotropic spectra are recorded.

Nitroxide mobility might differ in different environments of heterogeneous systems. It will also change with time due to water penetration or degradation of the delivery matrix. Figure 7.9 shows simulated spectra and their integrative forms of different ratios of samples where immobilized and mobile nitroxide molecules

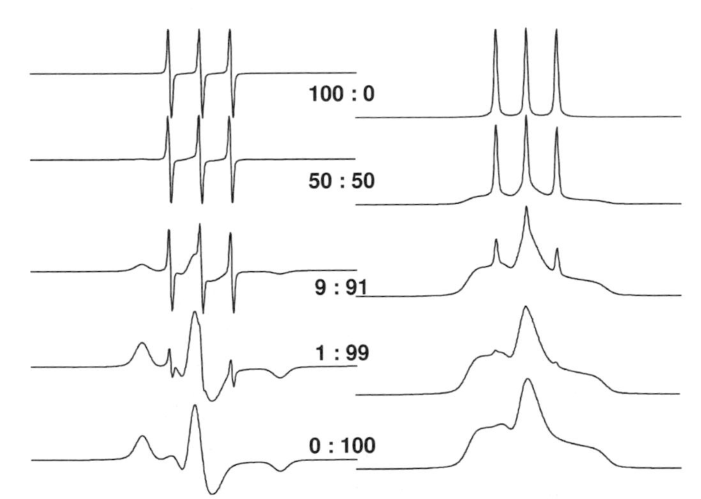

Fig. 7.9. Simulated ESR spectra of nitroxides with high and low mobility. The ratio indicates the ratio of mobile to immobile nitroxide molecules. Simulated ESR spectra (left) and their integrals (right) of different ratios of both components are represented.

coexist. The ESR spectra (= 1st derivative) are very sensitive to narrow lines and, therefore, mobile (= solubilized) nitroxides are detected below 1%. Immobilized nitroxides are more difficult to detect due to their broad lines and low signal amplitudes. They are easily overlooked in the ESR spectra if their concentration is below 50% of the total content. It is recommended that ESR spectra are integrated to make them more visible (Fig. 7.9).

The surface properties of nanoparticles play a significant role *in vivo*. Chauvierre et al. have used covalent ESR spin labeling of the dextran coat to investigate the surface properties of poly(isobutylcyanoacrylate) nanoparticles. They found a superposition of different nitroxide mobilities (similar to Fig. 7.9) and a dependency on the method used for particle synthesis [18]. The authors measured the mobile and immobile compartments in a quantitative manner. Italian scientists have used ESR to characterize the influence of lipid composition and the chain length of PEG on the characteristics of pegylated liposomes [19].

ESR can not only be used to characterize drug delivery systems before their use, it is also a method to shed more light on the release mechanisms themselves. The method can give unique and important information because different release mechanisms lead to different changes in spectral intensity and shape. Pure surface erosion will lead to a decrease in signal intensity, but no changes in spectral shape. In contrast, diffusion-controlled drug release can be recognized by changes in spectral shape due to increased mobility of the incorporated nitroxide.

The next example investigates the influence of temperature on the polarity of Poloxamer 188 solutions. Figure 7.10 shows ESR spectra of the lipophilic spin probe TEMPOLBENZOATE in 2% Poloxamer solutions. The ESR spectrum recorded at 25 °C indicates that the nitroxide is localized in a polar environment. The spectrum can be simulated with one species with a Hfs constant of $a_N = 1.66$ mT. Heating the sample to 60 °C leads to significant changes in spectral shape, which are most visible in the region of the third line. Obviously, the experimental spectrum is now a superposition of at least two ESR spectra from TEMPOLBENZOATE localized in different microenvironments. Simulation results for the ESR spectrum show that about 25% of the probe is localized in a polar environment ($a_N = 1.66$ mT, third spectrum from top from Fig. 7.10). This environment is highly polar and very similar to the one and only environment observed at 25 °C. About 75% of the spin probe, however, is localized in a less polar, more viscous environment ($a_N = 1.568$ mT, bottom spectrum in Fig. 7.10). This environment was not observed at 25 °C. Therefore, heating Poloxamer 188 solutions results in a new hydrophobic environment, in which lipophilic molecules accumulate. Experimental results demonstrate the thermal reversibility of this process. The temperature-induced formation of this hydrophobic environment has to be kept in mind if heating of Poloxamer solutions takes place (e.g., DSC).

The usefulness of ESR to get more insights has also been demonstrated for colloidal lipid carriers. The results indicate that the loading capacity of colloidal solid lipid dispersions is very low due to the crystalline nature of the lipid. They also showed that solid lipid nanoparticles (SLN) and nanostructured lipid carriers (NLC, composed of solid and liquid lipids) do not offer the proposed advantages in

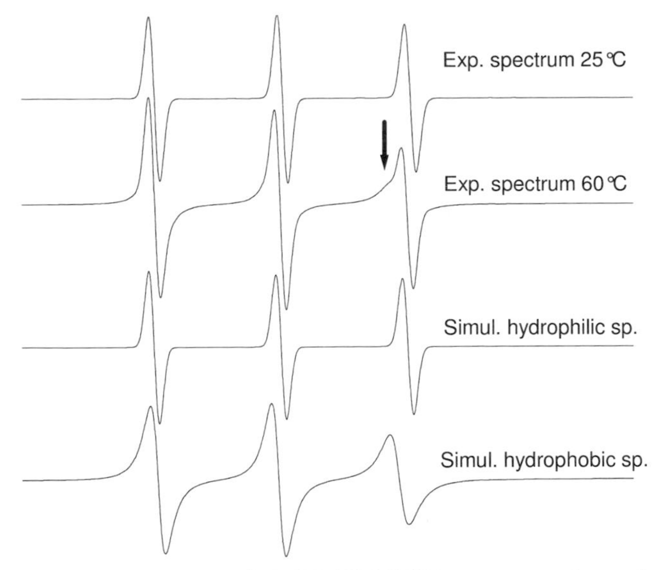

Fig. 7.10. ESR spectra of TEMPOLBENZOATE in 2% (w/w) Poloxamer 188 solutions at 25 and 60 °C. Note the different line shape at 60 °C, as indicated by the arrow. The bottom spectra are the result of simulating the experimental 60 °C spectrum, indicating the distribution of the nitroxide in two environments with different polarity.

comparison with a submicron emulsion, because lipid crystallization leads to expulsion of the lipophilic nitroxide into the aqueous phase. Detailed analysis of the ESR spectra shows that about $\frac{2}{3}$ of TEMPOLBENZOATE were exposed to a polar environment in a 10% colloidal dispersion of Compritol (a mixture of different glycerolbehanates) – which demonstrates the poor performance of SLN (Fig. 7.11) [20]. Incorporation into the lipid is even worse if the lipid crystal is more perfect. The incorporation rate of lipophilic nitroxides in cetylpalmitate is close to zero (own unpublished results).

Lipid crystallization might be retarded in nanosized systems due to the small size. Therefore, sample storage might greatly influence the lipid crystallization and localization of the incorporated drug. Figure 7.12 shows that the nitroxide TEMPO distributes in a ratio of 1:3 between the polar (aqueous) and nonpolar (lipid) environment, if the sample is stored at room temperature. Data from NMR, DSC and X-ray measurements show that the lipid matrix does not crystallize under such conditions. Storage at low temperatures induces crystallization of the lipid, at least partially. The lipid crystallization is connected with a redistribution of the nitroxide into a more polar environment and a 1:1 distribution is observed. Clearly, solid lipids can accommodate only traces of foreign molecules, because the concentration of the spin probe is rather low (3 mmol kg^{-1}).

An important aspect of colloidal drug delivery systems is dynamic phenomena, e.g., the kinetics of distribution processes. The dilution assay monitors the change in spectral shape after further addition of the outer phase (Fig. 7.13). Current re-

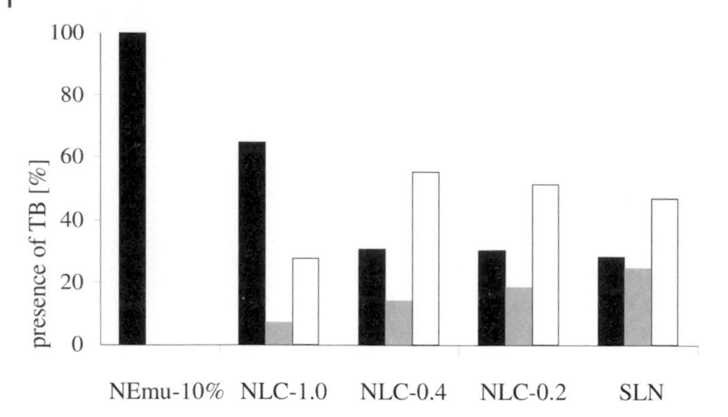

■ I: lipophilic compartment
■ II: very hydrophilic compartment (TB molecular dissolved in water, mobile)
□ III: compartment of moderate hydrophilicity (TB connected on the particle surface)

Fig. 7.11. Localization of the lipophilic spin probe TEMPOLBENZOATE (TB) in different colloidal lipid carriers. Note that the nitroxide is almost exclusively localized in liquid lipids and only partially in solid lipid nanoparticles (SLN) and nanostructured lipid carriers (NLC).

sults obtained from different colloidal carriers (SLN, NLC, nanoemulsions, nanocapsules and polymeric nanoparticles) suggest that distribution processes take place in the range from seconds to minutes for nitroxides with moderate lipophilicity (log $P < 4$). This is not surprising with respect to the very short diffusion length within the particles and the very small thickness (e.g., 10 nm) of the shell of nanocapsules [21]. Nitroxides with very high hydrophobicity such as HDPMI (Fig. 7.3) can be simulated with one component before and after dilution. They do

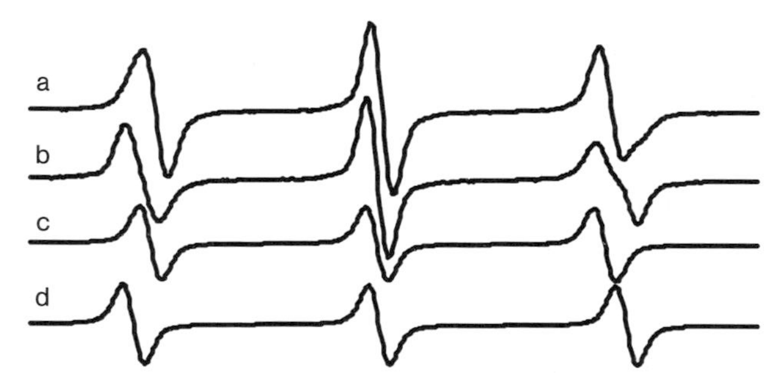

Fig. 7.12. 2.1 GHz ESR spectra of TEMPO loaded (20% w/w) WITEPSOL (mixtures of triglycerides and partial glycerides) dispersions, (a) experiment: sample stored at room temperature, (b) experiment: stored at refrigerator, (c) simulation: lipophilic component, (d) simulation: hydrophilic component.

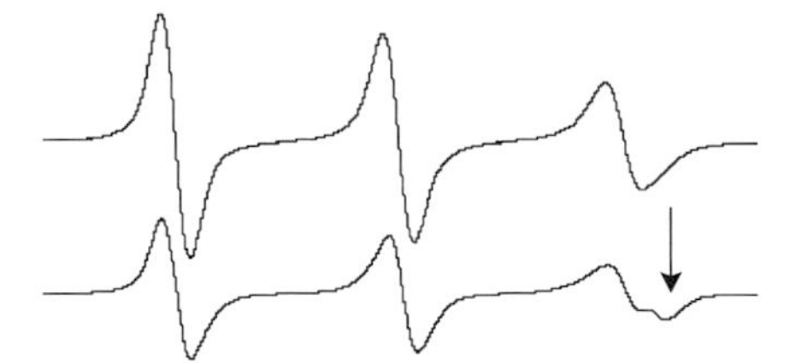

Fig. 7.13. ESR spectra of TEMPOLBENZOATE loaded nanocapsules before (top) and after (bottom) 1:4 dilution with the outer phase. The change in spectral shape is caused by increased localization of the probe in the aqueous environment.

not change their spectral shape due to their high preferential localization in the lipophilic environment.

ESR has been used to follow the pH-induced micelle-to-vesicle transformation of oleic acid in more detail [22]. Changes in the ESR spectra can also be used to follow the interaction between colloidal drug carriers and colloidal lipid dispersions with cells. Slovenian scientists found that the transfer depends on the spin probe, colloid matrix, and surfactant [23].

7.3.3 Monitoring of Microacidity

The development of pH-sensitive nitroxides offers the opportunity to monitor microacidity in drug delivery systems *in vitro* and *in vivo*. Figure 7.14 illustrates the principle.

Kroll et al. reported the first pharmaceutical application, monitoring drug degradation induced pH changes inside nontransparent w/o-ointments [24]. The pH is of crucial importance in the field of biodegradable polymers, because commonly used materials such as polyhydroxyesters, polyanhydrides and polyorthoesters are hydrolyzed to acids. The pH might affect the stability and solubility of incorporated drugs and, furthermore, also the degradation rate of many polymers to a large extent. The first *in vivo* measurement of the pH inside degrading PLGA implants was published in 1996 [ ]. The pH drops to as low as 2 inside subcutaneous PLGA implants in mice.

EPR has also been used to determine the acidity in PLGA-PEG microparticles designed for protein release [26]. For this purpose, albumin was covalently linked to a pH-sensitive label. Another study describes the influence of drugs and buffer substances on the acidity within the microparticles. Gentamicine base and sodium acetate are able to increase the pH inside the particles [27]. ESR is, therefore, a useful tool to monitor the success of a formulation measure in influencing the pH inside drug delivery systems. The principles of pH-measurement by ESR can also be applied to colloidal drug carriers and I am sure that we will soon witness further applications in this field.

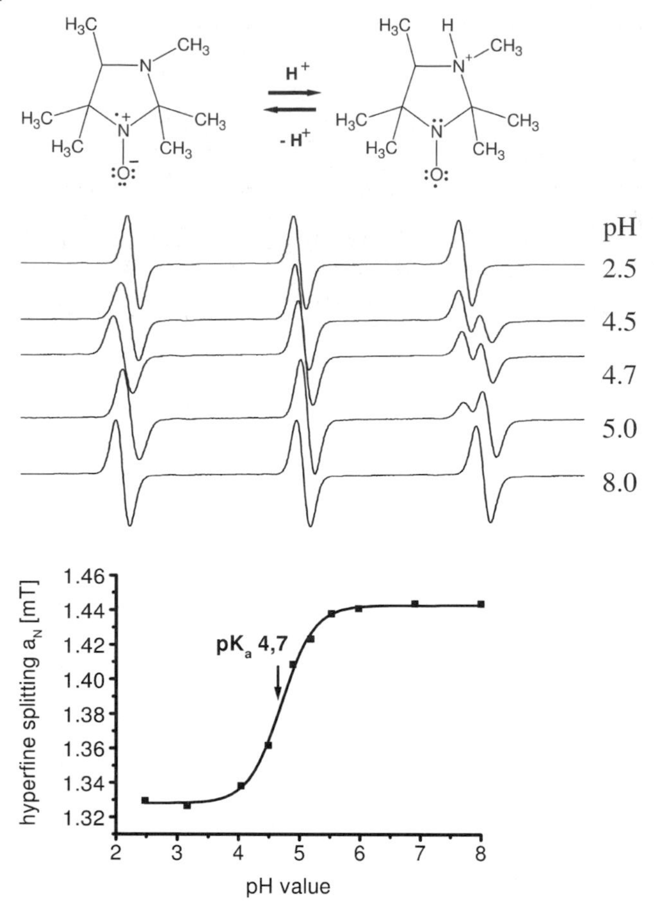

Fig. 7.14. Principle of pH measurement by imidazolidine nitroxides. Protonation of the nitrogen at position 3 leads to significant changes in the spin density and hyperfine coupling constant. At pK_a 4.7, the ESR spectrum results from equal contributions of protonated (small Hfs) and unprotonated (larger Hfs) forms of the nitroxide. The influence of pH on Hfs is shown in the lower part of the figure.

7.3.4
ESR Imaging

From the ESR spectrum, conclusions can be drawn on the nitroxide concentration, micropolarity, microviscosity and (using appropriate nitroxides) pH. Deeper insight into drug delivery processes is possible if the information can be attributed to a certain layer of the delivery system. Spectral–spatial information can be obtained by ESR imaging techniques. For example, pH gradients of almost two units within a distance of a few micrometers have been found in degrading polyanhydrides [28]. A more recent study showed the formation of pH gradients in polymers composed of polyorthoester and polylactide units [29]. Another study demonstrated that commercial ointments can modify the microacidity in human skin *in vitro*, leading to pH gradients in the skin [30]. The resolution of ESR imaging, in the range of a few microns, is insufficient to monitor distributions inside nano-scaled systems. However, it can be used to follow the fate of nanostructures with a macroscopic controlled release system (e.g., tablets that release nanoparticles) or, even more interestingly, to follow the fate of nanocarriers in mice and rats.

7.4
In Vivo ESR

The development of low-frequency spectrometers and imagers has made noninvasive ESR studies on mice and rats possible. *In vivo* applications include the characterization of antacid induced pH changes of the stomach in mice [31].

Correlation of *in vitro* and *in vivo* data is very important for the development and quality control of delivery systems. Major differences may arise from decreased water accessibility *in vivo*, different pH or buffer capacities or the involvement of enzymes. EPR is a useful tool to figure out whether the general release mechanism *in vitro* corresponds to the *in vivo* situation in a small mammal. For example, incomplete release of a water-soluble nitroxide has been found *in vivo* due to encapsulation of the polymer implant [32].

EPR observations show that release processes from biodegradable polyesters and polyanhydrides are rather complex [25, 33]. A surface erosion front mechanism with zero-order kinetics has been observed for P(CPP-SA) 20:80 polyanhydrides. The *in vivo* release was 1.5× slower than for *in vitro* experiments [33]. Substitution of the aromatic dicarboxylic acid CPP by long-chain aliphatic dicarboxylic acids (FAD) leads to increased mobility inside the manufactured polyanhydride and a more complex release pattern with partially diffusion-controlled processes [32]. For P(FAD-SA) implants, an intermediate increase in viscosity inside the degrading polymer has been observed, which was caused by precipitation of the dicarboxylic acids. EPR spectra of nitroxide-loaded poly(lactide-*co*-glycolide) implants did not change in shape or intensity during the first week, indicating an initial lag time. During the second week, a small portion of the nitroxides became solubilized within the polymer. The ratio of solubilized/immobilized nitroxides increased with time and a rapid, almost pulse-like release was observed after four weeks. The authors were able to distinguish release processes of the core from the outer layers by manufacturing sandwich implants that contained ^{14}N-nitroxides in the outer layers and ^{15}N-nitroxides in the core [33]. These studies show the possibility of obtaining unique information about drug delivery processes *in vivo*.

7.5
Summary and Outlook

ESR can be applied to samples that are otherwise difficult to assess, e.g., nontransparent, solid or multiphase samples, or samples with heterogeneities in the nanometer range. This allows monitoring of the distribution kinetics and protection capacity of nanoscaled drug delivery systems.

Technical developments now permit sensitive ESR spectroscopy and imaging of mice and rats. An increasing number of *in vivo* spectrometers are being built, and I am sure that we shall soon see many papers that focus on the *in vivo* characterization of nanoscaled drug delivery systems by ESR spectroscopy and imaging.

Acknowledgment

I thank the former and present members of my group, S. Liedtke, K. Jores, A. Rübe, C. Augsten, and R. K. Narayanan for the measurement of samples and discussions.

References

1 C.P. Poole, *Electron Spin Resonance: a Comprehensive Treatise on Experimental Techniques*, 2nd Edn. New York: John Wiley & Sons, **1983**.

2 L.J. Berliner ed., *In Vivo EPR (ESR): Theory and Applications*, New York: Kluwer Academic/Plenum Publishers, **2003**.

3 S.S. Velan, R.G.S. Spencer, J.L. Zweier, P. Kuppusamy. Electron paramagnetic resonance oxygen mapping (EPROM): Direct visualization of oxygen concentration in tissue, *Magn. Reson. Med.* 43 (**2000**) 804–809.

4 T. Yoshimura, S. Fujii, H. Yokoyama, H. Kamada. In-vivo electron-paramagnetic-resonance imaging of NO-bound iron complex in a rat head, *Chem. Lett.* (**1995**) 309–310.

5 K. Mäder, G. Bacic, H.M. Swartz. In vivo detection of anthralin derived free radicals in the skin of hairless mice by low frequency electron paramagnetic resonance spectroscopy, *J. Investigative Dermatol.* 104 (**1995**) 514–517.

6 R. Mason, M.B. Kadiiska. Ex vivo detection of free radical metabolites of toxic chemicals and drugs by spin trapping, in L.J. Berliner (ed) *In Vivo EPR(ESR): Theory and Applications, Biological Magnetic Resonance* 18, New York: Kluwer Academic/Plenum Publishing, pp. 309–323, **2003**.

7 M.J. Burkitt, R.P. Mason. Direct evidence for in vivo hydroxyl-radical generation in experimental iron overload: an ESR spin-trapping investigation. *Proc. Natl. Acad. Sci. U.S.A.* 88, (**1991**) 8440–8444.

8 G.S. Timmins, K.J. Liu. Spin trapping: facts and artifacts, in *In Vivo EPR: Theory and Applications*, L.J. Berliner (ed), New York: Kluwer Academic/Plenum Publishers, pp. 285–308, **2003**.

9 T.A. Konovalova, J. Lawrence, L.D. Kispert. Generation of superoxide anion and most likely singlet oxygen in irradiated TiO_2 nanoparticles modified by carotenoids, *J. Photochem. Photobiol. A: Chem.*, 162 (**2004**) 1–8.

10 X. Gao, J. Zhang, L. Zhang. Hollow sphere selenium nanoparticles: their in-vitro anti hydroxyl radical effect. *Adv. Mater. (Weinheim)*, 14(4) (**2002**) 290–293.

11 D. Teomim, K. Mäder, A. Bentolila, A. Magora, A.J. Domb. Gamma-irradiation stability of saturated and unsaturated aliphatic polyanhydrides-ricinoleic acid based polymers. *Biomacromolecules* 2 (**2001**) 1015–1022.

12 J. Raffi, S. Gelly, L. Barral, F. Burger, P. Piccerelle, P. Prinderre, M. Baron, A. Chamayou. Electron paramagnetic resonance of

radicals induced in drugs and excipients by radiation or mechanical treatments, *Spectrochim. Acta Part A* 58 (**2002**) 1313–1320.
13 M.B. Sintzel, K. Schwach-Abdellaoui, K. Mäder, R. Stösser, J. Heller, C. Tabatabay, R. Gurny. Influence of irradiation sterilization on semi-solid poly(ortho ester). *Int. J. Pharm.* 175 (**1998**) 165–176.
14 K. Mäder, A. Domb, H.M. Swartz. Gamma sterilization induced radicals in biodegradable drug delivery systems. *Appl. Radiation Isotopes*, 47 (**1996**) 1669–1674.
15 O. Mykhaylyk, N. Dudchenko, A. Dudchenko. Doxorubicin magnetic conjugate targeting upon intravenous injection into mice: High gradient magnetic field inhibits the clearance of nanoparticles from the blood. *J. Magn. Magn. Mat.*, 293 (**2005**) 473–482.
16 P. Smirnov, F. Gazeau, M. Lewin, J.C. Bacri, N. Siauve, C. Vayssettes, C.A. Cuenod, O. Clement. In vivo cellular imaging of magnetically labeled hybridomas in the spleen with a 1.5-T clinical MRI system, *Magn. Reson. Med.* 52 (**2004**) 73–79.
17 C. Wilhelm, F. Gazeau, J. Roger, J.N. Pons, J.-C. Bacri. Interaction of anionic superparamagnetic nanoparticles with cells: Kinetic analyses of membrane adsorption and subsequent internalization, *Langmuir* 18 (**2002**) 8148–8155.
18 C. Chauvierre, C. Vauthier, D. Labarre, H. Hommel. Evaluation of the surface properties of dextran-coated poly(isobutylcyanoacrylate) nanoparticles by spin-labelling coupled with electron resonance spectroscopy, *Colloid Polym. Sci.* 282 (**2004**) 1016–1025.
19 R. Bartucci, S. Belsito, L. Sportelli. Spin-label electron spin resonance studies of micellar dispersions of PEGs-PEs polymer-lipids, *Chem. Phys. Lipids* 124 (**2003**) 111–122.
20 K. Jores, W. Mehnert, K. Mäder. Physicochemical investigations on solid lipid nanoparticles (SLN) and on oil-loaded solid lipid nanoparticles: a NMR- and ESR-study, *Pharm. Res.* 20(8) (**2003**) 1274–1283.
21 A. Rübe, K. Mäder. An electron spin resonance study on the dynamics of polymeric nanocapsules, *J. Biomed. Nanotechnol.* 1 (**2005**) 208–213.
22 H. Fukuda, A. Goto, H. Yoshioka, R. Goto, K. Morigaki, P. Walde. Electron spin resonance study of the pH-induced transformation of micelles to vesicles in an aqueous oleic acid/oleate system, *Langmuir* 17 (**2001**) 4223–4231.
23 J. Kristl, B. Volk, P. Ahlin, K. Gombac, M. Sentjurc. Interactions of solid lipid nanoparticles with model membranes and leukocytes studied by EPR, *Int. J. Pharm.* 256 (**2003**) 133–140.
24 C. Kroll, K. Mäder, R. Stoesser, H.-H. Borchert. Nondestructive determination of pH values in nontransparent W/O systems by means of EPR spectroscopy, *Eur. J. Pharm. Sci.* 3 (**1995**) 21–26.
25 K. Mäder, B. Gallez, K.J. Liu, H.M. Swartz. Noninvasive in vivo characterization of release processes in biodegradable polymers by low frequency electron paramagnetic resonance spectroscopy, *Biomaterials* 17 (**1996**) 459–463.
26 K. Mäder, B. Bittner, Y. Li, W. Wohlauf, T. Kissel. Monitoring microviscosity and microacidity of the albumin microenvironment inside degrading microparticles from polylactide-co-glycolide (PLG) or ABA-triblock polymers containing hydrophobic poly(lactide-co-glycolide) A blocks and hydrophilic poly(ethylenoxide) B blocks, *Pharm. Res.* 15 (**1998**) 787–793.
27 A. Brunner, K. Mäder, A. Göpferich. The microenvironment inside biodegradable microspheres: changes in pH and osmotic pressure, *Pharm. Res.* 16 (**1999**) 847–853.
28 K. Mäder, S. Nitschke, R. Stösser, H.-H. Borchert, A. Domb. Non-destructive and localised assessment of acidic microenvironments inside biodegradable polyanhydrides by spectral spatial electron paramagnetic

resonance imaging (EPRI), *Polymer* 38 (**1997**) 4785–4794.

29 S. Capancioni, K. Schwach-Abdellaoui, W. Kloeti, W. Herrmann, H. Brosig, H.H. Borchert, J. Heller, R. Gurny. In vitro monitoring of poly(ortho ester) degradation by electron paramagnetic resonance imaging, *Macromolecules* 36 (**2003**) 6135–6141.

30 C. Kroll, W. Herrmann, R. Stösser, H.H. Borchert, K. Mäder. Influence of drug treatment on the microacidity in rat and human skin – an in vitro electron spin resonance imaging study, *Pharm. Res.* 18 (**2001**) 525–530.

31 B. Gallez, K. Mäder, H.M. Swartz. Non-invasive measurement of the pH inside the gut using pH-sensitive nitroxides. An in vivo EPR study, *Magn. Reson. Med.* 36 (**1996**) 694–697.

32 K. Mäder, Y. Cremmilleux, A. Domb, J.F. Dunn, H.M. Swartz. In vitro/in vivo comparison of drug release and polymer erosion from biodegradable P(FAD-SA) polyanhydrides – a non-invasive approach by the combined use of electron paramagnetic resonance spectroscopy and nuclear magnetic resonance imaging, *Pharm. Res.* 14 (**1997**) 820–826.

33 K. Mäder, G. Bacic, A. Domb, O. Elmalak, R. Langer, H.M. Swartz. Noninvasive in vivo monitoring of drug release and polymer erosion from biodegradable polymers by EPR spectroscopy and NMR imaging, *J. Pharm. Sci.* 86 (**1997**) 126–134.

8
X-ray Absorption and Emission Spectroscopy in Nanoscience and Lifesciences

Jinghua Guo

8.1
Introduction

The properties of matter at nanoscale dimensions can be dramatically different from the bulk or the constituent molecules. The differences arise through quantum confinement, altered thermodynamics or changed chemical reactivity. In general, electronic structure ultimately determines the properties of matter, and it is therefore natural to anticipate that a complete understanding of the electronic structure of nanoscale systems will lead to progress in nanoscience and bioscience, not inferior to that seen in recent years.

The ability to control the particle size and morphology of nanoparticles is of crucial importance, both from a fundamental and industrial point of view, considering the tremendous amount of high-tech applications of nanostructured metal oxide materials devices such as dye-sensitized solar cells, displays and smart windows, chemical, gas and biosensors, lithium batteries, supercapacitors. Controlling the crystallographic structure and the arrangement of atoms along the surface of nanostructured material will determine most of its physical properties.

So far the electronic structure and dynamics in biological systems have been little investigated. Synchrotron-based X-ray diffraction has been used to determine the molecular geometry that in many cases can determine the biological function of the molecule. Unfortunately, as crystallography was unable to address the local structural geometry of a chemisorbed molecule on surface, disordered biological systems may be a challenge too. When studying biological systems it is necessary to perform the experiments in a natural environment, i.e., in water/liquid solutions. We show later in this chapter that X-ray photon-in and photon-out spectroscopy can be used for *in situ* spectroscopic studies of wet systems.

X-rays originate from an electronic transition between a localized core state and a valence state. By the second half of the twentieth century, technological adavances in detector and X-ray spectrometry produced a renaissance of interest in X-ray spectroscopy as an analytical tool in chemical analysis and as a tool for studying the solid state in general [1]. The parallel development of computational methods in solid-state physics and quantum chemistry, abetted by modern high-speed com-

Nanotechnologies for the Life Sciences Vol. 3
Nanosystem Characterization Tools in the Life Sciences. Edited by Challa S. S. R. Kumar

ISBN: 3-527-31383-4

puters, enabled theoreticians to return to the many unrevolved problems in X-ray spectroscopy. Thus, the last decade has seen a rebirth of effort in expanding the methods of X-ray spectroscopy as well as applications to a growing lists of studies in various fields of science.

Soft X-ray spectroscopy has some basic features that are important to consider [2]. As a core state is involved, elemental selectivity is obtained because the core levels of different elements are well separated in energy, meaning that the involvement of the inner level makes this probe localized to one specific atomic site, around which the electronic structure is reflected as a partial density-of-states contribution. The participation of valence electrons gives the method chemical state sensitivity and, further, the dipole nature of the transitions gives particular symmetry information.

An introduction to soft X-ray absorption spectroscopy (XAS) along with soft X-ray emission spectroscopy (XES) is given in Section 8.2, and their chemical sensitivity in nanostructured and molecular materials is discussed in Section 8.3. A number of examples, including some recent findings, then illustrate the potential of XAS and XES applications in the nanoscience (Section 8.4) and bioscience (Section 8.5).

8.2
Soft X-ray Spectroscopy

Synchrotron radiation has become an indispensible tool in many areas of science. Synchrotron radiation is the light emitted by electrons as they circulate around a high-energy accelerator. This light covers the spectrum from hard (short wavelength) X-rays through soft (long wavelength) X-rays, ultraviolet (UV), visible, and infrared. The use of synchrotron radiation has grown to a powerful research tool in science in the last decade. Vacuum ultraviolet and X-ray photons emerging from storage rings are now among the most frequently used probes for advanced investigation of the electronic and geometric structure of matter.

The newly developed theoretical methods in band theory and quantum chemistry, as well as the use of different instrumentation, set the soft X-ray community apart from hard X-ray spectroscopists [3]. This separation has long been symbolized by the existence of separate triannual international conference series on Vacuum Ultraviolet Radiation Physics [4] and X-ray Physics and Inner Shell Ionization [5].

The utility of hard X-ray absorption spectroscopy as a probe of chemical identity in the biological, environmental and biomedical sciences is well documented [6–12]. Metals in enzymes can be probed in all oxidation states and spin states, and environmental samples can be examined in a chemically undisturbed form. In appropriate cases, intact tissues or even an entire organism can be examined. For example, a 23 amino-acid synthetic lytic peptide (Hecate) has been covalently linked to magnetite nanoparticles, and the lytic peptide-bound nanoparticles characterized by X-ray absorption near-edge structure spectroscopy [13]. In recent years, the advent of synchrotron sources as the best source for both X-ray regions has greatly reduced the barriers between the two communities [14]. Soft X-ray absorption fine structure, or NEXAFS technique, has become a powful tool to study molecules bonded on surfaces [15].

Current developments in synchrotron radiation techniques have led to extremely bright undulator sources. The use of synchrotron radiation for the excitation of soft X-ray emission spectra adds several important qualities to this spectroscopic method. Firstly, it provides a very intense photon-excitation source. Secondly, monochromatized soft X-ray photons offer a higher degree of energy selectivity than do electrons. Thirdly, synchrotron radiation offers the possibility of exciting soft X-ray spectroscopy by polarized light.

8.2.1 Soft X-ray Absorption Edges

The X-ray absorption spectrum of a given element consists of series of edges, each corresponding to excitation of a different core electron. Electron levels with the quantum number n equal to 1–4 are named as K, L, M, N levels, respectively. K-edges arise from the innermost 1s orbital, L-edges from the second shell, M-edges from the third and N-edges from the fourth. While there is only one K-edge, the higher order edges are further divided by the angular momentum of the resulting core electron vacancy. Hence, there are three L-edges, denoted L_1, L_2 and L_3, arising from 2s, $2p_{1/2}$ and $2p_{3/2}$ final states, respectively. Similarly, there are five M-edges and seven N-edges. These edges are well separated in energy. With Fe_2O_3, for example, the O K-edge is at 530 eV, and the Fe L_2 and L_3-edges are close together at 720 and 707 eV, while the observable M-edge has a very low energy of 52 eV.

The soft X-ray region ranges from about 50 to 1200 eV. Much of biological and environmental science can be based on soft X-ray spectroscopy measurements. This is because the X-ray edges in this region offer routes to chemical information not readily obtained by other techniques. Edges in the soft X-ray region include: the light element K-edges, including C, N, O, F, Si; the first transition metal L-edges, including Ti, Cr, Mn, Fe, Co, Ni, Cu, Zn; the M-edges of Fe through to the lanthanides, and N-edges of the lanthanides etc. All the absorption edges can be found on a website of the Center for X-ray Optics, Lawrence Berkeley National Laboratory [16], and some of the edges are illustrated in Fig. 8.1.

8.2.2 Soft X-ray Emission Spectroscopy

In soft X-ray emission, the core vacancy left by the excited 1s electron is filled by a valence-orbital electron, thereby also giving direct information about the chemical bonding. Characteristic X-ray emission spectra consist of spectral series with lines having a common initial state with the vacancy in the inner level. The emission intensity from a disordered sample is given by the formula

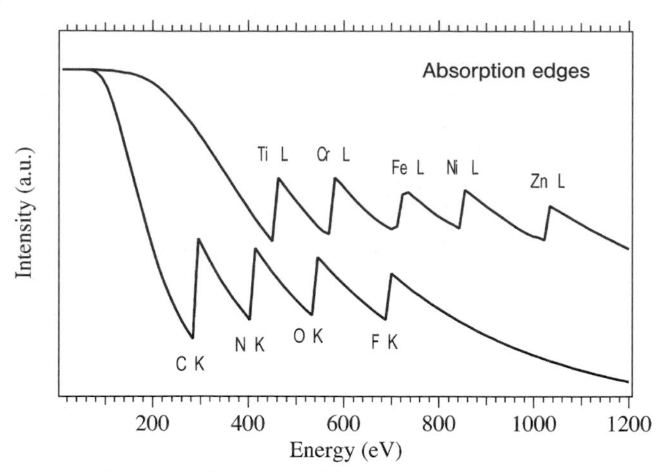

Fig. 8.1. Schematic illustration of X-ray absorption edges from the K-edges of some light elements and L-edges from some transition metals.

$$I \propto \sum_f \omega_{if}^3 |\mathbf{r}_{if}|^2 \Delta(\omega' - \omega_{if}, \Gamma) \tag{1}$$

Eq. (1) can be further simplified in the one-electron approximation. To be specific let us consider the K-emission of atom n in a molecule. The transition matrix element $\mathbf{r}_{if}$ describes now the one-electron transition between the $1s_n$ core orbital and occupied molecular orbital (MO) i, the effective wavefunction of which can be written as

$$\phi_i = \sum_n \chi_n c_{in} \tag{2}$$

where χ_n is the p-orbital of the atom n. Thus, the intensity Eq. (1) becomes

$$I \propto \sum_i c_{ni}^2 \Delta(\omega' - \omega_{i,1s}, \Gamma). \tag{3}$$

where $\omega_{i,1s} \approx \varepsilon_i - \varepsilon_{1s}$, ε_i and ε_{1s} are the energies of the 1s electron and MO i.

Further simplification could be made by retaining only the first term in the expansion of the exponential function, i.e., the so-called *dipole approximation*. This important approximation assumes $kr \ll 1$. Such an approximation works well for soft X-ray radiation. As result of the dipole approximation, K-emission probes the p-character of a molecular orbital ϕ_i, while L-emission probes the contributions of s and d atomic orbitals in ϕ. One can map the MO ϕ by measuring spectra of all series (K, L, M, ...) of all atoms.

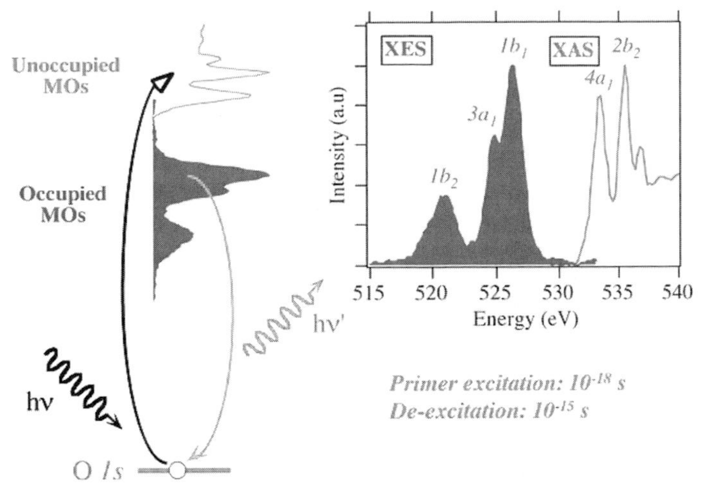

Fig. 8.2. Left: schematic illustration of X-ray absorption and emission spectroscopy. Right: X-ray absorption and emission of a water molecule.

8.2.3
Soft X-ray Absorption Spectroscopy

A soft X-ray absorption spectrum provides information about the unoccupied states. For example, in the oxygen K-edge absorption, the oxygen 1s electron is excited to empty electronic states in the water unoccupied molecular orbitals, and the dipole selection rule provides a tool to study locally the O 2p character of these unoccupied valence band (Fig. 8.2). The atomic nature of the core hole implies elemental and site selectivity. The probability of such a transition is related to the X-ray absorption cross section. The intensity of these secondary electrons or the photons can be measured as a function of incoming photon energy. This will reflect the absorption cross section as the intensity of the secondary electrons/emitted photons are proportional to the absorbed intensity. Because of the short mean free path of electrons, the electron-yield-detection method is very surface sensitive. If the out-coming photons are detected (fluorescence yield), the X-ray absorption becomes bulk probing (about 100–200 nm) due to the comparatively larger attenuation lengths.

The photoabsorption transition probability for a transition from an initial state $|i\rangle$ to a final state $|f\rangle$ is governed by Fermi's "Golden Rule":

$$P_{if} = \frac{2\pi}{\hbar}|\langle f|V|i\rangle|^2 \Delta(\omega - \omega_{fi}, \Gamma_{if}), \quad \omega_{fi} = \varepsilon_{\mathrm{f}} - \varepsilon_{\mathrm{i}} \tag{4}$$

where ε_{f} and ε_{i} are energies of final and initial states of the system, and Γ_{if} is the width of the spectral transition, V describes the interaction between molecule and light, and $\Delta(\varepsilon, \Gamma)$ is the Lorentzian function

$$\Delta(\varepsilon, \Gamma) = \frac{\Gamma}{\pi(\varepsilon^2 + \Gamma^2)} \tag{5}$$

The final electron state $|f\rangle$ could be a bound or a continuum state, depending on the photon frequency ω.

Thus, the expression for the X-ray absorption (XAS) cross section is obtained as Eq. (6).

$$\sigma = \frac{4\pi^2}{m^2}\frac{e^2}{c}\frac{1}{\omega}|\langle f|\mathbf{e}\cdot\mathbf{p}|i\rangle|^2\Delta(\omega - \omega_{fi}, \Gamma_{if}) \tag{6}$$

8.2.4 Resonant Soft X-ray Emission Spectroscopy

The big improvement in the performance of synchrotron radiation based soft X-ray spectroscopy during the last decade is the high brightness of the third generation source combined with high quality optical systems for refocusing the monochromatized soft X-ray beam. The new generation synchrotron radiation sources producing intense tunable monochromatized X-ray beams has opened up new possibilities for soft X-ray spectroscopy. X-ray absorption and emission have been traditionally treated as two independent processes, with the absorption and emission spectra providing information on the unoccupied and occupied electronic states, respectively.

The possibility to select the energy of the excitation has created an extra degree of freedom compared with traditional spectroscopy pursued with high-energy electron or characteristic X-ray excitation. The energy selectivity makes it possible to perform resonant excitation, i.e., exciting to particular empty states [28–30]. The introduction of selectively excited soft X-ray emission has opened a new field of study by disclosing many new possibilities of soft X-ray resonant inelastic scattering (RIXS). Among the new tools available with this technique are site selectivity in high-T_c superconducting materials [31], femtosecond dynamics [32] and chemical bonding mechanism [33] by detuning from resonance, etc.

Resonant inelastic X-ray scattering at core resonances has become a new tool for probing the optical transitions in transition metal oxides. The scheme of RIXS process is presented in Fig. 8.3. For the hard X-ray K-edge RIXS process, the intermediate state in a transition metal is the same as the final state in a K absorption measurement, while the RIXS final state (when scanning the Kα region) is the same as in L absorption spectroscopy [21, 22]. For L-edge and M-edge RIXS, the final states are typical d–d or f–f excitations [23, 24].

Final states probed via such a channel, RIXS or XRS, are related to eigenvalues of the ground state Hamiltonian. The core–hole lifetime is not a limit on the resolution in this spectroscopy. According to the many-body picture, an energy of a photon, scattered on a certain low-energy excitation, should change by the same amount as a change in an excitation energy of the incident beam, so that inelastic

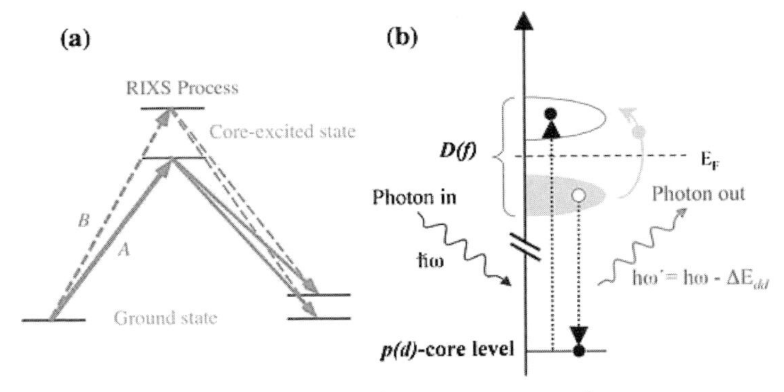

Fig. 8.3. Schematic representation of RIXS process (a), soft X-ray RIXS in study of dd (ff) excitations (b).

scattering structures have constant energy losses and follow the elastic peak on the emitted-photon energy scale.

The formulations of RIXS lead to a Kramers–Heisenberg type dispersion formula for the cross section, with generally only the resonant part of the scattering process taken into account [17–20]. Second-order perturbation theory for the RIXS process leads to the Kramers–Heisenberg formula for the resonant X-ray scattering amplitude. Using this starting point RIXS has been analyzed in periodic solids as a momentum conserving process, suggesting that it can be used as a novel "band-mapping" technique in diamond and graphite [25–27].

The spectral and polarization properties of the RIXS process can be described by a double differential cross section

$$\frac{d^2\sigma}{d\omega'\,d\Omega} = \sum_{v}\sum_{n}\frac{\omega'}{\omega}|F_{vn}(\omega)|^2\Delta(\omega-\omega'-\omega_{vn},\Gamma_{vn})\Phi(\omega-\omega_0) \tag{7}$$

where ω and ω' are the frequencies of incident and scattered X-ray photons, respectively. Γ_{vn} is the final-state lifetime broadening. $\Phi(\omega-\omega_0)$ is the spectral profile of the incoming beam. The broadening of the core excitations Γ_{vk} is much larger than the lifetime broadening Γ_{vn} of optical transitions $n \to v$. It is thus reasonable to assume that $\Gamma_{vn} = 0$, replacing the Δ function by a Dirac δ-function. This is the basis for super-highly resolved X-ray spectroscopy, which allows us to record X-ray fluorescence resonances without the lifetime broadening [18]. Further discussion on resonant soft X-ray emission spectroscopy is beyond the scope of this chapter, but it can be found in earlier reviews [34–37].

8.2.5
Experimental Details

An outline of an undulator beamline, such as beamline 7.0.1 of the Advanced Light Source [38], is presented in Fig. 8.4, showing its principle elements. It consists of a

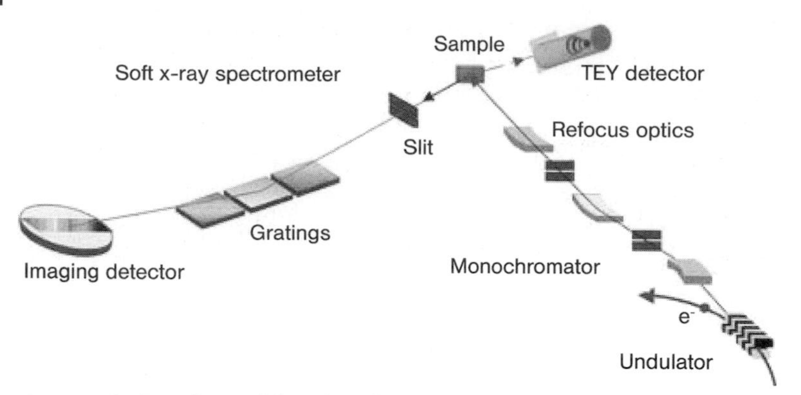

Fig. 8.4. Outline of an undulator beamline, including an undulator, monochromator and a Rowland circle geometry grating spectrometer and channeltron for soft X-ray fluorescence detection.

5 m, 5-cm period undulator and a spherical grating monochromator (SGM) covering the spectral range from 60 to 1200 eV. It is designed for high resolution operation with maximum photon flux and a small spot size (typically, 50–100 mm) at the sample, matched to the acceptance of the experiment spectrometers.

The experimental set-up at a synchrotron radiation beamline is called the end-station, since it constitutes an interchangable experiment at the end of the beamline. The end-station consists of three main parts: analysis chamber, preparation chamber and loadlock chamber, separated by valves [39].

X-ray absorption spectra are recorded in total-electron-yield (TEY) mode or fluorescence-yield (FY) mode. TEY is measured from sample drain current, and the FY is obtained from a channeltron. X-ray emission spectra are recorded by using a grazing incidence grating spectrometer. Figure 8.5 shows a common experimental set-up for soft X-ray spectroscopy.

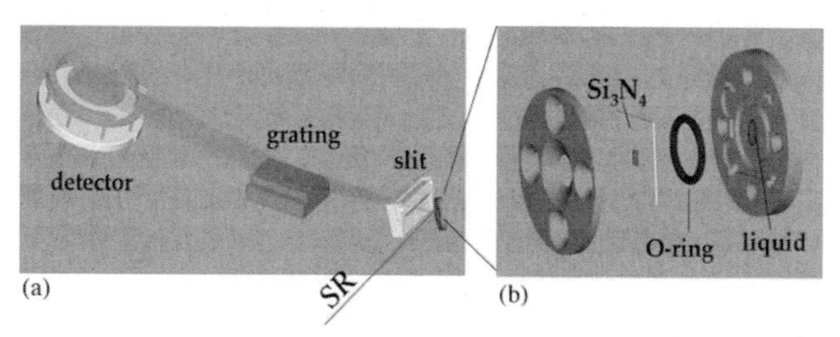

Fig. 8.5. A Rowland circle geometry grating spectrometer (a) and a static liquid cell (b) for soft X-ray spectroscopic study of liquid samples.

The high-resolution grating spectrometer was mounted parallel to the polarization vector of the incident photon beam to detect the emitted X-ray (Fig. 8.5a). The grazing-incidence grating spectrometer is based on Rowland geometry. It consists of slit, gratings and detector. Three spherical gratings are optimized to cover an operation range of 50–1000 eV. The gratings are mounted at angles of incidence to have a joint fixed slit, and the emitted X-rays are detected using a two-dimensional detector that can be positioned and oriented tangentially to the pertained Rowland circle. The detector consists of multi-channel plates and a resistive anode with a four-electrode readout.

Solid-state samples can be studied by placing the solid sample in front of the spectrometer slit, as shown in Fig. 8.5(a). While for the liquid (or wet) samples, a liquid cell has to be used. The liquid cell has a window to attain compatibility with UHV conditions of the spectrometer and beamline. The synchrotron radiation enters the liquid cell through a 100 nm thick silicon nitride window and the emitted X-rays exit through the same window. The thin silicon nitride window is commercially available [42]. The test showed that a 100 nm thin Si_3N_4 membrane of 2.25×2.25 mm^2 could hold one atmosphere pressure. The liquid cell (Fig. 8.5b) consists of a metal container and a 1×1 mm^2 and 100 nm thin Si_3N_4 membrane, which can withstand the differential pressure between the liquid on one side and UHV on another side. The transmission of X-rays at the C, O K-edge, and Fe L-edge for a 100 nm thick window is about 46%, 66%, and 82%, respectively. Silicon nitride membranes are not the best choice when compared to other materials with respect to their transmission of X-rays in the energy region 80–1000 eV, but their mechanical properties and fabrication process make them the only candidate at present. They also allow experiments to be performed at a vacuum pressure of less than 1×10^{-9} Torr.

8.3
Chemical Sensitivity of Soft X-ray Spectroscopy

The region from 200 to 700 eV covers the K-edges of the biologically relevant light elements carbon, nitrogen, oxygen and fluorine. The useful properties of soft X-ray spectroscopy stem from the electric dipole nature of the transitions involved. The emission originates from an electronic transition between a localized core state and a valence state. The energy of the emitted X-ray is equal to the difference in energy of the two states. Involvement of the inner level makes this probe localized to one specific atomic site, around which the electronic structure is reflected as a partial density-of-states (DOS) contribution. Chemical sensitivity is obtained when the resolution of the detected emission lines is high enough to resolve fine structure. The line shapes are determined by the valence electron distribution and the transitions are governed by dipole selection rules. For solids, essentially a partial DOS mapping is obtained. This is exemplified by the carbon Kα emission spectra of the carbon solids shown in Fig. 8.6 (see also Fig. 8.8a).

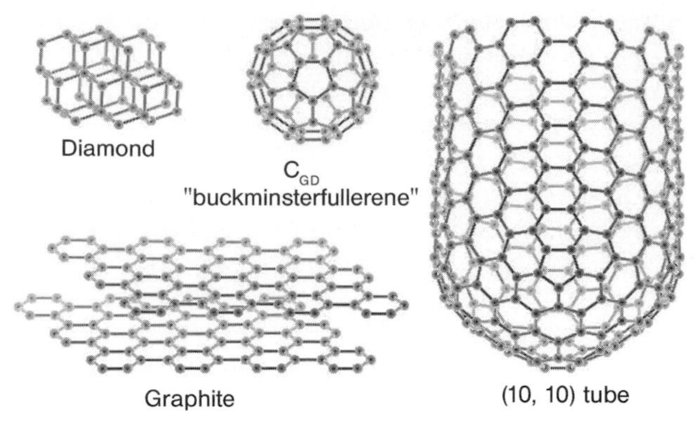

Fig. 8.6. Schematic representation of carbon in different allotropic forms: diamond, graphite, C_{60}, and nanotubes. (From Ref. [41].)

8.3.1
Electronic Structure and Geometrical Structure

The difference in structural arrangement of these allotropic forms (Fig. 8.6) of carbon gives rise to the wide differences in their physical properties. Carbon has an atomic number of 6 and has a $1s^2 2s^2 2p^2$ configuration in the electronic ground state. The atoms in diamond are tetrahedrally bonded to their four nearest-neighbors using linear combinations of 2s, $2p_x$, $2p_y$, and $2p_z$ orbitals in an sp^3 configuration. In contrast, in graphite, strong in-plane bonds are formed between a carbon atom and its three nearest-neighbors from 2s, $2p_x$, and $2p_y$ orbitals; this bonding arrangement is denoted by sp^2. The remaining electron with a p_z orbital provides only weak interplanar bonding, but it is responsible for the semi-metallic electronic behavior of graphite.

The sp-orbital, and the sp^3-hybrid orbital indicate rotational symmetry. Bonds of this kind are call *σ-bonds*. The electrons involved in such bonds are called σ-electrons. With double bonds, so-called *π-orbitals* occur with corresponding π-electrons. Such orbitals are not symmetrical with regard to their bonding orientation. Figure 8.7 displays X-ray absorption spectra of highly oriented pyrolytic graphite (HOPG) recorded at different incidence angles [40]. When the polarization vector of the incident beam is parallel to the basal plane only excitations to σ states are possible. Excitations to π states become more likely the more perpendicular to the basal plane the polarization vector is. Thus, one can see that the main feature below 292 eV has mainly π character, while the σ states are observed at energies above 292 eV. Notably, the sharp absorption feature appearing at 292.0 eV is not due to the band structure but to an excitonic state, which has been discussed in detail elsewhere [26].

The normal carbon K_α XES spectra of diamond, graphitic carbon, C_{70}, C_{60}, and benzene are presented in Fig. 8.8(a), where large differences in spectral profile are observed. The spectra of diamond and graphitic carbon show a wide band with

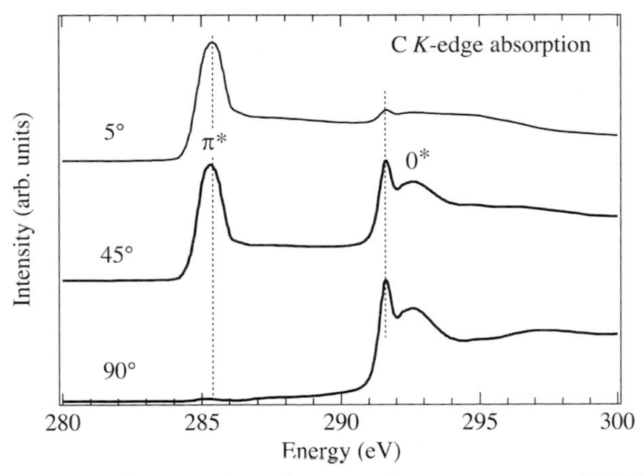

Fig. 8.7. Polarization dependent X-ray absorption spectra of HOPG.

some shoulder structures, where the energy positions of the peak maximum and band shapes are largely different. In some studies related to vapor deposition the XES spectral profile has been used as a means to identify certain chemical states [34]. In contrast, the spectra of benzene, C_{60}, and C_{70} exhibit clearly resolved emission peaks, indicating strong molecular character in their solid phases. The marked resolved emission features in benzene can find their countpart in the emission spectra of C_{60} and C_{70}, and all the fine structures are washed out in diamond and graphite.

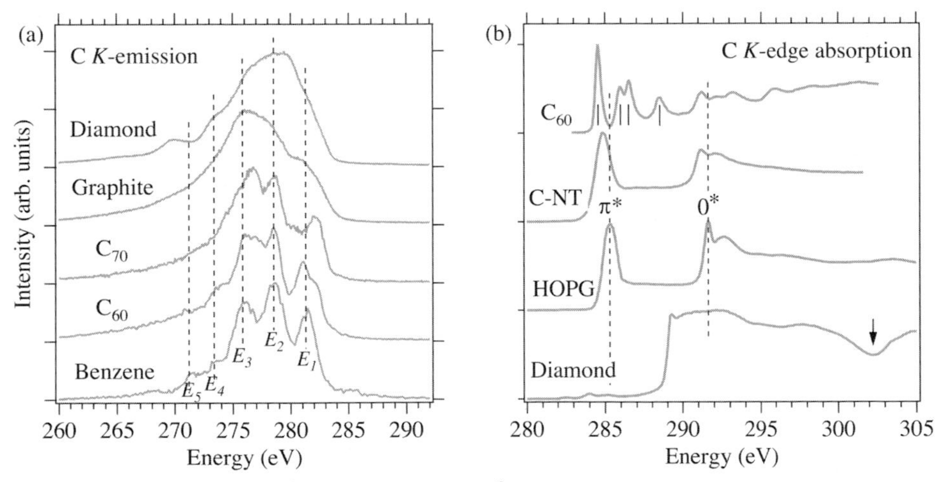

Fig. 8.8. (a) Non-resonant carbon Kα SXES spectra of diamond, amorphous carbon (graphitic), C_{70}, C_{60}, and benzene; (b) X-ray absorption spectra of C_{60}, carbon nanotubes, HOPG, and diamond.

Figure 8.8(b) shows the X-ray absorption spectra of C_{60}, carbon nanotubes (C-NT), HOPG, and diamond. The X-ray absorption spectral shape of carbon nanotubes is similar to that of HOPG, which is different to that of C_{60}. The π^* shows a shift towards low photon energy in the X-ray absorption of C-NT. C_{60} gives more resolved absorption features, which indicate a strong molecular character of C_{60}. Non-resonant carbon Kα XES spectra of carbon nanotubes and C_{60} show a similar shape, but resonantly excited XES spectra reveal large differences in the electronic structure of these two systems. The XAS spectrum of diamond shows no π^* contribution, and the absorption feature at 288.85 eV is the diamond exciton. The 2nd bandgap at 302 eV is also clearly observed.

8.3.2
Hydrogen Bonding Effect

A very important but hitherto less addressed question is that of the hydrogen bond effect on the electronic structure. This effect is essential for understanding the physical and chemical properties of many chemical and biological systems. The reason for this neglect, however, is the limited experimental access to the electronic structure of liquids. The application of spectroscopic methods to study the electronic structure of liquids has been hampered by the incompatibility of wet samples and high-vacuum conditions. Hence, even understanding the properties of pure liquid water remains a challenge.

Water is the most abundant substance on our planet, and it is the principal constituent of all living organisms. Chemical reactions taking place in liquid water are essential for many important processes in electrochemistry, environmental science, pharmaceutical science, and biology in general. Many models have been proposed to view the details of how liquid water is geometrically organized by a hydrogen bond network. Hydrogen bonding is an attractive interaction from a link of a hydrogen atom and one of the highly electronegative and non-metallic elements that contain a lone pair of electrons. Although H-bonds are much weaker than conventional chemical bonds, they have important consequences on the properties of water. Diffraction of X-rays [43] and neutrons [44] provides strong evidence that tetrahedral hydrogen-bond order persists beyond the melting transition, but with substantial disorder present [45]. Important questions remain about the precise nature of the disorder and how it is spatially manifested.

A network with four hydrogen bonds connects most of the water molecules in the liquid phase. Such a network can often be terminated by local structures with three or even two hydrogen bonds.

Soft-X-ray emission spectroscopy is essentially bulk sensitive, since the attenuation length of photons in this energy range is typically hundreds of nanometers in solid matter. The penetration depth offers a few experimental opportunities not present in electron-based spectroscopy. Figure 8.9 shows XAS and XES spectra of water in the gas phase and liquid phase at room temperature.

Another X-ray absorption study of liquid water [46] suggests that the four hydrogen-bonding networks mainly contribute to a single broad feature, while a shoulder located at 534.7 eV suggests the presence of the broken hydrogen bonds.

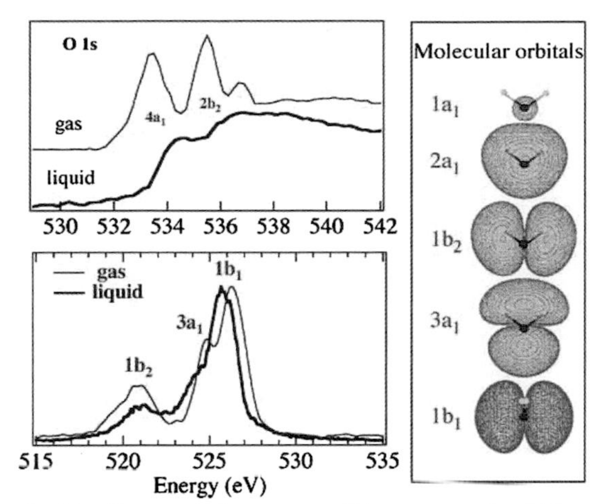

Fig. 8.9. X-ray emission spectra of the water molecules and liquid water, formed as electrons from the three outermost occupied molecular orbitals, schematically depicted in the right panel, fill a vacancy in the $1a_1$ core-level. The excitation-energy is 543 eV, well above the ionization limit but sufficiently low so that the emission from multiply excited states can be neglected.

Indeed, recent theoretical simulations assign this pre-edge structure to a particular three-hydrogen bond structure with one missing hydrogen bond at the hydrogen site.

8.3.3
Charge and Spin States of Transition Metals

Many metals that play important roles in biological and environmental sciences lack the partially filled d-shells used for UV–visible spectroscopy. Still, there are always soft X-ray resonances that can be exploited.

First transition metal L-edge spectroscopy is a relatively new technique for biological samples. L-edge spectra can be analyzed through theoretical simulation using ligand field multiplet theory (LFMT). LFMT is a multi-electron approach that describes the initial and final states as multiplets that are mixed and split by the symmetry of the ligand field [47–51].

The experimental Fe-$L_{2,3}$ absorption spectra for $K_4[Fe(CN)_6]$ (Fe^{2+}) and $K_3[Fe(CN)_6]$ (Fe^{3+}) are displayed in Fig. 8.10. The first peak at 704.5 eV is observed in the (Fe^{3+}) derivative but not in the (Fe^{2+}) one. It corresponds to the excitation $2p^6t_{2g}^5 \leftrightarrow 2p^5t_{2g}^6$ which is absent in the Fe^{2+} configuration (t_{2g}^6 ground state). This peak is a signature of the Fe^{3+} configuration [52]. The other peaks between 707 and 715 eV correspond to multiplet structures arising from transitions from 2p to e_g orbitals.

The second peak, which corresponds to the transition from $2p_{3/2}$ to empty e_g orbitals, exhibits at least five structures, which have to be analyzed with a multiplet model, taking into account the exchange interaction between the 2p hole and 3d electrons [51]. Note that the splitting of the L_3 line is related, but not directly scaled, to the lODq crystal field strength parameter (energy separation between

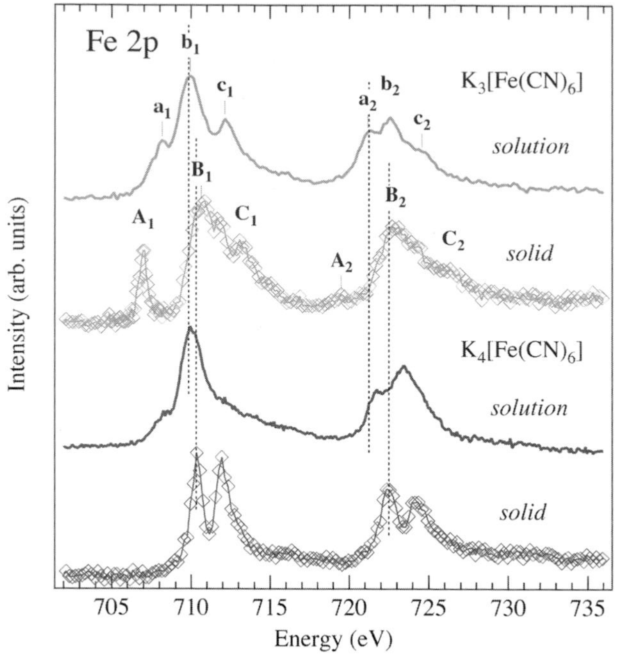

Fig. 8.10. Fe $L_{2,3}$-edges absorption from $K_3[Fe(CN)_6]$ and $K_4[Fe(CN)_6]$ in both solid state and water solutions. The peak labeled "A_1" corresponds to a singlet excitation.

the levels t_{2g} and the center of mass of the e_g levels). The Fe L-edge of both complexes, $K_3[Fe(CN)_6]$ and $K_4[Fe(CN)_6]$ presented the significant changes in the spectral profile when they were disvolved in water solutions.

The Ni L-edge XAS and RIXS spectra of NiO and $(Ph_4As)_2Ni[S_2C_2(CF_3)_2]$ are shown in Fig. 8.11 [53]. NiO is a typical high-spin Ni^{2+}, and the complex of $(Ph_4As)_2Ni[S_2C_2(CF_3)_2]$ is a Ni dithiolene, which has a low-spin Ni^{2+}. The dd and charge-transfer excitations can be observed along with some contributions of normal emission. The transition of dd excitations in the high-spin compound NiO was within 3.0 eV, while the low-spin system shows dd excitations with larger energies. Examination of these Ni models will serve as a first step in measuring real biological nickels, and one can create a spectral data base for use in future biological spectra analysis. Preliminary experiments on a small number of Ni compounds also confirm the feasibility.

8.4
Electronic Structure and Nanostructure

In the nano-regime, two effects dominate the chemical and physical properties: (a) increasing contribution of surface atoms (surface effect). The surface atom to bulk

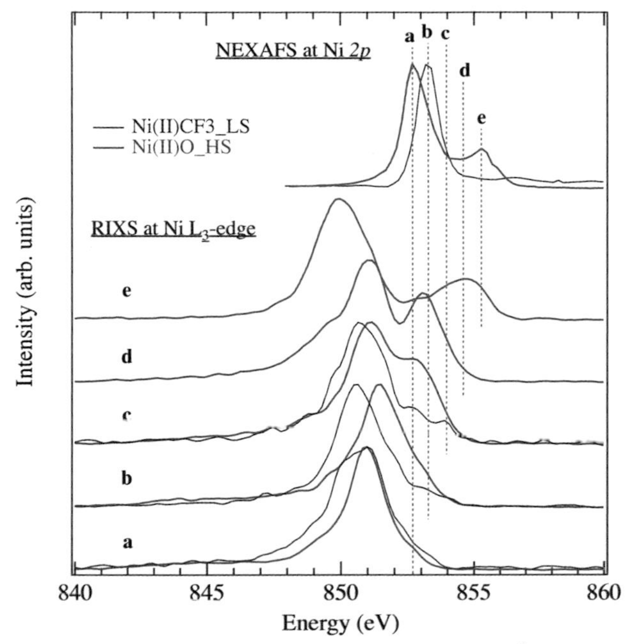

Fig. 8.11. X-ray absorption and RIXS spectra of NiO (Ni^{2+} in high spin) and $(Ph_4As)_2Ni[S_2C_2(CF_3)_2]$ (Ni^{2+} in low spin).

atom ratio increases dramatically as the particles size decreases. For a 30-nm particle, about 5% of all the atoms are on the surface, while for 3 nm particle, up to 50% of the atoms are on the surface, (b) quantum confinement as nanoclusters decrease down to a few nanometers (size effect), where electrons begin to feel the effects of quantum confinement when the size of the structure becomes comparable to the electron wavelengths.

Synchrotron radiation with photon energy at or below 1 keV is giving new insight into such areas as wet cell biology, condensed matter physics and extreme ultraviolet optics technology. In the soft X-ray region, the question tends to be, what are the electrons doing as they migrate between the atoms? [54].

8.4.1
Wide Bandgap Nanostructured Semiconductors

Soft X-ray spectroscopy has proved useful in probing the electronic structure of nanostructured solids [55, 56]. The development of the electronic band structure of solids from discrete localized atomic states is one of the key points of interest in cluster physics. As the size of nanocrystal extends into a truly molecular regime, such as cadimium thiolate [$Cd_8(SR)_{16}$ units], the optical excitation energies for the dissolved particles increase from 2.5 eV for CdS bulk to around 4.7 eV for Cd thiolate [57–59].

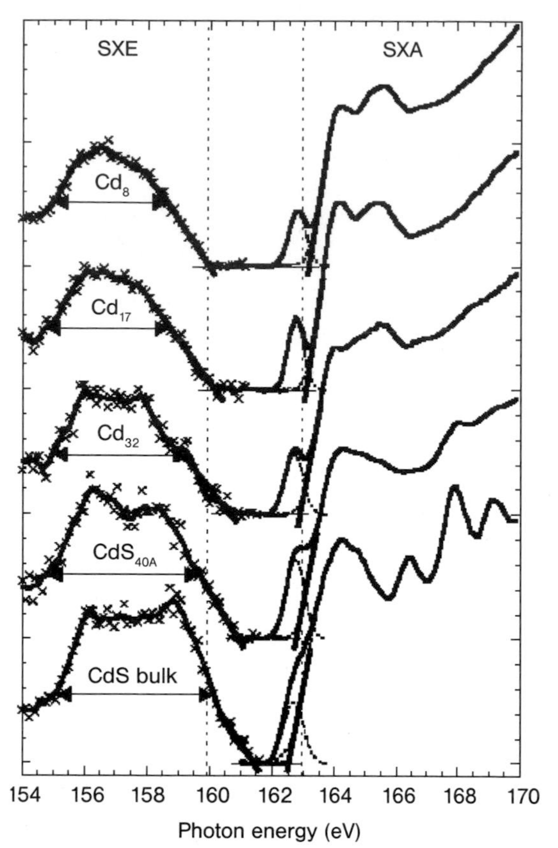

Fig. 8.12. Right: S L-edge XAS spectra of CdS bulk and Cd nanocrystallites. The dotted lines show the separation into excitonic and LPDOS excitation. Left: S L3 SXE spectra mapping the local DOS of the upper VB. Gray bars indicate the linear extrapolation of the emission cutoffs and the LPDOS absorption onsets, respectively. Courtesy of [60].

Figure 8.12 shows the S L-edge XAS and XES spectra for different samples (labeled by the number of Cd atoms per cluster) [60]. For decreasing particle size, an increase of the bandgap was observed. This opening of the bandgap can be separated into valence band and conduction band contributions.

X-ray absorption spectra have also used to measure the band edges of silicon nanocrystals, with average diameters ranging from 1 to 5 nm [61]. The conduction band (CB) edge shift (from XAS measurements) and the valence band (VB) edge shift (from photon emission measurements) vary as a function of nanocrystal diameter. If the observed VB or CB shifts are due to quantum confinement, one would expect the size of the band shifts to increase as the particle size of the nanocrystalline Si is decreased. Other X-ray absorption studies of quantum confinement have examined CdSe and InAs nanocrystals [62], nanodiamonds [63], and porous silicon [64], etc.

8.4.2
Cu Nanoclusters

Copper nanoclusters have been studied due to potential applications in optics, magneto-electronics and catalyst systems [65–67]. Silicon carbide provides a structurally and chemically stable support for such nanoclusters due to its stability against copper [68]. Although there are some studies on copper island formation on silicon carbide, the reports relied on the indirect spectroscopic results, mainly focusing on the stability of Cu/SiC interfaces [69–71]. Recently, we have observed the morphology of copper nanoclusters by high-resolution transmission electron microscopy (TEM) where copper nanoclusters had an ellipsoid shape inside a SiC matrix [72]. XAS studies can give a detailed description of the size dependence of the electronic structure of the copper nanoclusters in order to understand the nature of chemical bonds due to size effects.

Figure 8.13 shows the Cu L-edge XAS spectra of a series of Si/Ta(100 nm)/[Cu(0.5–3 nm)/SiC(2 nm)]10/Ta(5 nm) and the reference Cu. The bulk Cu sample shows three distinct features in the L_3-edge region, which are associated with the unoccupied 3d states in Cu. In a free Cu atom, the 3d orbital is fully occupied. In contrast, a solid Cu is characterized by a re-mixing of the valence wave functions in the ground state, gaining a small amount of non-vanishing 3d hole. According to dipole-selection rules, the dominant transition in XAS of Cu is from the Cu $2p_{3/2}$ and $2p_{1/2}$ states to the unoccupied Cu 3d states [73]. Thus, the peak at the onset of the L_3-edge is associated with the Cu 3d-derived character due to 3d–4s hybridization, giving a sharp $2p \rightarrow 3d$ transition superimposed on the smooth $2p \rightarrow 4s$ transitions. The peak at 939 eV and the peak at 943 eV are attributed to transitions

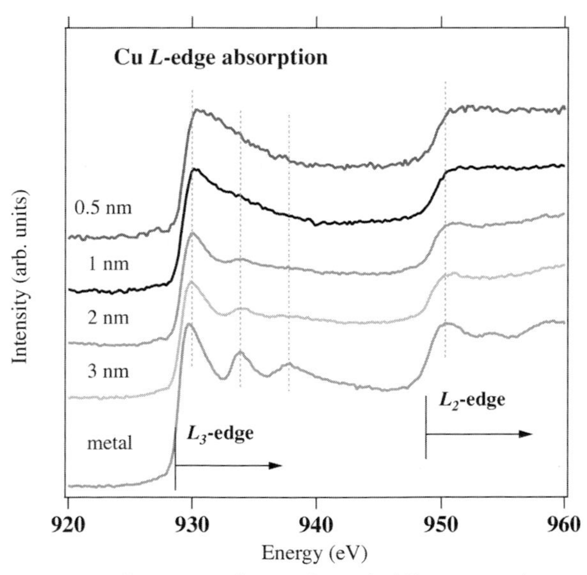

Fig. 8.13. XAS spectra of Cu L-edge with different nanocluster sizes.

towards empty states, mainly mixed with s, p characters. As the nanocluster size decreases, two distinct changes occur. First, the peaks move slightly to higher energy. Second, the peaks at 939 and 943 eV drastically decrease in intensity. The decrease of these peaks indicates that the s–p–d hybridization becomes weaker. This behavior can be attributed to the surface effect. The Cu atoms at surface have lower coordination number and see fewer Cu neighboring atoms, giving rise to a reduction in Cu–Cu interaction. The Cu–Cu interaction is mainly determined by d–d interaction and stronger Cu d–d interaction favors s–p–d re-hybridization [12, 13]; reduction of the d electron will weaken the hybridization.

8.4.3
ZnO Nanocrystals

ZnO, a wide band-gap semiconductor, has attracted considerable attention recently due to its potential technological applications such as, for instance, highly efficient vacuum fluorescent displays (VFD) and field-emission displays [76]. ZnO has also been used for short wavelength laser devices [77], high power and high frequency electronic devices [78], and light-emitting diodes (LED) [79, 80]. ZnO shows many advantages: (a) it has a larger exciton energy (60 meV) than GaN (23 meV); (b) the band-gap is tunable from 2.8 to 4 eV [81, 82]; (c) wet chemical synthesis is possible; (d) low power threshold at room temperature; (e) dilute Mn-doped ZnO shows room temperature ferromagnetism [83]. Recently, quantum size effects on the exciton and band-gap energies were observed in semiconductor nanocrystals [61, 84].

The controlled synthesis of ZnO nanostructures and in-depth understanding of their chemical/physical properties and electronic structure are the key issues for the future development of ZnO based nanodevices.

The XES spectra of bulk and nanotructured ZnO are displayed together with the corresponding XAS spectrum in Fig. 8.14 [85]. The O K-edge emission spectrum reflects the O 2p occupied states (valence band), and the O K-edge absorption spec-

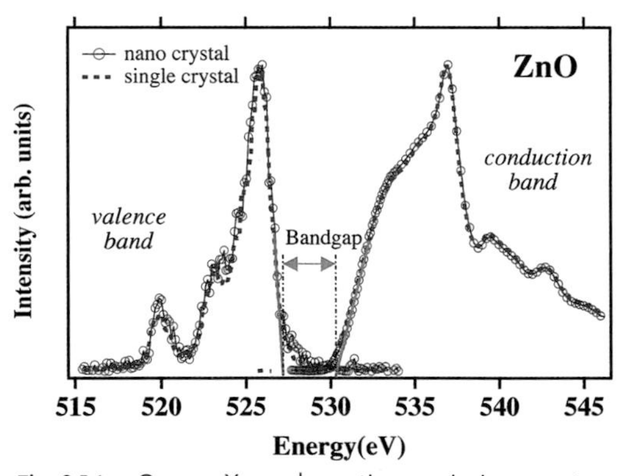

Fig. 8.14. Oxygen X-ray absorption–emission spectrum reflected conduction band and valence band near the Fermi-level of ZnO nanoparticles in comparison with bulk ZnO.

trum reflects the O 2p unoccupied states (conduction band). In the photon energy region of 530–539 eV, the X-ray absorption can be mainly assigned to the O 2p hybridized with Zn 4s states. In the region 539–550 eV the spectrum is mainly attributed to O 2p hybridized with Zn 4p states. Above 550 eV, the contribution comes mainly from O 2p–Zn 4d mixed states [86]. Stronger s–p–d hybridization was revealed in nanostructured ZnO since the contributions of features at 520 and 523 eV are enhanced. A well-defined band-gap can be observed between the valence-band maximum and conduction-band minimum. Our absorption–emission spectrum yields a fundamental band-gap energy of 3.3 eV.

8.5 Electronic Structure and Molecular Structure

Our microscopic understanding of a liquid is based very much on the study of spatial and of spatio-temporal correlation functions. This is the main bequest by the liquid state to the whole of physics: think in terms of correlations, be these between molecules in water or between galaxies in the cosmos. The study of correlations allows us to appreciate the local organization of one molecule around others, and to unravel the microscopic dynamics. Neutron diffraction and scattering have and continue to play a major role in these studies. X-ray diffraction is important as well, and very recently X-ray inelastic scattering has become available. We should add to these the soft X-ray spectroscopy expeiments.

Using X-ray absorption and selectively excited X-ray emission spectroscopy to probe unoccupied and occupied electronic states, one can establish a firm interpretation for the unusual thermodynamic properties of molecular liquids. Furthermore, one can elucidate finer details of their structural properties. XAS and XES spectra reflect the local electronic structure of the various conformations; in this case, the oxygen lineshape is sensitive to the hydrogen bonding configurations.

8.5.1 Hydrogen Bonding in Liquid Water

The local structure of liquid water is still under debate. Soft X-ray emission spectra, emanating from the radiative decay, subsequent to core excitation, can be useful in assigning structures in XAS spectra [87–90], and especially we have, earlier, shown that resonantly excited XES spectra of liquid water are compatible with the traditional view that three and four hydrogen bonds dominate in the structure [89]. Here we show that a theory that assumes that most water molecules in liquid water only make two hydrogen bonds fails to reproduce the experimental XES spectra.

Both the experiment and calculation have been described earlier [89], and the salient results are shown in Fig. 8.15. At excitation high above threshold (545.5 eV), the XES spectrum is well described by a calculation both with symmetric four-fold coordination, and conformations with one broken hydrogen bond. Thus, at high

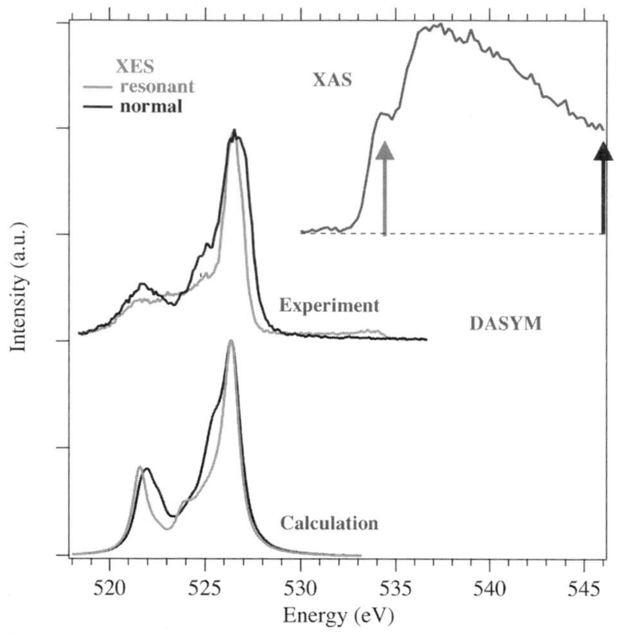

Fig. 8.15. Resonantly (excitation energy at 534.7 eV) and non-resonantly (at 545.5 eV) excited XES spectra of liquid water, compared to calculations for the SYM and DASYM species. Excitation energies are marked in the XAS spectrum.

excitation energy the spectra are not predicted to be sensitive to the breaking of a single hydrogen bond. When tuning the excitation energy to the pre-peak at 534.7 eV, the spectral changes comply with what is expected when structures with one broken hydrogen bond at the hydrogen donor site (DASYM) [46, 89] are resonantly excited. We have shown elsewhere that a thorough analysis of XES spectra excited in the threshold region can give further details on the contributions by various structures to the XAS spectrum [91].

8.5.2 Molecular Structure in Liquid Alcohol and Water Mixture

Near-edge X-ray absorption spectra of water, methanol and a mixture of the two are displayed in Fig. 8.16. The X-ray absorption spectra of all samples present a similar shape: a strong pre-edge at 534.5 eV (B) and a broad main absorption threshold arising from 537.0 eV (C). The pre-edge (B) in liquid water has been the fingerprint of the particular water cluster with a broken hydrogen bond at the hydrogen side [89]. In methanol, the ring-structure configuration contributes mostly to the pre-edge (B) and the chain-structure configuration contributes mostly to the main threshold (C) of the absorption spectra [92]. Upon mixing water and methanol, a small X-ray absorption peak (A) appears at 531.5 eV, below the known X-ray ab-

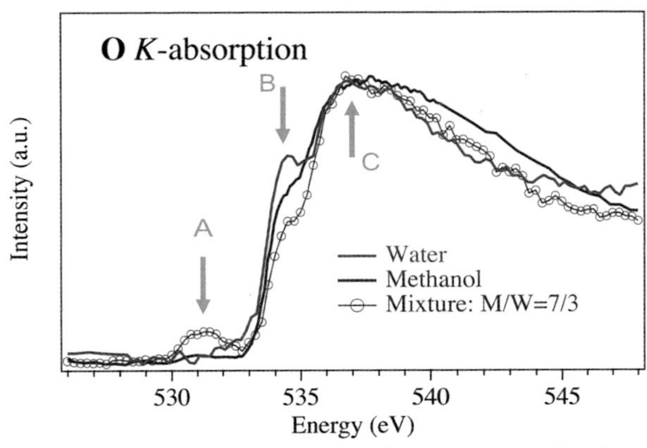

Fig. 8.16. X-ray absorption spectra of liquid water, methanol and a mixture of molar concentration 7:3.

sorption pre-edge (B). The new absorption peak (A) suggests direct interaction between water molecules and methanol molecules.

To understand the origin of the pre-peaks in X-ray absorption spectra, resonant X-ray emission measurements were performed (Fig. 8.17). Structures of rings and chains for methanol and mixtures of water clusters and methanol are fully optimized at the hybrid density functional theory B3LYP level with 6-31G basis set using GAUSSIAN 98 [92]. The theoretical X-ray emission spectra of methanol were generated by the group theory formulation [20], using the adiabatic approximation (ground state electronic structure) for intensities as established for molecules and clusters [93]. All transition moments and orbital energies were calculated at the canonical Hartree–Fock level with the Sadlej basis set by using the DALTON program [94].

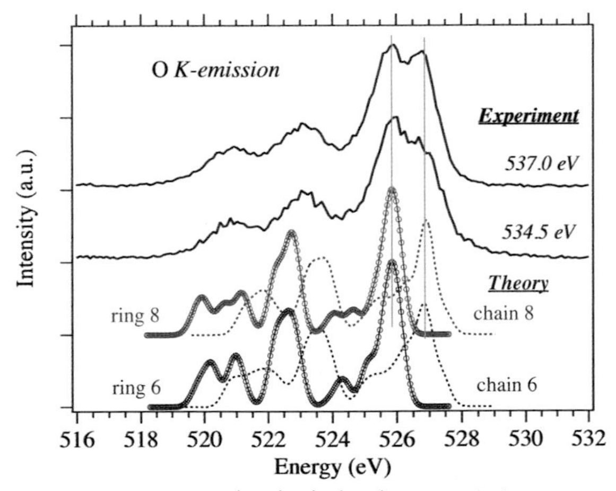

Fig. 8.17. Experimental and calculated X-ray emission spectra of liquid methanol.

The double structure emission band centered at 527 eV indicates two major different chemical species (molecular structure) in the alcohol and water mixtures. Comparison between experiment and theory shows that the two species originate from the combination of methanol rings and chains. These two peaks clearly have different origins: the one at higher energy side is from the chains, and the other from the rings. From the intensity distribution, it was found that the chains and rings have a similar size with same portion.

As to the mixtures of liquid water and methanol, when selecting the excitation energies around the pre-peak (A) in the X-ray absorption, the main emission band becomes very narrow and shifts towards higher photon energy, opposite to the shifting of the elastic peak. Is the interaction between water molecules and these chains or rings be responsible for the sharp spectral features in the resonantly excited X-ray emission spectra? We learnt previously [90] that the dominant chain and ring structures in methanol are those with 6 and 8 molecules. Of course, we do not know if this is the answer or just one of the possibilities. Notably, when methanol mixes with water, the ring and chain become the same. The results from alcohol–water mixtures of different concentrations are prerequisites to support the experimental findings from the equimolar study [11]. Thus, we know that the exact mixing level is not critical to our conclusions.

This study shows that, when a few water molecules interact with chains, the chains start to bend over to form open-ring structures. By adding water, the alcohol structures become more compact. The formation of such ordered molecular structure upon mixing provides evidence that the entropy of the alcohol and water mixture is not less than expected for ideal solutions.

We find a strong involvement of hydrogen bonding in the mixing of water and methanol molecules. The local electronic structure of water and methanol clusters, where water cluster is bridging within a six-membered open-ring structured methanol cluster, is separately determined. The experimental findings suggest an incomplete mixing of water–alcohol systems and a strong self-association between methanol chain and water cluster through hydrogen bonding. The enhancement of joint water–methanol-ring structure explains the loss of entropy of the aqueous solutions.

The answer to this specific question is likely to be of interest to a much broader scientific community of readers. The novelty of this work is that it relies on a spectroscopic probe to investigate problems that have, largely, been the territory of scattering methods. This technique could have great impact to the extent that this approach can be generalized to investigate other solvent mixtures.

8.5.3 Electronic Structure and Ion Solvations

Ion solvation is a phenomenon that is of fundamental interest in many chemical contexts because solvated ions are almost omnipresent on Earth. Hydrated ions occur in aqueous solution in many chemical and biological systems [95, 96]. Metal-ion transport in both aqueous- and polymer-solvent media involves continual

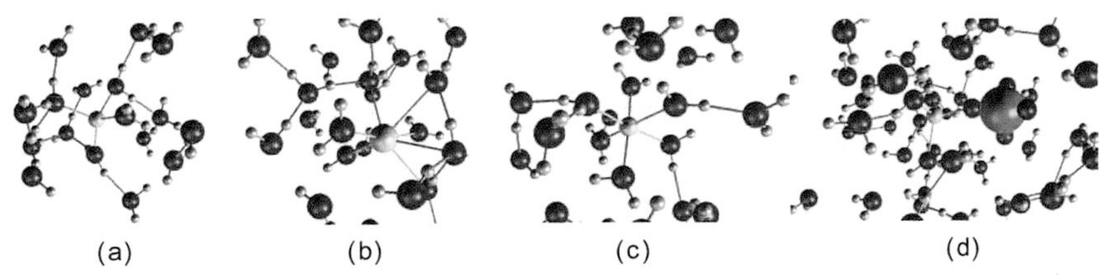

Fig. 8.18. Sample snapshots showing configurations with water molecules in the first and second solvation shells for (a) Li^{+}, (b) Na^{+}, (c) Mg^{2+}, and (d) ion pair water solutions.

solvent–ligand exchange. Metal-ion coordination chemistry is therefore fundamental to these phenomena where a dramatic exchange of ligand occurs. Alternatively, the effects of solvated cations could be monitored via examining the spectra of solvent. The great advantage of this approach is that the restriction to cations no longer applies. For example, it can be used in studying what happens when salts of the alkali and alkaline earth metals are added to aqueous solvent.

An ion in solution disturbs the local solvent structure. The local structure around a Li^{+} ion central water molecule, a Na^{+} ion, and a Mg^{2+} ion in water solutions as obtained using the polarizable MD model is shown in Fig. 8.18. The influences of cations on the water molecular structure can be seen as the threshold shifts towards higher energy in the X-ray absorption spectra; the mixing of molecular orbital in $3a_1$ symmetry is reinforced as the intensity of $3a_1$ is further reduced. We find that the charge difference of the cations may not be the only factor that accounts for the interactions between the cations and water molecules.

Figure 8.19 reports the O K-edge absorption spectra of liquid water and NaCl, $MgCl_2$, $AlCl_3$ water solutions. The pre-peak (marked as α) in XAS spectrum of

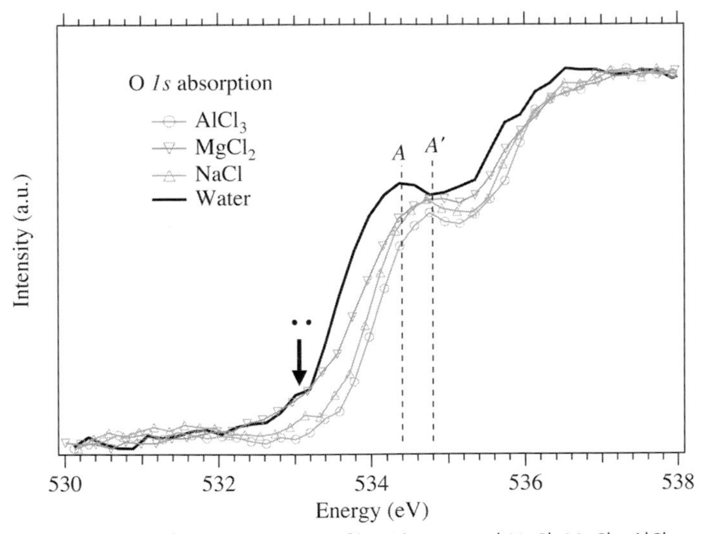

Fig. 8.19. X-ray absorption spectra of liquid water and NaCl, $MgCl_2$, $AlCl_3$ water solutions.

water has partly $4a_1$ symmetry. The major difference on going from pure water to ion–water solutions can be seen as a shift of 0.4 eV for the absorption pre-peak (A), which is probably a result of a higher degree ionic character of the chemical bonding between cations and oxygen. The front slope of the absorption threshold has shifted most away from that of pure water for Al^{3+}-water, followed by Na^{+}-water, while Mg^{2+}-water shows a tendency to resemble that of pure water in the region marked by A. For the monovalent Na^{+}, the distance between the water molecules in the first and second solvation shells is long, 2.40–2.50 Å [97], and rather unaffected by the ion. For Mg^{2+} and Al^{3+}, the distances are substantially shorter, 2.00–2.14 Å and 1.87–1.90 Å, respectively. Thus, the changes observed XAS spectra are not in line with the charge order of the cations: Na^{+}, Mg^{2+}, and Al^{3+}.

8.5.4
Drugs in Water Solution

All orally administered drugs must be dissolved, absorbed, and transported by the blood stream to the site of action. It is a pharmaceutical challenge to establish predictive models for this complex process. One essential aspect of the problem is the lack of understanding of solubility on the microscopic level. Especially when hydrogen bonding is involved, the situation is often too complex for reliable predictions.

It is possible to understand drug solubility by applying X-ray spectroscopy to substances in solid phases, in aqueous solution, and in the gas-phase (Fig. 8.20). The influence of the molecular surrounding on the local electronic structure is reflected in soft X-ray absorption fine structure, and in site-selectively excited X-ray emission spectra.

The overall shapes of the XAS spectra recorded at the nitrogen and oxygen K edges of atenolol and nadolol are similar (Fig. 8.21), apart from some remarkable exceptions. In atenolol there is an additional sharp resonance that is virtually absent in nadolol, both at the oxygen and nitrogen edge. These resonances can be assigned to excitations at specific sites by considering the structure of the molecules (Fig. 8.21). Nadolol has a single nitrogen site, (–NH–), whereas atenolol contains an additional amino-group nitrogen ($-NH_2$). It is, therefore, straightforward to assign the additional peak at 400.5 eV to excitations at this site. The smaller resonance at 398.5 eV can be assigned to the (–NH–) site, as it is present in both molecules. Similarly, the atenolol resonance at around 532 eV in the oxygen spectrum can be assigned to the (=O) site, which is unique to this molecule. Regions in the spectra with less pronounced features cannot be assigned without theory, and, especially at higher energies, overlapping intensity from several non-equivalent sites is difficult to disentangle.

8.5.5
Electronic Structure of Bases in DNA Duplexes

Characterization of electronic states near the Fermi level of DNA duplexes has been desired to clarify the mechanisms of long-range charge migration in DNA [98–103].

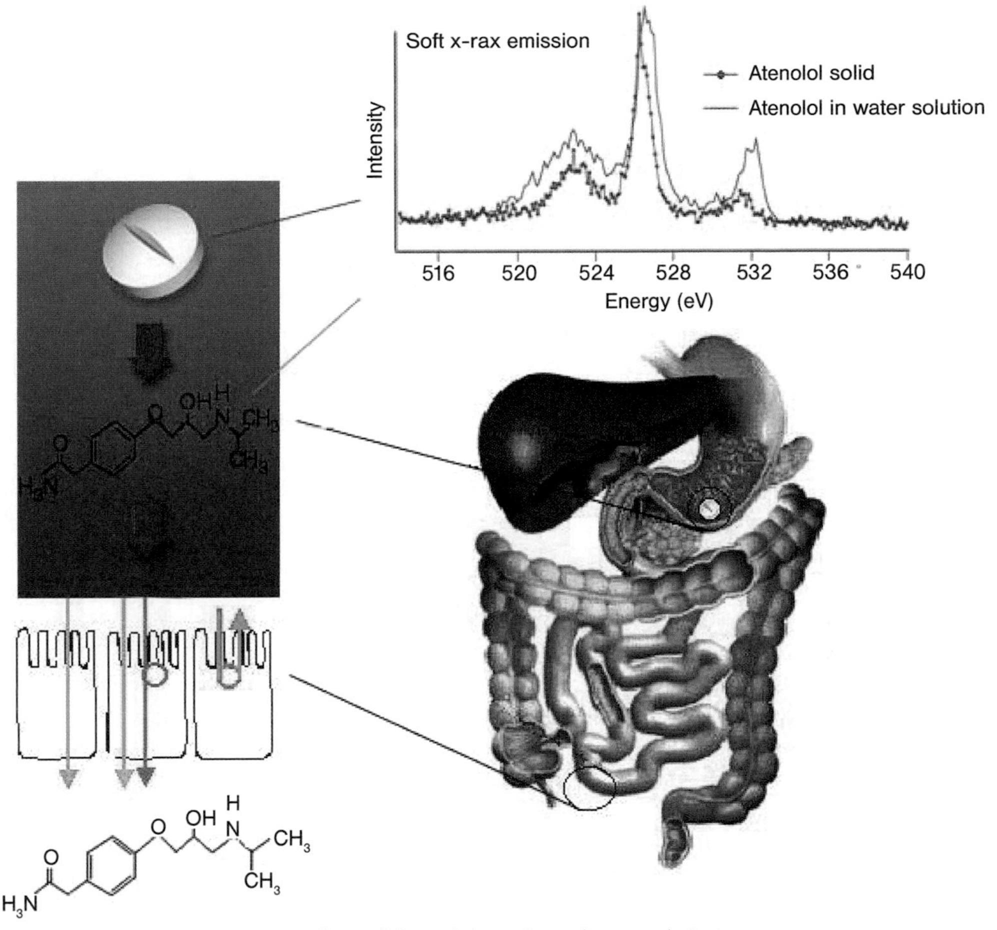

Fig. 8.20. X-ray emission spectra of atenolol in solid powder and water solution.

It was recently reported that the N K-edge XAS and Resonant Photoemission (RPE) spectroscopy characterized the electronic structure near the Fermi level of DNA duplexes to specify the charge migration mechanism [104]. The samples were thick GC- and AT-DNA films on SiO_2/p-Si(111) substrates. Since N atoms are included in only bases in DNA duplexes, the RPE spectra excited from N 1s to unoccupied states purely extract the electronic orbital features of the bases in DNA. It was concluded that the charge-hopping model is suitable for electric conduction in DNA duplexes rather than the charge-transfer model via delocalized states when electrons pass through the π^* states of DNA bases.

Soft X-ray absorption and emission spectroscopies have been applied to study the nitrogen bonding structure in poly(dC) $\cdots$ poly(dG) [105]. The three sharp peaks at 398.5, 399.5, and 400.9 eV (Fig. 8.22a) indicate that absorption features originate from the well-defined structures. From previous studies of carbon nitride films [88, 106], the nitrile structure (N_1-site) aligns very well with the peak of 399.5 eV.

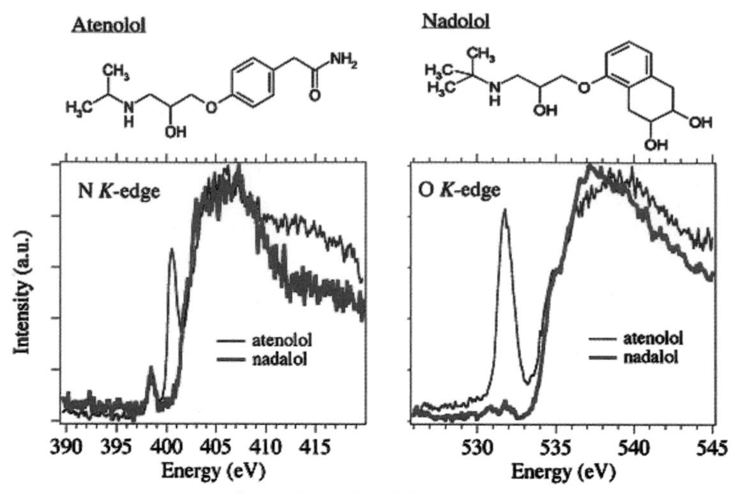

Fig. 8.21. XAS spectra of atenolol and nadolol, measured at the nitrogen and oxygen K edges.

Since the nitrile bond predominantly has p-character, this peak appears strong in the π^* region. The 398.5 eV peak corresponds to pyridine-like N (N_2-site). The spectra for graphite-like N_3-site shows a peak at 401.7 eV, with a shoulder to the low energy side. This could be a low-energy σ^* resonance; however, with the complex structure in DNA it is difficult to separate the π and σ bonds. This structure together with pyridine-like N are, possibly, the origin of the absorption peak at 400.9 eV in XAS spectrum.

Nitrogen K-emission spectra are shown in Fig. 8.22(b). The excitation energies were selected to correspond to the absorption features in the XAS spectrum. The fact that N atoms in different bonding environments are excited depending on the photon energy is clearly reflected by the differences in XES spectra. For the lowest excitation energy (398.5 eV), presumably corresponding to the $1s \leftrightarrow \pi^*$ transition of pyridine-like N, the main emission line is centered at 393.5 eV. The overall spectral shape is similar to that of pyridine. When the excitation energy is increased to 399.5 eV, mainly nitrogen in the N_1 structure should be excited. However, the emission spectra indicate that a large fraction of the pyridine-like N atoms are also excited at this energy. This can be explained by a relatively broad signal corresponding to pyridine-like N due to shifts that depend on the second nearest neighbors. Thus the emission spectrum can be modeled by a superposition of the N_1 and N_2 spectra. The excitation energy of 400.9 eV, however, mainly excites the N_1 structure.

X-ray transitions, where a core-level vacancy is filled by a valence-orbital electron, give direct information about the chemical bonding. While such transitions have been analyzed using X-ray spectrometers since the late 1920s, interest in the technique is presently booming due to the advent of third generation synchrotron radiation sources. Today the method is frequently applied in research fields ranging

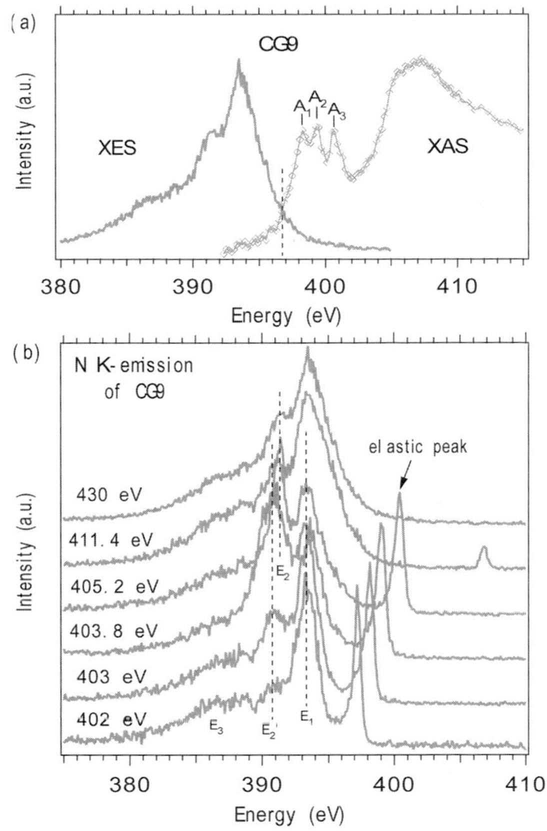

Fig. 8.22. (a) N K-edge absorption and emission spectra of poly(dC) · · · poly(dG). (b) N K-edge emission spectra recorded at selected excitation energies.

from atomic and molecular physics to materials research. Understanding protein functionality is of fundamental importance in biochemistry. Soft-X-ray absorption and emission study of poly(dC) · · · poly(dG) can elucidate the relation between the structure and functionality of proteins.

Acknowledgments

This chapter would not exist without the contributions from my collaborators. I acknowledge some of them for their support over the years, to name a few: Y. Luo, J. Nordgren, J.-E. Rubensson, A. Augustsson, C. L. Dong, S. Kashtanov, C. L. Chang, L. Vaysseries, S. Butorin, L. Duda, C.-J. Englund, S. H. Yang, D. W. Shin, J. Söderström, P.-A. Glans, T. Learmonth, K. Smith, etc. Preparation of the manuscript for this chapter was also supported by the office of Science, Basic Energy Sciences, and the Department of Energy under contract DE-AC03-76SF000098 at Lawrence Berkeley National Laboratory.

References

1 LEONID V. AZAROFF, ed. *X-ray Spectroscopy.* (McGraw-Hill Inc., **1974**).

2 TSUN-KONG SHAM, ed. *Chemical Applications of Synchrotron Radiation* (World Scientific, Singapore, **2002**). Chapter 10 Soft X-ray fluorescence spectroscopy for materials science and chemical physics, J. NORDGREN, S. M. BUTORIN, L. C. DUDA, J.-H. GUO, J. E. RUBENSSON, p. 517.

3 J. A. SAMSON, D. L. EDERER, ed. *Vacuum Ultraviolet Spectroscopy.* (Academic Press, London, **2000**). Chapter 13, Soft X-ray fluorescence spectroscopy, T. A. CALLCOTT, p. 279.

4 Twelve International Conferences on Vacuum Ultraviolet Radiation Physics have been held since 1965, the most recent being VUV12 in San Francisco in 1998.

5 Seventeen International Conferences on X-ray Physics and Inner Shell Ionization have been held, the most recent being X-96 in Prague, Czechoslovakia in 1996.

6 S. P. CRAMER, K. O. HODGSON, Bioinorganic applications of X-ray absorption spectroscopy, *Prog. Inorg. Chem.* 25, 1 (**1979**).

7 D. KONIGSBERGER, R. PRINS, ed. *Extended X-ray Absorption Fine Structure*, (Plenum: New York, **1988**). Chapter Biochemical Applications of X-ray Absorption Spectroscopy, S. P. CRAMER, pp. 257–320.

8 J. TELSER, ed. *Paramagnetic Resonance of Metallobiomolecules*, (American Chemical Society, Washington D.C., **2003**). Chapter X, X-Ray Magnetic Circular Dichroism – A Primer for Chemists, S. P. CRAMER.

9 M. V. ALDRICH, J. L. GARDEA-TORRESDEY, J. R. PERALTA-VIDEA, J. G. PARSONS, Uptake and Resuction of Cr(VI) to Cr(III) by mesquite (Prosopis spp.): chromate-plant interaction in hydroponics and solid media studied using XAS, *Environ. Sci. Technol.* 37, 1859 (**2003**).

10 JORGE L. GARDEA-TORRESDEY, EDUARDO GOMEZ, JOSE R. PERALTA-VIDEA, JASON G. PARSONS, HORACIO TROIANI, MIGUEL JOSE-YACAMAN, Alfalfa Sprouts: A Natural Source for the Synthesis of Silver Nanoparticles, *Langmuir* 19, 1357 (**2003**).

11 J. GOULON, A. ROGALEV, G. GOUJON, Ch. GAUTHIER, E. MOGUILINE, A. SOLE, S. FEITR, F. WHIHELM, N. JAOUEN, Ch. GOULON-GINET, P. DRESSLER, P. ROHR, M.-O. LAMPERT, R. HENCK, Advanced detection systems for X-ray fluorescence excitation spectroscopy, *J. Synchro. Rad.* 12, 57 (**2005**).

12 S. ROUX, B. GARCIA, J.-L. BRIDOT, M. SALOME, C. MARQUETTE, L. LEMELLE, P. GILLET, L. BLUM, Pascal Perriat, Olivier Tillement, Synthesis, characterization of dihydrolipoic acid capped gold nanoparticles, and functionalization by the electroluminescent luminol, *Langmuir* 21, 2526 (**2005**).

13 CHALLA S. S. R. KUMAR, C. LEUSCHNER, E. E. DOOMES, L. HENRY, M. JUBAN, J. HORMES, Efficacy of lytic peptide-bound magnetite nanoparticles in destroying breast cancer cells, *J. Nanosci. Nanotechnol.* 4, 245 (**2004**).

14 FRANK DE GROOT, High-resolution X-ray emission and X-ray absorption spectroscopy, *Chem. Rev.* 101, 1779 (**2001**).

15 JOACHIM STÖHR, *NEXAFS Spectroscopy.* (Springer-Verlag, Berlin Heidelberg, **1992**).

16 http://www.cxro.lbl.gov/opticalconstants/

17 F. Kh. GEL'MUKHANOV, L. N.

Mazalov, A. V. Kontratenko, A theory of vibrational structure in the X-ray spectra of molecules, *Chem. Phys. Lett.* 46, 133 (**1977**).

18 F. Kh. Gel'mukhanov, H. Ågren, Resonant inelastic X-ray scattering with symmetry-selective excitation, *Phys. Rev. A* 49, 4378 (**1994**).

19 F. Kh. Gel'mukhanov, H. Ågren, Channel interference in resonance elastic X-ray scattering, *Phys. Rev. A* 50, 1129 (**1994**).

20 Y. Luo, H. Ågren, F. Kh. Gel'mukhanov, Symmetry assignments of occupied and unoccupied molecular orbitals through spectra of polarized resonance inelastic X-ray scattering, *J. Phys. B: At. Mol. Phys.* 27, 4169 (**1994**).

21 K. Hamalainen, D. P. Siddons, J. B. Hastings, L. E. Berman, Elimination of the inner-shell lifetime broadening in X-ray-absorption spectroscopy, *Phys. Rev. Lett.* 67, 2850 (**1991**).

22 K. Hamalainen, C. C. Kao, J. B. Hastings, D. P. Siddons, L. E. Berman, V. Stojanoff, S. P. Cramer, Spin-dependent X-ray absorption of MnO and MnF_2, *Phys. Rev. B* 46, 14274 (**1992**).

23 S. M. Butorin, J.-H. Guo, M. Magnuson, P. Kuiper, J. Nordgren, Low-energy *d–d* excitations in MnO studied by resonant X-ray fluorescence spectroscopy, *Phys. Rev. B* 54, 4405 (**1996**).

24 P. Kuiper, J.-H. Guo, C. Såthe, L.-C. Duda, J. Nordgren, J. J. M. Pothuizen, F. M. F. de Groot, G. A. Sawatzky, Resonant X-ray Raman spectra of Cu *dd* excitations in $Sr_2CuO_2Cl_2$, *Phys. Rev. Lett.* 80, 5204 (**1998**).

25 Y. Ma, N. Wassdahl, P. Skytt, J.-H. Guo, J. Nordgren, P. D. Johnson, J.-E. Rubensson, T. Böske, W. Eberhardt, S. D. Kevan, Soft-X-ray resonant inelastic scattering at the C K edge of diamond, *Phys. Rev. Lett.* 69, 2598 (**1992**).

26 Y. Ma, P. Skytt, N. Wassdahl, P. Glans, D. C. Mancini, J.-H. Guo, J. Nordgren, Core excitons and vibronic coupling in diamond and graphite, *Phys. Rev. Lett.* 71, 3725 (**1993**).

27 J. A. Carlisle, E. L. Shirley, E. A. Hudson, L. J. Terminello, T. A. Calcott, J. J. Jia, D. L. Ederer, R. C. C. Perera, F. J. Himpsel, Probing the graphite band structure with resonant soft-X-ray fluorescence, *Phys. Rev. Lett.* 74, 1234 (**1995**).

28 J. Nordgren, N. Wassdahl, Soft X-ray fluorescence spectroscopy using tunable synchrotron radiation, *J. Electron Spectrosc. Relat. Phenom.* 72, 273 (**1995**).

29 E. J. Nordgren, Soft X-ray emission spectroscopy in the nineties, *J. Electron Spectrosc. Relat. Phenom.* 78, 25 (**1996**).

30 J. Nordgren, P. Glans, K. Gunnelin, J.-H. Guo, P. Skytt, C. Såthe, N. Wassdahl, Resonant soft X-ray fluorescence spectra of molecules, *Appl. Phys. A* 65, 97 (**1997**).

31 J.-H. Guo, S. M. Butorin, N. Wassdahl, P. Skytt, J. Nordgren, Y. Ma, Electronic structure of $La_{2-x}Sr_xCuO_4$ studied by soft-X-ray-fluorescence spectroscopy with tunable excitation, *Phys. Rev. B* 49, 1376 (**1994**).

32 P. Skytt, P. Glans, J.-H. Guo, K. Gunnelin, C. Såthe, J. Nordgren, F. Kh. Gel'mukhanov, A. Cesar, H. Ågren, Quenching of symmetry breaking in resonant inelastic X-ray scattering by detuned excitation, *Phys. Rev. Lett.* 77, 5035 (**1996**).

33 M. Nyberg, Yi Luo, L. Qian, J.-E. Rubensson, C. Såthe, D. Ding, J.-H. Guo, T. Käämbre, J. Nordgren, Bond formation in titanium fulleride compounds studied through X-ray emission spectroscopy, *Phys. Rev. B* 63, 115117 (**2001**).

34 J.-H. Guo, J. Nordgren, Resonant C Kα X-ray emission of some carbon allotropes and organic compounds, *J. Electron Spectrosc. Relat. Phenom.* 110–111, 105–134 (**2000**).

35 S. M. Butorin, Resonant inelastic X-ray scattering as a probe of optical scale excitations in strongly electron-correlated systems: quasi-localized view, *J. Electron Spectrosc. Relat. Phenom.* 110–111, 213–233 (**2000**).

36 S. M. Butorin, J.-H. Guo, N. Wassdahl, J. E. Nordgren, Tunable-excitation soft X-ray fluorescence spectroscopy of high-Tc superconductors: an inequivalent-site seeing story, *J. Electron Spectrosc. Relat. Phenom.* 110–111, 235–273 (**2000**).

37 A. Kotani, S. Shin, Resonant inelastic X-ray scattering spectra for electrons in solids, *Rev. Mod. Phys.* 73, 203 (**2001**).

38 T. Warwick, P. Heimann, D. Mossessian, W. McKinney, H. Padmore, Performance of a high resolution, high flux density SGM undullator beamline at the ALS, *Rev. Sci. Instrum.* 66, 2037 (**1995**).

39 J.-H. Guo, N. Wassdahl, P. Skytt, S. M. Butorin, L.-C. Duda, C. J. Englund, J. Nordgren, End station for polarization and excitation energy selective soft X-ray fluorescence spectroscopy, *Rev. Sci. Instrum.* 66, 1561 (**1995**).

40 P. Skytt, P. Glans, D. C. Mancini, J.-H. Guo, N. Wassdahl, J. Nordgren, Y. Ma, Angle-resolved soft-X-ray fluorescence and absorption study of graphite, *Phys. Rev. B* 50, 10457 (**1994**).

41 Allotropic figures are downloaded from http://cnst.rice.edu/pics.html.

42 Silson Ltd, JBJ Business Park, Northampton Road, Blisworth, Northampton, NN7 3DW, England. (Email: peter.anastasi@silson.com).

43 A. H. Narten, H. A. Levy, Liquid water: molecular correlation functions from X-ray diffraction, *J. Chem. Phys.* 55, 2263 (**1971**).

44 A. H. Narten, Liquid water: atom pair correlation functions from neutron and X-ray diffraction, *J. Chem. Phys.* 56, 5681 (**1972**).

45 F. H. Stillinger, Water revisited, *Science* 209, 451 (**1980**).

46 S. Myneni, Y. Luo, L. Å. Näslund, M. Cavalleri, L. Ojamäe, Ogasawara, A. Pelmenschikov, Ph. Wernet, P. Väterlein, C. Heske, Z. Hussain, L. G. M. Pettersson, A. Nilsson, Spectroscopic probing of local hydrogen-bonding structures in liquid water, *J. Phys. Condensed Matter* 14, L213 (**2002**).

47 T. Yamaguchi, S. Shibuya, S. Suga, S. Shin, Inner-core excitation spectra of transition-metal compounds: II. *p–d* absorption spectra, *J. Phys. C* 15, 2641 (**1982**).

48 B. T. Thole, G. van der Laan, P. H. Butler, Spin-mixed ground state of Fe phthalocyanine and the temperature-dependent branching ratio in X-ray absorption spectroscopy, *Chem. Phys. Lett.* 149, 295 (**1988**).

49 G. van der Laan, B. T. Thole, G. A. Sawatzky, M. Verdaguer, Multiplet structure in the $L_{2,3}$ X-ray-absorption spectra: A fingerprint for high- and low-spin Ni^{2+} compounds, *Phys. Rev. B* 37, 6587 (**1988**).

50 F. M. F. de Groot, J. C. Fuggle, B. T. Thole, G. A. Sawatzky, $L_{2,3}$ X-ray-absorption edges of d0 compounds: K^+, Ca^{2+}, Sc^{3+}, and Ti^{4+} in O_h (octahedral) symmetry, *Phys. Rev. B* 41, 928 (**1990**).

51 F. M. F. de Groot, J. C. Fuggle, B. T. Thole, G. A. Sawatzky, 2*p* X-ray absorption of 3*d* transition-metal compounds: An atomic multiplet description including the crystal field, *Phys. Rev. B* 42, 5459 (**1990**).

52 C. Cartier dit Moulin, P. Rudolf, A.-M. Flank, C. T. Chen, Spin transition evidenced by soft X-ray absorption spectroscopy, *J. Phys. Chem.* 96, 6196 (**1992**).

53 J.-H. Guo, H. Wang, S. Cramer, Resonant Soft X-ray Emission Spectroscopy of Some Ni Complexes, in preparation (**2005**).

54 N. Smith, Science with Soft X Rays, *Physics Today*, January, 29 (**2001**).

55 L. Pavesi, E. Buzaneva, ed., *Frontier of Nano-Optelectronic Systems*, (Kluwer Academic Publishers, Netherlands, **2000**). Soft X-ray spectroscopy as a probe of the electronic structure of nanostructured solids, S. Eisebitt and W. Eberhardt, p. 347–362.

56 J. Guo, Synchrotron radiation, soft-X-ray spectroscopy and nanomaterials, *Int. J. Nanotechnol.* 1–2, 193 (**2004**).

57 T. Vossmeyer, G. Reck, B. Schulz, L. Katsikas, H. Weller, Double-layer superlattice structure built up of $Cd_{32}S_{14}(SCH_2CH(OH)CH_3)_{36}{\cdot}4H_2O$

clusters, *J. Am. Chem. Soc.* 117, 12881 (**1995**).

58 T. Vossmeyer, G. Reck, L. Katsikas, E. T. K. Haupt, B. Schulz, H. Weller, A double diamond superlattice built up of $Cd_{17}S_4(SCH_2CH_2OH)_{26}$ clusters, *Science* 267, 1476 (**1995**).

59 T. Vossmeyer, G. Reck, L. Katsikas, E. T. K. Haupt, B. Schulz, H. Weller, A new three dimensional crystal structure of a cadmium thiolate, *Inorg. Chem.* 34, 4926 (**1995**).

60 J. Lüninga, J. Rockenberger, S. Eisebitt, J.-E. Rubensson, A. Karl, A. Kornowski, H. Weller, W. Eberhardt, Soft X-ray spectroscopy of single sized CdS nanocrystals: size confinement and electronic structure, *Solid State Commun.* 112, 5 (**1999**).

61 T. van Buuren, L. N. Dinh, L. L. Chase, W. J. Siekhaus, L. J. Terminello, Changes in the electronic properties of Si nanocrystals as a function of particle size, *Phys. Rev. Lett.* 80, 3803 (**1998**).

62 K. S. Hamad, R. Roth, J. Rockenberger, T. van Buuren, A. P. Alivisatos, Structural disorder in colloidal InAs and CdSe nanocrystals observed by X-ray absorption near-edge spectroscopy, *Phys. Rev. Lett.* 83, 3474 (**1999**).

63 J.-Y. Raty, Giulia Galli, C. Bostedt, T. van Buuren, L. J. Terminello, Quantum confinement and fullerenelike surface reconstructions in nanodiamonds, *Phys. Rev. Lett.* 90, 37401 (**2003**).

64 T. van Buuren, T. Tiedje, S. N. Patitsas, W. Weydanz, Effect of thermal annealing on the conduction- and valence-band quantum shifts in porous silicon, *Phys. Rev. B* 50, 2719 (**1994**).

65 R. L. Zimmerman, D. Ila, E. K. Williams, B. Gasic, A. Elsamadicy, A. L. Evelyn, D. B. Poker, D. K. Hensley, D. J. Larkin, Gold, silver and copper nanocrystal formation in SiC by MeV implantation, *Nucl. Instrum. Meth. B* 166/167, 892 (**2000**).

66 H. Fukuzawa, H. Yuasa, S. Hashimoto, K. Koi, H. Iwasaki, M. Takagishi, Y. Tanaka, M. Sahashi, MR ratio enhancement by NOL current-confined-path structures in CPP spin valves, *IEEE, Trans. Magn.* 40, 2236 (**2004**).

67 P. L. Hansen, J. B. Wagner, S. Helveg, J. R. Rostrup-Nielsen, B. S. Clausen, H. Topse, Atom-resolved imaging of dynamic shape changes in supported copper nanocrystals, *Science* 295, 2053 (**2002**).

68 K. Nishimori, H. Tokutaka, S. Nakanishi, S. Kishida, N. Ishihara, Off-Angle SiC(0001) Surface and Cu/SiC Interface Reaction, *Jpn. J. Appl. Phys. Lett.* 28, L1345 (**1989**).

69 Z. An, A. Ohi, M. Hirai, M. Kusaka, M. Iwami, Study of the reaction at Cu/3CPSiC interface, *Surf. Sci.* 493, 182 (**2001**).

70 Z. An, M. Hirai, M. Kusaka, T. Saitoh, M. Iwami, Analysis of nanostructure formation using photon/electron spectroscopies: Cu on SiC substrates, *Jpn. J. Appl. Phys. Part 1* 40/3B, 1927 (**2001**).

71 I. Dontas, S. Ladas, S. Kennou, Study of the early stages of Cu/6HPSiC(000-1) interface formation, *Diamond Relat. Mater.* 12, 1209 (**2003**).

72 D. W. Shin, S. X. Wang, A. Marshall, W. Kimura, C. L. Dong, A. Augustsson, J.-H. Guo, Growth and characterization of copper nanoclusters embedded in SiC matrix, *Thin Solid Films* 473, 267 (**2005**).

73 H. H. Hsieh, Y. K. Chang, W. F. Pong, J. Y. Pieh, P. K. Tseng, T. K. Sham, I. Coulthard, S. J. Naftel, J. F. Lee, S. C. Chung, K. L. Tsang, Electronic structure of Ni-Cu alloys: The *d*-electron charge distribution, *Phys. Rev. B* 57, 15204 (**1998**).

74 S. Shionoya, W. M Yen, eds. *Phosphor Handbook*, (CRC Press, Boca Raton, FL, **1999**), p. 255.

75 J. Nause, S. Ganesan, B. Nemeth, V. Munne, A. Valencia, P. Keisel, H. Morkoc, D. Look, From sunscreen and diaper ointment to wide band gap optoelectronics: ZnO-based homoepitaxial device technology, *III-Vs Rev.* 12, 28 (**1999**).

76 M. W. Shin, R. J. Trew, GaN

MESFETs for high-power and high-temperature microwave applications, *Electron. Lett.* 31, 498 (**1995**).

77 F. Hamdani, A. E. Botchkarev, H. Tang, W. Kim, H. Morkoc, Effect of buffer layer and substrate surface polarity on the growth by molecular beam epitaxy of GaN and ZnO, *Appl. Phys. Lett.* 71, 3111 (**1997**).

78 T. Detchprohm, K. Hiramatsu, H. Amano, I. Akasaki, Hydride vapor phase epitaxial growth of a high quality GaN film using a ZnO buffer layer, *Appl. Phys. Lett.* 61, 2688 (**1992**).

79 R. D. Vispute, V. Talyansky, S. Choopun, R. P. Sharma, T. Venkatesan, M. He, X. Tang, J. B. Halpern, M. G. Spencer, Y. X. Li, L. G. Salamanca-Riba, A. A. Iliadis, K. A. Jones, Heteroepitaxy of ZnO on GaN and its implications for fabraication of hybrid optoelectronic devices, *Appl. Phys. Lett.* 73, 348 (**1998**).

80 A. Ohtomo, M. Kawasaki, T. Koida, H. Koinuma, Y. Sakurai, Y. Yoshida, M. Sumiya, S. Fuke, T. Yasuda, Y. Segawa, Double heterostructure based on ZnO and $Mg_xZn_{1-x}O$, *Mater. Sci. Forum* 264, 1463 (**1998**).

81 P. Sharma, A. Gupta, K. V. Rao, F. J. Owens, R. Sharma, R. Ahuja, J. M. Osorio Guillen, B. Johansson, G. A. Gehring, Ferromagnetism above room temperature in bulk and transparent thin films of Mn-doped ZnO, *Nat. Mater.* 2, 673 (**2003**).

82 Y. K. Chang, H. H. Hsieh, W. F. Pong, M.-H. Tsai, F. Z. Chien, P. K. Tseng, L. C. Chen, T. Y. Wang, K. H. Chen, D. M. Bhusari, J. R. Yang, S. T. Lin, Quantum confinement effect in diamond nanocrystals studied by X-ray-absorption spectroscopy, *Phys. Rev. Lett.* 82, 5377 (**1999**).

83 C. L. Dong, C. Persson, L. Vayssieres, A. Augustsson, T. Schmitt, M. Mattesini, R. Ahuja, C. L. Chang, J.-H. Guo, Electronic structure of nanostructured ZnO from X-ray absorption and emission spectroscopy and the local density approximation, *Phys. Rev. B* 70, 195325 (**2004**).

84 J.-H. Guo, L. Vayssieres, C. Persson, R. Ahuja, B. Johansson, J. Nordgren, Polarization-dependent soft-X-ray absorption of highly oriented ZnO microrod arrays, *J. Phys: Condens. Matter* 14, 6969 (**2002**).

85 K. Gunnelin, P. Glans, P. Skytt, J.-H. Guo, J. Nordgren, H. Ågren, Assigning X-ray absorption spectra by means of soft-X-ray emission spectroscopy, *Phys. Rev. A* 57, 864 (**1998**).

86 N. Hellgren, J.-H. Guo, C. Såthe, A. Agui, J. Nordgren, Y. Luo, H. Ågren, J.-E. Sundgren, Nitrogen bonding structure in carbon nitride thin films studied by soft X-ray spectroscopy, *Appl. Phys. Lett.* 79, 4348 (**2001**).

87 J.-H. Guo, Y. Luo, A. Augustsson, J.-E. Rubensson, C. Såthe, H. Ågren, H. Siegbahn, J. Nordgren, X-ray emission spectroscopy of hydrogen bonding and electronic structure of liquid water, *Phys. Rev. Lett.* 89, 137402 (**2002**).

88 J.-H. Guo, Y. Luo, A. Augustsson, S. Kashtanov, J.-E. Rubensson, D. Shuh, H. Ågren, J. Nordgren, The molecular structure of alcohol-water mixtures, *Phys. Rev. Lett.* 93, 157401 (**2003**).

89 A. Augustsson, S. Kashtanov, Y. Luo, J.-H. Guo, C.-L. Dong, C. L. Chang, H. Ågren, J.-E. Rubensson, J. Nordgren, Conformations and core-excitation dynamics in liquid water, in preparation (**2005**).

90 M. J. Frisch et al., Gaussian 98, 1998, Gaussian Inc., Pittsburgh PA, 1998. See http://www.gaussian.com.

91 T. Privalov, F. Gel'mukhanov, H. Ågren, Role of relaxation and time-dependent formation of X-ray spectra, *Phys. Rev. B* 64, 165 115 (**2001**).

92 T. Helgaker et al., Dalton, An ab initio electronic structure program, Release 1.0 (1997). See http://www.kjemi.uio.no/software/dalton/dalton.html.

93 D. T. Richens, *The Chemistry of Aqua Ions*. (Wiley, Chichester, **1997**).

94 W. Kaim, B. Schwederski, *Bioinorganic Chemistry: Inorganic*

Elements in the Chemistry of Life: An Introduction and Guide. (Wiley, Chichester, **1994**).

95 H. Ohtaki, T. Radnai, Structure and dynamics of hydrated ions, *Chem. Rev.* 93, 1157 (**1993**).

96 E. Braun, Y. Eichen, U. Sivan, G. Ben-Yoseph, DNA-templated assembly and electrode attachment of a conducting silver wire, *Nature (London)* 391, 775 (**1998**).

97 P. J. de Pablo, F. Moreno-Herrero, J. Colchero, J. Gómez Herrero, P. Herrero, A. M. Baró, P. Ordejón, J. M. Soler, E. Artacho, Absence of dc-Conductivity in λ-DNA, *Phys. Rev. Lett.* 85, 4992 (**2000**).

98 D. Porath, A. Bezryadin, S. de Vries, C. Dekker, Direct measurement of electrical transport through DNA molecules, *Nature (London)* 403, 635 (**2000**).

99 K.-H. Yoo, D. H. Ha, J.-O. Lee, J. W. Park, J. Kim, J. J. Kim, H.-Y. Lee, T. Kawai, H. Y. Choi, Electrical conduction through Poly(dA)-Poly(dT) and Poly(dG)-Poly(dC) DNA molecules, *Phys. Rev. Lett.* 87, 198 102 (**2001**).

100 H. Fink, C. Schönenberger, Electrical conduction through DNA molecules, *Nature (London)* 398, 407 (**1999**).

101 A. Yu. Kasumov, M. Kociak, S. Guéron, B. Reulet, V. T. Volkov, D. V. Klinov, H. Bouchiat, Proximity-induced superconductivity in DNA, *Science* 291, 280 (**2001**).

102 H. S. Kato, M. Furukawa, M. Kawai, M. Taniguchi, T. Kawai, T. Hatsui, N. Kosugi, Electronic structure of bases in DNA duplexes characterized by resonant photoemission spectroscopy near the Fermi level, Phys. *Rev. Lett.* 93, 86403 (**2004**).

103 J.-H. Guo, S. Kashtanov, J. Söderström, H. Cheng, P.-A. Glans, T. Learmonth, K. Smith, J. Nordgren, Y. Luo, The electronic structure of Bases in DNA Duplexes studied by soft-X-ray absorption and emission spectroscopy, in preparation (**2005**).

104 N. Hellgren, J. Guo, Y. Luo, C. Såthe, A. A. S. Kashtanov, J. Nordgren, H. Ågren, J.-E. Sundgren, Electronic structure of carbon nitride thin films studied by X-ray spectroscopy techniques, *Thin Solid Films* 471, 19 (**2005**).

9
Some New Advances and Challenges in Biological and Biomedical Materials Characterization

Filip Braet, Lilian Soon, Thomas F. Kelly, David J. Larson, and Simon P. Ringer

9.1
Introduction

The three sections of this chapter identify new and recent advances as well as challenges in the microscopy of selected biological and biomedical materials using (1) atom probe tomography, (2) atomic force microscopy and (3) cryo-transmission electron microscopy.

Section 9.2 describes the fundamental principles of atom probe tomography, a time-of-flight mass spectroscopy method that exploits the quantum-mechanical tunneling phenomenon that occurs when many conducting solids are raised to very high electric fields. The resultant ionization associated with the tunneling liberates an ion with a specific mass-to-charge ratio that is subsequently accelerated towards a position-sensing detector. Time-of-flight data is transformed into mass-to-charge ratios in a way that provides identification of the individual isotopes of the elements, as well as molecular species. This section also provides the results of recent applications of atom probe to biomedical materials that are used as human-prosthetic devices and components. These applications demonstrate the core competency of the atom probe technique in analyzing three-dimensional (3D) atomic-level compositional information of materials. Probing interfaces using compositional profiles, identifying the presence, or otherwise, of chemical heterogeneities within phases and measuring the local composition near and within nanoscale precipitate phases are key areas where this technique can contribute to the science and technology of biomedical materials. Applications data are also presented for biological materials after a brief review of work to date. This section concludes with some of the challenges facing the field in terms of sample preparation issues, specimen stability issues under high fields and the challenge of interpreting mass spectra from complex molecules.

Sections 9.3 and 9.4 provide a correlative microscopy context for fenestral studies in the hepatic endothelial cell model. Section 9.3 focuses on atomic force methods and Section 9.4 focuses on electron microscopy approaches and data. The result is a new insight into the morpho-function and structure of the liver sieve.

Nanotechnologies for the Life Sciences Vol. 3
Nanosystem Characterization Tools in the Life Sciences. Edited by Challa S. S. R. Kumar

ISBN: 3-527-31383-4

9.2 Modern Atom Probe Tomography: Principles, Applications in Biomaterials and Potential Applications for Biology

9.2.1 The Need for an Ideal Microscope

The length scale at which human knowledge and endeavor operates is directly linked to the scale of our microscopies. Though people have long been able to conjecture about the nature and significance of length scales beyond our ability to observe, be they microscopical or astronomical, practical technologies always require observations and measurements that can transform such conjecture into fact. Examples include the postulation of the very concept of the atom [1], the potential for a double helix DNA structure [2] and the notion that crystal defects (dislocations) were the origin of the discrepancy between the observed and theoretical strength of metals and alloys [3]. As we enter the century of nanotechnology, the need for the "ideal" microscope will grow because the critical dimensions of materials and devices trend towards the truly atomic. Modern atom probe microscopy is coming closer to this ideal than any other microscopy. In this section, the modern instrument is reviewed and some of the significant challenges and potential opportunities of the technique and its application in both biomaterials and in biological science are discussed.

9.2.1.1 Field Ion Microscopy and the Modern Atom Probe Instrument

As introduced above, Greek philosophers were, interestingly, able to hypothesize the existence of atoms, but it took 25 centuries before they were first observed by humans with, in fact, a field ion microscope (FIM) pioneered by Erwin Müller [4]. The technique is mentioned here because of its potential significance in imaging molecules and the close relation to the atom probe technique. Miller et al. [5] provide a detailed explanation of the FIM, and Panitz [6] has provided a detailed review of the technique as applied to molecules. Essentially, a high positive potential is applied as a standing voltage to a cryogenically cooled specimen-tip in a controlled ultra-high vacuum (UHV) chamber. An inert gas such as He, Ne or Ar is leaked into the chamber to fill a back pressure of $\sim 10^{-3}$ Pa and, in the presence of the applied field, the gas atoms are field polarized and so are gradually drawn towards the tip. The gas atoms eventually make contact with the specimen tip and migrate, though a series of collisions towards positions of low potential energy. Once these gas atoms have lost most of their kinetic energy and are trapped by the very high field of the specimen, they may be field ionized by electron tunneling processes. Such ions are repelled from the tip and projected in an almost radial direction towards a fluorescent screen. Field ionization occurs preferentially at the atomic steps and terrace positions and so forms a stereographic-like projection image of the material surface with atomic resolution. A tungsten (W) FIM image is

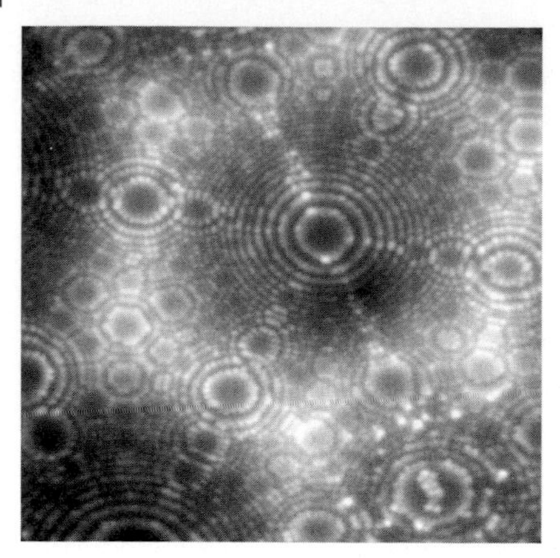

Fig. 9.1. Field ion micrograph of W atoms imaged using Ne. Each bright dot is an individual atom.

provided in Fig. 9.1, where each bright dot corresponds to an individual atom. The concentric circles around low-index crystallographic poles correspond to the outer atoms of atomic planes, one atop another.

Figure 9.2(a) is a simplified schematic of the atom probe principle, showing how the application of a positive high voltage pulse, V_{pulse}, to a sharp needle-shaped specimen under a standing positive voltage, V_{dc}, can raise the local field on the sample so as to render some of the surface atoms as ions, through the quantum mechanical process of electron tunneling. This field-induced ionization of the surface atoms results in their "evaporation" from the specimen along trajectories close to normal to the tangent to the sample surface. These ions eventually strike a single-ion detector and this *stop* signal is compared to the *start* of the initial pulse to generate an ion flight-time. The mass-to-charge (m/n) ratio of the ion is determined by equating the potential energy of the ion just prior to field evaporation to the kinetic energy just after, such that

$$neV = \frac{1}{2}mv^2$$

Fig. 9.2. (a) Atom probe principle: individual atoms are field ionized from the sample surface and accelerated towards a detector. Time-of-flight mass spectroscopy is used to identify the atom species and this is correlated to spatial orientation (*x*, *y*) by the position-sensitive detector. Tomographic data is generated when the 2D image slices are correctly reconstructed along *z*. (b) Atom probe tomography from a spinodal alloy, revealing a spinodal interface with separate Fe- and Ni-rich regions [7, 8]. Isoconcentrational surfaces for 10 at% Ni are charted in the upper 3D atom map (30 × 30 × 25 nm). Blue = Mn, red = Fe, yellow = Ni and teal = Al.

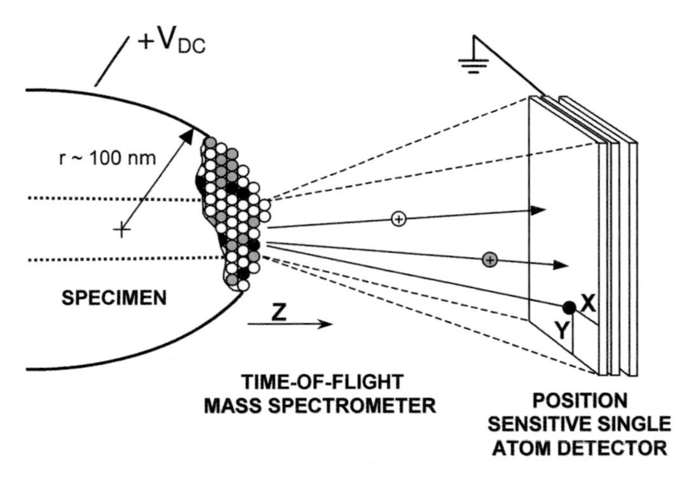
$+V_{DC}$
r ~ 100 nm
SPECIMEN
Z
X
Y
TIME-OF-FLIGHT
MASS SPECTROMETER
POSITION
SENSITIVE SINGLE
ATOM DETECTOR
a

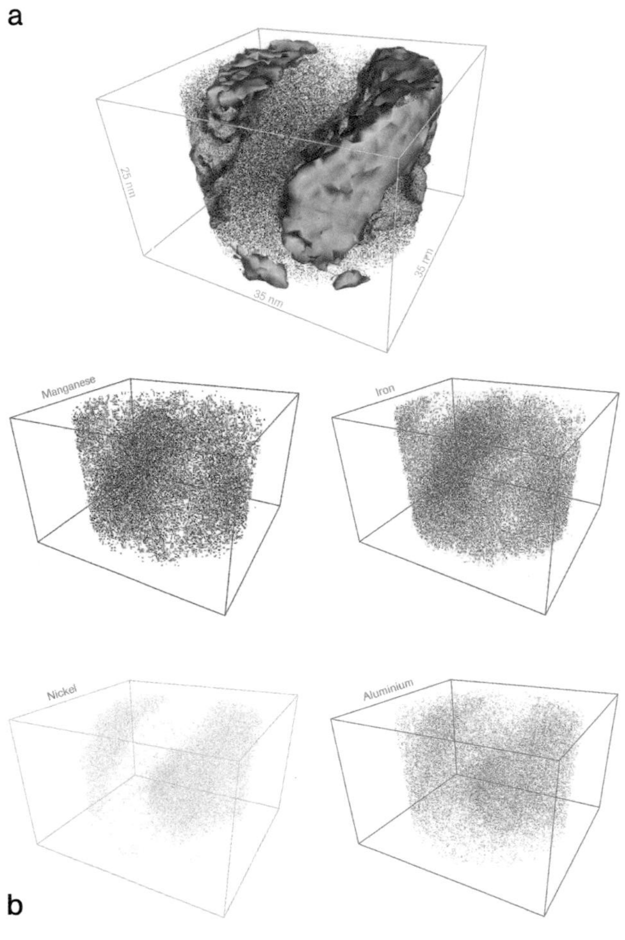
25 nm
35 nm
35 nm
Manganese
Iron
Nickel
Aluminium
b

and so

$$\frac{m}{n} = KV\frac{t^2}{d^2} \tag{1}$$

where n is the number of electrons removed in the field ionization, e is the elementary charge, V is the total electric field on the specimen, m and v are the mass and velocity of the ion, respectively, d is the flight distance from the specimen to the detector and K is a constant. The identity of the ion is then correlated spatially to the hit-position (x, y) on the detector and the depth z within the sample. Figure 9.2(b) is an example of 3D atom probe tomographic data. Within this analysis volume is a spinodal interface decorated with solute atoms of Fe, Ni, Mn and Al, revealing nanoscale phase separation into the Fe-rich and Ni-rich phases. In this way, researchers can view a tomographic reconstruction of materials, since 3D images of the internal structure are generated by the reconstruction along z of thousands of 2D x–y slices. This approach to nanostructural analysis has taken microscopy and materials characterization into a new era and is opening up new directions for materials research and development [5–9]. Excellent texts on the workings, instrumentation, historical development and applications of conventional 1D and 3D atom probe techniques are available [5, 6, 9].

The current state-of-the art of atom probe employs a local electrode ahead of the specimen to mediate extraction of ions (Fig. 9.3). Figure 9.3(a) is drawn so as to emphasize that the sample may be an individual needle or tip or may be composed of an array of many tips (Fig. 9.3b). Such arrays have the advantage that they can present geometrically similar tips to the local electrode and so effectively normalize many of the stochastic field evaporation issues. Arrays will also allow statistical evaluations of structures in the same way that cryo-electron microscopy enables single-particle image analysis, via thousands of images of randomly-oriented individual protein molecules. Figure 9.3(b) is a schematic that emphasizes this approach and conveys the importance of the proximity or "local" effects of the counter electrode with the specimen [10–12]: hence the name, local electrode atom probe or LEAP®. Here, the cryogenically cooled needle-shaped specimen, Fig. 9.3(c) – or planar array specimen (Fig. 9.3d) – is mounted on a nanopositioning stage and pointed towards a funnel-shaped local electrode. The specimen is aligned to the aperture of the local electrode with the aid of a pair of orthogonal long-range optical microscopes. The initially detected field evaporated ions are used to finalize the alignment.

For a specimen positioned close to a local electrode aperture of ~20–30 μm in diameter, the field at the specimen apex is very high even at relatively low voltages. It is viable to build suitable pulsers (2 ns fwhm pulse with amplitudes of up to 2000 V) that can operate at pulse repetition rates of up to 200 kHz. This rate is two orders of magnitude faster than previous instruments. In addition, the ions that are field evaporated from the specimen reach the local electrode in a significantly shorter time (tens of picoseconds) and thereafter are shielded from time-varying fields as the V_{pulse} decays. This eliminates the major source of energy defi-

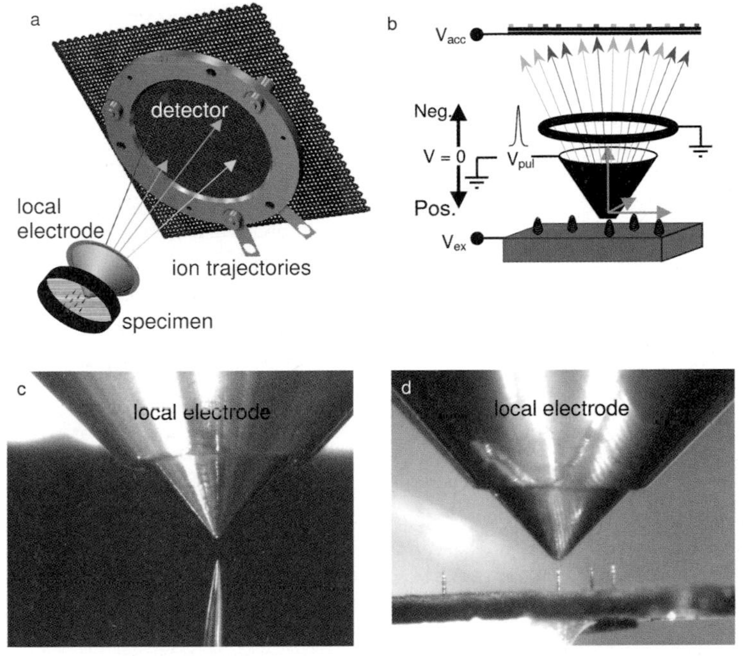

Fig. 9.3. (a) LEAP™ principle: individual atoms are field ionized from the sample surface and accelerated towards a detector. (b) The local electrode enhances the field on the specimen, permitting lower applied voltages V_{ex} and V_{pulse}, which allow a much higher pulsing frequency. In addition, having a small V_{ex} and a large V_{accel} effects an acceleration of the ions through the local electrode that provides for highly sensitive mass separation (high mass resolution time-of-flight mass spectroscopy). The time-of-flight mass spectroscopy is used to identify the atom species and this is correlated to spatial orientation (x, y) by the position-sensitive detector. Tomographic data is generated when the 2D image slices are correctly reconstructed along z. (c) Close-up optical image of a sharpened tip in the vicinity of the local electrode. (d) Demonstration application of an array of tips.

cits. Therefore, the local electrode atom probe can achieve high mass resolution (better than $\Delta m/m$ of $1/500$) over a 1.5 steradian field of view. This mass resolution is sufficient to separate the individual isotopes of all the elements. The wide field of view permits up to $\sim 10^8$ atoms to be collected routinely from a specimen.

To position correctly ions at these higher pulse rates, a novel crossed delay line (CDL) detector is used. The CDL detector features a pair of delay lines, placed perpendicularly, in place of the phosphor screen or anodes used in the other variants of 3DAP. When electrons from the microchannel plate strike the delay line at some location, they travel to each end of the line. The position of the ion's impact is therefore determined from the arrival times measured at each end of the delay line. These times are converted into true x and y distances on the detector. A crossed delay line detector and high speed digital timing system that provides

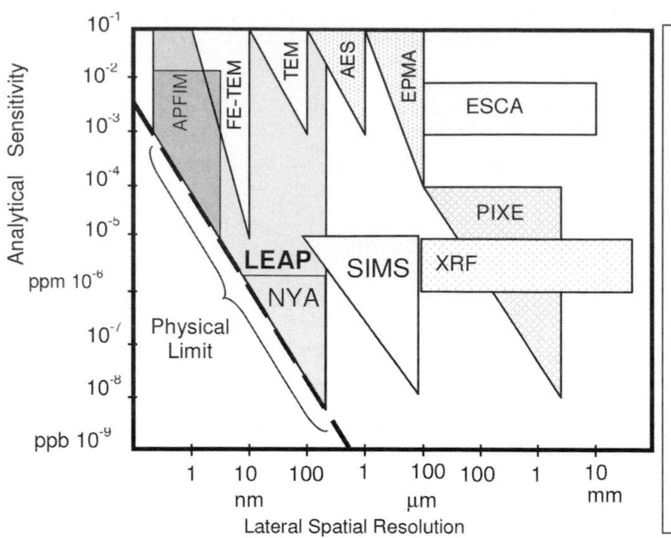

LEGEND
Analytical Sensitivity measured in terms of minimum detectable mass fraction.

LEAP: Local electrode atom probe
APFIM: Conventional 3-D atom probe field ion microscopy
FE-TEM: Field emission transmission electron microscopy
TEM: transmission electron microscopy
AES: Auger electron spectroscopy
EPMA: Electron probe microanalysis
ESCA: Electron spectroscopic chemical analysis (X-ray Photoelectron Spectroscopy)
PIXE: Proton induced X-ray emission
XRF: X-ray fluorescence
SIMS: Secondary ions mass spectroscopy
NYA: Not yet available

Fig. 9.4. Potential for a modern atom probe (LEAP) to be combined with other analytical techniques so as to probe the structure and, in particular, chemical composition of materials.

600 × 600 pixels at rates up to 25 000 ions per second has been developed by Imago.

The new features – local electrode, high speed pulser and high speed single atom sensitive CDL detector – are incorporated in a commercial instrument [12] that achieves the high-performance gains originally envisioned for this geometry. The Imago LEAP® microscope can collect data at sustained rates of up to 20 000 correctly positioned ions per second (one million ions per minute). These developments have dramatically improved the overall performance of atom probe tomography for a wide range of materials. In particular, the characterization of planar specimens in addition to traditional needle-shaped specimens is now practical [13].

Figure 9.4 is a survey of a range of analytical microscopy techniques and shows the analytical capacity of the modern atom probe in the context of related techniques. The strengths of the LEAP are clear and the potential for solving problems over various length-scales by a correlative microscopy approach, such as is advocated in this chapter, is very high. There are interesting opportunities and significant challenges for those working in biological science and technology and some of these will now be introduced.

9.2.1.2 **Applications in Biomaterials**

As other chapters in this volume affirm, the term "biomaterials" is very broad. One important class of biomaterials is those used as implants that replace or augment components of the human body such as tissue, organs, skeletal or the like. The most common examples include the use of medical grade austenitic 316 stainless steels, which are used as implanted spinal fixation devices, bone screws, cardiovascular and neurological stents, and as critical components of minimally invasive surgical devices. These applications are made possible due to suitable physical and mechanical properties, good corrosion resistance in biological environments, reasonable biocompatibility, and ease of manufacture [14].

These steels are thermomechanically processed so as to ensure the thermodynamic stability of the austenite (γ) phase at high temperatures and to suppress the formation of martensite at lower (near ambient) temperatures. The typical composition of these alloys is (wt%): Fe, <0.03 C, 16–18.5 Cr, 10–14 Ni, 2–3 Mo, <2 Mn, <1 Si, <0.045 P, <0.03 S and they possess yield strengths of at least 170 MPa. Elements that stabilize the γ-phase include Ni, Mn and N, usually in some combination [15, 16]. The relatively common allergy affects attributed to Ni and the potential for carcinogenic effects associated with this element have driven the development of Ni-free 316 grade stainless steel. In more general terms, the conventional theory of Cr-oxide passivation applies to these steels and is the main protection mechanism that results in their high corrosion resistance. However, other corrosion processes, such as crevice corrosion and pitting, can be very damaging in biomaterials applications. Therefore, much of the design of composition and thermomechanical processing for these alloys aims to mitigate or overcome these damage mechanisms by minimizing precipitation of a second phase that is electrochemically active with respect to the γ-phase. This requires suppression of precipitation of carbides, nitrides, or even body-centered-cubic α or δ ferrite. This is effected by lowering the amount of C and N that remains in the γ-phase. This, in turn, can be achieved by alloying with strong carbo-nitride forming elements such as Ti and Nb, so that these elements preferentially scavenge the interstitials and form highly stable, electrochemically inactive carbo-nitride M (CN) precipitates. This approach is particularly effective in avoiding the deleterious process of sensitization where Cr-based carbo-nitrides nucleate heterogeneously on γ/γ grain boundaries and so deplete the capacity of Cr to protect the surface of the component (resulting in severe local attack). Therefore, the electrochemical activity of the grain boundaries must be kept low and it is essential to minimize any localized changes in alloy chemistry that can produce electrochemical interaction with body fluids.

This is a formidable materials requirement and effectively requires the control of nanostructure and composition of the material. For example, the surface of such alloys will be especially critical since biological responses and corrosion occur at the material–environment interface. In addition, precise control of the chemistry and structure of buried interfaces such as grain boundaries and other chemical heterogeneities is critically important. Such characterization requires microscopy in three-dimensions (3D) at high resolution: reference to the foregoing discussion and to Fig. 9.2(b) reveals that the modern atom probe can provide key insights into these issues.

Figure 9.5(a–d) presents recent research on both the nanostructure of surfaces and buried interfaces in medical device 316L stainless steel. Figure 9.5(a) and (b) show a 3D atom map reconstruction of a 316L grade, chemically modified at Med-

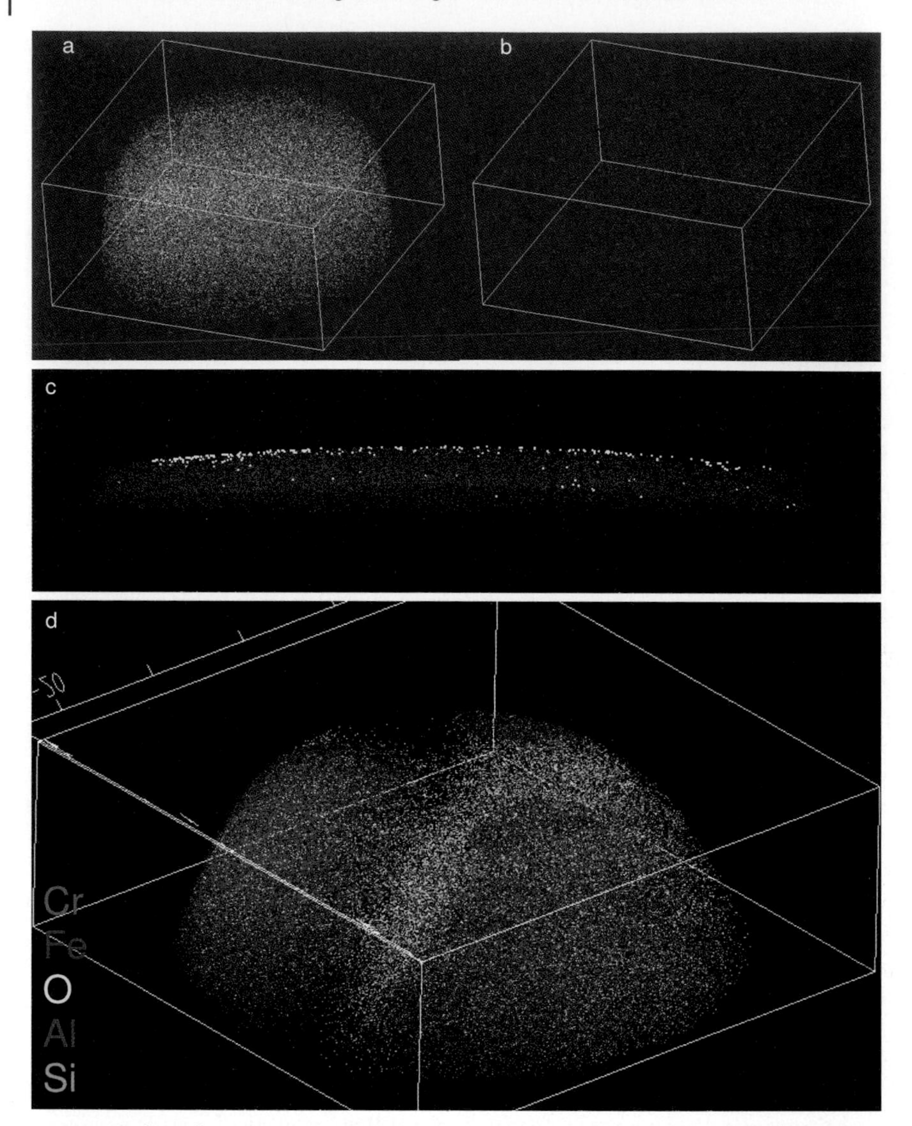

Fig. 9.5. (a, b) Bulk structure of a Medtronic 316L specimen. Each color-coded dot shows the location of an individual atom. Cr is in yellow, Ni in light blue, Si in blue and Cu in red. Image is 20 nm wide and 10 nm high. (c) Near-surface structure of a 316L wire specimen; Cr in red, oxygen in gold. The image is about 25 nm wide. (d) Region that is thought to be a grain boundary triple point in a 316L wire specimen, showing clear enrichment of O.

tronic [14]. To determine the bulk structure of the Medtronic sheet specimens, segments were cut from the interior volume and polished. These were then mounted in holders, electropolished with 10% perchloric acid in acetic acid at 8–20 V DC to a nominal end radius of 20–50 nm, to provide the necessary needle-shape for atom probe analysis. The bulk atomic composition and structure of this material was then studied with a LEAP microscope. All anticipated elements were located in the specimen, including Fe, Cr, Ni, C, P, Si, Mo, and Cu, with a uniform elemental distribution and without precipitates or grain boundaries in the analyzed regions. This is highly significant because of the relation between biocompatibility, related performance as a biomaterial and the uniformity of the solid solution γ-phase. Figure 9.5(c) provides 3D atom map reconstructions from a commercial 316L grade procured as wire 0.38 mm in diameter, 1/8 hard, from a commodity supplier. The commodity 316L wires were electropolished as described above. During electropolishing and subsequent air exposure, a thin oxide naturally forms on the surface of the sharpened wire. This oxide was used to develop specimen preparation methods for analysis of oxides on medical-grade 316L samples. Notably, there is relatively little atom probe work examining metal oxide surfaces with atom probe microscopes. To prepare the surface oxide for LEAP analysis, a ~15 nm thick capping layer of Ni was applied to the sharpened wire specimens by argon (Ar) ion sputtering to protect the surface during the first stages of atom probe analysis. Analyses demonstrated that such coatings provided some preservation of the oxide layer, enabling the imaging of the oxide–stainless steel interface, Fig. 9.5(c). Note the presence of oxygen (O) at the stainless steel surface (for clarity, Ni atoms species are not shown in this projection). Although the volume and total number of atoms in this image are limited, there is a clear oxygen-rich layer on the specimen surface.

Significantly, this data gives access to the body fluid/oxide/γ-phase interface chemistry, since it is the first 20–50 nm of the biomaterial (in this case 316L stainless steel) over which ionic mass transport occurs and is almost always mediated via an oxide layer. An understanding of surface roughness, oxide thickness and elemental compositional profiles near these oxides is essential for thorough characterization of biomaterials interfaces. In Fig. 9.5(c), the oxide is seen to be ~1 nm thick, and [O] atoms are also observed deeper into the γ-phase. Roughness indices on both sides of the oxide interface can be developed, as can concentration profiles.

Figure 9.5(d) provides a 3D atom map reconstruction from the commodity 316L wire-stock, revealing a significant oxide-based chemical heterogeneity within the analyzed volume, which is thought to be a grain boundary. Such chemical heterogeneities at the grain boundaries of biomaterials are very serious for reasons explained above, since they have the potential to increase the potency of these sites for heterogeneous nucleation, leading ultimately to sensitization of the steel. Moreover, depending on the proximity of these O/Cr/Si enriched boundaries to the body fluids, there is the risk of developing a localized electro-active region that leaches toxic elements into the body and stimulates a corrosion reaction. The regions of 316L such as presented in Fig. 9.5(d) are inferior to those such as in Fig. 9.5(a) and (b), where the degree of chemical homogeneity is much higher.

9.2.1.3 **Applications and Challenges for Biological Science**

It is almost 50 years since Erwin Müller first observed the flicker of individual atoms on a phosphor (P) screen from field ion microscopy, along with his work shortly after in pioneering the atom probe to select individual atoms from the

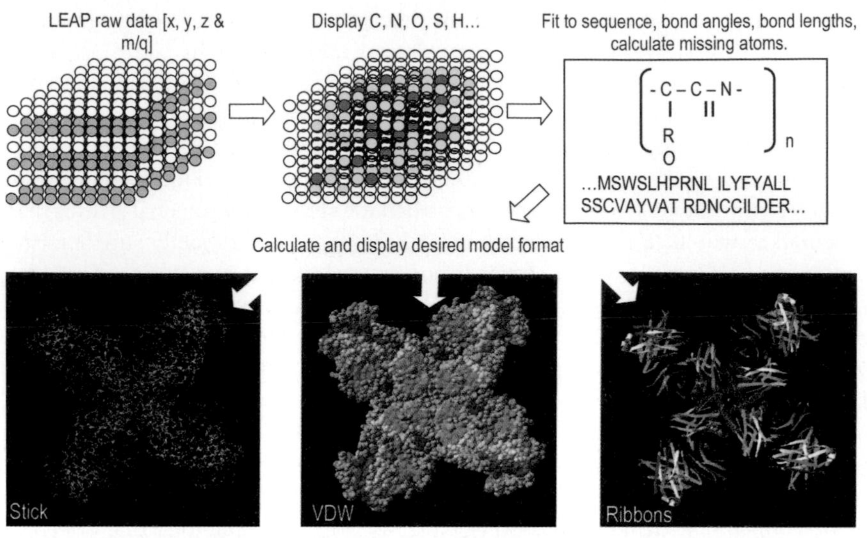

Fig. 9.6. Idealized atom probe experiment of biological material. The LEAP acquires positional (*x*, *y* & *z*) and atom identity data, via *m*/*q*. Specific individual atoms are identified and displayed, and an informatics-like algorithm would fit and identify the molecular structure, which could be displayed in the appropriate format.

FIM image and analyze them [17]. Today, we see the core-competence of the atom probe in analyzing multi-million atom data-sets and representing a premium high-resolution analytical technique for materials metrology and characterization. The examples provided in Figs. 9.2 and 9.5 demonstrate this.

Having discussed the capacity of the atom probe to analyze individual atoms in 3D in the inorganic context of biomaterials, it is appropriate to also examine the capacity of this instrument to solve complex problems in organic biological materials. From the preceding discussion on the analysis of alloy biomaterials, where individual atoms can be mapped in 3D, lattice-plane by lattice-plane, the biologist might be tempted to view the instrument as something akin to an atom-level confocal microscope. The possibility of reconstructing molecular conformation in 3D, such as is performed, albeit with some experimental difficulty using cryo-electron microscopy (see, e.g., Ref. [18]), is highly attractive. Questions related to the structure of proteins and macromolecular complexes, proteins at interfaces, virus structures, ligand binding, protein membrane structure and so on represent seminal questions in modern structural biology. The concept here is outlined in Fig. 9.6. In this experiment, individual light elements are displayed and mathematical filters applied to the data that display selected atoms (C, N, O, S, H etc.) with specific coordination in terms of angles, density or simple proximity [19]. These could then be fitted to the molecular sequence and ultimately display complete biological structures [20]. This approach is not unlike the computational methods applied to X-ray synchrotron and cryo-electron microscopy data and, in particular, to the computational data-fitting that has recently become so successful in nuclear magnetic

Tab. 9.1. Biological specimen preparation methods and relevance to atom probe.

Method	*Comment for atom probe*
Chemical fixation	Not suitable for all biological samples
Dry specimens	Use of low-tension solvents, freeze drying
Embedded and crosslinked samples	Offer possibilities for stabilizing molecules under the electric field
Provide electrical conductivity	Metal coatings, OsO_4, RuO_4 or conductive embedding agents can assist the molecules to field ionize via electrically pulsed atom probe

resonance (NMR). However, there are access, sample preparation and cost issues with these types of characterization, which require significant equipment installations, and the experiments are non-trivial. The idealized atom probe experiment described above requires the following research and development in at least three specific areas.

Sample Fabrication Issues Biomolecules need to be stabilized for imaging under high vacuum conditions, since these conditions are far from their natural state. Fortunately, electron microscopists have developed numerous technologies to address this issue, many of which seem relevant here. Indeed, such has been the recent progress in biological specimen preparation that this is no longer the limiting factor for achieving high resolution. In fact, destructive electron-beam specimen interactions are more limiting.

Table 9.1 lists various biological specimen preparation methods and mentions some of the issues that require attention in assessing these methods for atom probe. Certainly, with respect to stabilizing the specimens for UHV, there seem to be excellent prospects of adapting a method for atom probe tips or tip arrays.

Stability of the Specimen under High Electric Fields The response of biomolecules to the very high applied electric fields that occur with an atom probe is one of the most complex and problematic aspects of these experiments. However, Panitz has examined Poly(GC) DNA using FIM [21] and Machlin has published FIM images that are reported to come from tRNA [22]. Other work has appeared on synthetic polymers, including polypyrrole [23] and octacyanophthalocyanine metal complexes [24]. These pioneering experiments have shown that the stability of the molecules under the field can be resolution-limiting. Figure 9.7(a) provides schematic images that describe the preparation of a specimen of human immunoglobin G via FIM: a W tip is Au-coated before application of the IgG. This was chemically fixed and examined in the FIM (Fig. 9.7b shows the results). Individual bright dots are seen in the images that may correlate to individual molecules or groups thereof. Progressive evaporation of the sample revealed systematic changes in the FIM image until, eventually, the IgG material was all evaporated and the substrate Au became visible (Fig. 9.7b, final two images). These FIM images are

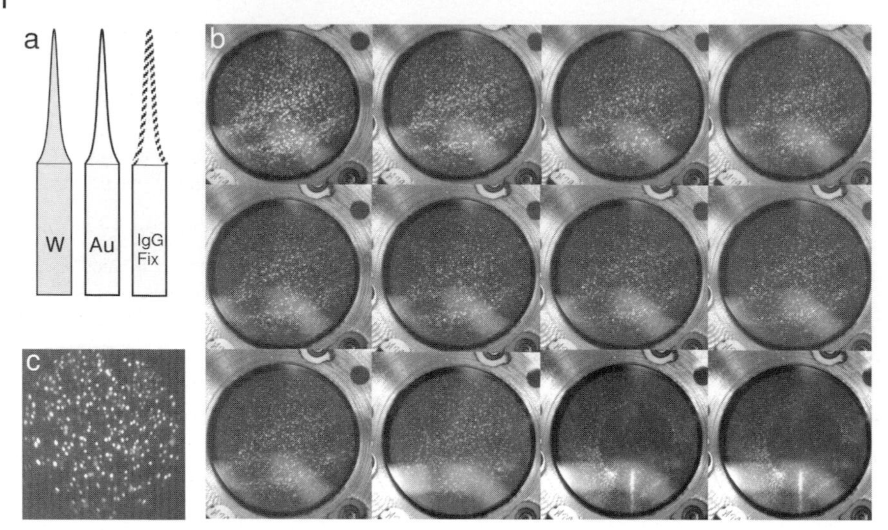

Fig. 9.7. (a) Schematic showing the three stages of specimen preparation for a human IgG specimen: a polished W tip is sputter-coated with a few nm of Au, offering excellent bonding for the IgG-fixed molecule. (b) Series of FIM images recorded with Ne imaging gas. The series show self-consistent and uniform contrast: progressive evaporation of what are thought to be the biomolecules continues until the substrate Au is revealed. (c) Ne FIM image of amorphized alloy. The uniform contrast of amorphous alloys provides similar contrast.

very similar to the image (Fig. 9.7c) to a Cu-Co multi-layer film device amorphized by a Ga focused ion beam [25]. In this respect, amorphous inorganic materials represent closer analogies to organic molecular samples than their crystalline counter-parts and the so comparison is of interest. Whereas the bright contrast in the image in Fig. 9.7(c) arises, almost certainly, from individual atoms on the alloy surface, a contrast theory for the series of images in Fig. 9.7(b) is not available: bright spots may be individual atoms or could be particular functional groups or the domains thereof. There are numerous other artifacts that could, in principle, also explain the contrast observed in these early experiments. Nevertheless, the systematic character of the images suggests that there is some local equilibrium at the surface of these tips and that we are observing molecular species. It is unlikely that the images are from Au metal or any oxide thereof. The challenge is to reproduce such images and to understand them in terms of the primitive molecular structure of the specimen. The recent discussion by Panitz [26] also supports the notion that stable images may be formed from biomolecules in FIM. As powerful as the atom probe technique is, it is suggested that a significantly better understanding of FIM images such as those provided in Fig. 9.7 is required and that stable FIM imaging may be a necessary precursor to sound atom probe analysis.

Mode of Field Evaporation: Notion of the Molecular Probe As described earlier, atom probe experiments usually involve the application of energy pulses to a sample under an applied standing voltage so that the surface atoms are field evaporated as ions that can be identified by time-of-flight mass spectroscopy. The energy is usually transmitted via a high voltage pulse although it can be transmitted via a laser to effect, e.g., a thermally-induced evaporation process. In any case, there are at least three modalities by which one can imagine the field evaporation of a biomolecule may occur.

Type I field evaporation involves, predominantly, the liberation of individual atoms from the molecule surface. In this case, the mass spectra are closely similar to those obtained for inorganic materials.

Type II field evaporation involves, predominantly, the liberation of molecular fragments. An enormous range of molecular bonds may be sheared from the sample, providing C, C–C, C–C–C, C–H, C–H–H etc. In this case, the mass spectra will be very complex. Clearly, a major challenge in developing atom probe science to solve problems in biology is to provide a system for understanding the mass spectra.

Type III field evaporation involves predominantly the liberation of rational molecular fragments, as recognizable functional groups. In this scenario, an enormous range of molecular bonds may also be sheared from the sample. However, conceivably, the proposed *molecular probe* could apply look-up tables of known molecular species that correspond to specific mass-charge ratios, in the same way that *atom probe* uses the charge state of the isotope abundances to develop range files that window signals from the individual elements in the specimen.

Clearly, the exact effect of the energy pulse on biomolecules is a fertile area for new research. Such efforts will need to use both modeling and experiments and also utilize the rapidly developing knowledge-base arising from the success of other spectroscopic techniques, particularly SIMS and nanoSIMS, XPS and MALDI (matrix-assisted laser desorption/ionization). Figure 9.8 provides one representation of these techniques, where the different modalities of signal generation are emphasized. Whereas the high-energy ion beam in SIMS generates variable fragments of the initial molecular specimen (closer to type II field evaporation), XPS causes an emission of characteristics electrons, and MALDI causes ionization of whole molecules (closer to type III). We expect an atom probe to cause field evaporation across all three types listed above.

Figure 9.9 is the mass spectrum for a C_{18} self-assembled-monolayer (SAM). Here, the raw number of counts of detector-events is plotted as a function of m/q. The mass spectrum is "indexed" by attributing the known or expected molecular fragments to the significant peaks. In the same way that we have presented indicative early-stage FIM data in Fig. 9.7 from quite new experiments that demonstrate the challenges in comprehensive interpretation of FIM images, Fig. 9.9 reveals the opportunities and challenges for molecular probing of these sorts of samples. An encouraging aspect of Fig. 9.9 is the clear success of the instrument in probing the molecular species of the specimen, since discrete peaks of organic matter are identified in the mass spectra, as opposed to Au peaks etc. Prima facie, most of the peaks can be plausibly identified and labels are provided. Moreover, the data can be reconstructed in 3D (Fig. 9.9c). Here, the outer surface is displayed in the 3D

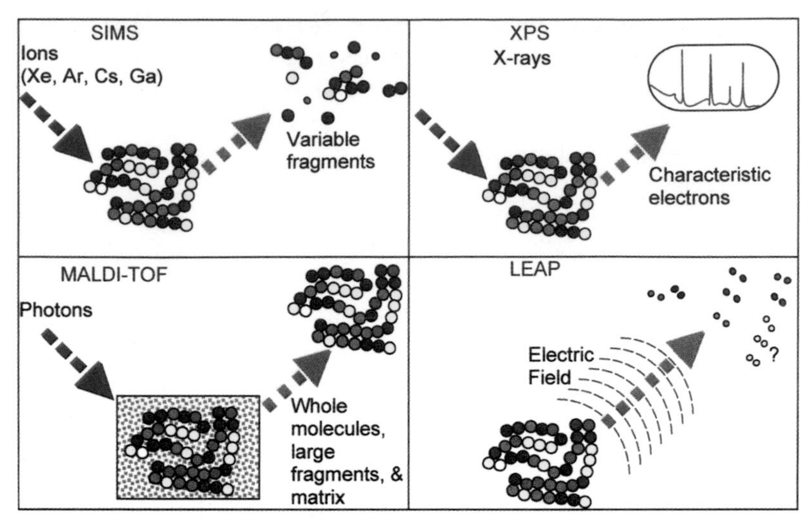

Fig. 9.8. Schematic representation of signal modalities from a selection of molecular analysis techniques. The SIMS technique is a mixture of type II and type III evaporation, liberating variable fragments of the original molecule. XPS uses characteristic electrons. MALDI (matrix-assisted laser desorption/ionization) TOF is capable of ionizing whole molecules.

reconstruction-cube, and the species are clearly located at the surface. As a method of probing biointerfaces, this approach seems promising. However, a key uncertainty in such data, which will require unequivocal resolution, is the specific identity of mass spectra peaks. In Fig. 9.9, for example, we have left a significant peak at $m/q = 22$ unidentified. Also, the significance of the particular type of molecular species that occur in the mass spectra is uncertain, since certain species clearly evaporate more readily than others. Another important question relates to the effect of molecular size and asymmetric charge distribution on trajectory aberrations. Since the ultimate molecular properties depend on coordination, orientation and bond angles, it is strongly desired that these characteristics are preserved when the molecules hit the position-sensing detector. Nevertheless, the results of initial experimentation with these ideas using the LEAP, taken together with other pioneering work attempting FIM and atom probe of organics referred to here, suggests that, under certain circumstances, biomolecules may be imaged and analyzed using these techniques. Clearly, much experimentation and development remains.

Future Developments Figure 9.10 reveals the common range of bond energies, comparing ionic, covalent and the dispersion bond forces. One of the intriguing characteristics of biomolecular atom probe specimens is the way that they can, potentially, present an enormously diverse range of bond energies to the instrument. The fact that the field evaporation process must occur across such a range of interatomic or inter-molecular forces would seem to imply that a substantial range of fragmentation will occur in response to the pulse. Figure 9.10 also lists some of

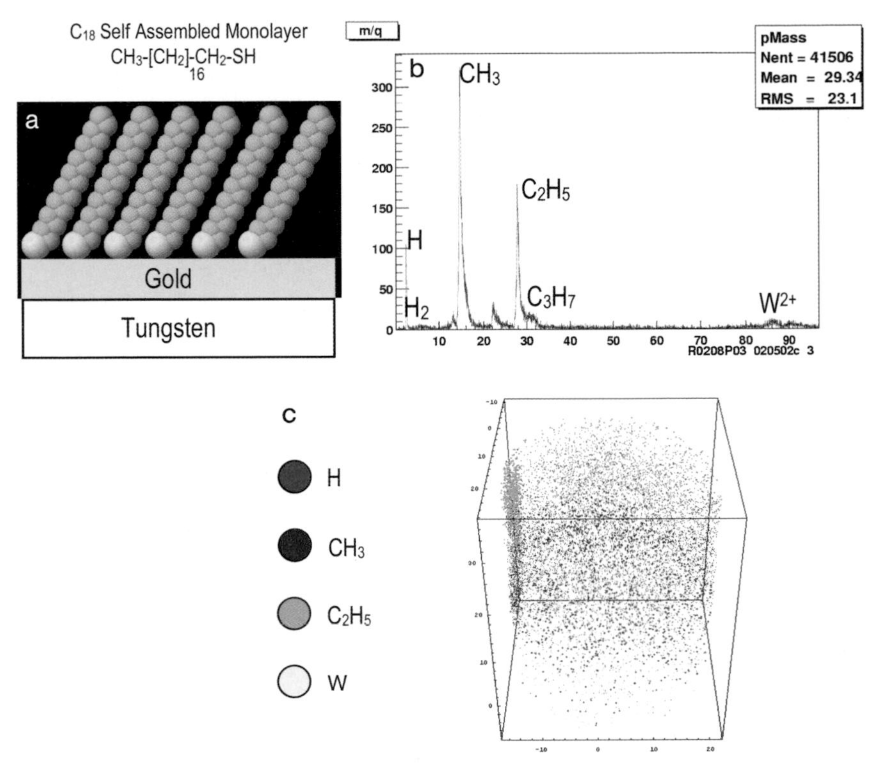

Fig. 9.9. (a) Schematic of a C_{18} self-assembled monolayer chemisorbed on an Au-coated W-tip via the thiol headgroup. (b) A LEAP mass spectrum. (c) 3D atom map reconstructed from the mass spectrum in (b).

the key interatomic bond energies that are so prevalent in biomolecular materials. As shown, the probe pulse must rupture significant interatomic bond energies before individual atoms are ionized and detected as single events on the detector. A theme of developments in this area will be to mitigate the effects of the variety of bond-energies through specimen preparation strategies.

9.3 Atomic Force Microscopy

9.3.1 Introduction

In 1986, Binnig et al. [28] revolutionized microscopy through the invention of the atomic force microscope (AFM). Marketable instruments of this new imaging technique began to appear in the five years following its discovery. In the early 1990s,

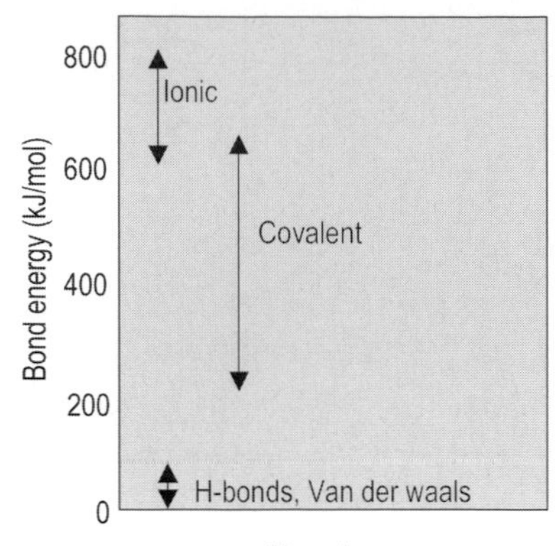

Bond	Bond Energy (kJ/mol)	Bond	Bond Energy (kJ/mol)
C-C	348	S-S	214
C-H	414	P-O	419
C-O	352	P=O	502
C=O	712	S-Au	184
C-N	293	Si-O	368
C-S	260	Si-Si	176
O-H	465	C-Si	289

Fig. 9.10. Range of bond energies in organic and inorganic materials, together with a table summarizing the more significant interatomic bond energies in organic materials such as biomolecules [27].

Henderson et al. [29] and Radmacher et al. [30] both illustrated the potential of the instrument to image biological preparations in real time under near physiological conditions with nanometer resolution. From then onwards the AFM has fascinated biologists and the number of publications describing biological applications of AFM has grown swiftly [31].

Atomic force microscopy is becoming a valuable tool for determining biological structure and function [32]. It can be operated *in vitro* on live cells without the necessity for further specimen preparation such as fixation or staining [33]. Atomic force microscopy potentially provides for nanometer-scale resolution [34, 35]. Recently, it has been combined with simultaneous confocal laser scanning microscopic imaging [35]. It also can be used to image individual isolated cellular complexes (i.e., proteins, DNA, organelles, etc.) [36]. In addition, biomolecular probing with functionalized tips can be used to generate force versus displacement curves, providing the capability to obtain single-molecule bond-strength information [37]. Furthermore, the elasticity or softness or compliance of biological samples can be assessed [38].

9.3.2
Instrumentation

A range of materials, methods and notes have appeared with regard to imaging cultured cells with the AFM [39]. In general, maintenance of steady-state culture conditions involves a high degree of thermal stability (37 ± 0.5 °C), continual renewal of the culture medium (osmolarity of ± 320 mOsmol per kg-H_2O) and a neutral pH (7.4 ± 0.3), stabilized by the use of 20 mM HEPES buffer and/or a flow of CO_2 through the AFM cell chamber to maintain a steady-state concentration of 5%. Fluctuations in temperature, osmotic pressure and pH have severe affects on cell viability and structure, making consistent AFM imaging difficult [40]. Commercial liquid cells for studying biological samples under controlled conditions are available, but are only useful for a limited number of special application problems. Therefore, biologists have designed their own set-ups and, consequently, various home-made systems, which differ from laboratory to laboratory, are found when reviewing the literature.

Our AFM studies used (a) the Topometrix Explorer^{TM} TMX equipped with a 100 μm XY/12 μm Z TrueMetrix Linearized Liquid Scanner installed on a Zeiss IM 35 inverted microscope with a home-made XY specimen stage adaptation; and (b) a home-made fluid cell in combination with a heating stage. This design allows positioning of the cantilever in the optical axis of the inverted microscope, movement of the sample via the inverted microscope independently of the AFM, and minimizes cantilever drift by controlling temperature-induced variations. Time-lapse images of living cells in contact [33, 41–43] or non-contact mode [43, 44] have been obtained over 2–3 h, before peripheral parts of the cytoplasm started to detach from the substrate or before cell viability started to decrease as determined by the trypan blue exclusion test [38].

9.3.2.1 **Live Cell Imaging**

The major advantages of AFM over scanning electron microscopy (SEM) for imaging cells are that no coating and no vacuum are required, electrons are avoided and imaging can be carried out in an aqueous environment. As a result, living cell studies under near physiological conditions can be performed. Evidence for the successful application of AFM for biological imaging comes from abundant studies in the past decade, where the dynamic behavior of living cells at a resolution comparable with SEM has been be imaged and analyzed (for reviews see Refs. [31] and [32]). In our set-up, images of living cells in contact mode could be obtained repeatedly over a period of 1 h (Fig. 9.11) and more [33, 43] depending on the cell type used, and at the same time we observed no scanning-induced artifacts such as lateral deformation. Instead, attention-grabbing cell biological processes such as moving membrane sheets at the rim of the cytoplasm of rat hepatocytes could be observed, illustrating lamellipodial activity (Fig. 9.11, 0–60 min). In some instances, prolonged AFM-imaging of cells may result in removal of parts of cells or even cells *in toto* from the substrate due to repeated tip contact (Fig. 9.11, 75 min). Therefore, to assure optimal viability of the cells, scanning of the sample should be carried out for a maximum of 2(–3) hours, after which the sample should be replaced. At the end of the experiment, the viability should be checked routinely with the aid of the trypan blue and/or the propidium iodide test. In our combined AFM-light microscope set-up the overall viability usually drops by ~6% with every hour. In addition, the combined AFM/inverted microscope allows the cell and AFM tip to be seen by the optical microscope at all times during the scanning process. By doing so, the morphology of the cells during AFM imaging can be easily judged and tip-induced alterations such as detachment or removal of the peripheral parts of the cytoplasm can be easily observed. These tip-induced changes are typical morphological signs for the onset of decreased cell viability.

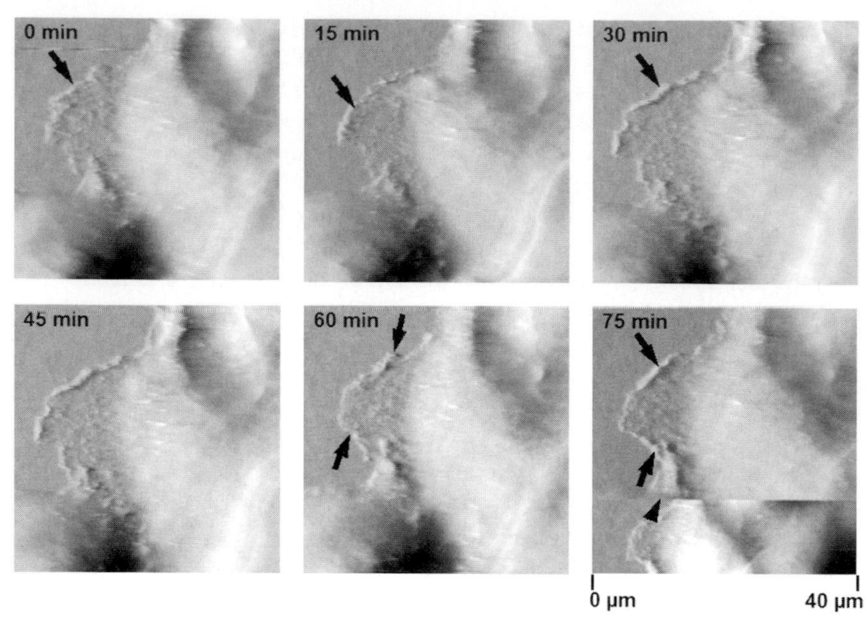

Fig. 9.11. Rat liver parenchymal cells. Time-lapse AFM series of moving membrane sheets at the rim of the cytoplasm in living rat liver parenchymal cells (hepatocytes). Arrows denote the lamellipodial activity observed at the rim of the cytoplasm of the hepatocytes over time. Note that prolonged scanning resulted in artifactual smearing and even removal of cell parts (arrowhead).

The AFM can gather, simultaneously, correlative topology and submembranous structures on the same cell at high resolution [33]. Sample deformation is an important component of the contrast mechanism in the visualization of living cells by AFM, and originates from local variations in stiffness when the tip palpates the cell membrane [29]. The cell stiffness or elasticity is determined mainly by the various organelles lying in the cytoplasm of the cell, and high-resolution imaging of submembranous cell compartments in the past was only possible when cells underwent detergent-extraction, fixation and/or immunocytochemical staining, thereby precluding dynamic studies [38, 40]. Therefore, indirect AFM imaging of organelles underneath the plasma membrane is a powerful tool to probe subcellular dynamics at nanometer resolution in intact living cells without the necessity of further preparative steps. The most prominent submembranous structure that can be probed with the AFM is the cytoskeleton of cells. An example is given in Fig. 9.12(A), illustrating the presence of long actin fibers with a straight outline in rat liver fibroblasts. We previously reported cytoskeletal changes in living rat skin fibroblasts for up to one hour after applying the microfilament-disrupting drug latrunculin A [33]. Interestingly, when cells underwent short fixation with 0.1% glutaraldehyde the morphology of the cell surface changed drastically (Fig. 9.12B); i.e., (i) there was an increase in cell height (Fig. 9.12C versus 12D); (ii) underlying

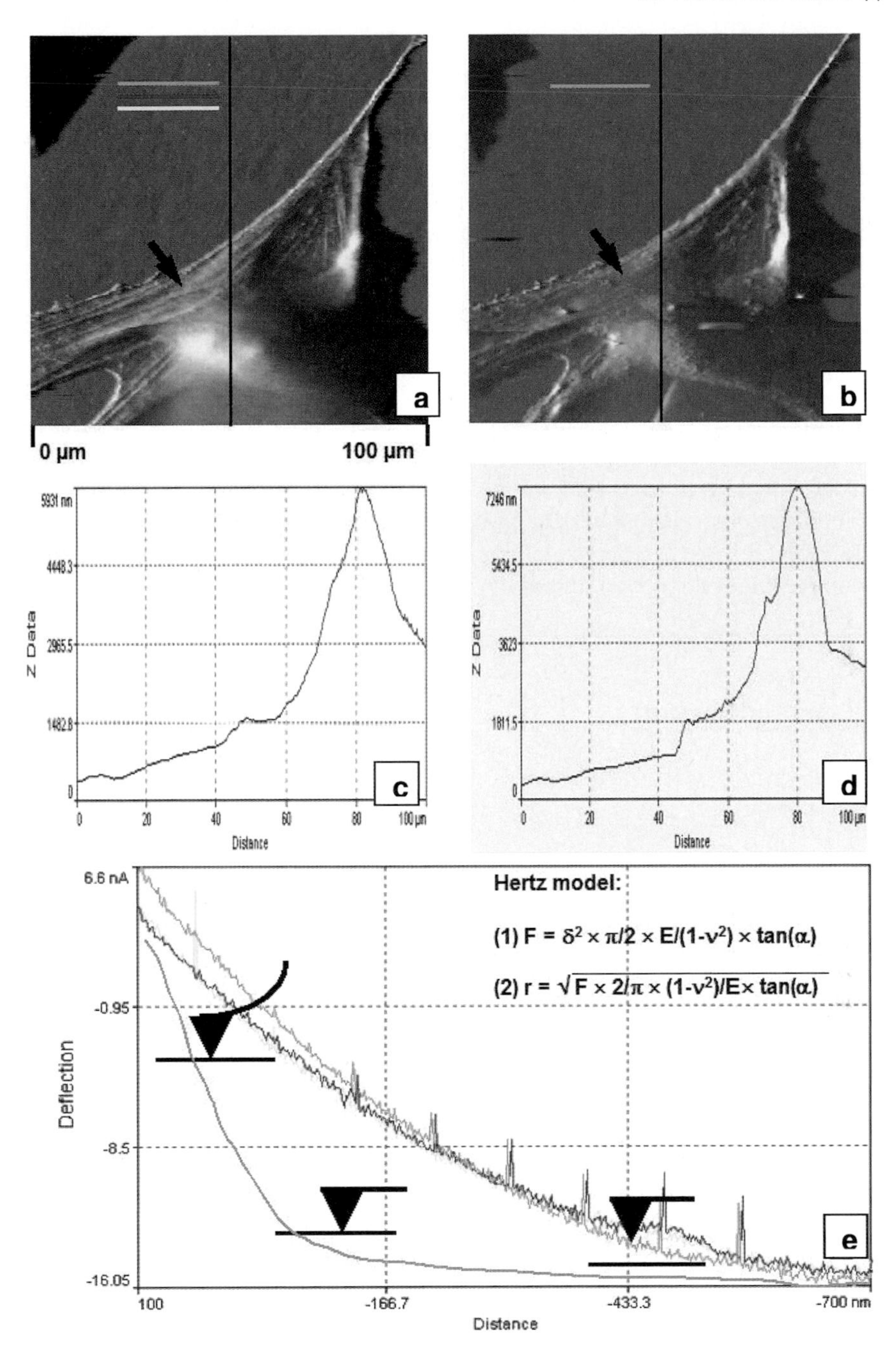

a
b
0 µm
100 µm
5931 nm
4448.3
2965.5
1482.8
Z Data
Distance
c
7246 nm
5434.5
3623
1811.5
Z Data
Distance
d
6.6 nA
-0.95
-8.5
-16.05
Deflection
100
-166.7
-433.3
-700 nm
Distance
Hertz model:
(1) F = δ² × π/2 × E/(1-ν²) × tan(α)
(2) r = √F × 2/π × (1-ν²)/E× tan(α)
e

cytoskeleton structures and clearly depicted cell contours could no longer be observed, even when the imaging force was increased 10-fold; and (iii) images were dominated by topographic information (Fig. 9.12A versus 9.12B). The latter two points can be explained by the fact that the stiffness of the cell membrane is much higher after fixation than the spring constant of the AFM-cantilever used, resulting in less deformation of the membrane around rigid submembranous structures. This is confirmed by the softness (i.e., elasticity) measurements performed on living versus fixed fibroblasts (Fig. 9.12E). The elasticity of those cells, as calculated by the Hertz model, increased ~12-fold, i.e., from 10 kPa for the living cell to more than 120 kPa for the fixed status.

9.3.3 Summary

Finally, much can be expected from the recent integration of atomic force and confocal fluorescence microscopies in one instrument, combining the high-resolution topographical imaging of probe microscopy with the reliable biomolecular identification capabilities of optical microscopy [35].

9.4 Cryo-electron Microscopy

9.4.1 Introduction

Cryo-electron microscopy has seen an increase in biological applications in the past fifteen years with the advent of significant technical innovation [45]. This microscopy method incorporates both vitrification, a freezing process that subverts

Fig. 9.12. Rat liver fibroblasts. AFM data on living (a, c) and glutaraldehyde fixed (b, d) rat liver fibroblasts. (a) Low magnification AFM-image of a living fibroblast, revealing cytoskeletal fibers traversing along the long axis of the cell (arrow). (b) AFM-image of the same cell as depicted in (a) after glutaraldehyde fixation. Notice that the earlier visualized cytoskeleton fibers in part disappear and instead the image is dominated by topographical information. (c, d) Height data of the corresponding AFM-images, illustrating the increase in height (~20%) after the cells underwent fixation. Height curves were obtained based on measurements along the black solid line depicted in parts (a and b). (e) Force curves taken on living (green, red and yellow lines) versus fixed (blue line) fibroblasts. These force curves have been plotted on top of each other to depict the differences. With a living cell, shallow force curves are observed, indicating the softness of the sample, whereas under fixed conditions the curve shows a very linear response, illustrating that the deflection is almost proportional to the force applied (see also the illustrative black tip drawings). From the force curves we have determined the elastic modulus of the cells, using the Hertz model, and calculated that the elastic modulus increases from around 10 kPa for the living condition to about 120 kPa for the fixed cells.

crystalline ice formation, and image acquisition under cryo-conditions. The cryo-fixation process is rapid, of the order of 0.1 ms, which is 10^4 times faster than conventional infiltration methods [46, 47], and preserves spatial as well as temporal biological states. In addition, sample preparation methods such as high-pressure freezing allow vitrification of specimens up to 200 μm thick. Significant achievements have also been made in the understanding and technology of freezing thin samples such as hepatic endothelial monolayers that are vulnerable to osmotic and temperature effects, as described in following section.

9.4.2 Instrumentation

Chemical fixation, dehydration and drying or embedding/sectioning of cells can induce image artifacts, resulting in different observations when different preparation techniques are applied [48]. In cryo-electron microscopy, living cells are physically fixed by rapid cooling, enabling their study as whole mounts without the necessity of further preparation steps. However, transmission electron microscopy of thin cells *in toto* (whole mount) has long been tried and considered impossible or extremely difficult due to the mass-thickness of the specimen. This view is usually substantiated by failures reported in the literature. Culturing cells as a single layer thick on grids, beam–specimen interaction resulting in specimen damage, and problems associated with cryo-specimen preparation have all been held responsible for this failure, depending on the spirit of the time. Over ten years ago the possibilities of cryo-electron microscopy on whole-mounted cells was discussed. By that time, the isolation and culture of hepatic endothelial cells on electron microscopy grids had become established [49], and cryo-electron microscopy had become an accepted approach for cellular imaging, as supported by the cryo-observations on intact blood platelets [50] and bacteria [51], revealing subcellular details such as organelles, membranous structures and cytoskeleton elements. It was argued, based on preliminary atomic force microscopy data [52], that the thickness of hepatic endothelial cells was of the same dimensional order as blood platelets and bacteria, and thus cryo-electron microscopy imaging could be expected using intermediate voltages. This holds especially for the thin fenestrated areas of cultured hepatic endothelial cells, which are less than 100 nm thick. As a result, electron beam-related problems (damage) or electron optical limitations were not expected. By that time, the development of an automated, computer-controlled vitrification system had begun that ultimately culminated in the Vitrobot™ [53–55] and a temperature and humidity controlled glove box in conjunction with a Vitrobot™ [55–57].

9.4.2.1 Cryo-electron Microscopy Imaging

The use of a temperature and humidity controlled glove box in conjunction with a Vitrobot™ is essential for the manipulation of whole-mount hepatic endothelial cells from culture conditions in preparation for cryo-electron microscopy investigation [56, 57]. The fenestrae and surrounding cytoskeleton elements and different

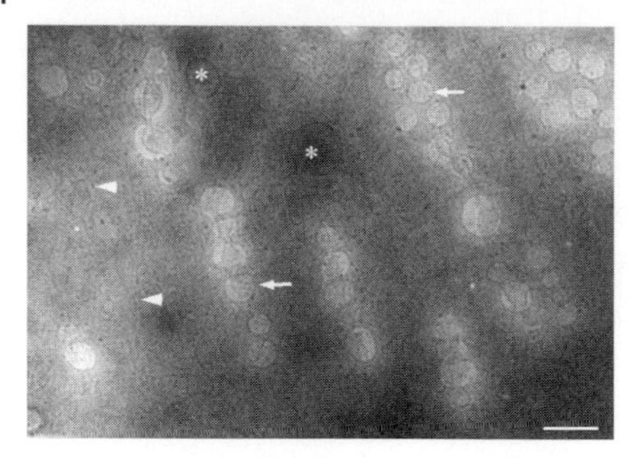

Fig. 9.13. High magnification image of the fenestrated cytoplasm of a vitrified hydrated whole-mounted hepatic endothelial cell obtained under controlled sample handling conditions by using a temperature and humidity controlled glove box in combination with the Vitrobot^{TM}, as described previously [35, 36], showing fenestrae and the associated cytoskeleton rings (arrow). Note the cytoskeleton elements (arrowhead) running next to the fenestrae and membrane-bound vesicles (asterisk). Scale bar: 500 nm. (Courtesy of Dr Peter Frederik & Paul Bomans, University of Maastricht, The Netherlands).

membrane-bound organelles are easily observed in these types of cryo-images without the presence of preparation-induced artifactual gaps (Fig. 9.13), indicating that our earlier observations on the fenestrae-associated cytoskeleton are not an artifact introduced by chemical fixation, partial extraction or other preparation procedure [57]. Further improvements in cryo-imaging may be expected when higher accelerating voltages are used (300 instead of 120 kV) in conjunction with electron tomography [45].

9.4.3
Summary

Once a biological sample is vitrified correctly the challenge becomes to extract all the available 3D information and in the time-domain. Tomography at the cryo-electron microscopy level can resolve details at a 3D resolution better than 3 nm in a specimen of 100 nm thick, providing a solid basis for tomography as an emerging 3D technology that may greatly contribute to the 3D study of intact cells at the (supra-)molecular level [58].

9.5
Conclusions

Imaging in biology has seen increasingly improved resolutions over the decades with the advent of new microscopical techniques. Applications of high-resolution microscopy span different sectors, from medical to biotechnology to fundamental-type research; analytical surveys serve as measures of quality control and materials improvement in medical technology; and the nano-world of basic biology bridges the structure–function relationship at several levels, from macromolecular to cellular and physiological, as seen with the wide use of AFM and cryo-EM. Soon, atomic resolutions from the atom probe will allow better device diagnostics and the generation of new questions and fields in biology appropriate for this new level of analysis. There will also be a continued need for correlative microscopy to develop control measures and for cross-referencing of methodologies from sample preparation to software development for data interpretation and rendering.

Acknowledgments

The facilities, scientific and technical assistance from staff in the NANO Major National Research Facility at the Electron Microscope Unit, the University of Sydney are gratefully appreciated. We gratefully acknowledge the staff of Imago Scientific Instruments, in particular Mr Tom Kunicki for helpful discussions and insights, and Dr Steve Goodman now at 10H, Inc. Consulting (www.10htech.com) for contributions to experimental data and illustrations. With respect to the cryo-electron microscopy data, the authors gratefully acknowledge the work of Drs Peter Frederik and Paul Bomans (University of Maastricht, The Netherlands).

References

1 Allen, R.E., Furley, D. *Studies in Presocratic Philosophy*. Routledge, New York, **1975**.

2 Watson, J.D. *The Double Helix: A Personal Account of the Discovery of the Structure of DNA*. Simon & Schuster Adult Publishing Group, **2001**.

3 Hirth, J.P., Lothe, J. *Theory of Dislocations*. Wiley, New York, 1982.

4 Müller, E.W., Field ion microscopy. *Science*, **1965**, *149*, 591.

5 Miller, M.K., Cerezo, A., Hetheringto, M.G., Smith, G.D.W. *Atom Probe Field Ion Microscopy*. Oxford University Press, Oxford, **1996**.

6 Panitz, J. Point-projection microscopy of macromolecular contours. *J. Microsc.*, **1982**, *125*, 3.

7 Saxey, D.W., Hanna, J., Zheng, R.K., Marceau, R.K.W., Baker, I., Ringer, S.P. Nanonstructural analysis of advanced alloys in a local electrode atom probe. Proceedings of Conference: Microsc. Microanal., Honolulu, Hawaii, USA, Supplement to *J. Microsc. Microanal.*, **2005**, *11*, 872CD.

8 Ringer, S.P., Ratinac, K. On the role of characterisation in the design of interfaces in nanoscale materials technology. *Microsc. Microanal.*, **2004**, *10*, 324–335.

9 Miller, M.K. *Atom Probe Tomography*. Kluwer Academic/Plenum Press, New York, **2001**.

10 Nishikawa, O., Ohtani, Y., Maeda, K., Watanable, M., Tanaka, K. Development of the scanning atom probe and atomic level analysis. *Mater. Char.* **2000**, *44*, 29–57.

11 Kelly, T.F., Larson, D.J. Local electrode atom probes. *Mater. Char.* **2000**, *44*, 59–85.

12 Kelly, T.F., Gribb, T.T., Olson, J.D. Oltman, E., Wiener, S.A., Lenz, D.R., Shepard, J.D., Martens, R.L., Ulfig, R.M., Strennen, E.M., Bunton, J.H., Strait, D.R., Kunicki, T.C., Payne, T. and Watson, J. Configuration and performance of a local electrode atom probe. *Microsc. Microanal.* **2003**, *9*, 564–565.

13 Larson, D.J., Petford-Long, A.K., Ma, Y.Q., Cerezo, A. Overview No. 138: Information storage materials: nanoscale characterisation by three-dimensional atom probe analysis. *Acta Mater.*, **2004**, *52*, 2847–2862.

14 Goodman, S.L., Mengelt, T.J., Ali, M., Ulfig, R.M., Istephanous, N., Kelly, T.F. Atomic structure and compositional analysis of 316L stainless steel medical device materials with the local electrode atom probe. Proceedings of Conference: Microsc. Microanal., Savannah, Georgia, USA, Supplement. *J. Microsc. Microanal.*, **2004**, *10*, Suppl. 2, 541.

15 Pekner, D., Bernstein, I.M. *Handbook of Stainless Steel*. McGraw-Hill, New York, **1977**.

16 See: www.azom.com.

17 Müller, E.W., Panitz, J.A., McLane, S.B. The atom-probe field ion microscope. *Rev. Sci. Instrum.*, **1968**, *39*, 83–86.

18 Zhou, Y., Morais-Cabral, J.H., Kaufman, A., Mackinnon, R. Chemistry of ion coordination and hydration revealed by a K+ channel-Fab complex at 2.0 A resolution. *Nature*, **2001**, *414*, 43–48.

19 Guex, N., Peitsch, M.C. SWISS-MODEL and the Swiss-PdbViewer: An environment for comparative protein modeling. *Electrophoresis*, **1997**, *18*, 2714–2723.

20 Unger, V.M., Kumar, N.M., Gilula, N.B., Yeager, M. Three-dimensional structure of a recombinant gap junction membrane channel. *Science*, **1999**, *283*, 1176–1180.

21 Panitz, J. Point-projection imaging of unstained ferritin clusters. *Ultramicroscopy*, **1982**, *7*, 3.

22 Machlin, E.S., Freilich, A., Agrawal, D.C., Burton, J.J., Briant, C.L. Field ion microscopy of biomolecules. *J. Microsc.* **1975**, *104*, 127–168.

23 Maruyama, T., Nishi, T., Hasegawa, Y., Sakurai, T. *Interfaces in Polymer, Ceramic and Metal Matrix Composites*. ed.: Ishida, M., Elsevier, Amsterdam, **1988**, p. 73.

24 Iwatsu, F., Morikaw, H., Terao, T. FIM of phthalocyanines. *J. Phys.*, **1984**, *45–C9*, 471.

25 Larson, D.J., Foord, D.T., Petford-Long, A.K., Anthony, T.C., Rozdilsky, I.M., Cerezo, A., Smith, G.W.D. Focused ion beam milling for atom probe field ion microscopy specimen preparation: Preliminary experiments. *Ultramicroscopy*, **1998**, *75*, 147–159.

26 Panitz, J. In search of the chimera: molecular imaging in the atom-probe. Proceedings of Conference: Microsc. Microanal. Honolulu, Hawaii, USA, Supplement to *J. Microsc. Microanal.*, **2005**, *11*, 92.

27 *Handbook of Chemistry and Physics*, 85th edition, CRC Press, Boca Raton, FL, **2004–2005**.

28 Binnig, G., Quate, C.F., Gerber, C.H. Atomic force microscope. *Phys. Rev. Lett.* **1986**, *56*, 930–933.

29 Henderson, E., Haydon, P.G., Sakaguchi, D.S. Actin filament dynamics in living glial cells imaged by atomic force microscopy. *Science*, **1992**, *257*, 1944–1946.

30 Radmacher, M., Tillman, R.W., Fritz, M., Gaub, H.E. From molecules to cells: imaging soft samples with the atomic force microscope. *Science*, **1992**, *257*, 1900–1905.

31 Horber, J.K., Miles, M.J. Scanning probe evolution in biology. *Science*, **2003**, *302*, 1002–1005.

32 Santos, N.C., Castanho, M.A. An overview of the biophysical applications of atomic force microscopy. *Biophys. Chem.* **2004**, *107*, 133–149.

33 Braet, F., Seynaeve, C., De Zanger, R., Wisse, E. Imaging surface and submembranous structures with the atomic force microscope: a study on living cancer cells, fibroblasts and macrophages. *J. Microsc.* **1998**, *190*, 328–338.

34 Frederix, P., Akiyama, T., Staufer, U., Gerber, C.H., Fotiadis, D., Muller, D.J., Engel, A. Atomic force bio-analytics. *Curr. Opin. Chem. Biol.* **2003**, *7*, 641–647.

35 Kassies, R., van der Werf, K.O., Lenferink, A., Hunter, C.N., Olsen,

J.D., Subramaniam, V., Otto, C. Combined AFM and confocal fluorescence microscope for applications in bio-nanotechnology. *J. Microsc.* **2005**, *217*, 109–116.

36 Ikai, A., Afrin, R. Toward mechanical manipulations of cell membranes and membrane proteins using an atomic force microscope: an invited review. *Cell Biochem. Biophys.* **2003**, *39*, 257–277.

37 Rounsevell, R., Forman, J.R., Clarke, J. Atomic force microscopy: mechanical unfolding of proteins. *Methods Enzymol.* **2004**, *34*, 100–111.

38 Radmacher, M. Measuring the elastic properties of living cells by the atomic force microscope. *Methods Cell Biol.* **2002**, *68*, 67–90.

39 Nagao, E. and Dvorak, J.A. An integrated approach to the study of living cells by atomic force microscopy. *J. Microsc.* **1998**, *191*, 8–19.

40 Radmacher, M. Measuring the elastic properties of biological samples with the AFM. *IEEE Eng. Med. Biol. Mag.* **1997**, *16*, 47–57.

41 Braet, F., Rotsch, C., Wisse, E., Radmacher, M. AFM imaging and elasticity measurements on living rat liver macrophages. *Appl. Phys.* **1998**, *66*, S575–S578.

42 Braet, F., De Zanger, R., Seynaeve, C., Baekeland, M., Wisse, E. A comparative atomic force microscopy study on living skin fibroblasts and liver endothelial cells. *J. Electron Microsc.* **2001**, *50*, 283–290.

43 Braet, F., Vermijlen, D., Bossuyt, V., De Zanger, R., Wisse, E. Early detection of cytotoxic events between hepatic natural killer cells and colon carcinoma cells as probed with the atomic force microscope. *Ultramicroscopy*, **2001**, *89*, 265–273.

44 Braet, F., De Zanger, R., Kämmer, S., Wisse, E. Noncontact versus contact imaging: An atomic force microscopic study on hepatic endothelial cells *in vitro*. *Int. J. Imaging Syst. Technol.* **1997**, *8*, 162–167.

45 Koster, A.J., Klumperman, J. Electron microscopy in cell biology: integrating structure and function. *Nat. Rev. Mol. Cell Biol.* **2003**, Supp, SS6–SS10.

46 Sitte, H., Edelman L., Neumann, K. *Cryotechniques in Biological Electron Microscopy.* eds.: Steinbrecht, R.A., Zierold, K., Springer-Verlag, Berlin, **1987**, pp. 87–113.

47 Sitte, H., Edelman, L., Neumann, K. Cryofixation without pretreatment at ambient pressure. *Cryotechniques in Biological Electron Microscopy*, ed. R.A. Steinbrecht, K. Zierold, Springer-Verlag, Berlin, **1987**, pp. 87–113.

48 King, M. Dimensional changes in cells and tissues during specimen preparation for the electron microscope. *Cell Biophys.* **1991**, *18*, 31–55.

49 Braet, F., De Zanger, R., Baekeland, M., Crabbé, E., Van Der Smissen, P., Wisse, E. Structure and dynamics of the fenestrae-associated cytoskeleton of rat liver sinusoidal endothelial cells. *Hepatology*, **1995**, *21*, 180–189.

50 Frederik, P., Stuart, M.C., Bomans, P., Busing, W., Burger, K., Verkleij, A. Perspective and limitations of cryo-electron microscopy. From model systems to biological specimens. *J. Microsc.* **1991**, *161*, 253–262.

51 Frederik, P., Bomans, P., Stuart, M. The ultrastructure of cryo-sections and intact vitrified cells – the effects of cryoprotectants and acceleration voltage on beam induced bubbling. *Scanning Microsc.* **1991**, *5*, S43–S51.

52 Braet, F., Kalle, W., De Zanger, R., De Grooth, B., Raap, A., Tanke, H., Wisse, E. Comparative atomic force and scanning electron microscopy: an investigation on fenestrated endothelial cells in vitro. *J. Microsc.* **1996**, *181*, 10–17.

53 Frederik, P., Bomans, P., Braet, F., Wisse, E. *Cells of the Hepatic Sinusoid* eds.: Wisse, E., Knook, D.L., Balabaud, C. Kupffer Cell Foundation, Leiden, **1997**, 476–478.

54 Frederik, P., Bomans, P., Franssen, V., Laeven, P. *Proceedings of the 12th European Congress on Electron Microscopy*, ed. Janish, R., Lech, S., Brno, Reklamní Atelier Kupa, **2000**, pp. B383–B384.

55 Frederik, P., Hubert, D.H. Cryo-electron microscopy of liposomes. *Methods Enzymol.* **2005**, *391*, 431–448.

56 Braet, F., Bomans, P., Wisse, E., Frederik, P. The observation of intact hepatic endothelial cells by cryo-electron microscopy. *J. Microsc.* **2003**, *212*, 175–185.

57 Braet, F., Wisse, E., Frederik, P., Geerts, W., Koster, A., Soon, L., Ringer, S. Contribution of high-resolution correlative imaging techniques in the study of the hepatic sieve in three-dimensions. *Microsc. Res. Tech.* **2005**, in press.

58 Baumeister, W. From proteomic inventory to architecture. *FEBS Lett.*, **2005**, *579*, 933–937.

10
Dynamic Light Scattering Microscopy

Rhonda Dzakpasu and Daniel Axelrod

10.1
Introduction

We describe here a novel imaging technique for optical microscopy based on dynamic light scattering. Conventional dynamic light scattering (DLS, also known as quasielastic light scattering, QELS) is a well-established laser-based nonmicroscopic, nonimaging technique commonly used to measure diffusion coefficients of proteins in solution [1, 2]. DLS is based on the time-dependent interference among electric fields emanating from scattering centers in relative motion and is sensitive to relative motions that are six times smaller than the optical resolution of the microscope. We have adapted DLS to microscopy (DLSM) and, with the use of a CCD camera in a line-streaking mode, we use the rate of data readout transfer as the rate of data collection. This allows us to image rapid motions over submicroscopic distances. Rather than creating images based on static index of refraction variations, spatial maps based on the rates of motions are produced. This can provide valuable insight into the processes imaged in biological systems.

The theory of DLSM is presented here, followed by a description of a detection system that uses a slow scan (a few Hz exposure rate) CCD to record the rapid (kHz range) fluctuations of DLS, and finally followed by tests on a model system (small polystyrene beads in suspension) and a living cell system (macrophages). The goal of the tests on macrophages is not to answer any particular biological question at this point, but rather to verify the applicability of DLSM to cell biological systems.

Previously reported applications of DLS in a microscope have been limited to a single point (rather than spatial mapping) measurement of diffusion coefficients and flow rates (velocimetry) [3–15]. These works did not address the theoretical question as to how intensity fluctuations can occur from scattering centers mutually close enough to fall within the optical resolution distance of the microscope; this question is addressed here.

DLSM in its present form is chemically nonspecific, as are dark field, phase contrast, and differential interference contrast. Unlike those techniques that detect spatial but temporally static variations in local refractive index, DLSM detects tem-

Nanotechnologies for the Life Sciences Vol. 3
Nanosystem Characterization Tools in the Life Sciences. Edited by Challa S. S. R. Kumar

ISBN: 3-527-31383-4

porally dynamic variations in a spatially resolved format. We discuss possible extensions of the technique that might increase its chemical specificity while preserving its ability to detect rapid motions in submicroscopic regions.

This chapter is divided into five further sections: Section 10.2 presents the theoretical solution for the functional form of the scattered electric field from a single and then a collection of scattering centers. The temporal and spatial autocorrelation function is derived for the intensity at the image plane of the microscope. We show there can be significant intensity fluctuations due to the relative motion of scattering centers within the optical resolution of the microscope. In addition, we show that fast time scale fluctuations can be measured with this technique.

Section 10.3 describes the experimental realization of the technique for our control system of polystyrene beads and the first application to cell biology, macrophage cells.

Section 10.4 describes the data analysis performed on the intensity fluctuations obtained from the experimental system.

Section 10.5 presents results for the two experimental systems. We show that the rates of motion obtained for different bead sizes agree with those expected from hydrodynamic theory for such beads and are in relative agreement based on their size ratio. In addition, we also show that we can observe spatially resolved differences in the decay rates in macrophages, both untreated and treated by a pharmacological agent.

Section 10.6 discusses the experimental and theoretical results stemming from this work. It also presents possible applications of the technique and suggests potential improvements.

10.2 Theory

This section presents a theoretical derivation of the functional form of the scattered electric field at the image plane of a microscope (which is modeled as a simple lens), first from a single scattering center and then from a collection of centers. From the resulting intensity, the temporal and spatial autocorrelation functions are derived. The theory is essentially a combination of scalar diffraction theory for a simple lens and a generalization of the conventional DLS theory as presented in Ref. [16].

The optical resolution of a microscope specifies the minimum separation of two objects in object space required to form distinctly separated images. If the separation is greater than the resolution distance, then clearly little interference can occur in the image plane. We show here that for scattering centers spaced closer than the optical resolution distance, sufficient phase variations still exist to create intensity fluctuations in the image plane.

As individual scattering centers can enter and/or leave an imaged region monitored by a single pixel in the detector, additional intensity fluctuations are created due to the change in particle number. These intensity fluctuations would occur even for incoherent light scattering (e.g., fluorescence). We derive an expression that includes the effect of intensity fluctuations due to both phase and number variations.

The characteristic decay time of the temporal intensity autocorrelation function guides what is the minimum sample time that should be employed in the detection system. The characteristic decay distance of the spatial autocorrelation function guides what is the maximum pixel size that should be employed in the detection system.

10.2.1
Single Scattering Center

We first consider the electric field as it scatters from a single particle toward the objective lens, refracts through the lens, and propagates to the image plane. The scattering center is assumed to be much smaller than the wavelength of the incident light λ (the Raleigh scattering limit); the polarization of the incident light is assumed to be linear and the polarization of the scattered light is assumed to be the same as the incident light. The electric fields are thereby represented as (complex) scalars rather than vectors. All of the light (incident and scattered) is monochromatic so the common factor $\exp(i\omega t)$ is everywhere suppressed.

The coordinate systems are depicted in Fig. 10.1. Scattering center (object) space, objective lens space, and image space coordinates are unprimed, primed, and double primed, respectively. The optical axis of the microscope is defined as the z-axis. The incident light is formed from a collimated laser beam propagating in the y–z plane and passing through a cylindrical lens. This lens focuses in the x-dimension only and leaves the incident light as a thin stripe along the y-direction, although still much larger in every dimension than the optical resolution. Therefore, the illumination electric field amplitude E_0 can be considered constant over a region competent for mutual interference at one detector pixel. The direction of propagation and focusing is such that no direct incident light reaches the objective (i.e., essentially dark field). If $\mathbf{k}_0$ is the incident wave vector and $\mathbf{r}$ is the location of a

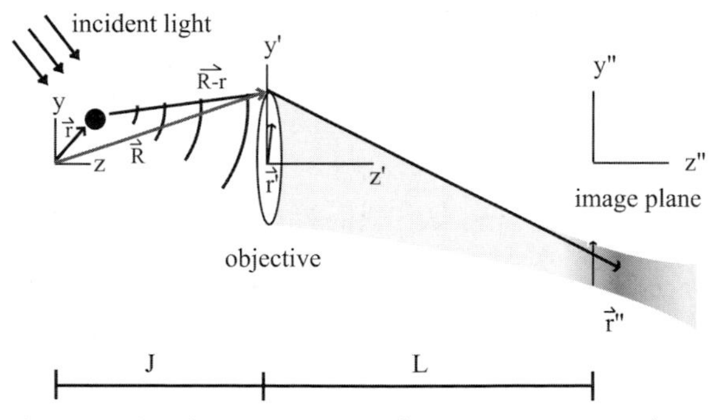

Fig. 10.1. Coordinate systems used in Section 10.2. The object is shown as a discrete black dot, and the point spread intensity in the image region as a blur. The origin in image space is located in the plane of the detector. The origin in object space is chosen at a point on the optical axis such that its focused image is centered at the origin in image space.

particular scattering center in object space, then the incident electric field $E(\mathbf{r})$ at a single scattering center is:

$$E(\mathbf{r}) = E_0 \exp(\mathbf{i}\mathbf{k}_0\mathbf{r}) \tag{1}$$

At the position of the objective lens (before propagating through it), the scattered light will produce an electric field $E'(\mathbf{r}')$ with amplitude proportional to both $E(\mathbf{r})$ and some scattering efficiency factor dependent upon the polarizability of the scattering center. However, this scattering efficiency factor is assumed constant among all scattering centers and also isotropic over the range of angles gathered by the lens, so its appearance will be suppressed in the expression for $E'(\mathbf{r}')$:

$$E'(\mathbf{r}') = E(\mathbf{r}) \exp(ik_s|\mathbf{R} - \mathbf{r}|) \tag{2}$$

where $\mathbf{R}$ is the vector from the origin in $\mathbf{r}$-space to a point on the objective lens represented by two-dimensional vector $\mathbf{r}'$ in the plane of the lens; $|\mathbf{R} - \mathbf{r}|$ is the distance from the scattering center to that point on the objective; and $k_s = |\mathbf{k}_s|$ is the amplitude of the scattered wave vector.

The electric field $E''(\mathbf{r}'', z'')$ in the image region (with positions denoted in cylindrical coordinates) is given in Ref. [17]:

$$E''(\mathbf{r}'', z'') = \frac{\exp(ik_0 L)}{i\lambda L} \iint E'(\mathbf{r}') \exp\left(\frac{-ik_0 r'^2}{2f}\right) \times \exp\left\{\frac{ik_0 r'^2}{2}\left[\frac{1}{L}\left(1 - \frac{z''}{L}\right)\right] - \frac{ik_0 \mathbf{r}' \cdot \mathbf{r}''}{L}\right\} \mathrm{d}^2\mathbf{r}' \tag{3}$$

where f is the focal length of the objective lens, $k_0 \equiv |\mathbf{k}_0|$ and L is the distance from the objective to the image plane. Assuming a small angle (i.e., small numerical aperture) approximation, the term $\exp(-ik_0 r'^2/2f)$ corresponds to the phase shift imposed by the objective lens. The factor

$$\exp\left\{\frac{ik_0 r'^2}{2}\left[\frac{1}{L}\left(1 - \frac{z''}{L}\right)\right] - \frac{ik_0 \mathbf{r}' \cdot \mathbf{r}''}{L}\right\}$$

describes the phase alteration as the field propagates in the empty space from the objective lens to the image region.

Combining Eqs. (1–3), noting that $\mathbf{k}_s$ is oriented in the same direction as $\mathbf{R} - \mathbf{r}$ with $k_s \approx k_0$, and regrouping, we obtain:

$$E''(\mathbf{r}'', z'') = E_0 \frac{\exp(ik_0 L)}{i\lambda L} \exp(\mathbf{i}\mathbf{k}_0 \cdot \mathbf{r}) \iint \exp\left\{ik_0|\mathbf{R} - \mathbf{r}| - \frac{ik_0 \mathbf{r}' \cdot \mathbf{r}''}{L} - \frac{ik_0 \mathbf{r}'^2}{2}\left[\frac{1}{f} - \frac{1}{L}\left(1 - \frac{z''}{L}\right)\right]\right\} \mathrm{d}^2\mathbf{r}' \tag{4}$$

For $r \ll R$, we can substitute an approximation for $|\mathbf{R} - \mathbf{r}|$:

$$|\mathbf{R} - \mathbf{r}| = \left\{ R^2 \left[1 + \left(\frac{r}{R}\right)^2 - \frac{2\mathbf{r} \cdot \mathbf{R}}{R^2} \right] \right\}^{1/2} \cong R + \frac{1}{2}\left(\frac{r^2}{R}\right) - \frac{\mathbf{r} \cdot \mathbf{R}}{R} \tag{5}$$

Since $\mathbf{R} = \mathbf{r}' + J\hat{\mathbf{z}}$, where J is the object distance, $1/R$ can be written in the small aperture approximation ($r' \ll J$) as:

$$\frac{1}{R} = (r'^2 + J^2)^{-1/2} \cong \frac{1}{J}\left[1 - \frac{1}{2}\left(\frac{r'}{J}\right)^2 + \cdots\right] \tag{6}$$

Substituting Eqs. (5) and (6) into Eq. (4), and noting that $\frac{1}{f} = \frac{1}{J} + \frac{1}{L}$, the electric field in the image region becomes:

$$E''(\mathbf{r}'', z'') = E_0 \frac{\exp ik_0(L+J)}{i\lambda L} \int \exp ik_0 \left(-\mathbf{r} \cdot \mathbf{Q}(\mathbf{r}') + \frac{r^2\gamma(r'')}{2J} - \frac{\mathbf{r}' \cdot \mathbf{r}''}{L} - \frac{z'' r'^2}{2L^2}\right) d^2\mathbf{r}' \tag{7}$$

where

$$\gamma(r') \equiv 1 - \frac{r'^2}{2J^2} \tag{8}$$

and

$$\mathbf{Q}(\mathbf{r}') \equiv \gamma\left(\frac{\mathbf{r}'}{J} + \hat{\mathbf{z}}\right) - \hat{\mathbf{k}}_0 \tag{9}$$

Vector $k_0\mathbf{Q}$ is a generalized conventional scattering vector, analogous to $\mathbf{q}$ ($\equiv \mathbf{k}_s - \mathbf{k}_0$) in conventional nonimaging DLS. Here, $\mathbf{Q}$ additionally takes into account the range of scattering angles gathered by the microscope objective lens.

To simplify Eq. (7) further, we assume that the detector is located in the image plane ($z'' = 0$). The scattering centers imaged within the same optical resolution area are then very close to the origin in object space so that $r \ll r'$ for almost the entire range of the integral. Therefore, the exponent term $\mathbf{r} \cdot \mathbf{Q} \gg r^2\gamma/2J$, implying that the factor $\exp(-ik_0\mathbf{r} \cdot \mathbf{Q})$ in Eq. (7) varies much more rapidly than $\exp(-ik_0 r^2\gamma/2J)$ over the range of the r', so that the latter factor can be assumed constant and close to unity. Eq. (7) then becomes:

$$E''_{z''=0}(\mathbf{r}, \mathbf{r}'') = E_0 \frac{\exp ik_0(L+J)}{i\lambda L} \int \exp ik_0 \left(-\mathbf{r} \cdot \mathbf{Q} - \frac{\mathbf{r}' \cdot \mathbf{r}''}{L}\right) d^2\mathbf{r}' \tag{10}$$

10.2.2 Multiple Scattering Centers

Each scattering center i located at position $\mathbf{r}_i$ produces an electric field at $z'' = 0$ according to Eq. (10). The total electric field E and the consequent intensity I at the image plane depend on the set of all the $\mathbf{r}_i$ positions $(I = 1, \ldots N)$ as follows:

$$E(\{\mathbf{r}_i\}, \mathbf{r}'', t) = \sum_i^N b_i(t) E''_{z''=0}(\mathbf{r}_i, \mathbf{r}'') \tag{11}$$

$$I(\{\mathbf{r}_i\}, \mathbf{r}'', t) = E * E \tag{12}$$

To understand the meaning of the $b_i(t)$ parameters, we define an "equivalent volume" v_{pix} in object space that contains all of the $\mathbf{r}$ positions that contribute to the intensity observed by a single CCD camera pixel at one position in the image plane. (Of course, the actual region from which scattered light is gathered has graded rather than sharp edges.) We also define an arbitrarily larger volume V that subsumes v_{pix} and contains the N scattering centers included in the sum in Eq. (11). The occupation number $b_i(t)$ equals unity if scattering center i is in v_{pix} at time t and zero otherwise. We assume that the positions $\mathbf{r}_i$ are statistically independent and randomly time-dependent (e.g., due to Brownian motion). These random motions cause E to fluctuate in both phase and amplitude, and the resulting intensity to fluctuate in amplitude. The temporal and spatial behavior of the intensity fluctuations can be investigated through autocorrelation functions.

10.2.3 Temporal Autocorrelation of Intensity

The temporal autocorrelation is defined as

$$\Gamma(\tau) = \langle I(t_1) I(t_2) \rangle \tag{13}$$

where $\tau \equiv t_2 - t_1$, the intensities at the two times are measured at the same $\mathbf{r}''$ position in the image plane, and the ensemble average indicated by the brackets is taken over all possible $\{\mathbf{r}_i\}$ configurations. Because the system is assumed to be in equilibrium, Γ depends only on the time difference τ and not the absolute times. After substituting Eqs. (10–12) into Eq. (13), we get Eq. (14).

$$\Gamma(\tau) = \left(\frac{|E_0|}{i\lambda L}\right)^4 \sum_{i,j,k,l}^N \langle b_i(t_1) b_j(t_1) b_k(t_2) b_l(t_2) \rangle$$
$$\times \iiiint \left\langle \begin{array}{l} \exp(-ik_0 \mathbf{r}_i(t_1) \cdot \mathbf{Q}_a) \exp(ik_0 \mathbf{r}_j(t_1) \cdot \mathbf{Q}_b) \\ \exp(ik_0 \mathbf{r}_k(t_2) \cdot \mathbf{Q}_c) \exp(-ik_0 \mathbf{r}_l(t_2) \cdot \mathbf{Q}_d) \end{array} \right\rangle$$

$$\times \exp\left(-\frac{ik_0\mathbf{r}'_a \cdot \mathbf{r}''}{L}\right) \exp\left(\frac{ik_0\mathbf{r}'_b \cdot \mathbf{r}''}{L}\right) \exp\left(\frac{ik_0\mathbf{r}'_c \cdot \mathbf{r}''}{L}\right)$$

$$\times \exp\left(-\frac{ik_0\mathbf{r}'_d \cdot \mathbf{r}''}{L}\right) d^2\mathbf{r}'_a\, d^2\mathbf{r}'_b\, d^2\mathbf{r}'_c\, d^2\mathbf{r}'_d \tag{14}$$

The phase fluctuations (arising from the complex exponential factors) are uncorrelated with number fluctuations (arising from the b factors); this is why the single ensemble average in Eq. (13) can be separated into a product of two ensemble averages (number and phase) in Eq. (14).

The summation in Eq. (14) can be separated according to the relationships among the summation indices i, j, k, l such that

$$\Gamma = \left(\frac{|E_0|}{i\lambda L}\right)^4 \sum_{m=1}^{6} (\Sigma_m \Gamma_m^{\text{num}} \Gamma_m^{\text{ph}}) \tag{15}$$

where Γ_m^{num} and Γ_m^{ph} are the number and phase fluctuation factors, respectively, and Σ_m represents sums over i, j, k, l restricted as shown in Table 10.1.

10.2.4 Phase Fluctuation Factors

In the first three cases ($m = 1, 2, 3$) at least one index is unique from all of the others. In such cases, a factor $\langle \exp(-ik_0\mathbf{r} \cdot \mathbf{Q}) \rangle$ with the unique index on the $\mathbf{r}$ vector can be factored out from the overall ensemble average since the motions of the scattering centers are mutually independent. That factor can be handled as follows (written here for a particular scattering index i):

$$\langle \exp(-ik_0\mathbf{r}_i(t_1) \cdot \mathbf{Q}_a) \rangle = \int \chi_i(\mathbf{r}_i) \exp(-ik_0\mathbf{r}_i \cdot \mathbf{Q}_a)\, d^3\mathbf{r}_i \tag{16}$$

where χ_i is the probability density that the particle is located in the vicinity of position $\mathbf{r}_i$. The scattering centers i are assumed to be uniformly distributed over the volume v_{pix} imaged by an individual pixel in r-space so that $\chi_i(\mathbf{r}_i) = \chi = 1/v_{\text{pix}}$. Eq. (16) becomes:

Tab. 10.1. Number of unique indices.

m Index	*Scattering center indices*	*Number of unique indices*
1	$i \neq j \neq k \neq l$	4
2	$i \neq j \neq k = l$	2
3	$i \neq j = k = l$	1
4	$i = j \neq k = l$	0
5	$i = k \neq j = l$	0
6	$i = j = k = l$	0

$$\langle \exp(-ik_0 \mathbf{r}_i(t_1) \cdot \mathbf{Q}_a) \rangle = \frac{1}{v_{\text{pix}}} \int \exp(-ik_0 \mathbf{r}_i \cdot \mathbf{Q}_a)\, \mathrm{d}^3\mathbf{r}_i = \frac{(2\pi)^{3/2}}{v_{\text{pix}}} \delta(k_0 \mathbf{Q}_a) \tag{17}$$

The integral over $\mathbf{r}'_a$ in Eq. (14) then becomes:

$$\begin{aligned}
&\int \langle \exp(-ik_0 \mathbf{r}_i(t_1) \cdot \mathbf{Q}_a) \rangle \exp\left(-\frac{ik_0 \mathbf{r}'_a \cdot \mathbf{r}''}{L}\right) \mathrm{d}^2\mathbf{r}'_a \\
&= \frac{(2\pi)^{3/2}}{v_{\text{pix}}} \int \delta(k_0 \mathbf{Q}_a) \exp\left(-\frac{ik_0 \mathbf{r}'_a \cdot \mathbf{r}''}{L}\right) \mathrm{d}^2\mathbf{r}'_a \\
&= \frac{(2\pi)^{3/2}}{v_{\text{pix}}} \left(\frac{\mathrm{J}}{k_0 \gamma}\right)^2 \int \delta(k_0 \mathbf{Q}_a) \exp\left(-\frac{ik_0 \mathbf{r}'_a \cdot \mathbf{r}''}{L}\right) \mathrm{d}^2(k_0 \mathbf{Q}_a)
\end{aligned} \tag{18}$$

The integral has a nonzero value only when $k\mathbf{Q}_a = 0$. From the definition of $\mathbf{Q}_a$ in Eq. (9), we can obtain the $\mathbf{r}'_a$ for which this condition is satisfied:

$$\mathbf{r}'_a|_{\mathbf{Q}_a=0} = \frac{\mathrm{J}}{\gamma} \hat{\mathbf{k}}_0 - \mathrm{J}\hat{\mathbf{z}} \tag{19}$$

This particular $\mathbf{r}'_a$ is located where the extension of the incident beam crosses the plane of the objective lens. Since our experimental setup was designed so that the incident light misses the objective, the integral over $\mathbf{r}'_a$ in Eq. (18) (which is limited to the area of the objective) does not include $\mathbf{r}'_a|_{\mathbf{Q}_a=0}$. Thus, those terms in the sum of Eq. (15) with at least one unique summation index (i.e., $m = 1, 2, 3$) are zero.

The phase term for the $m = 4$ case from Eq. (15) is given by Eq. (20).

$$\begin{aligned}
\Gamma_4^{\text{ph}} &= \iint \langle \exp ik_0[\mathbf{r}_i(t_1) \cdot \Delta\mathbf{Q}_{ab}] \rangle \exp\left(\frac{ik_0 \Delta\mathbf{r}'_{ab} \cdot \mathbf{r}''}{L}\right) \mathrm{d}^2\mathbf{r}'_a\, \mathrm{d}^2\mathbf{r}'_b \\
&\quad \times \iint \langle \exp -ik_0[\mathbf{r}_k(t_2) \cdot \Delta\mathbf{Q}_{cd}] \rangle \exp\left(\frac{-ik_0 \Delta\mathbf{r}'_{cd} \cdot \mathbf{r}''}{L}\right) \mathrm{d}^2\mathbf{r}'_c\, \mathrm{d}^2\mathbf{r}'_d \\
&= \left| \iint \langle \exp ik_0[\mathbf{r}_i(t_1) \cdot \Delta\mathbf{Q}_{ab}] \rangle \exp\left(\frac{ik_0 \Delta\mathbf{r}'_{ab} \cdot \mathbf{r}''}{L}\right) \mathrm{d}^2\mathbf{r}'_a\, \mathrm{d}^2\mathbf{r}'_b \right|^2
\end{aligned} \tag{20}$$

where $\Delta\mathbf{Q}_{\alpha\beta} \equiv \mathbf{Q}_\beta - \mathbf{Q}_\alpha$ and $\Delta\mathbf{r}'_{\alpha\beta} \equiv \mathbf{r}'_\beta - \mathbf{r}'_\alpha$. The terms of the form $\langle \exp ik_0[\mathbf{r}(t_1) \cdot \Delta\mathbf{Q}] \rangle$ are similar to that in Eq. (17), except for the factor $\Delta\mathbf{Q}$ instead of $\mathbf{Q}$. Therefore, the integral in Eq. (20) can be reduced to:

$$\begin{aligned}
&\frac{1}{v_{\text{pix}}} \iint \left(\int \exp ik_0[\mathbf{r}_i(t_1) \cdot \Delta\mathbf{Q}_{ab}]\, \mathrm{d}x_i\, \mathrm{d}y_i\, \mathrm{d}z_i \right) \exp\left(\frac{ik_0 \Delta\mathbf{r}'_{ab} \cdot \mathbf{r}''}{L}\right) \mathrm{d}^2\mathbf{r}'_a\, \mathrm{d}^2\mathbf{r}'_b \\
&= \frac{l_{\text{pix}}(2\pi)}{v_{\text{pix}}} \iint \delta(k_0 \Delta\mathbf{Q}_{ab}) \exp\left(\frac{ik_0 \Delta\mathbf{r}'_{ab} \cdot \mathbf{r}''}{L}\right) \mathrm{d}^2\mathbf{r}'_a\, \mathrm{d}^2\mathbf{r}'_b
\end{aligned}$$

$$
\begin{aligned}
&= \frac{(2\pi)}{s_{\text{pix}}}\left(\frac{\mathrm{J}}{k_0\gamma}\right)^4 \iint \delta(k_0\Delta\mathbf{Q}_{ab}) \exp\left(\frac{\mathrm{i}k_0\Delta\mathbf{r}'_{ab}\cdot\mathbf{r}''}{L}\right) \mathrm{d}^2(k_0\mathbf{Q}_a)\,\mathrm{d}^2(k_0\mathbf{Q}_b) \\
&= \frac{(2\pi)}{s_{\text{pix}}}\left(\frac{\mathrm{J}}{k_0\gamma}\right)^4 \int\left[\int \exp\left(\frac{\mathrm{i}k_0\Delta\mathbf{r}'_{ab}\cdot\mathbf{r}''}{L}\right)\right]\Bigg|_{a=b} \mathrm{d}^2(k_0\mathbf{Q}_b) \\
&= \frac{(2\pi)}{s_{\text{pix}}}\left(\frac{\mathrm{J}}{k_0\gamma}\right)^4 \int \mathrm{d}^2(k_0\mathbf{Q}) = \frac{(2\pi)}{s_{\text{pix}}}\left(\frac{\mathrm{J}}{k_0\gamma}\right)^2 \int \mathrm{d}^2\mathbf{r}' = \frac{(2\pi)}{s_{\text{pix}}}\left(\frac{\mathrm{J}}{k_0\gamma}\right)^2 A
\end{aligned} \tag{21}
$$

where l_{pix} is the z-dimension of the observed volume v_{pix}; s_{pix} is the area of the observed volume; and A is the area of the objective. Therefore,

$$
\Gamma_4^{\text{ph}} = \frac{(2\pi)^2}{{s_{\text{pix}}}^2}\left(\frac{\mathrm{J}}{k_0\gamma}\right)^4 A^2 \tag{22}
$$

For the $m = 5$ term in Eq. (15),

$$
\begin{aligned}
\Gamma_5^{\text{ph}} &= \Bigg|\iint \langle \exp \mathrm{i}k_0[\mathbf{r}_i(t_2)\cdot\mathbf{Q}_c - \mathbf{r}_i(t_1)\cdot\mathbf{Q}_a]\rangle \\
&\quad \times \exp\left(-\frac{\mathrm{i}k_0\mathbf{r}'_a\cdot\mathbf{r}''}{L}\right) \exp\left(\frac{\mathrm{i}k_0\mathbf{r}'_c\cdot\mathbf{r}''}{L}\right) \mathrm{d}^2\mathbf{r}'_a\,\mathrm{d}^2\mathbf{r}'_c\Bigg|^2 \\
&= \Bigg|\iint \langle \exp(-\mathrm{i}k_0\Delta\mathbf{r}_i\cdot\mathbf{Q}_c) \exp(-\mathrm{i}k_0\mathbf{r}_i(t_2)\cdot\Delta\mathbf{Q}_{ac})\rangle \\
&\quad \times \exp\left(-\frac{\mathrm{i}k_0\Delta\mathbf{r}'_{ac}\cdot\mathbf{r}''}{L}\right) \mathrm{d}^2\mathbf{r}'_a\,\mathrm{d}^2\mathbf{r}'_c\Bigg|^2 \\
&= \Bigg|\iint \langle \exp(-\mathrm{i}k_0\Delta\mathbf{r}_i\cdot\mathbf{Q}_c)\rangle\langle \exp(-\mathrm{i}k_0\mathbf{r}_i(t_2)\cdot\Delta\mathbf{Q}_{ac})\rangle \\
&\quad \times \exp\left(-\frac{\mathrm{i}k_0\Delta\mathbf{r}'_{ac}\cdot\mathbf{r}''}{L}\right) \mathrm{d}^2\mathbf{r}'_a\,\mathrm{d}^2\mathbf{r}'_c\Bigg|^2
\end{aligned} \tag{23}
$$

where $\Delta\mathbf{r}_i \equiv \mathbf{r}_i(t_2) - \mathbf{r}_i(t_1)$. The term $\langle \exp(-\mathrm{i}k_0\mathbf{r}_i(t_2)\cdot\Delta\mathbf{Q}_{ac})\rangle$ reduces to a δ-function, so we obtain:

$$
\begin{aligned}
\Gamma_5^{\text{ph}} &= \Bigg|\frac{(2\pi)}{s_{\text{pix}}}\left(\frac{\mathrm{J}}{k_0\gamma}\right)^4 \int\left[\int \delta(k_0\Delta\mathbf{Q}_{ac}) \exp\left(-\frac{\mathrm{i}k_0\Delta\mathbf{r}'_{ac}\cdot\mathbf{r}''}{L}\right) \mathrm{d}^2(k_0\mathbf{Q}_a)\right] \\
&\quad \times \langle \exp(-\mathrm{i}k_0\Delta\mathbf{r}_i\cdot\mathbf{Q}_a)\rangle\,\mathrm{d}^2(k_0\mathbf{Q}_c)\Bigg|^2 \\
&= \Bigg|\frac{(2\pi)}{s_{\text{pix}}}\left(\frac{\mathrm{J}}{k_0\gamma}\right)^2 \int \langle \exp(-\mathrm{i}k_0\Delta\mathbf{r}_i\cdot\mathbf{Q})\rangle\,\mathrm{d}^2\mathbf{r}'\Bigg|^2
\end{aligned} \tag{24}
$$

where the subscript "c" on the $\mathbf{r}'_c$ factors has been suppressed. The ensemble average in Eq. (24) can be rewritten as:

$$\langle \exp(-ik_0\Delta\mathbf{r}_i \cdot \mathbf{Q}) \rangle = \int p[\Delta\mathbf{r}_i(\tau)|0] \exp(-ik_0\Delta\mathbf{r}_i \cdot \mathbf{Q})\, \mathrm{d}^3\Delta\mathbf{r}_i \tag{25}$$

where $p[\Delta\mathbf{r}_i(\tau)|0]$ is the conditional probability of finding a particle at position $\Delta\mathbf{r}_i$ at time τ given that it was at the origin ($\Delta\mathbf{r}_i = 0$) at $\tau = 0$. The right-hand side of Eq. (25) is the Fourier transform of $p[\Delta\mathbf{r}_i(\tau)|0]$:

$$\langle \exp(-ik_0\Delta\mathbf{r}_i \cdot \mathbf{Q}) \rangle = (2\pi)^{3/2} \tilde{p}(\mathbf{Q}) \tag{26}$$

where $\tilde{p}(\mathbf{Q})$ is the Fourier transform of $p[\Delta\mathbf{r}_i(\tau)|0]$ into $\mathbf{Q}$-space.

We assume that the scattering centers are undergoing random diffusive motion. Taking the Fourier transform of the diffusion equation $\partial p/\partial\tau = D\nabla^2 p$ from $\Delta\mathbf{r}_i$-space to $\mathbf{Q}$-space gives:

$$\frac{\partial\tilde{p}}{\partial\tau} = -D{k_0}^2\mathbf{Q}^2\tilde{p} \tag{27}$$

so that

$$\tilde{p}(\mathbf{Q}) = (2\pi)^{-3/2} \exp(-D\mathbf{Q}^2{k_0}^2\tau) \tag{28}$$

and therefore, in combination with Eq. (26):

$$\langle \exp(-ik_0\Delta\mathbf{r}_i \cdot \mathbf{Q}) \rangle = \exp(-D\mathbf{Q}^2{k_0}^2\tau) \tag{29}$$

Eq. (24) then becomes:

$$\Gamma_5^{\mathrm{ph}} = \frac{(2\pi)^2}{s_{\mathrm{pix}}^2} \left(\frac{J}{k_0\gamma}\right)^4 A^2 \left| \frac{1}{A} \int \exp(-D\mathbf{Q}^2{k_0}^2\tau)\, \mathrm{d}^2\mathbf{r}' \right|^2 \tag{30}$$

For the $m = 6$ term in Eq. (15),

$$\begin{aligned}
\Gamma_6^{\mathrm{ph}} &= \iiiint \langle \exp ik_0[\mathbf{r}_i(t_1) \cdot \Delta\mathbf{Q}_{ab}] \exp ik_0[\mathbf{r}_i(t_2) \cdot \Delta\mathbf{Q}_{cd}] \rangle \\
&\quad \times \exp\left(\frac{ik_0\Delta\mathbf{r}'_{ab} \cdot \mathbf{r}''}{L}\right) \exp\left(\frac{ik_0\Delta\mathbf{r}'_{cd} \cdot \mathbf{r}''}{L}\right) \mathrm{d}^2\mathbf{r}'_a\, \mathrm{d}^2\mathbf{r}'_b\, \mathrm{d}^2\mathbf{r}'_c\, \mathrm{d}^2\mathbf{r}'_d \\
&= \iiiint \langle \exp ik_0[\mathbf{r}_i(t_1) \cdot (\Delta\mathbf{Q}_{ab} + \Delta\mathbf{Q}_{cd})] \exp ik_0[\Delta\mathbf{r}_i(\tau) \cdot \Delta\mathbf{Q}_{cd}] \rangle \\
&\quad \times \exp\left(\frac{ik_0\Delta\mathbf{r}'_{ab} \cdot \mathbf{r}''}{L}\right) \exp\left(\frac{ik_0\Delta\mathbf{r}'_{cd} \cdot \mathbf{r}''}{L}\right) \mathrm{d}^2\mathbf{r}'_a\, \mathrm{d}^2\mathbf{r}'_b\, \mathrm{d}^2\mathbf{r}'_c\, \mathrm{d}^2\mathbf{r}'_d \\
&= \frac{1}{v_{\mathrm{pix}}} \iiiint \left(\int \exp ik_0[\mathbf{r}_i(t_1) \cdot (\Delta\mathbf{Q}_{ab} + \Delta\mathbf{Q}_{cd})]\, \mathrm{d}x_i\, \mathrm{d}y_i\, \mathrm{d}z_i \right)
\end{aligned}$$

$$\times \langle \exp ik_0[\Delta \mathbf{r}_i(\tau) \cdot \Delta \mathbf{Q}_{cd}] \rangle \exp\left(\frac{ik_0 \Delta \mathbf{r}'_{ab} \cdot \mathbf{r}''}{L}\right)$$

$$\times \exp\left(\frac{ik_0 \Delta \mathbf{r}'_{cd} \cdot \mathbf{r}''}{L}\right) d^2\mathbf{r}'_a\, d^2\mathbf{r}'_b\, d^2\mathbf{r}'_c\, d^2\mathbf{r}'_d$$

$$= \frac{(2\pi) l_{\text{pix}}}{v_{\text{pix}}} \left(\frac{\text{J}}{k_0 \gamma}\right)^2 \iiint \left[\int \delta(\mathbf{Q}_b - \mathbf{Q}_a + \mathbf{Q}_d - \mathbf{Q}_c)\right.$$

$$\left. \times \exp\left(\frac{ik_0 \mathbf{r}'' \cdot (\Delta \mathbf{r}'_{ab} + \Delta \mathbf{r}'_{cd})}{L}\right) d^2(k_0 \mathbf{Q}_a)\right]$$

$$\times \langle \exp ik_0[\Delta \mathbf{r}_i(\tau) \cdot \Delta \mathbf{Q}_{cd}] \rangle\, d^2\mathbf{r}'_b\, d^2\mathbf{r}'_c\, d^2\mathbf{r}'_d$$

$$= \frac{(2\pi)}{s_{\text{pix}}} \left(\frac{\text{J}}{k_0 \gamma}\right)^2 \iint \langle \exp ik_0[\Delta \mathbf{r}_i(\tau) \cdot \Delta \mathbf{Q}_{cd}] \rangle\, d^2\mathbf{r}'_c\, d^2\mathbf{r}'_d \int d^2\mathbf{r}'_b$$

$$= \frac{(2\pi)}{s_{\text{pix}}} \left(\frac{\text{J}}{k_0 \gamma}\right)^2 A \iint \exp(-Dk_0^2 \Delta Q^2 \tau)\, d^2\mathbf{r}'_c\, d^2\mathbf{r}'_d \qquad (31)$$

10.2.5 Number Fluctuation Factors

The total volume of the imaged sample is V and the total number of scattering centers in that volume is N. The volume "imaged" by a single pixel is v_{pix}. The number of particles $M(t)$ (assumed $\ll N$) and its expectation value in the volume v_{pix} can be written:

$$M(t) = \sum_i^N b_i(t) \qquad (32)$$

$$\langle M \rangle = \left\langle \sum_i b_i(t) \right\rangle = \sum_i \langle b_i(t) \rangle = N \langle b_i \rangle \qquad (33)$$

The variance of M can be derived from Eqs. (32) and (33):

$$\text{var}\ M = N\ \text{var}\ b_i \qquad (34)$$

The assumption that $\langle b_i \rangle \ll 1$ implies that M follows a Poisson distribution and, therefore, var M equals $\langle M \rangle$. Therefore,

$$\text{var}\ b_i = \langle M \rangle / N = \langle b_i \rangle \qquad (35)$$

and the temporal autocorrelation function for b_i can be written as:

$$\langle b_i(t_1) b_i(t_1 + \tau) \rangle = (\text{var}\ b_i) g_{\text{num}}(\tau) + \langle b_i \rangle^2 \approx \frac{\langle M \rangle}{N} g_{\text{num}}(\tau) \qquad (36)$$

where $g_{num}(\tau)$ is a normalized number fluctuation autocorrelation function such that

$$g(0) = 1 \quad \text{and} \quad g(\infty) = 0$$

We are now set to consider the terms Γ_m^{num} that appear in Eq. (15). Since the $m = 1, 2, 3$ terms in Eq. (15) are forced to zero by their phase fluctuation factors, we need consider only $\Gamma_{4,5,6}^{num}$. We make the approximations that $N \gg \langle M \rangle \gg 1$.

For the $m = 4$ term of Eq. (15),

$$\begin{aligned}\Gamma_4^{num} &= \sum_{i \neq k}^{N} \langle b_i^2(t_1) b_k^2(t_1 + \tau) \rangle = \sum_{i \neq k}^{N} \langle b_i(t_1) b_k(t_1 + \tau) \rangle = \sum_{i \neq k}^{N} \langle b_i(t_1) \rangle \langle b_k(t_1 + \tau) \rangle \\ &= (N^2 - N) \langle b_i \rangle^2 \approx \langle M \rangle^2 \end{aligned} \tag{37}$$

For the $m = 5$ term,

$$\begin{aligned}\Gamma_5^{num} &= \sum_{i \neq j}^{N} \langle b_i(t_1) b_i(t_1 + \tau) b_j(t_1) b_j(t_1 + \tau) \rangle \\ &= \sum_{i \neq j}^{N} \langle b_i(t_1) b_i(t_1 + \tau) \rangle \langle b_j(t_1) b_j(t_1 + \tau) \rangle \\ &= (N^2 - N) \langle b_i(t_1) b_i(t_1 + \tau) \rangle^2 \approx \langle M \rangle^2 g_{num}^2(\tau) \end{aligned} \tag{38}$$

For the $m = 6$ term,

$$\begin{aligned}\Gamma_6^{num} &= \sum_{i}^{N} \langle b_i^2(t_1) b_i^2(t_1 + \tau) \rangle = \sum_{i}^{N} \langle b_i(t_1) b_i(t_1 + \tau) \rangle \\ &= N \langle b_i(t_1) b_i(t_1 + \tau) \rangle \approx \langle M \rangle g_{num}(\tau) \end{aligned} \tag{39}$$

The $m = 6$ term is smaller than the $m = 4$ and $m = 5$ terms by a factor of $\langle M \rangle$; therefore, it will be neglected. The complete temporal autocorrelation function thereby becomes:

$$\Gamma(\tau) = \langle I \rangle^2 \left\{ 1 + g_{num}^2(\tau) \left| \frac{1}{A} \int \exp(-DQ^2 k_0^2 \tau) \, d^2\mathbf{r}' \right|^2 \right\} \tag{40}$$

where $\langle I \rangle$ is the mean intensity observed at a pixel from the $\langle M \rangle$ scattering centers in its view:

$$\langle I \rangle = \left(\frac{|E_0|}{i \lambda L} \right)^2 \frac{(2\pi)}{s_{pix}} \left(\frac{J}{k_0 \gamma} \right)^2 A \langle M \rangle \tag{41}$$

To compare the result given in Eq. (40) with experimentally obtained autocorrelation functions, we construct the normalized temporal autocorrelation function:

$$g_T(\tau) \equiv \frac{\Gamma(\tau) - \langle I \rangle^2}{\langle I \rangle^2} \tag{42}$$

Combining Eqs. (40) and (42) shows that $g_T(\tau)$ monotonically decays to zero:

$$g_T(\tau) = g_{\text{num}}^2(\tau) \left[\frac{1}{A} \int \exp(-DQ^2 k_0 \tau)\, \mathrm{d}^2\mathbf{r}' \right]^2 \tag{43}$$

10.2.6
Characteristic Times and Distances

Figure 10.2(a) shows $g_T(\tau)$ [Eq. (43)] plotted as a function of the unitless time variable $Dk_0{}^2\tau$ with the indicated integration performed numerically for the particular case where the objective numerical aperture NA equals 0.4 (as used in the experimental setup).

The characteristic decay time $Dk_0^2\tau_c$, defined as the time required for $g_T(\tau)$ to reach its e^{-1} value, is $Dk_0^2\tau_c \approx 0.52$ for this particular numerical aperture. In that time, the mean distance r_c the particle travels laterally by three-dimensional diffusion is:

$$r_c \equiv (4D\tau_c)^{1/2} = (4 \cdot 0.52/k_0^2)^{1/2} = 0.23\lambda \tag{44}$$

This characteristic distance is about a factor of six smaller than the resolution of the microscope, which according to the Raleigh criterion is $r_{\text{res}} = 0.61\lambda/\text{NA} = 1.5\lambda$ for NA $= 0.4$. This proves that dynamic light scattering intensity fluctuations of significant amplitude do occur amongst scattering centers within a resolution distance of each other.

The actual characteristic time τ_c can be estimated for an aqueous suspension of 200 nm diameter polystyrene nanospheres as used in some of our experiments. Hydrodynamics predict $D = 2.2 \times 10^{-8}$ cm^2 s^{-1} for such spheres. For $\lambda = 632.8$ nm, the characteristic time of the temporal intensity autocorrelation function would be $\tau_c = 2.4$ ms. The experimental detection system must be able to observe these fast time-scale fluctuations.

10.2.7
Spatial Autocorrelation of Intensity

We define the spatial correlation region to be the spatial extent of the intensity fluctuations at the image plane. It is measured as the characteristic distance of the spatial autocorrelation function and determines the maximum pixel size allowable for measuring the temporal behavior of the intensity fluctuations. For example, if a pixel is larger than several characteristic spatial correlation regions, then the rela-

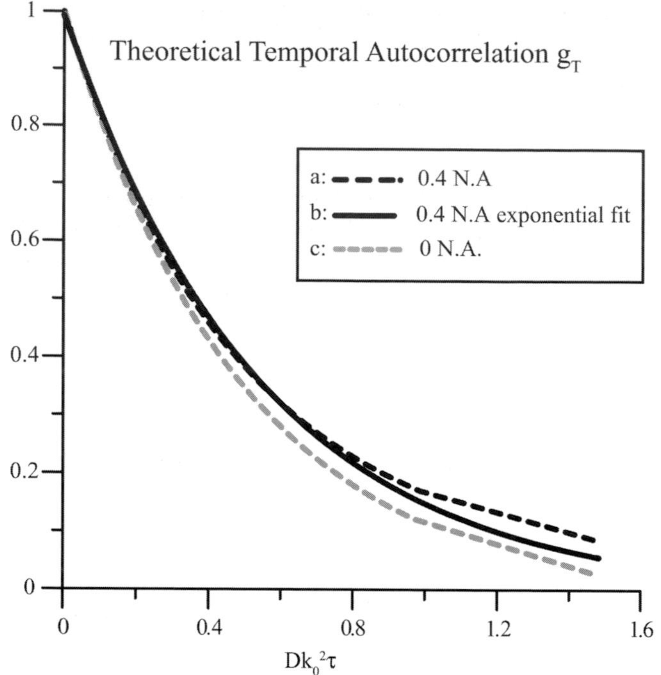

Fig. 10.2. Theoretical temporal autocorrelation function of the scattered light intensity [as calculated from Eq. (43)] vs. the unitless time parameter $Dk_0{}^2\tau$. Three curves are plotted. (a) The black dashed line curve is obtained from a numerical integration of Eq. (43) for a 0.4 numerical aperture objective. (b) The black solid curve is a single exponential decay, fitted to the points obtained from the numerical integration. (c) The grey dashed line is the pure exponential decay obtained from the zero aperture limit.

tive size of the observed fluctuations will be greatly reduced, compromising the signal-to-noise ratio. Ideally, a pixel should cover less than one spatial correlation region.

We start the calculation of the spatial autocorrelation function in a manner similar to the temporal autocorrelation function [Eq. (14)] except here using the intensities at two different *positions* $\mathbf{r}''_{1,2}$, recorded at the same time. (Because of the similarity of the mathematical procedures, we will skip most of the details here.) We count only those scattering centers that are actually present in the illuminated region at one snapshot of time, so all the $b_i(t)$ factors can be set equal to unity.

$$\Gamma_S(\Delta\mathbf{r}'') = \left(\frac{|E_0|}{i\lambda L}\right)^4 \left\langle \sum_{i,j,k,l}^{N} \int \exp(-ik_0\mathbf{r}_i(t_1)\cdot\mathbf{Q}_a)\exp(ik_0\mathbf{r}_j(t_1)\cdot\mathbf{Q}_b) \right.$$
$$\left. \times \exp(ik_0\mathbf{r}_k(t_1)\cdot\mathbf{Q}_c)\exp(-ik_0\mathbf{r}_l(t_1)\cdot\mathbf{Q}_d) \right\rangle$$

$$\times \exp\left(-\frac{\mathrm{i}k_0\mathbf{r}'_a \cdot \mathbf{r}''_1}{L}\right) \exp\left(\frac{\mathrm{i}k_0\mathbf{r}'_b \cdot \mathbf{r}''_1}{L}\right) \exp\left(\frac{\mathrm{i}k_0\mathbf{r}'_c \cdot \mathbf{r}''_2}{L}\right)$$

$$\times \exp\left(-\frac{\mathrm{i}k_0\mathbf{r}'_d \cdot \mathbf{r}''_2}{L}\right) \mathrm{d}^2\mathbf{r}'_a\,\mathrm{d}^2\mathbf{r}'_b\,\mathrm{d}^2\mathbf{r}'_c\,\mathrm{d}^2\mathbf{r}'_d \tag{45}$$

As in the calculation for the temporal autocorrelation function, the spatial autocorrelation function terms corresponding to $m = 1, 2, 3$ (see Table 10.1) produce zero values and the $m = 4, 5, 6$ terms produce non-zero values. In the latter terms, an integral appears that can be related to a first-order Bessel function:

$$\int \exp\left(\frac{-\mathrm{i}k_0\mathbf{r}' \cdot \Delta\mathbf{r}''}{L}\right) \mathrm{d}^2\mathbf{r}' = \frac{2J_1(\mu)}{\mu} \tag{46}$$

where $\Delta\mathbf{r}'' \equiv \mathbf{r}''_2 - \mathbf{r}''_1$ and $\mu = k_0 r_o{}'\Delta\mathbf{r}''/L$ and $r_o{}'$ is the radius of the objective.
The final form of the spatial autocorrelation function becomes:

$$\Gamma_S(\Delta\mathbf{r}'') = \langle I\rangle^2 \left\{ \left[1 + \left(\frac{2J_1(\mu)}{\mu}\right)^2\right] + \langle M\rangle^{-1} \left[\left(\frac{2J_1(\mu)}{\mu}\right)^4 - \left(\frac{2J_1(\mu)}{\mu}\right)^2 - 1\right] \right\} \tag{47}$$

For large $\langle M\rangle$, the $\langle M\rangle^{-1}$ term is small and is not included in further calculations. The leading term of the spatial autocorrelation function has the same distance dependence as the point spread function of the microscope objective at the image plane (an Airy disk). In analogy with Eq. (42), a normalized form of $\Gamma_s(\Delta\mathbf{r}'')$ can be written as

$$g_S(\Delta\mathbf{r}'') = \frac{\Gamma_S(\Delta\mathbf{r}'') - \langle I(\mathbf{r}'')\rangle^2}{\langle I(\mathbf{r}'')\rangle^2} \tag{48}$$

The characteristic spatial correlation distance is qualitatively the average distance in the image plane from constructive to destructive interference [18, 19]. It can be defined quantitatively as the distance l_c in $\mathbf{r}''$-space corresponding to $\mu = 1$. Parameter μ [see after Eq. (46)] can be rewritten in terms of the numerical aperture (NA) and magnification (mag) of the objective (in the low aperture, air immersion case) as:

$$\mu = \left(\frac{2\pi}{\lambda}\right)\left(\frac{\mathrm{NA}}{\mathrm{mag}}\right)\Delta\mathbf{r}'' \tag{49}$$

Thus the spatial correlation distance l_c is:

$$l_c = \Delta\mathbf{r}''_c = \left(\frac{\lambda}{2\pi}\right)\left(\frac{\mathrm{mag}}{\mathrm{NA}}\right) \cong 8\ \mu\mathrm{m} \tag{50}$$

The size of the CCD camera pixel in the experiments reported here is 6.3 μm, so that approximately one spatial correlation distance is observed in each pixel.

10.2.8
Variance of Intensity Fluctuations: Mobile Fraction

We define the mobile fraction β to be the ratio of the scattered intensity from the mobile scattering centers $\langle I\rangle_{\mathrm{mob}}$ to the total scattering intensity $\langle I\rangle$ in the collection volume of each pixel:

$$\beta = \frac{\langle I\rangle_{\mathrm{mob}}}{\langle I\rangle} \tag{51}$$

where

$$\langle I\rangle = \langle I\rangle_{\mathrm{mob}} + \langle I\rangle_{\mathrm{fix}} \tag{52}$$

and $\langle I\rangle_{\mathrm{fix}}$ arises from fixed scattering centers (such as the sample substrate). The mobile fraction β can be estimated from the variance of the intensity fluctuations, which is the difference between the extrapolated values of the temporal autocorrelation function values at $\tau = 0$ and at $\tau = \infty$. Combining the definition of Γ in Eq. (13) with Eq. (52), we get

$$\begin{aligned}\Gamma(\tau) &= \langle I_{\mathrm{mob}}(t)I_{\mathrm{mob}}(t+\tau)\rangle + \langle I_{\mathrm{mob}}(t)I_{\mathrm{fix}}(t+\tau)\rangle \\ &\quad + \langle I_{\mathrm{fix}}(t)I_{\mathrm{mob}}(t+\tau)\rangle + \langle I_{\mathrm{fix}}(t)I_{\mathrm{fix}}(t+\tau)\rangle\end{aligned} \tag{53}$$

The first term in the above equation can be reduced to $\langle I\rangle^2_{\mathrm{mob}} + g(\tau)\langle I\rangle^2_{\mathrm{mob}}$ where $g(0) = 1$ and $g(\infty) = 0$. The next two terms are each $\langle I\rangle_{\mathrm{mob}}\langle I\rangle_{\mathrm{fix}}$ and the last term is $\langle I^2\rangle_{\mathrm{fix}} = \langle I\rangle^2_{\mathrm{fix}}$ since the scattering due to the immobile intensity does not fluctuate. Therefore,

$$\Gamma(0) = 2\langle I\rangle^2_{\mathrm{mob}} + 2\langle I\rangle_{\mathrm{mob}}\langle I\rangle_{\mathrm{fix}} + \langle I\rangle^2_{\mathrm{fix}} \tag{54}$$

and

$$\Gamma(\infty) = \langle I\rangle^2_{\mathrm{mob}} + 2\langle I\rangle_{\mathrm{mob}}\langle I\rangle_{\mathrm{fix}} + \langle I\rangle^2_{\mathrm{fix}} = \langle I\rangle^2 \tag{55}$$

and therefore

$$\beta^2 = \frac{\langle I\rangle^2_{\mathrm{mob}}}{\langle I\rangle^2} = \frac{\Gamma(0) - \Gamma(\infty)}{\Gamma(\infty)} \tag{56}$$

10.3
Experimental Design

10.3.1
Optical Setup

Light from a CW helium neon laser (15 mW, 632.8 nm) was focused through an 8-mm focal length cylindrical lens and propagated down towards the sample (at ~60° from the vertical) such that the direct incident light missed the objective (Fig. 10.3a). The resulting illumination region was a thin line, 2 μm full-width at half-maximum and oriented so that the stripe was in the plane of incidence.

A Leitz Diavert inverted microscope with a 0.4 NA, 32X long working distance objective collected the scattered light from the sample. The objective focused the scattered light at an image plane above the microscope at which a 5 μm wide slit was located. Lens L in the path between the image plane and the face of the CCD camera refocused the scattered light and image plane slit onto the CCD array of a digital camera with no additional magnification. The image plane slit served to select a clean-edged line of scattered light that illuminated only a single column of pixels in the CCD camera (Fig. 10.3b). The image plane slit also created a confocal-like effect, reducing the out-of-focus light at the camera.

The sample chamber containing either cells or beads (see below) was placed on a motorized stage (Maerzhaeuser Wetzlar, Germany) that could be translated stepwise in the direction transverse to the illumination stripe.

Emission path filters (Chroma Technology, Corp., Brattleboro, VT) were chosen depending upon whether light scattering (for DLSM experiments) or fluorescence (for number fluctuation experiments) was performed. For DLSM, a 632 ± 10 nm laser narrow bandwidth filter was used. For the number fluctuation experiments, a long pass 650 nm filter was used to block scattered excitation light and transmit fluorescence from dye-labeled beads.

10.3.2
Data Acquisition

During an image acquisition period, the camera shutter remained open. The progressive scan cooled CCD camera (Pentamax-KAF-1400, Roper Scientific, Trenton, NJ, pixel size 6.8 × 6.8 μm) reads out data column-by-column such that the intensities recorded in each pixel column shifts one column (Fig. 10.4) every 300 μs. The shifting continues until the full frame of the CCD array is read. Therefore, each pixel along the column of illumination formed by the image plane slit in our optical setup gave rise to a temporal streak of intensity fluctuations with a "bin time" of 300 μs.

Custom PC software written in LabView and interfaced to a data acquisition board (National Instruments, Austin, TX) coordinated the image acquisition with the microscope stage controller (Lang MCL-2, Huttenberg, Germany). Images

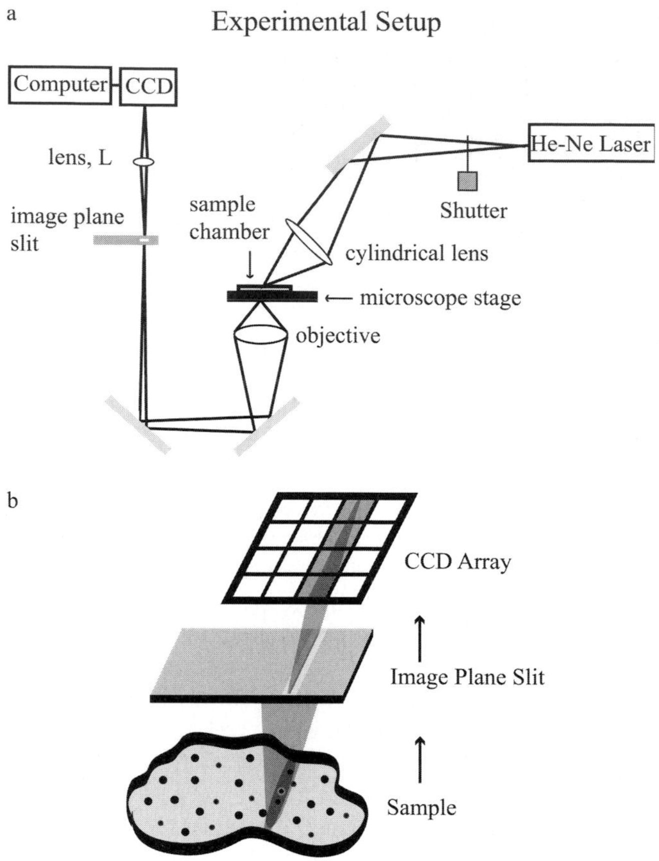

Fig. 10.3. Schematic drawing of experimental setup. (a) A cylindrical lens focuses the beam from a He-Ne laser to create a thin stripe of illumination (oriented in the page plane here) on the sample. A motorized stage controls the motion of the sample and the sample is moved in the *x*-direction only (normal to the page plane). A 32X, 0.4 NA, objective gathers the scattered light from the sample and focuses it onto a 5 µm wide slit placed in the image plane and oriented in the direction of the illumination stripe. A lens (L) re-images the slit onto the face of a CCD camera without additional magnification. (b) Detail of the imaging on the CCD array, with lenses omitted from the drawing. The slit excludes out-of-focus light and creates a well-defined focused line of scattered illumination that gets mapped onto a single column of the CCD camera array. Other columns of the CCD camera array are not illuminated.

were taken every 500 ms, followed by the advancement of the microscope stage to a new position on the sample. The sequence was repeated until the entire sample had been stepped through, resulting in a stack of images containing the intensity fluctuations from every point in the sample. Images were stored using a Windows 98, 600 MHz Intel P-III computer running WinView (Roper Scientific, Trenton, NJ).

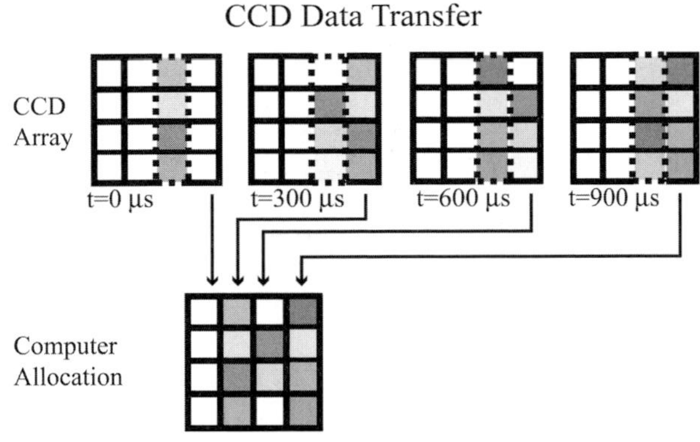

Fig. 10.4. Schematic of CCD readout mechanism. The dashed rectangle represents the column of the CCD array that is constantly exposed to the scattered light. The CCD array readout shift register is represented by the column of pixels at the extreme right. (Between the illuminated columns and the shift register there may be intermediate columns but these are not shown.) The (progressive scan) camera is operated in its normal readout mode, except with the camera shutter always open. At the end of every 300 μs interval, accumulated intensity data in the illuminated column of pixels is advanced rightward column-by-column toward the shift register. At the same time, blank counts from the pixel columns to the left are shifted into the illuminated column, effectively clearing it, and data in the shift register is read into an array in the computer memory. This process continues until the shift register has read every column of the CCD array, thereby completing image acquisition for a single stage position of the sample. The final "image" as recorded in the computer consists of a time sequence ($\Delta t = 300$ μs) of 800 intensity readings at each pixel along the illuminated pixel column. The next streak-like image is acquired after the motorized microscope stage has advanced to the next position on the sample. The whole image acquisition process is complete when the stage has stepped through the entire sample.

10.3.3
Sample Preparation: Polystyrene Beads

For dynamic light scattering experiments, 200 and 500 nm diameter polystyrene nanospheres (Duke Scientific, Palo Alto, CA) were tip-sonicated (to break up large clusters) and loaded at their undiluted aqueous suspension concentrations of 2.1×10^{12} particles mL^{-1} (200 nm) and 1.4×10^{11} particles mL^{-1} (500 nm) into separate rectangular cross-section glass microcapillary tubes (inner thickness 0.05×0.5 mm wide, Wilmad Specialty Glass, Buena, NJ) by capillary action. The two microcapillary tubes containing the two bead sizes were placed side-by-side on a plastic holder and oriented at a diagonal across the microscope field of view so that laser light line illuminated both microcapillary tubes simultaneously. Vacuum grease was used to seal the ends of the microcapillary tubes and to adhere them to the holder. For number fluctuations measurements, 200 nm carboxylate-modified fluorescent dark red FluoSpheres were obtained from Molecular Probes (Eugene, OR). FluoSpheres were drawn into microcapillary tubes at their undiluted concentration of 5.3×10^{12} particles mL^{-1} and mounted similarly.

The CCD camera acquired 40 images of the polystyrene beads assembly, each of which represented a line position 10 μm apart on the sample.

10.3.4
Sample Preparation: Living Macrophages

Coverslips (25 mm diameter, #2 thickness, and 32.5 mm diameter, #1.5 thickness) were treated for several hours in concentrated H_2SO_4 in a porcelain holder and then rinsed for 2 h in distilled water and oven dried overnight at 130 °F. This treatment greatly reduced the light scattering from the coverslips.

Two aqueous buffers, Ringer (RB) (155 mM NaCl, 5 mM KCl, 2 mM $CaCl_2$, 1 mM $MgCl_2$, 2 mM NaH_2PO_4, 10 mM HEPES, pH 7.2, 10 mM glucose) and Ringers with acetate (ARB) (80 mM NaCl, 70 mM sodium acetate, 5 mM KCl, 2 mM $CaCl_2$, 1 mM $MgCl_2$, 2 mM NaH_2PO_4, 10 mM HEPES pH 6.8, 10 mM glucose) were prepared for use during the experiment.

Monoclonal mouse macrophage cultures (RAW 264.7, American Type Tissue Culture, Manassas, VA) were obtained from the laboratory of Dr Joel Swanson (Department of Immunology, University of Michigan Medical School). The cultures were maintained in 60 mm diameter polystyrene tissue culture dishes in 10 mL of a 0.2 μm filtered (Millipore, Bedford MA) solution of Dulbecco's modified Eagle's medium containing 10% heat-inactivated fetal bovine serum, high glucose, L-glutamate and sodium pyruvate (DMEM, Grand Island Biological Company, Grand Island, NY) and penicillin–streptomycin. For replating onto the glass coverslips, cells were dislodged by vigorous trituration. A 1:30 dilution of the resulting suspension in DMEM was prepared, plated onto 32.5 mm coverslips (pre-cleaned as described above) in polystyrene dishes, and incubated for two days at 37 °C. Macrophage cells on a glass coverslip can spread to ~80 μm diameter circular shapes, with thicknesses of a couple of microns at the periphery and about 5 μm in the center.

The cell sample chamber consisted of a 25 mm diameter, 50 μm thick Teflon spacer ring, cut into two slightly separated halves to provide a channel for easy fluid exchange by capillary action. These spacers were placed on top of the 32.5 mm coverslip with adherent cells. A 25 mm diameter coverslip was then placed over the spacers to create a cell sandwich chamber, which was clamped with plastic clips over a drilled hole in a plastic plate. The cells in the chamber were gently rinsed with 1 mL of RB.

10.3.5
Buffer Changes during Data Acquisition

The CCD camera first took 120 successive streak images of the cell in RB with a 5 μm step stage motion between each image. Subsequently, 1 mL of ARB was exchanged into the cell coverslip sandwich and incubated for 20 min before a second similar round of streak image recording was initiated over the same cell area. A final 1 mL of RB buffer was exchanged into the sample chamber, another 20 min of incubation time elapsed, and the camera acquired a final similar sequence of 120 streak images over the same area.

10.4 Data Analysis

The two types of intensity autocorrelation functions – temporal and spatial – for each pixel along an illuminated line can be calculated from the width and height, respectively, of the same streak image (Fig. 10.5a).

10.4.1 Temporal Intensity Autocorrelation Function

Normalized temporal autocorrelation functions $g_T(\tau)$ of the experimental intensity fluctuations [see Eq. (42)] were calculated along the rows for each point on the line of illumination.

The theoretically expected temporal autocorrelation function [Eq. (43)] can be integrated numerically and fit to a single exponential (Fig. 10.2b). The close agreement between the fitted function and the numerical integration justifies approximating the experimental data with a single exponential. However, DLSM is very sensitive to extraneous sources of noise, such as table and acoustic vibrations and laser source fluctuations. To remove these largely periodic noise effects, the obtained temporal autocorrelation functions were fit to a sum of a single exponential, three sine waves, and a constant:

$$g_{\text{fit}}(\tau) = \mathrm{A}_0 e^{-\mathrm{A}_1\tau} + \mathrm{A}_2 \cos(2\mathrm{A}_3\pi\upsilon\tau) + \mathrm{A}_4 \cos(2\mathrm{A}_5\pi\upsilon\tau) + \mathrm{A}_6 \cos(2\mathrm{A}_7\pi\upsilon\tau) + \mathrm{A}_8 \tag{57}$$

where A_{0-8} are the fitting parameters. The A_1 decay rate parameter was assigned a pseudocolor for each pixel to create a spatial map of the fluctuation decay rates. The mean value of the decay rates for each bead size was used to calculate the apparent diffusion coefficients using $Dk_0^2\tau_c \approx 0.52$ where $\tau_c = 1/A_1$.

10.4.2 Spatial Intensity Autocorrelation Function

We calculated the normalized spatial autocorrelation function $g_s(\Delta\mathbf{r}'')$ [see Eq. (48)] along each of the columns of a streak image of the intensity fluctuations from polystyrene bead suspensions. To obtain a single spatial autocorrelation function for comparison with theory, all of the $g_S(\Delta\mathbf{r}'')$ were averaged over all of the columns. The result was not fit to the theoretical function, but the characteristic decay distance was compared with theory [see Eqs. (49) and (50)].

a

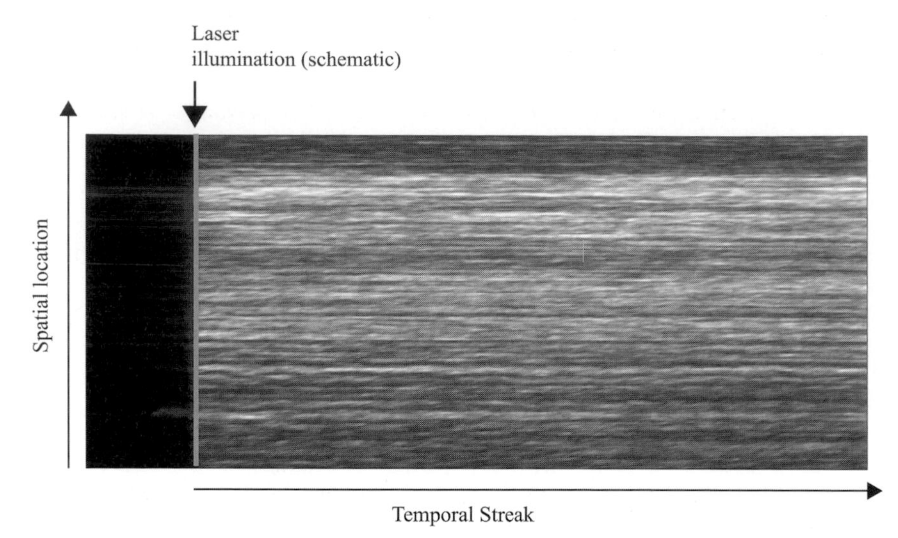

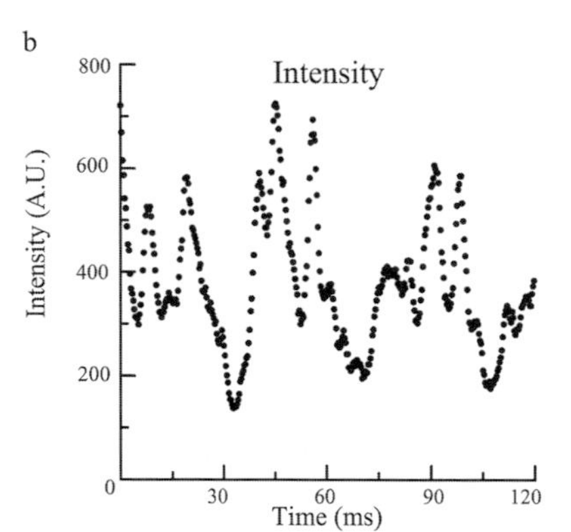

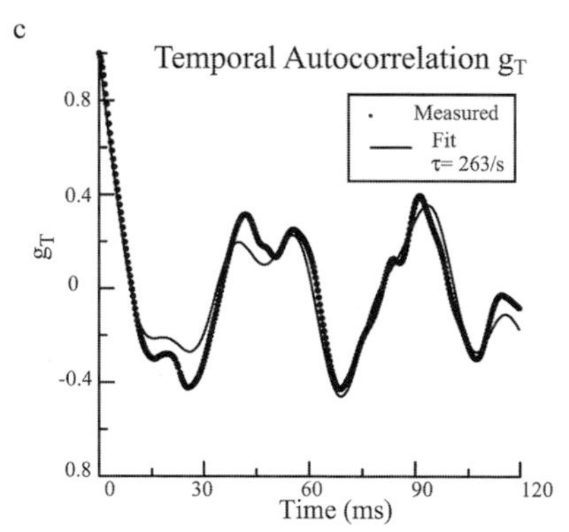

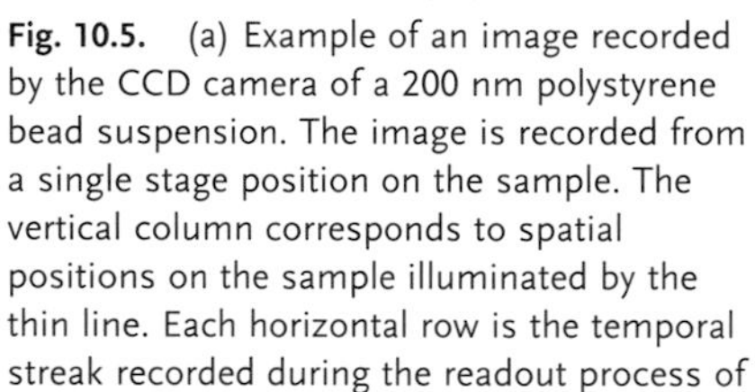

Fig. 10.5. (a) Example of an image recorded by the CCD camera of a 200 nm polystyrene bead suspension. The image is recorded from a single stage position on the sample. The vertical column corresponds to spatial positions on the sample illuminated by the thin line. Each horizontal row is the temporal streak recorded during the readout process of the CCD camera at this location. (b) An example of a recorded intensity fluctuation along one of the rows. (c) The autocorrelation function of the intensity time course shown in (b), based directly on the experimental intensities (dotted line) and then subsequently fitted with the form of Eq. (43) (solid line).

10.4.3
Mobile Fraction

Mobile fraction β was calculated pixel-by-pixel from $\Gamma(0)$ and $\Gamma(\infty)$ [Eq. (56)], and the calculated values were assigned different pseudocolors for display. Eq. (56) assumes that phase fluctuations are the only contributors to $\Gamma(\tau)$. However, in an actual experiment, $\Gamma(\tau)$ can also be affected by other factors, as follows.

(a) *Photon shot noise* contributes to the amplitude of $\Gamma(\tau)$, but only at $\tau = 0$ since the photon arrival times are uncorrelated. Thus, to obtain an accurate estimate of $\Gamma(0)$ apart from shot noise, a linear extrapolation was performed from the values of $\Gamma(1)$ and $\Gamma(2)$.
(b) *Number fluctuations* were recorded on a sample that produces no coherent phase fluctuations: fluorescent beads. Here, the variance was very small, only marginally above the noise level in $\Gamma(\tau)$, and $\Gamma(0)$ was estimated as $\Gamma(1)$.
(c) *Overly large pixels size* can decrease the recorded variance; pixels should be smaller than the spatial correlation distance, as indeed occurs here (see Section 10.2).

10.5 Experimental Results

Polystyrene beads were used to quantitatively test the DLSM technique and compare the experimental values of the mean decay rates and mobile fractions with values predicted from theory. DLSM was then applied to macrophage cells as a test on a living biological system. Macrophages are very motile and their motility can be altered pharmacologically.

10.5.1 Polystyrene Beads: Temporal Phase Autocorrelation

An example of scattered light intensity fluctuations along with the corresponding autocorrelation function is displayed in Figs. 10.5(b) and (c). The temporal autocorrelation functions were fit to a linear combination of a single exponential and three sine waves [Eq. (57)]. The sine waves were added to the fitting function because the obtained autocorrelation functions exhibited some pseudo-oscillatory behavior with varying frequencies in addition to the expected exponential decay. An average of the (~1000) temporal autocorrelation functions obtained from the rows of a single streaked image of 200 nm polystyrene beads is displayed in Fig. 10.6. This average exhibits only an exponential decay, indicating that the pseudo-oscillations in some individual rows are random noise arising from the low number of correlation times in the time represented by a streak (240 ms).

Figure 10.7(b) shows a DLSM decay rate pseudocolor spatial map of a section of two adjacent microcapillary tubes containing 200 and 500 nm diameter beads, respectively. Although the range of colors suggests a wide distribution of decay rates in each tube, there are more pixels with shorter decay rates for the system of 200 nm beads than for 500 nm beads, as can be seen in normalized histograms of the decay rates for both sonicated and unsonicated beads of both sizes (Fig. 10.8). Fluctuation decay rates are generally much faster for the 200 nm beads. In addition, sonication resulted in an increase of the average decay rates towards the theoretical values. Average values of decay rates measured with our technique were 209 and 256 s^{-1} (non-sonicated and sonicated, respectively) for the 200 nm beads and

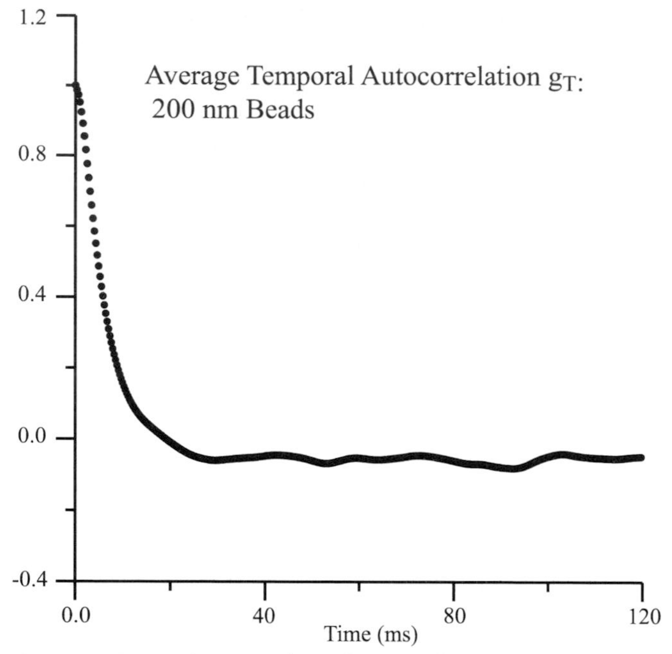

Fig. 10.6. Temporal autocorrelation function of 200 nm polystyrene beads, averaged over all the rows of the streaked image shown in Fig. 10.5(a).

107 and 144 s^{-1} (non-sonicated and sonicated, respectively) for the 500 nm beads. These decay rates correspond to diffusion coefficients of 1.35×10^{-8} and 8.75×10^{-9} cm^2 s^{-1} for the sonicated 200 and 500 nm beads, respectively. Table 10.2 compares the experimentally obtained results with the theoretically predicted values. The experimental results are in good agreement with the hydrodynamic theory values for the 500 nm sonicated beads and about 40% lower than the theoretical expectation for the 200 nm sonicated beads. There is no obvious correlation between variations in the local decay rates and the local mean scattered light intensity (data not shown).

10.5.2
Variance of Intensity Fluctuations on Beads: Phase Fluctuations

A qualitative representation for the size of the intensity fluctuations can be obtained by visual inspection of an image from the CCD camera. Figure 10.9 shows a raw streaked image recorded by the CCD camera for scattered light (DLSM fluctuations, Fig. 10.9a) and for fluorescence (number fluctuations, Fig. 10.9c), both recorded from 200 nm carboxylate-modified dark red FluoSpheres. Figure 10.9(b) shows the normalized temporal autocorrelation function, $g_T(\tau)$, for a typical row in the DLSM streaked image depicted in Fig. 10.9(a). Figure 10.9(d) shows a corre-

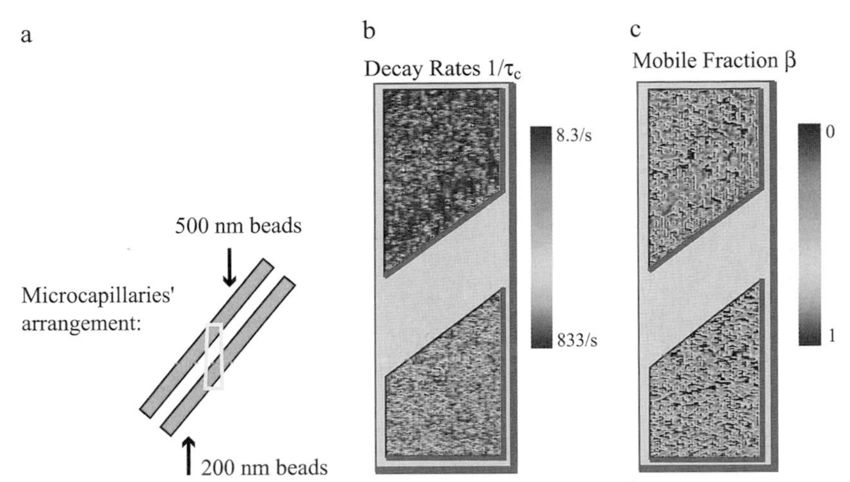

Fig. 10.7. Experimental spatial maps for the polystyrene beads. (a) Two different sizes of polystyrene beads were used in a single experiment: 200 and 500 nm. The beads were in separate microcapillary tubes oriented at an angle with respect to the optical axis and mounted on a homemade microscope sample chamber. The yellow box represents the field of view of the microscope. (b) Spatial map of the decay rates in the field of view. Note the faster diffusion of the 200 nm beads, as expected. (c) Spatial map of the mobile fractions, obtained using Eq. (56), in the same field of view. A zero on the scale implies that most of the scattering centers are immobile whereas unity implies all of the scattering centers are mobile. Both bead sizes show a large mobile fraction: for the 200 nm beads $\beta_{avg} = 0.63$, and for the 500 nm beads $\beta_{avg} = 0.42$. The 200 nm beads show a smaller spread of the degree of mobility than the 500 nm beads.

sponding normalized temporal autocorrelation function for the number fluctuation experiment of Fig. 10.9(c). The relative variance of phase fluctuations is much larger than the relative variance of number fluctuations.

In principle, the amplitude of the normalized autocorrelation function for an entirely mobile DLSM sample should be unity. But, as can be seen from Fig. 10.9(b), the experimental value is ~0.6. This decreased amplitude is probably due to a fixed scattering background and a consequent "mobile fraction" that is less than unity.

10.5.3 Polystyrene Beads: Number Fluctuations

Number fluctuations without contamination by phase interference effects could be autocorrelated from streaked images of the fluorescence from labeled polystyrene beads. The average value of the normalized variance for all of the temporal autocorrelation functions from number fluctuations was only 10^{-3}. Thus, we conclude that the fluctuations observed at 632 nm were due almost exclusively to dynamic light scattering phase fluctuations rather than number fluctuations of particles dif-

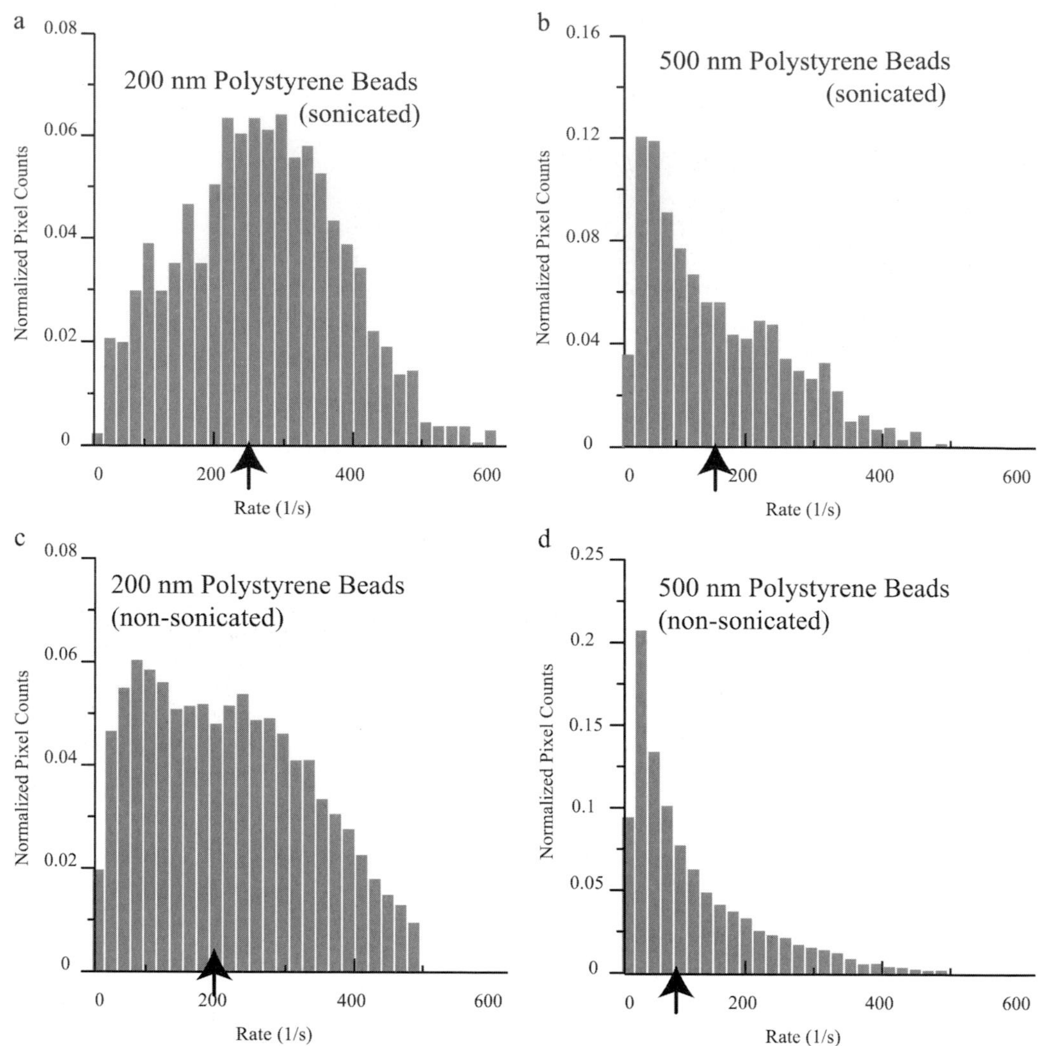

Fig. 10.8. Histograms of the decay rates for the polystyrene beads. (a, c) 200 nm beads (sonicated and unsonicated). (b, d) 500 nm beads (sonicated and unsonicated). Arrows indicate mean values.

Tab. 10.2. Mean diffusion coefficients of polystyrene beads (average experimentally measured diffusion coefficients for 200 and 500 nm beads along with expected values from hydrodynamic theory).

Diffusion coefficient (10^{-8} cm^2 s^{-1}) $\pm$ S.E.	*200 nm sonicated beads (2.1×10^{12} particles mL^{-1})*	*500 nm sonicated beads (1.4×10^{11} particles mL^{-1})*
Experimental	1.35 ± 0.03	0.875 ± 0.03
Theoretical	2.23	8.76

a

Raw CCD Image: Phase Fluctuations

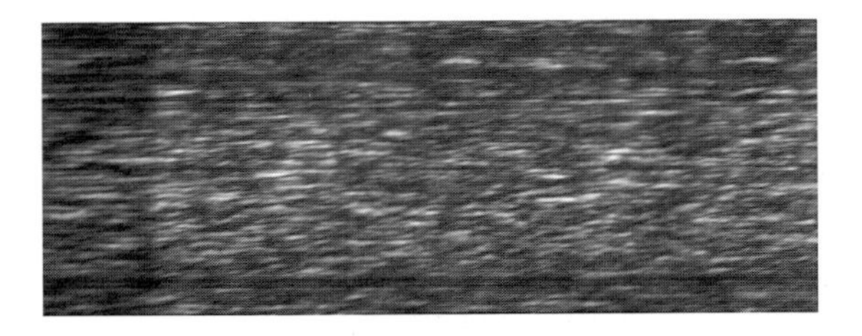

b

Autocorrelation g_T

c

Raw CCD Image: Number Fluctuations

d

Autocorrelation g_T

Fig. 10.9. CCD camera images and their corresponding autocorrelation functions of intensity fluctuations from 200 nm carboxylate-modified dark red FluoSpheres. (a) Raw streak image from dynamic light scattering measurement. (b) Normalized temporal autocorrelation function for panel a, calculated using Eq. (43). The large variations in intensity across the raw image imply a large amplitude in the temporal autocorrelation function. (c) Image from the number fluctuation experiment is taken from the same sample as in panel a, but with a long pass 650 nm filter placed in the emission path. The image is very smooth, indicating small intensity variations. (d) Normalized temporal autocorrelation function for panel c. The function has a very small amplitude, implying that the contribution from particle number variations to the DLSM intensity fluctuations seen in panels a and b is very small.

fusing in and out of the volume observed by a pixel. The intensity fluctuations resulting from the fluorescence emission were so small that decay rates could not be reliably extracted from the temporal autocorrelations.

10.5.4
Polystyrene Beads: Spatial Autocorrelation

Figure 10.10 shows the spatial autocorrelation function obtained for the 200 nm sonicated polystyrene beads along with the theoretically predicted one [Eq. (48)].

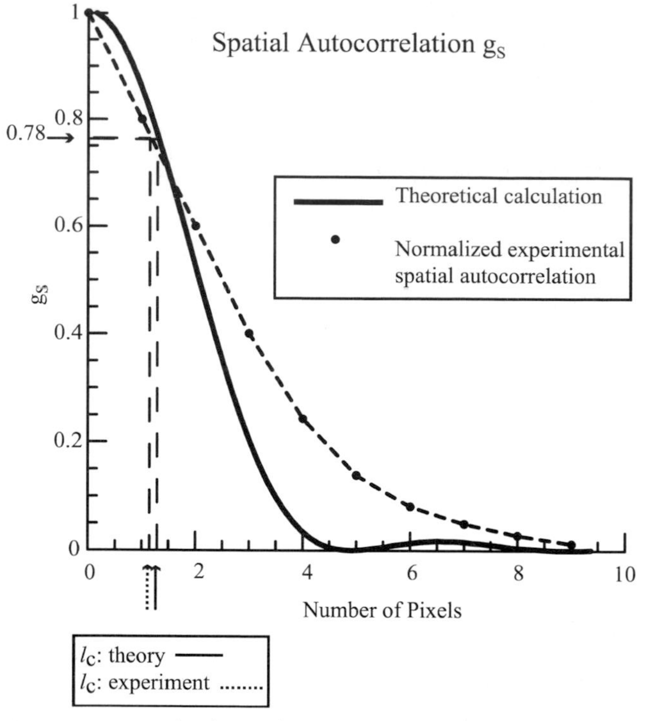

Fig. 10.10. Normalized spatial autocorrelation function for the 200 nm polystyrene beads. The experimentally obtained spatial autocorrelation function is depicted by black dots connected by dashed lines. The theoretical spatial autocorrelation function [based on Eq. (48)] is depicted by the solid line. The characteristic decay distance of the spatial autocorrelation function (where $\mu = 1$) is marked for both the experimentally and theoretically obtained curves.

The amplitude at which the characteristic distance is defined ($\mu = 1$) is indicated, from which the spatial correlation distance l_c can be derived [Eq. (50)]. The spatial correlation distance obtained from the theoretical curve is approximately $l_c = 1.27$ pixels, which corresponds to 8 μm, whereas the value calculated from the experimentally obtained spatial autocorrelation function is $l_c = 1.1$ pixels, which corresponds to 6.93 μm. Thus, the experimentally obtained value of the spatial correlation distance agrees well with the theoretically derived one. In addition, since the size of the CCD camera pixel is 6.3 μm, each pixel will observe intensity fluctuations from approximately one spatial correlation area.

10.5.5
Polystyrene Beads: Mobile Fractions

Mobile fractions were calculated using Eq. (56). Figure 10.7(c) shows the spatial map of the mobile fractions obtained from the dynamic light scattering intensity fluctuation measurements for the 200 and 500 nm beads. The 200 nm beads show a higher overall relative mobility than the 500 nm beads. This is because the concentration of the 200 nm beads is an order of magnitude larger than that of the 500 nm beads, which caused the mean light scattering intensity to be larger from the smaller beads. Thus, scattering from the smaller beads constituted a larger fraction of the total scattering in the experimental system, which includes background scattering from the microcapillary tubes and optics.

10.5.6 Living Macrophage Cells: Temporal Autocorrelation

DLSM experiments were performed on a total of 21 macrophage cells as follows: (a) 10 min after the cell was removed from the incubator and washed with Ringers

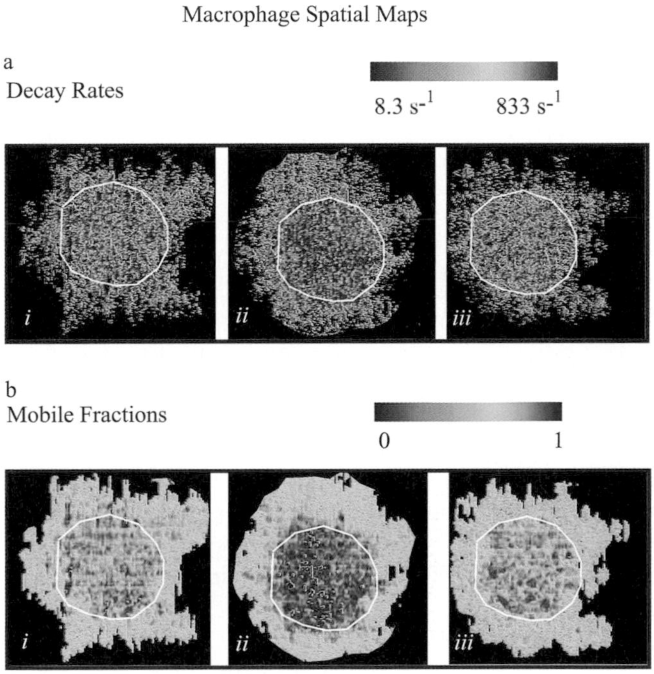

Fig. 10.11. Experimental spatial maps of a macrophage cell. For both panels a and b, each map depicts one of the three treatments on the cell: (i) Cell was flushed with 1 mL of Ringer's buffer (RB); (ii) cell was flushed with 1 mL of an acetate-containing RB (ARB) and incubated for 20 min; (iii) the ARB was washed out with RB and the cell was incubated for 20 min. (a) Spatial map of the temporal autocorrelation intensity fluctuation decay rates. There is a noticeable decrease in motility of the acetate-treated cell. Washing out the acetate reverses this attenuated motility, restoring the cell to its initial state. Black pixels occur where the scattering intensity was too low to obtain useful data. (b) Spatial map of the mobile fractions β of macrophages. The mobile fraction data are plotted on a logarithmic scale. Most of the motion occurs in the central interior portion of the cells.

buffer; (b) 20 min after the acetate-containing Ringers buffer was introduced; and (c) 20 min after Ringers buffer with acetate was replaced with plain Ringers buffer to remove the acetate. Figure 10.11(a) shows pseudocolor spatial maps of the fluctuation decay rates created for the three conditions for a typical cell.

A white overlay line divides the cell image into outer and inner regions. In all three maps, the outer region of the cell shows faster motility than the inner region. Additionally, the center of the cell shows a noticeable decrease in the macrophage motility when acetate is added. The original motility seems to be restored when acetate is washed out. The black areas within the periphery of the cell region represent the locations at which the decay rates could not be reliably calculated because the scattered intensity was too low.

10.5.7 Living Macrophage Cells: Mobile Fraction

Figure 10.11(b) shows the spatial maps of the mobile fractions on a log scale for the three treatments described above. Overall, the mobile fraction is quite low, especially in the outer region where the mean scattering intensity from the cell is a smaller fraction of the background scattering.

10.6 Discussion

We have demonstrated that the dynamic light scattering imaging technique presented here is a viable method to observe and image small relative motions among nearby scattering centers in a microscope [20, 21]. These relative motions, visualized as intensity fluctuations, are detected in both non-biological and biological systems using a slow scan CCD camera. Fluctuations as fast as 250 s^{-1} were measured and relative motions between scattering centers on the order of a fraction of a wavelength were spatially mapped. The technique provides information about the relative motion of the scattering centers (rather than their relative location as with conventional microscopy) without sacrificing optical resolution.

10.6.1 Polystyrene Beads

The experimental diffusion coefficients were relatively correct for the ratio of 200 vs. 500 nm polystyrene beads with the values for the 200 nm beads slower than expected. However, both were somewhat lower than theoretically predicted from hydrodynamics. This discrepancy can be explained by the fact that aggregates were most likely present in both samples, forming particles with larger effective sizes and therefore lowering their diffusion rate. Sonication of the bead suspensions increased the diffusion rates towards the theoretical value.

Because of the high concentration of beads in these samples, we would expect that number fluctuations should be small, as was experimentally confirmed. Also, because the relevant characteristic volume for number fluctuations (i.e., the effective the volume observed by a pixel) is typically larger than the characteristic volume leading to phase fluctuations, number fluctuations are likely to be slower than phase fluctuations, as was also experimentally confirmed. Number fluctuations do not contribute much to the DLSM streaked images observed here.

Although not relevant to our experimental conditions, extreme cases could show interesting number fluctuation effects, according to the theory. In the limit of one scattering center, the only contribution to the intensity autocorrelation function will arise entirely from number fluctuations and so there will not be a contribution due to phase fluctuations. If the number fluctuations occur on a faster time scale than the phase fluctuations, there would be an appreciable decay of the autocorrelation function before a phase fluctuation took place. Thus, only if the phase fluctuations occur on a faster time scale than the number fluctuations will any phase fluctuations be detected.

The spread in diffusion rates is in part due to the noisiness of the temporal autocorrelation function. This effect arises from the fact that the temporal autocorrelation function is computed over a relatively small number of coherence times. The coherence time of the intensity fluctuations for the 200 nm polystyrene beads (τ_c) was approximately 3 ms and the intensity fluctuations were collected over a 240 ms period (T, the time duration of a streak in the CCD record). The signal-to-noise in the autocorrelation function is thereby expected to be

$$(T/\tau_c)^{1/2} \approx 9$$

As an alternative approach to estimating the spread of autocorrelation decay rates arising solely from a finite experimental time T, we calculated the temporal autocorrelation function from an ensemble of numerically generated "telegraph" signals, over the same number of coherence times as in the experiment. A telegraph signal has transitions between $+1$ and -1 at completely uncorrelated times (but at a certain average rate) and has an exponentially decaying autocorrelation function [22].

Figure 10.12(a) shows the experimentally obtained distribution of decay rates for 200 nm polystyrene beads. Figure 10.12(b) is the distribution of decay rates from the autocorrelation function of the telegraph signal. The mean decay rate of the telegraph signal was scaled to match that of the experimental data. Comparison of the two confirms that much of the spread of experimental characteristic rates on the beads is due to an insufficient number of correlation times in T. However, the width of the distribution of the decay rates for the beads is still somewhat wider than that for the noise signal and it is skewed towards the slower decay rates. This indicates that the width of the distribution for the beads is in due in part to spatial heterogeneity of different sized aggregates in the 240 ms time scale of the experiments.

The theoretical calculation indicates that the amplitude of the normalized tem-

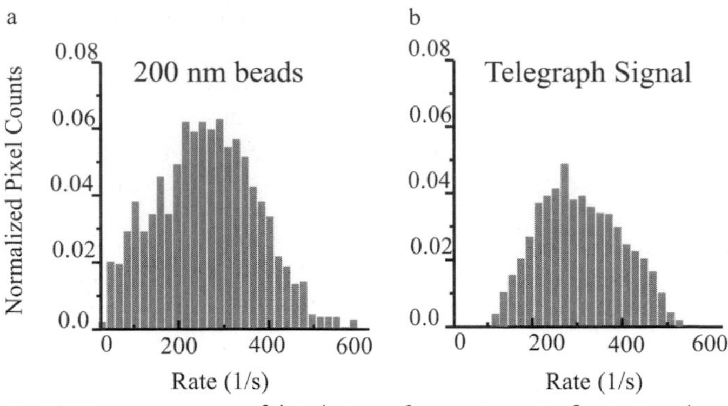

Fig. 10.12. Comparison of distribution of decay rates for the 200 nm polystyrene beads and for a numerically generated telegraph signal with the same average rate and number of coherence times for calculation. (a) Normalized intensity fluctuation decay rates for 200 nm polystyrene beads. (b) Normalized intensity fluctuation decay rates for the telegraph signal. The finite number of coherence times results in a spread in the intensity fluctuation decay rates for both the experimental data as well as the numerically generated noise signal.

poral autocorrelation function should be unity. However, the amplitude of the experimentally obtained autocorrelation function for polystyrene beads is significantly less. This amplitude reduction could arise for two reasons. First, there is a large background of immobile scattering emanating from the collection optics. This background does not contribute to the intensity fluctuation decay of the autocorrelation function, but does increase in the overall light intensity. Secondly, since the spatial correlation distance covered a region less than two pixels in size, not all the parts of a single pixel will see the same intensity, thereby decreasing the pixel-to-pixel variance of the recorded intensities.

The "proof-of-concept" experiments were performed on 200 and 500 nm polystyrene beads. However, notably, this technique is limited not by the size of the scattering center, but by the "scattering power". If the scattering center is large but has the same index of refraction as the surrounding environment it will not be detected, but if the scattering center is extremely small but index mismatched to the surrounding environment it will be detected. If, in a microscope resolution region, the scattering center emits less light than the Raleigh scattering of the water, it will not produce much of a DLSM effect.

10.6.2
Macrophages

Macrophages are immune system cells well known for their high motility. They are found in all tissues and in the blood. They are scavenger cells, acting as the first line of defense in recognizing and killing foreign microorganisms through phagocytosis [23]. During phagocytosis, the macrophage invaginates foreign material and breaks it down enzymatically.

Previous work on macrophages showed there are two distinct regions of the cell, a thin outer section and a thicker central section [24]. The DLSM spatial maps of the fluctuation decay rates of the imaged cells confirm this finding. Those maps indicate that the outer zone has faster fluctuations than the center of the cell. This result is particularly interesting because previous experiments from this lab, using polarized total internal reflection fluorescence microscopy, [25] showed the plasma membrane in the outer zone to be rather flat and inert whereas the plasma membrane of the central zone was quite bumpy and dynamic. Evidently, other intracellular scattering centers in that outer zone are moving rapidly.

Treatment with acetate-containing buffer reduced the rate of motion of scattering centers in the interior zone (Fig. 10.11). Previous studies have shown that exposure of the macrophage to acetate-containing Ringer's buffer acidifies the cell and causes the lysosomes concentrated in the center of the cell to migrate radially outward to the cell edge [26]. This causes certain regions within the cell to switch from high activity to quiescence.

The studies on macrophages here show that DLSM is feasible on living cells. It essentially provides contrast in unlabeled cells based on rapid rates of submicroscopic motion rather than static refractive index gradients. However, background scattering can be significant and can make some of the least-scattering regions of a cell unusable for DLSM.

10.6.3 Improvements for DLSM

Further improvements in DLSM should aim to increase its spatial and chemical specificity. Our current optical setup works in an epi-illumination mode. Since the depth of the focal region is approximately 2 μm, the entire thickness of the cell is illuminated. Therefore, the scattered light intensity comes from multiple layers of scattering centers in the region of illumination. To begin to understand the various physical processes contributing to the intracellular motion causing the phase fluctuations, the illuminated region should be better defined to a thin optical slice, such as obtainable with total internal reflection (TIR) illumination [27]. With TIR illumination, the cell membrane and submembrane structure would be the major component in the region of illumination.

Another improvement to DLSM would be to increase its chemical specificity while maintaining the coherent scattering necessary for intensity phase fluctuations. One possibility is to combine DLSM with second harmonic generation imaging microscopy (SHIM) [28]. Second harmonic generation (SHG) is a second-order optical process such that the monochromatically scattered light is coherent but is double the frequency of the incident light. SHG signals can only arise from scattering centers lacking a center of symmetry. Cell membranes and many highly ordered structural proteins intrinsically lack this symmetry center and can be imaged via SHIM. In addition, chemical dyes are available having the required second-order optical nonlinearities. The dyes can be specifically bound to target organelles and have been used to increase the SHG resolution and contrast within the membrane of the cell [29].

Another technique that could be combined with DLSM is a modification of resonance light scattering microscopy (RLS) [30, 31]. RLS in its current form produces highly monochromatic scattered light from gold particles in suspension illuminated by a xenon lamp, with the scattered color dependent upon the size of the particles. In an application to cell biology, these gold particles could be biochemically modified to specifically label organelles within a cell. The gold particles have a high light-scattering power and are unbleachable. Illumination with a wavelength-tunable laser tuned to the resonance of the gold particles would produce coherent scattered light emanating mainly from the gold particles. Phase intensity fluctuations would occur between neighboring gold particles; these fluctuations could then be autocorrelated to measure relative motions among specifically constituents in the cell.

Acknowledgments

We thank Drs Kenneth Christensen, Adam Hoppe, and Joel Swanson for macrophage cell culturing assistance.

References

1 Pecora, R., Doppler shifts in light scattering from pure liquids and polymer solutions. *J. Chem. Phys.* **1964**, *40*, 1604–1614.
2 Cummins, H. Z., Knable, N., Yeh, Y., Observation of diffusion broadening of Rayleigh scattered light. *Phys. Rev. Lett.* **1964**, *12*, 150–153.
3 Maeda, T. and Fujime, S., Quasi-elastic light scattering under optical microscope. *Rev. Sci. Instrum.* **1972**, *43*, 566–567.
4 Mishina, H., Asakura, T., Nagai, S., A laser Doppler microscope. *Opt. Commun.* **1974**, *11*, 99–102.
5 Cochrane, T., Earnshaw, J. C., Practical laser Doppler microscopes. *J. Phys. E: Sci. Instrum.* **1978**, *11*, 196–198.
6 Herbert, T. J., Acton, J. D., Photon correlation spectroscopy of light scattered from microscopic regions. *Appl. Opt.* **1979**, *18*, 588–590.
7 Nishio, I., Tanaka, T., Imanishi, Y., Ohnishi, S. T., Hemoglobin aggregation in single red blood cells of sickle cell anemia. *Science* **1983**, *220*, 1173–1174.
8 Blank, P. S., Tishler, R. B., Carlson, F. D., Quasielastic light scattering microscope spectrometer. *Appl. Opt.* **1987**, *26*, 351–356.
9 Peetermans, J., Nishio, I., Ohnishi, S. T., Tanaka, T., Light-scattering study of depolymerization kinetics of sickle hemoglobin polymers inside single erythrocytes. *Proc. Natl. Acad. Sci. U.S.A.* **1986**, *83*, 352–356.
10 Peetermans, J. A., Foy, B. D., Tanaka, T., Accumulation and diffusion of crystallin inside single fiber cells in intact chicken embryo lenses. *Proc. Natl. Acad. Sci. U.S.A.* **1987**, *84*, 1727–1730.
11 Peetermans, J. A., Matthews, E. K., Nishio, I., Tanaka, T., Particle

motion in single acinar cells observed by microscope laser light scattering spectroscopy. *Eur. Biophys. J.* **1987**, *15*, 65–69.

12 Peetermans, J. A., Nishio, I., Ohnishi, T., Tanaka, T., Single cell laser light scattering spectroscopy in a flow cell: repeated sickling of sickle red blood cells. *Biochim. Biophys. Acta* **1987**, *931*, 320–325.

13 Tishler, R. B., Carlson, F. D., A study of the dynamic properties of the human red blood cell membrane using quasi-elastic light scattering spectroscopy. *Biophys. J.* **1993**, *65*, 2586–2600.

14 Wong, A., Wiltzius, P., Dynamic light scattering with a CCD camera, *Rev. Sci. Instrum.* **1993**, *64*, 2547–2549.

15 Kaplan, P. D., Trappe, V., Weitz, D. A., Light scattering microscope, *Appl. Opt.* **1999**, *38*, 4151–4157.

16 Cummins, H. Z., Carlson, F. D., Herbert, T. J., Woods, G., Translational and rotational diffusion constants of tobacco mosaic virus from Rayleigh linewidths. *Biophys. J.* **1969**, *9*, 518–546.

17 Klein, M. V., *Optics*. Wiley, New York, **1970**.

18 Jakeman, E., Oliver, C. J., Pike, E. R., The effects of spatial coherence on intensity fluctuation distributions of Gaussian light. *J. Phys. A* **1970**, *3*, L45–L48.

19 Cantrell, C. D., Fields, J. R., Effect of spatial coherence on the photoelectric counting statistics of Gaussian light. *Phys. Rev. A* **1973**, *7*, 2063–2069.

20 Dzakpasu, R., Axelrod, D., Dynamic light scattering microscopy: A novel optical technique to image submicroscopy motions I: Theory, *Biophys. J.* **2004**, *87*, 1279–1287.

21 Dzakpasu, R., Axelrod, D., Dynamic light scattering microscopy: A novel optical technique to image submicroscopy motions II: Experimental applications, *Biophys. J.* **2004**, *87*, 1288–1297.

22 Davenport, W. and Root, W., *An Introduction to the Theory of Random Signals and Noise*. McGraw Hill, New York, **1958**.

23 Cannon, G. J. and Swanson, J. A., The macrophage capacity for phagocytosis. *J. Cell Sci.* **1992**, *101*, 907–913.

24 Swanson, J. A., Locke, A., Ansel, P., Hollenbeck, P. J., Radial movement of lysosomes along microtubules in permeabilized macrophages. *J. Cell Sci.* **1992**, *103*, 201–209.

25 Sund, S. E., Swanson, J. A., Axelrod, D., Cell membrane orientation visualized by polarized total internal reflection fluorescence. *Biophys. J.* **1999**, *77*, 2266–2283.

26 Heuser, J., Changes in lysosome shape and distribution correlated with changes in cytoplasmic pH. *J. Cell Biol.* **1989**, *108*, 855–864.

27 Axelrod, D., Total internal reflection fluorescence microscopy in cell biology. *Methods Enzymol.* **2003**, *361*, 1–33.

28 Campagnola, P. J., Clark, H. A., Mohler, W. A., Lewis, A., Loew, L. M., Second harmonic imaging microscopy of living cells. *J. Biomed. Opt.* **2001**, *6*, 277–286.

29 Campagnola, P., Millard, A., Terasaki, M., Hoppe, P., Malone, C., Mohler, W., Three-dimensional high-resolution second-harmonic generation imaging of endogenous structural proteins in biological tissues. *Biophys. J.* **2002**, *82*, 493–508.

30 Yguerabide, J. and Yguerabide, E., Light-scattering submicroscopic particles as highly fluorescent analogs and their use as tracer labels in clinical and biological applications I. theory. *Anal. Biochem.* **1998**, *262*, 137–156.

31 Yguerabide, J. and Yguerabide, E., Light-scattering submicroscopic particles as highly fluorescent analogs and their use as tracer labels in clinical and biological applications II. Experimental characterization. *Anal. Biochem.* **1998**, *262*, 157–176.

11
X-ray Scattering Techniques for Characterization of Nanosystems in Lifesciences

Cheng K. Saw

11.1
Introduction

This chapter aims to provide the basics of using X-ray diffraction techniques to obtain information on the structure and morphology of nanosystems, and also to point out some of its strengths and weaknesses when compared to other characterization techniques. X-ray scattering techniques cover a wide range of density domains, from a tenth to a thousandth of an angstrom. Essentially, this covers a whole range of condensed matter, including the structure and morphology of nanosystems, which is particularly useful for examining nanostructures in the lifesciences. This range of domain size requires both wide-angle X-ray scattering (WAXS) and small-angle X-ray scattering (SAXS) techniques. Roughly, WAXS covers from 2 nm down, and SAXS covers from 0.5 to 100 nm and possibly 1000 nm for a finely tuned instrument. A brief theoretical description of both WAXS and SAXS is given in this chapter. WAXS is a powerful technique in providing information on the crystallographic structure, or lack of structure, atomic positions and sizes in a unit cell and, to some extent, chemical compositions and chemical stoichiometry. Examples of such experiments will also be given. To describe the technique of X-ray scattering, some historical and theoretical background will be given in the hope of making this subject both interesting and simple.

Over the past 10 to 20 years, the major development in this scattering technique is in the instrumentation. Better and faster detectors have been developed. Solid-state detectors and position sensitive detectors clearly play an important role in energy discrimination and high-speed data acquisition. The X-ray beams are conditioned with the latest technology in monochromators, mirrors and multi-layers and they can also be made to focus or collimate to the desired probe sizes without significant reduction in X-ray flux. X-ray optics have been improved drastically. Improvements in data quality have also been demonstrated. The basic theory regarding X-ray diffraction and interactions of X-ray with matter developed in the early nineteen hundreds remains. The arrival of synchrotron radiation for materials probe afforded a quantum leap in all aspects of diffraction capabilities (Section 11.13). Some examples are given in the later part of this chapter.

Nanotechnologies for the Life Sciences Vol. 3
Nanosystem Characterization Tools in the Life Sciences. Edited by Challa S. S. R. Kumar

ISBN: 3-527-31383-4

By and large, X-ray diffraction capability is a major, and necessary, component in any modern characterization laboratory. In most cases, several diffraction instruments are needed for numerous reasons, e.g., high through-put, different X-ray optics for high resolution requirements, different incident energies may be needed, different scan types and so on. Some of these will be more apparent in the later part of this chapter. In general, X-ray scattering is the quickest and cheapest method to obtain a great deal of structural information. For example, identification of crystalline phases, phase impurities, additional disordered phases and the quality of the phases as well as lack of phases. It is also non-destructive and requires very little, or no, major sample preparation. In today's technological world, structural information is required on very different sample types, e.g., thin films, multilayers, very small amounts of samples, specific crystalline orientations, residual stresses and so on. X-ray diffraction has also played an important part in protein crystallography where high quality data with an enormous number reflections are needed for analysis to determine the structure of the protein. Clearly, with such sophisticated demands the experimental setups have to be somewhat unique for each investigation. This chapter also points out the experimental difficulties and ways to compensate and optimize the scattering properties when carrying out these types of experiments on different nanosystems. These small and unique changes in the set-up determine the quality of data and, thus, the success of the experiment.

X-ray scattering techniques also play a major role in obtaining information on both the structure and morphology of materials in nanosystems in the lifesciences. Clearly, the understanding of structure and morphology of materials is a basic requirement for one to design and generate materials with particular desired functions. Other characterization techniques, well established over the years for materials science, also play major roles in characterizing nanosystems for lifescience applications.

Experiments requiring higher resolution or sensitivity can be carried out using more sophisticated national facilities, like the synchrotron and neutron sources at several major research facilities. In fact, the development of synchrotron sources for characterization purposes represents a major advancement for X-ray scattering techniques. Experiments that are either not possible or very difficult to perform using in-house sources can now be easily carried out on one of the beamlines at a synchrotron source. It has greatly enhanced X-ray scattering capabilities. Most national facilities are open to general users around the world. Synchrotron radiations are also X-rays. In this case, the X-rays are highly collimated, polarized, tunable and also have enormous brightness. Many new techniques, e.g., absorption spectroscopy, anomalous scattering, elastic scattering and diffraction at different energies and high spatial resolution techniques (microprobes), to name a few, are being developed. To enhance atomic speciation, anomalous WAXS and SAXS have also been developed and are often used in advanced laboratories. This arises from the tunability of the X-ray to the desired energies close to the absorption edges to enhance the scattering power by taking advantage of anomalous scattering factors. These techniques are covered elsewhere in this volume (e.g. Chapter 8).

In general, nanostructures are crystalline structures with very small crystallite sizes in the nanometer range. However, they can also be disordered in the angstrom domains but highly ordered in the nanometer range. If the crystallites are small enough, the diffraction peaks are essentially very broad. As the size decreases, the material goes from paracrystalline to, eventually, the amorphous state. This chapter will describe the concept of small crystallite size, paracrystallinity and, briefly, "structures" in an amorphous state.

Nanostructures in the application of nanostructure technology, very often, are formed with some kind of coating around each particle to prevent agglomeration or the fusing of the particles to form a bigger crystallite size. Characterization of this size domain is generally well suited and carried out using the SAXS technique. SAXS is also used to examine morphology and molecular ordering in lifesciences, particularly with proteins, micelles and lipids. Essentially, SAXS is used in two major cases: dilute and highly correlating systems. A dilute system refers to non-interacting nanostructure entities, e.g., proteins and micelles in solutions. Highly correlating systems, for example, can be lipids and collagen where the overall molecules are highly ordered. In other situations, particularly polymer chains, mass fractal analysis is required. This will not be discussed here. Readers are encouraged to examine foundation books listed in the reference section. Several fundamental X-ray diffraction books used in formulating this chapter are listed in the reference section [1–13].

11.2
Brief Historical Background and Unique Properties

In 1895, Wilhelm Konrad Röntgen, while experimenting with electric discharges in a tube noted that a barium platinocyanide treated paper lit up as a result of fluorescence induced by unknown so-called X-rays. The X-rays originated from the tube and showed great penetration power, which is inversely proportional to the atomic number. The earliest application of X-ray was in radiography in surgical operations, which clearly revolutionized the medical field. For this discovery, Röntgen won the Nobel Prize in 1901. Another prominent discovery was by C. G. Barkla, who noted that there is a homogeneous energy component in the emitted X-ray operating under certain conditions and is characteristic of the target elements. He also noted that there are two main groups of emission lines, known as K and L lines, which clearly play a major role in Niels Bohr atomic model. For this he won the Nobel Prize of 1917. As the theory developed, X-rays were found to be electromagnetic radiation with a wavelength in the range of atomic spacing. Friedrich and Knipping first recorded the diffraction diagram of a copper sulfate crystal. With this information Laue formulated the theory of diffraction, which won him the Nobel Prize in 1914.

W. L. Bragg explained and correlated the X-ray spots with the atomic spacing and suggested that the diffraction of X-rays is caused by planes of atoms arranged in a lattice. Together with his son, W. H. Bragg, he built the first X-ray spectrometer and carefully recorded the quantitative diffracted intensity and angular positions.

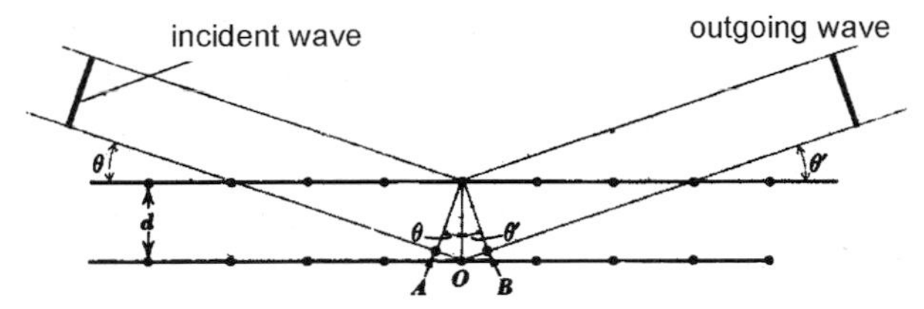

Fig. 11.1. Diagram showing the constructive interference of monochromatic X-rays.

As a result, the concept of diffraction was described and Bragg's law was introduced. These findings won them the Nobel Prize of 1915.

This phenomenon is explained by supposing that the wave property of X-rays is scattered by individual atoms, spherically, in the lattice (Fig. 11.1). The scattered waves then interfere either constructively or destructively at some distance away in space. Therefore, it is a phase issue. For the waves to add up constructively, the Bragg condition has to be satisfied, i.e.,

$$n\lambda = 2d \sin \theta \tag{1}$$

where n is the order of reflection, λ is the incident wavelength, θ is the observed angle and d is the interatomic spacings. The right-hand side of Eq. (1) is essentially the difference in path lengths when the waves are scattered at positions O and O′ (see Fig. 11.1). These d-spacings are generated by the atomic locations of positions in lattices. As a result, many crystal structures were found, lattice spacings were accurately measured and, also, characteristic X-rays from different elements were found and better defined. This is indeed a big advance in materials science and fundamental atomic physics.

Energy is often used to describe the X-ray, instead of wavelength, and is related by

$$E(\text{keV}) = h\nu = \frac{hc}{\lambda(A)} \tag{2}$$

where $hc = 12.4$; h is Planck's constant and c is the speed of light.

11.3 Scattering of X-rays

To see how scattering theory developed, it is important to start by describing the scattering of X-rays by a charged particle, say an electron. This result can then be carried over to represent the scattering intensity of an assembly of electrons, say in an atom, and then to a larger volume of mass. Because matter is essentially surrounded by electrons, X-ray scattering is accomplished by X-ray–electron interactions. Basically, there are two major components of scattered X-rays, coherent and incoherent, assuming monochromatic incident X-ray beam. Coherent refers to scattered X-rays that have the same wavelength (energy) as the incident X-rays. Incoherent refers to those X-rays that change wavelength (energy), implying that the incident X-rays imparted part of their energy to the charged particle. The incoherent component (also referred to as inelastic) is a slowly varying function over the scattering angles. The coherent component, which eventually interacts via diffraction, is the elastic component, and will be considered in this chapter.

Classically, an X-ray is described as a wave and is represented by the electric and magnetic field components. The wave is then incident on a charged electron, is then modified, either coherently or incoherently, and is then spherically scattered. The X-ray intensity is proportional to the square of the amplitude. The intensity of X-ray scattering by a charged particle is given by the classic Thomson formula [14],

$$I_{\mathrm{e}} = I_{\mathrm{o}} \frac{e^4}{m^2 c^4 R^2} \left(\frac{1 + \cos^2 2\theta}{2} \right) \quad (3)$$

where e is the electron charge, m is the mass of the electron, c is the velocity of light, and R is the distance between the observer or detector and the electron. The component in parenthesis is referred to as polarization factor. As Warren [10] has pointed out, the component to the left is the scaling number of the order of 10^{-26} for 1 electron; however, even in milligram amounts of sample there are 10^{20} electrons. The scattering by an individual atom with atomic number Z can be written as the summation of individual electron scattering.

This is not totally correct, since the electrons are not located at a point. There is an electronic distribution function around an atom. Hence, the structure factor f is introduced into the scattering equation. This factor is an integral of the electronic density distribution function, generally derived by empirical means. Without going into details of the derivation, the structure factor is given by Eq. (4), where $k = (4\pi \sin\theta)/\lambda$, λ is the X-ray wavelength, and $\rho(r)$ is the electronic density distribution.

$$f = \frac{4\pi}{\mathrm{e}} \int_0^\infty r^2 \rho(r) \frac{\sin kr}{kr} \mathrm{d}r \quad (4)$$

Thus, f is proportional to the atomic number Z at an angle of 0° and diminishes at higher angles. Notably, Eq. (4) is valid only if the distribution is truly spherical and the energy of the incident X-ray is not close to the absorption edge. Due to the anharmonicity of the electron distribution function and the wavelengths near the absorption edges, the structure factor is more complicated and has two additional modifying terms,

$$f = f_{\mathrm{o}} + \Delta f' + \mathrm{i}\Delta f'' \quad (5)$$

where f' and f'' are, respectively, the real and imaginary components, often referred to as the anomalous dispersion. Derivation of these factors will not be discussed here. These values are also tabulated in the *International Table of Crystallography* [15].

11.4
Crystallography

As the result of the development of X-ray diffraction and the complexity of Bragg's planes, a method of defining the X-ray peak positions and intensities is need. This is essentially accomplished by mathematical development through space groups. Again, the results of this work have been accurately tabulated, in the four-volume series *International Table for X-ray Crystallography*, a widely used reference. More detailed descriptions of crystallography appear in many fundamental X-ray diffraction books.

The crystallographic planes generated by lattices defined by the space groups are correlated to the X-ray scattering peaks. Hence, it is important to know where the atoms are located and what kind of planes they generate, in order to calculate the peak positions. Crystallography describes the atomic arrangements in a lattice, which are defined by atomic translations and rotations forming some kind of periodic arrangements. The smallest unique atomic arrangement used to generate the crystal by rotations and translations is called the unit cell. Unit cell parameters are defined by three axes and three angles defined by the axes. There are seven forms of large subgroups of crystallographic systems and 14 unique Bravais lattices for the atoms to arrange themselves in. The seven systems are triclinic, monoclinic, orthorhombic, tetragonal, hexagonal, rhombohedral and cubic. Essentially, they are defined by the relationship of the lattice parameters of three axes a, b and c, and three axial angles α, β and γ defined by the axes. Table 11.1 lists the crystal structures and characteristic parameters of the axes and angles.

With the atoms placed in the unit cell positions, 14 unique systems can then be described. In the cubic system, there are three lattices, namely simple, body-centered, and face-centered cubic. There are two tetragonal structures (simple and body-centered), four orthorhombic structures (simple, body-centered, end-centered

Tab. 11.1. Crystal systems and their axes [16].

System	*Axes*	*Axial angle*
Triclinic	$a \neq b \neq c$	$\alpha \neq \beta \neq \gamma \neq 90°$
Monoclinic	$a \neq b \neq c$	$\alpha = \gamma = 90°, \beta \neq 90°$
Orthorhombic	$a \neq b \neq c$	$\alpha = \beta = \gamma = 90°$
Tetragonal	$a = b \neq c$	$\alpha = \beta = \gamma = 90°$
Hexagonal	$a = b \neq c$	$\alpha = \beta = 90°, \gamma = 120°$
Rhombohedral	$a = b = c$	$\alpha = \beta = \gamma \neq 90°$
Cubic	$a = b = c$	$\alpha = \beta = \gamma = 90°$

and face-centered), and two monoclinic (simple and end-centered). Additional atoms can be placed in each of the unit cells; however, these positions are simple equivalent positions.

These lattices generally have a unique set of *d*-spacings; hence, in X-ray diffraction experiments, these spacings are reflected in the intensity of the scattered X-ray at specific Bragg angles by these unique planes. Eventually, X-ray diffraction from crystals progressed to such a level that a simpler way of identifying these planes had to be developed. This is accomplished by so-called Miller indices (hkl). X-ray peak positions can be correlated to the type of lattices or space group.

11.5
Scattering from a Powder Sample

The most common X-ray diffraction experiments are carried out on powder or bulk material with random crystalline orientation. Single-crystal diffraction experiments are also performed quite often; however, they have to be carried out using a four-circle goniometer. When the unit cell is complex with a large number of atoms, a single crystal experiment may be the only way to extract the desired structural information. Single-crystal experiments are generally far more complicated and difficult to perform without significant experience. Also, when the samples are prepared they are polycrystalline. Therefore, the selection of a single homogeneous crystal is often very tedious and difficult.

The total intensity from a powder pattern for a particular (hkl) at the angle 2θ is generally given by the scattering power [10],

$$P = I_{\text{o}}\left(\frac{e^4}{m^2c^4}\right)\frac{V\lambda^3 m|F_{hkl}|^2}{4v_{\text{a}}^2}\left(\frac{1+\cos^2 2\theta}{2\sin\theta}\right) \tag{6}$$

where I_{o} is the intensity of the primary beam, e, m, c, λ and θ take their usual meaning, V is the effective volume of the crystalline material in the powder sample, m is the multiplicity for that reflection and $|F_{hkl}|^2$ is the structure factor squared. The term in brackets on the right is known as the Lorentz polarization factor. For an actual experiment using a diffractometer, the scattering power is divided per unit length of the diffraction circle.

In an actual experiment, Compton scattering needs to be subtracted, a temperature factor needs to be added and the scattering power per unit detection length calculated. In most cases, with slits, the sample-detector distance is kept the same. Then, the scattering power is given by Eq. (7),

$$P = Km_{hkl}|F_{hkl}|^2(\text{LP})_{hkl} \tag{7}$$

where K is a constant with intensity proportional to the multiplicity m_{hkl}, structure factor square $|F_{hkl}|^2$ and Lorentz polarization factor $(\text{LP})_{hkl}$. The structure factor is essentially,

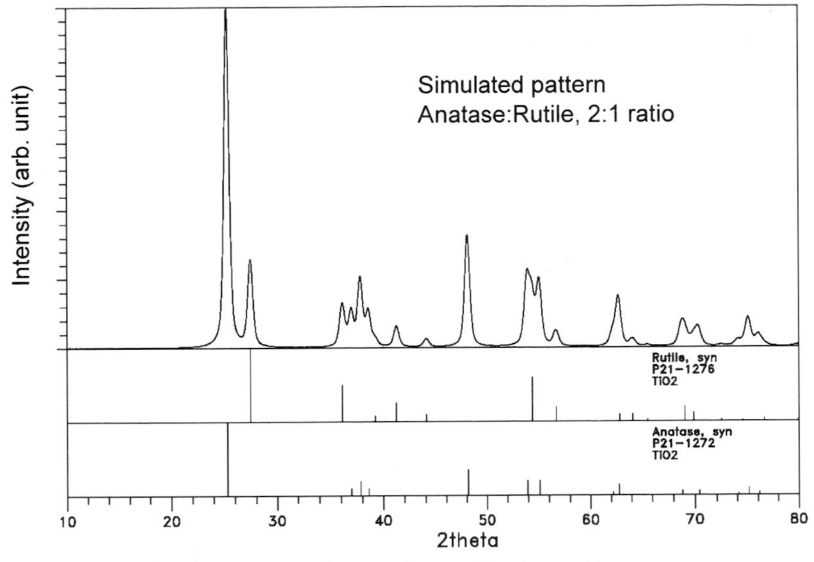

Fig. 11.2. Simulated XRD pattern for two phases of titanium oxide.

$$F_{hkl} = \sum_n f_n e^{2\pi i(hx_n + hy_n + lz_n)} \tag{8}$$

where f_n is the scattering factor, which is directly dependent on the atomic number, and $e^{2\pi i(hx_n + hy_n + lz_n)}$ is the form factor, which has the sinusoidal property and turns the intensity on or off. There are some exceptions to this rule, which will be described below.

Figure 11.2 shows a typical diffraction pattern for a simulated pattern of two phases of titanium oxide; the major phase is anatase and the minor is rutile. Along with the plots are the reported peak intensities and positions from the ICDD database. Both phases are tetragonal, with different cell parameters. Thus, phases can be identified even for the same compound. Such results are common.

The structure factor as described in Eq. (8) is summed over all the atoms in the unit cell. For example, in the face-centered cubic (FCC) cell, for simplicity, there are four atoms, located at $(0,0,0)$, $\left(0,\frac{1}{2},\frac{1}{2}\right)$, $\left(\frac{1}{2},0,\frac{1}{2}\right)$ and $\left(\frac{1}{2},\frac{1}{2},0\right)$. Eq. (8) will become,

$$F_{hkl} = [1 + e^{\pi i(h+k)} + e^{\pi i(h+l)} + e^{\pi i(k+l)}] \tag{9}$$

noting that $e^{\pi i m} = (-1)^m$ when m is an integer. Hence, it can be concluded that $F_{hkl} = 4$ if the indices hkl are all odds or evens, and $F_{hkl} = 0$ when the indices are mixed. Turning our attention to a body-centered cubic (BCC) structure, with two atoms $(0,0,0)$ and $\left(\frac{1}{2},\frac{1}{2},\frac{1}{2}\right)$ per unit cell. Eq. (8) becomes

$$F_{hkl} = [1 + e^{\pi i(h+k+l)}] \tag{10}$$

For F_{hkl} to be non-zero, $(h + k + l)$ must be even. Hence, in FCC, reflections occur when hkl are all even or all odd, but with a BCC structure, $(h + k + l)$ has to be even. For another FCC system, for example, the rock salt structure of sodium chloride, having four Na atoms and four Cl atoms per unit cell, the summation in Eq. (8) will result in slightly different F_{hkl}. Again, the sample rule applied here is that hkl have to be all evens or all odds. However, for the even case, F_{hkl} will be four times the sum of the scattering factors, whereas for the odd case F_{hkl} will be four times the difference of the scattering factors.

In the order–disordered structure of non-monatomic systems, it is possible to violate the above hkl requirements. This is because, even though the atomic positions are periodic, the atomic species may not be. The scattering factors do not completely cancel out, resulting in the observation of non-allowed peaks. A good example is the well-known Cu_3Au alloy [10]. Another system that needs to be considered is materials with antiphase domains where the atoms of different species have been interchanged, resulting in super structures and X-ray diffraction peaks not normally allowed.

11.6 Scattering by Atomic Aggregates

In practice, not all matter in life has atoms arranged in nice lattices. The most interesting of these materials are those with complex atomic connectivity and having long atomic chains and no kind of ordering in the unit cell range. Examples of such systems are given Section 11.11, which describes the component of amorphous and highly disordered scatterings. It is often very difficult to distinguish between an amorphous structure, similar to liquid or gas, and an extremely small crystallite state, also sometimes referred as a "paracrystalline" state. For nanostructures, it may be important to determine whether the structure is amorphous or simply paracrystalline.

For time average scattering of non-interacting scatterers like mono-atomic gases, the scattering is given by the Debye equation,

$$I_{eu} = \sum_m \sum_n f_m f_n \frac{\sin kr_{mn}}{kr_{mn}} \tag{11}$$

where f_m, f_n are the scattering factors and r_{mn} are the interatomic distances. This equation is represented in Fig. 11.3 for $r_{mn} = 1$. This result clearly shows that even for a non-interacting system there is a ripple effect, which declines very rapidly on going from low to high scattering angles. For polyatomic systems, the result is the same except that the ripples are much broader.

For a condensed system like a liquid, the summation from Eq. (11) can be simplified to

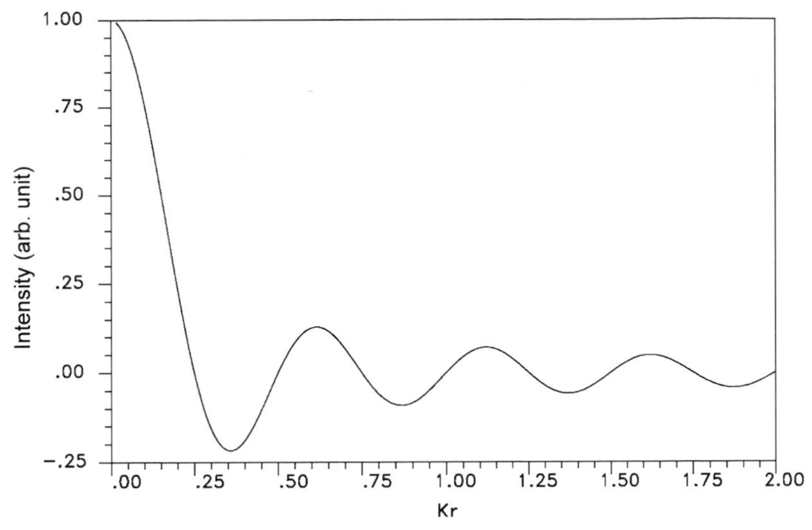

Fig. 11.3. Intensity function of a monatomic gas.

$$I_{eu} = Nf^2\left(1 + \sum_m \frac{\sin kr_{mn}}{kr_{mn}}\right) \tag{12}$$

where the 1st term is the summation onto itself and the 2nd is the interacting term. This equation can then be converted into an integral [Eq. (13)].

$$I_{eu} = Nf^2\left[1 + \int 4\pi\rho(r)\frac{\sin kr}{kr}\mathrm{d}r\right] \tag{13}$$

Using algebraic manipulation and defining $\rho(r) = [\rho(r) - \rho_0] + \rho_0$, Eq. (13) becomes

$$k[i(k)] = 4\pi\int_0^\infty r[\rho(r) - \rho_o] \sin kr\,\mathrm{d}r \tag{14}$$

Where $i(k) = [I_{eu}/(N - f^2)]/f^2$, for simplicity, and by using the theorem of Fourier's conversion we can write the radial distribution function as Eq. (15).

$$4\pi r^2\rho(r) = 4\pi r^2\rho_o + \frac{2r}{\pi}\int_0^\infty k[i(k)] \sin kr\,\mathrm{d}k \tag{15}$$

Eq. (15) provides a mean of converting the intensity function, which is in k space, into the radial distribution function in real space. In this formulation, the atoms are arranged in completely random fashion, giving an amorphous state. There are broad diffraction peaks, which belong to the amorphous structure. Unlike crystals, there is no crystallographic ordering. Several papers have demonstrated on model calculations from ideal random atomic arrangements and then determined the radial distribution functions. The density distribution function can be converted into the intensity or interference distribution function and be compared to the experiments [17]. For binary systems, with atoms of A and B species, for example, the intensity function $i(k)$ is a composite of the three partial functions $i_{\text{A-A}}(k)$, $i_{\text{A-B}}$ and $i_{\text{B-B}}(k)$. It is, therefore, impossible to extract each partial density distribution function without further experiments. Because of differences in scattering factors for X-ray, neutrons and electrons, partial distribution functions can be extracted using a combination of the three experiments. Anomalous scattering experiments can and have also been performed to extract the partial distribution functions. By using the dense random packed models and comparing the calculated $i_{\text{total}}(k)$ with the experimental observations, the partial functions can be determined [18]. This is particularly useful in the understanding of atomic and chemical short-range order. Clearly, for ternary systems the functions get very complex.

11.7 Crystallite Size and Paracrystallinity

Paracrystalline is a term used to describe small size ordering. Essentially, when the crystalline size is small enough, the resulting diffraction peaks will be broadened; however, the peak positions, if they can be deconvoluted, will fall precisely at the crystalline positions. Extraction of the crystallite size from the peak widths can never be accurate. Hence, in most cases crystallite size is generally an estimate. There are many factors that impact on the peak width, e.g., lattice distortions possibly due to the thermal vibrations, micro-strains, nature of the paracrystallinity. In general, to improve the precision of extracting the crystallite size, the instrumentation broadening component can be subtracted out by performing an exact experiment on a highly ordered structure like silicon powder. Instrument factors are due to misalignment, horizontal divergence (on a vertical goniometer), energy spread of the incident beam, interference of the α_1 and α_2 doublets, and also absorption of the X-rays by the sample. Notably, instrument broadening does not remain constant with all 2θ but, rather, increases with increasing angle. For this discussion, we are only concerned with determining the estimated crystallite size D_{hkl} of nanoparticles; it is generally given by the Scherrer equation,

$$D_{hkl} = \frac{K\lambda}{\beta \cos \theta_{hkl}} \tag{16}$$

where β is the peak broadening with the inherent instrument broadening subtracted out, θ_{hkl} is the Bragg angle measured at hkl reflection. K is normally taken to be 0.9 or unity, depending on the crystal shapes. For details on different values of K, refer to Refs. [3, 4]. Figure 11.4 shows a plot of crystal size against X-ray peak

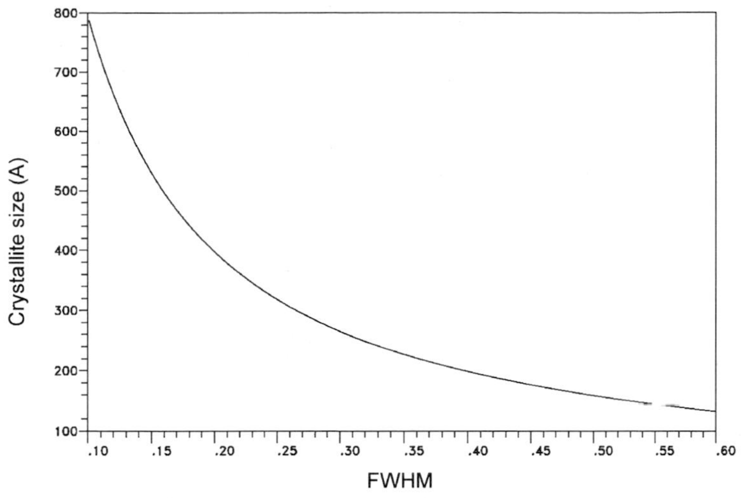

Fig. 11.4. Crystallite size versus FWHM.

broadening (fwhm – full-width at half-maximum) using Eq. (16) and assuming $2\theta_{hkl} = 20^\circ$, indicating the sensitivity in the measurement of small crystallites but not large ones. That is, a small change in fwhm translates into a large change in crystallite size in the small width region and a small change in crystallite size in the large width region. Very often, due to uncertainty in the measurements, "apparent crystallite size" is used.

As the crystallite size decreases, the peak broadens. Figure 11.5(a) and (b) compare the simulated diffraction spectra of crystalline hexagonal and cubic Co as a function of crystallite size, calculated using Jade software (MDI) for copper K_α radiation. Multiple peaks are observed in the hexagonal structure for crystallite sizes of 40 Å and above. Below 40 Å, the three distinct peaks between 40 and 50° (2θ) diminish; hence it is not so clear whether the Co is either cubic or hexagonal phase. Moreover, without significant features, it can be concluded that the Co may be essentially amorphous. In this case, above 40 Å should not be a problem in making the distinction.

11.8
Production of X-rays

It is important to understand the production and properties of the X-ray used in the probe. X-rays are generated when electrons emitted from a tungsten filament are accelerated with sufficiently high energy onto a specific target. Typically, X-ray tube voltages are set at 20 to 50 keV. Interaction between the electrons and the electrons of the target materials results in a broad continuous X-ray spectrum. This radiation is often referred to as white radiation or the Bremsstrahlung (German for braking) radiation. This radiation is useful in the energy dispersive type of diffrac-

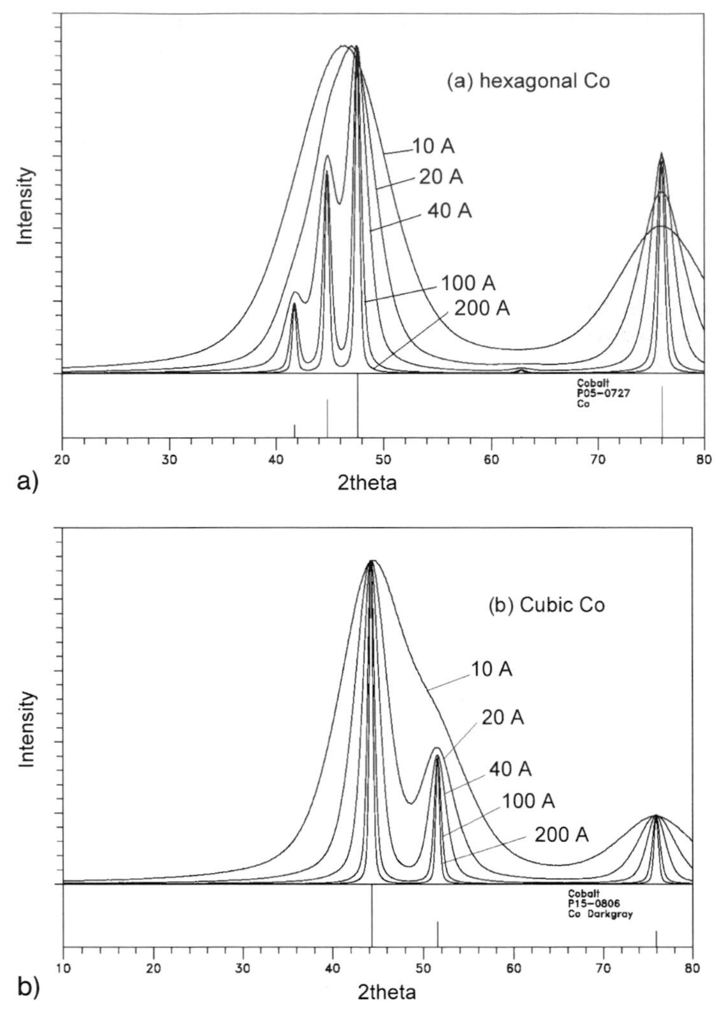

Fig. 11.5. Simulated spectra of crystalline hexagonal (a) and cubic (b) cobalt of different crystallite sizes.

tion experiment or a Laue experiment for single crystals but is a nuisance when monochromatic radiation is needed. The energies and intensity of the Bremsstrahlung radiation are affected by changing the tube voltage.

With sufficiently high kinetic energy, the electrons can kick out bound (orbital) electrons from the target. Subsequently, the decay of higher bound state electrons to the lower bound state results in the release of energy by emitting X-rays of characteristic energies and wavelengths, referred to as the K series, with $K_{\alpha 1}$, $K_{\alpha 2}$, $K_{\beta 1}$ and $K_{\beta 2}$ radiations. The population of the α series is much greater than that of the β series. In general, most diffraction experiments are carried out using K_{α} radiation, and K_{β} radiation is masked off by several techniques.

Less than 1% of the energy incident on the target is converted into X-rays. Hence, this is a very inefficient method. Most of the energy is dissipated as heat at the target. This heat needs to be removed immediately otherwise the targets can melt away. Generally, targets are designed to be on top of a highly thermal conducting heat sink, which is also under constant cooling. Because of this heating issue, there is a limiting power at which the X-ray generator can operate. This clearly depends on the tube design to achieve optimum cooling of the targets. Another target design, which allows for higher power setting, is the rotating anode X-ray generator. The target element is deposited on a highly thermal conducting rotating drum (usually made of copper), which is constantly cooled. The drum rotates at a very high speed, thus constantly changing the target position where the electrons hit. This design alleviates the heat dissipation problem and, therefore, a higher power setting can be achieved, resulting in higher X-ray flux. Unfortunately, maintaining vacuum conditions with a rotating shaft is a challenge.

11.9 Absorption of X-rays

It is also important to understand the absorption behavior of X-rays through opaque materials. In general, X-rays are absorbed whenever they pass through matter. The equation governing this rule is simply

$$I = I_o e^{-\mu\tau} \tag{17}$$

where I_o is the incident intensity, I is the transmitted intensity, μ is the linear absorption coefficient (cm^{-1}) and τ is the thickness (cm) of the absorber. This is assuming that the X-ray energy is far from the absorption edge. For convenience, the mass absorption coefficients μ/ρ ($cm^2\ g^{-1}$) are used and have been tabulated [15].

$$I = I_o e^{-(\mu/\rho)\rho\tau} \tag{18}$$

Hence, the mass absorption coefficient is independent of density, and the overall absorption coefficient will just be the summation of the coefficients weighted according to the composition. For X-ray energies close to the absorption edges, the coefficient of absorption changes abruptly. This is directly related to the work function required to knock electrons from their original shells. The absorption coefficient can also be calculated empirically [15].

11.10 Instrumentation: WAXS

Crystallographic structural information on nanoparticles has to be obtained using conventional X-ray diffraction techniques. In general, the most common diffraction

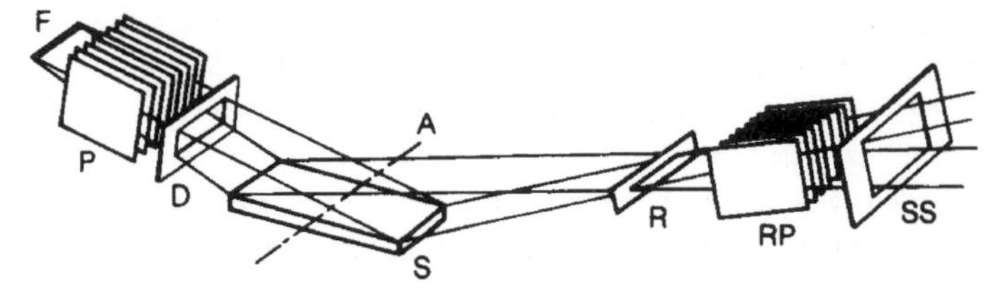

Fig. 11.6. Para-focusing Bragg–Brentano geometry.

experiments are carried out on bulk and powder samples. Again, most materials are polycrystalline. Occasionally, single-crystal experiments are very useful for obtaining detailed structural information, small lattice anisotropy and complex unit cell or cells with a large number of atoms. Single crystal experiments have to be carried out on a four-circle goniometer and will not be described here. Also, there are times when special scans are needed, e.g., to measure residual stress, pole figures, rocking curves etc. a four-circle X-ray goniometer is needed.

The most common and widely used X-ray powder diffraction technique is the Bragg–Brentano method. It is self-focusing and optimizes for intensity and resolution using the conventional in-house X-ray sources. Figure 11.6 shows the X-ray optics. Such diffractometers are essentially very versatile. The distance from the source F to the sample and from the sample to the receiving slit is always kept equal and constant at all scattering angles. Parallel vertical slits P and RP are added to improve the scattering signal. D is the defining slit and R is the receiving slit. Because of the focusing requirement, alignment of the sample is critical for accurate peak position measurements. To help in alignment and calibration, a known standard is used, e.g., a well-characterized silicon powder standard from NIST.

Traditionally, the X-rays are detected using high sensitivity film. In some cases, they are still used. The films are then processed and read by a digitizer, which is essentially a measurement of transmitted light over the spatial positions on the films. X-ray films are also used when examining two-dimensional scattering. Because of the difficulty in processing the film, the modern image plate was invented. Here, an image reader is needed.

For pulse counting techniques, a gas ionization counter, also known as a Geiger-Müller counter, is used. Essentially, it consists of exposing a gas-filled region, normally a noble gas like argon, krypton or xenon mixed with a small amount of quenching gas, held at high potential to the X-rays. Normally, this region is protected by an X-ray-transparent window. When an X-ray enters the compartment it ionizes the gas, producing electrons. These electrons are accelerated to the anode, thus creating a current pulse. The quantum efficiency of this detector greatly depends on the design and type of gas it uses. The energy resolution is poor and the detector can be saturated without going to very high count rates.

A more convenient scintillation has also been developed. This type of detector depends greatly on the fluorescence of certain crystals, e.g., sodium iodide or cesium iodide crystal doped with thallium. Other crystals are also used. These crystals emit light when exposed to X-rays. The light is then converted into an electron pulse via a set of multiplier plates that can then be counted by the electronics. In this case, the dead time is much shorter and counting for higher count rates can be achieved.

The energy resolution of the above two counts is not optimal for normal diffraction techniques. The K_β radiation and other background radiations, e.g., contamination from the X-ray tube, need to be discriminated. Traditionally, foil filters are used. Selection of the foils is based on the absorption edges, e.g., when using copper K_α radiation a nickel filter foil of 0.0158 mm is needed, which will absorb 98.4% of the K_β radiation [4].

Another way of removing background radiation is the commonly used analyzing crystal (also called diffracted beam monochromator), which is placed in front of the detector (which can either be the gas proportional counter or the scintillation counter) and after the receiving slit. The choice of crystal is based on the crystal mosaic, for energy selectivity and efficiency. The most widely used energy discriminator is usually graphite for efficiency without significantly sacrificing X-ray intensity. For experiments with more stringent energy discrimination, a LiH monochromator is used. In most cases, the analyzing crystal does not have sufficient resolution to remove the $K_{\alpha 2}$ radiation.

Another way of improving the scattering signal is to use a solid-state detector (SSD). Normally, the detector consists of Si(Li) (silicon-lithium drifted) or Ge(Li) (germanium-lithium drifted) held at liquid nitrogen temperature. When an X-ray impinges on the detector, an electron-hole pair is created and the number of electron–hole pairs is proportional to the energy. Hence, in this case, the energy resolution is roughly about 130 eV. K_β radiation can be eliminated as well as other background radiations and fluorescence from the sample.

The high energy-resolution of the solid-state detector permits energy dispersive diffraction. Essentially, this takes advantage of the full broad energy spectrum of X-rays coming from the tube because of the Bremsstrahlung radiation and, by having the detector at specific angle, a full range of Bragg scattering can be obtained from the energy dispersion. The advantage here is that there is no moving part when conducting the experiment. It makes use of the full range of X-rays and counting statistics can be greatly improved. The drawbacks are that Bragg peak is significantly broader than that of the conventional scanning technique. The detector is big and clumsy, and has to be constantly cooled with liquid nitrogen. However, a newer version of SSD known as a Peltier detector is electrically cooled. This is also rather heavy and clumsy.

A normal mono-energetic experiment makes use of the strong K_α radiations, which consist of $K_{\alpha 1}$ (8.04778 keV) and $K_{\alpha 2}$ (8.02783 keV) energies [19]. For copper K_α radiation, $K_{\alpha 2}$ intensity is about half that of $K_{\alpha 1}$. $K_{\alpha 2}$ radiation is an unwanted component of the incident beam. The difference in energies is about 20 eV. However, if a more stringent monochromator is used, the intensity will be drastically reduced. Most powder diffraction experiments simply ignore it. However, when one examines nanostructures where the crystallite size is very small, the presence of $K_{\alpha 2}$ will not be a factor. On the other hand, if one chose to use higher energetic radiation, like molybdenum, the difference is about 105 eV. Normally, the best solid-state detectors have a resolution of 130–150 eV.

Unfortunately, this setup is not favorable for samples with elements excitable by the incident X-ray with energy 8.04 keV for copper K_α radiation, e.g., magnetic nanoparticles Fe and Co-containing samples. In these situations, some form of tighter discrimination is needed to extract the coherently scattered X-rays and not the fluorescence from the sample, which is essentially noise for the diffraction experiment. A way around this problem is to change the probe radiation to that of CoK_α for cobalt-containing samples and FeK_α for iron-containing samples. In this way the X-ray probe energies will not be sufficient to cause fluorescence. By changing incident X-ray energies, one has to be aware that the penetration depth is decreased and the energy discrimination needs to be readjusted. Another way to carry out diffraction experiments on highly fluorescent samples is to ensure that the incident energies are sufficiently far from the fluorescence of the samples, e.g., using MoK_α radiation on iron- and cobalt-containing samples. In this way, the diffracted beam crystal can easily filter out the fluorescent signals.

11.11
Small Angle X-ray Scattering

The regime covered by small angle X-ray scattering techniques typically covers 0.5 to 100 nm and, according to Bragg's equation, this turns out to be only a few degrees in 2θ. This regime also turns out to be very useful for examining many types of nanostructures. In general, for the application of nanostructures, it is customary to have a layer of polymer coated on each particle. Information on size, shapes, interparticle correlations and density fluctuations of nanostructure are of importance and can be obtained using this technique. In lifesciences, for example, biological cells and cell tissues, proteins and protein folding, colloids, micelles, bacteria and viruses can also be investigated using SAXS. This technique has also been widely used in examining tumors and cancerous cells, and distinctions in the diffraction patterns between healthy and cancerous cells have been shown. It is the information on the macro-scale, the connectivity and arrangement of the local domains, that is critical in understanding the behavior of these systems. For example, a SAXS study examining healthy and cancerous human breast tissues found that the collagens have very different structures [20]. It has also been widely used to examine tumors and cancerous cells [21], and, again, distinctions have been drawn between healthy and cancerous cells. Further research by many groups is continuing to try to understand the causes of cancer. Even so, by combining SAXS and WAXS characterization techniques, experiments can be carried out on well chosen cell tissue samples that will provide information for the understanding of disease processes evolution. This represents a new form of diagnostic capability for health sciences.

Information on the nanostructure, size, shapes, interparticle correlations, density fluctuations are of importance and can be obtained. In the lifesciences, most systems, such as those listed above, are highly disordered at the local level (angstroms) and at times they are even semi-crystalline. Clearly, they are very complex and in most cases consist of interpenetrating components of the different make-up of molecules. These molecules, which can have odd shapes, essentially hinder the formation of unit cell arrangements. Again, the macro-crystalline arrangements are critical in the generation of properties.

Another example of X-ray scattering experiments in lifesciences is in understanding the structure and morphology of bone. Bones are made up of fibrillar collagens with nano-particles of crystalline hydroxyapatite distributed in the macro-structure. It is this macro-arrangement of the collagen and the crystalline component that provides the optimal compression strength for the load bearing of the anatomy [22]. A network of collagen fibers is deposited and grown within and between the collagen fibrils [23]. SAXS experiments can provide information on the macro-domains. One area in lifesciences that SAXS will play an important role is in understanding the cell membrane. Information on the structure of the lipid moieties in the cell membrane essentially defines the functionality of the membrane [24]. Even though lipids are considered to have bilayer structures, monolayer structures do exist. Of particular interest is understanding lipid behavior in solvent, which again can be carried out using SAXS. Depending on the acquisition time, *in situ* experiments as a function of temperature or time decay can also be carried out.

For nanostructures in biological materials SAXS analysis is used to extract information relating to the macro-lattice on highly correlating domains, which often can be periodic. A typical example for this is the tendons. Many biological systems also possess periodic structures, e.g., animal tail tendons and myelin membranes in the nerve. In polymers, the scattering signal arises from the stacking of lamellar crystals and block copolymers having segregating density domains.

Laboratory prepared nanoparticles (or crystals), e.g., cobalt [25] when coated with organic layers, can self-assemble into superstructures. The structure and dimension of these superstructures have also been examined using both WAXS and SAXS, thus providing a full description of the overall structure. The superstructure of biomaterials like colloidal particles of glycolipids has also been examined using SAXS. The technique was able to provide information on the structure of the dispersed particles [26].

The analysis of SAXS data can be divided into three groups: dilute systems, highly correlating systems and fractals. In dilute systems, when the structures are non-interacting, information on the shape, size and mass of the molecules can be obtained. For highly correlating systems, the basic formulation follows Bragg's law. The fractal component is generally useful for polymers and polymer chains and will not be discussed here. The reader is encouraged to refer to other sources [7, 12].

11.11.1
Dilute Systems

The scattering for dilute systems assumes that the scattering from each particulate is independent. Hence, it is critical for the sample to have an appropriate thickness so that no multiple scattering can occur. The X-ray interacts with the electron fluctuations surrounding the particulates and the total intensity is merely the sum of all the individual scattering components. Therefore, information on sizes, shapes and mass can be obtained. The generalized scattering function is essentially given as

$$I(k) = (\rho_o v)^2 F^2(k) \tag{19}$$

where ρ_o is the density difference over total volume v, and $F(k)$ is the structure factor. The form factor is simply represented by

$$F(k) = \int \rho(r) e^{ikr}\, dr \tag{20}$$

where the integral is over the electron density over each particulate and the exponential component is the form factor. By expanding the form factor and dropping out the higher order terms, Guinier has shown that the intensity function is

$$I(k) = \rho_o^2 v^2 \exp\left(-\frac{1}{3} k^2 R_g^2\right) \tag{21}$$

or

$$I(k) = I(0) \exp\left(-\frac{1}{3} k^2 R_g^2\right) \tag{22}$$

where $I(0)$ is the scattering at zero angle. Essentially, this function is Gaussian. Very often, ln $I(k)$ is plotted against k^2 and the radius of gyration R_g can be extracted from the resultant slope. Shape information is obtained from R_g; the value of R_g for many forms of particulate has been calculated and will not be discussed here. Actually, R_g is defined as in Eq. (23) and described as the root-mean-square distance of all points in the particle from the center of mass weighted accordingly.

$$R_g^2 = \frac{\int_0^\infty \rho(r) r^2\, dr}{\int_0^\infty \rho(r) r\, dr} \tag{23}$$

The size of the particle is conveniently expressed by R_g and, assuming the shape is known, the radius of gyration can be extracted accurately. The radius of gyration for some common shapes has been tabulated in many review articles. By knowing or assuming the shape of the particle examined, the dimension of the particle can be gotten by solving for the radius of gyration. Extrapolation of the Guinier plot to scattering at zero angle results in $\rho_o^2 v^2$, which provides information on the particle molecular weight and volume. However, in this case, accurate intensity of the main beam is needed.

For homogenous particles with distinct boundaries between the particles and the matrix, which can be a solution or another media with different density, e.g., in polymers with crystalline and amorphous components, Porod has indicated that the scattering $I(k)$ should decrease as k^{-4} for large k and the proportionality constant is related to the total area S occupied by the boundaries and is expressed as

$$I(k) \rightarrow \frac{2\pi\rho_0^2 S}{k^4} = \frac{P}{k^4} \tag{24}$$

By plotting log $(I(k))$ versus log (k) and extrapolating to $k = 0$, the P (Porod constant) can be extracted. When this parameter is divided by the "invariant", which is the scattering intensity for the total volume, the surface area to volume ratio can be gotten. When this parameter is divided by the density, the total surface area (m/g) resulted. The information is particularly useful, especially in the area of catalysis when the amount of surface area controls the property of the material.

11.11.2 Highly Correlating Systems

Most systems in lifesciences are highly correlated or rather densely packed with significant density fluctuation and are also highly oriented. For example, collagens are very well packed in a fibrillar fashion and the diffraction pattern shows many orders of reflections. The peaks are also very sharp, with a dimension of 67 nm for native rat tendon [26]. Obviously, for densely packed systems, the scattering theory is governed by Bragg's law.

Collagens are common in lifescience. It is a major constituent in tendon, bone, skin, cornea, cartilage and other parts that require strength and flexibility. Clearly, the packing and deformation of collagens will reveal the state of the matter when compared with healthy tissue. The arrangement of collagen is believed to be liquid-crystalline-like as well as having large scale triple helices arrangements for roughly 300 nm [27, 28]. The change in packing indicates some kind of mineralization that prevents proper function of the component. It is indeed a mystery that collagen, which is abundant in our anatomy, has so many functions. It is the arrangement on the macro-scale that changes its properties [29].

As described earlier [24], examination of protein bilayers is of great importance in biological science. Essentially, these are highly correlated systems. Clearly, the structure and makeup of these bilayers constitute the working of the membranes, peptides etc. There are enormous numbers of biological systems for which both SAXS and WAXS can be utilized for understanding their behavior. Biological systems are normally very complex.

11.12 SAXS Instrumentation

A SAXS experiment, in principle, has a very simple geometry (Fig. 11.7); however, improperly arranged and aligned components will result in all kinds of problems, e.g., parasitic scattering, calibrations, peak profiles and background noise. The

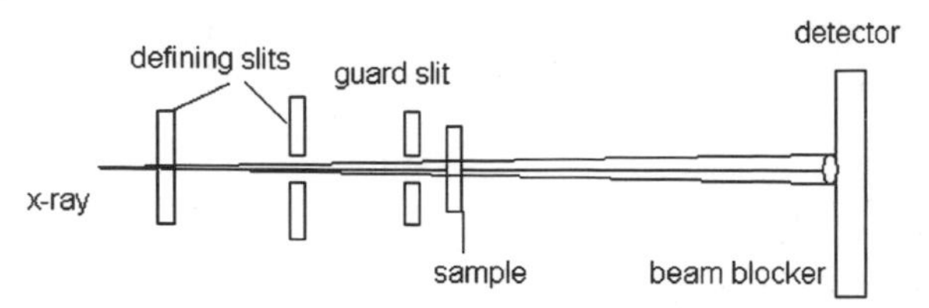

Fig. 11.7. Optics for carrying out SAXS experiment using the three-slit system.

quality of the instrument greatly depends on the level of noise because the diffuse scattering intensity for most samples can be very weak for different reasons. A very well-defined X-ray beam is needed for this application.

Numerous SAXS instrumentations are available commercially. Essentially, there are two types of configuration, pin-hole and the line source technique, each with its own strengths and weaknesses. The pin-hole technique is useful for examining anisotropic materials, e.g., fibers, films and any sample with orientations. However, there is an inherent lost in intensity, resulting in long data acquisition time. A two-dimensional detector system is required, e.g., conventional highly sensitive X-ray film, image plate, 2D proportional gas detector and even the CCD (charge coupled device) detector systems. Both the film and image plate require X-ray exposure and then have to be processed and read by some kind of digitizer or image reader. In this way, some level of spatial resolution and dynamic range will be lost. When small peaks are of interest or accurate intensity function is required, these are not the desired techniques. Moreover, real-time experiments or experiments performed as a function of temperature will be extremely tedious and inefficient.

A basic requirement for a good experimental setup for SAXS is in the collimation. This is because SAXS is a measurement of extremely small intensity at very small angle, next to a very strong incident main beam. Hence, the scattering around the incident beam has to be as clean as possible and protected against parasitic scattering cause by the defining slits and beam blocker. Commonly used in the pin-hole technique for the in-house systems is the three pin-hole slits system with different sizes (Fig. 11.7). They have to be properly arranged and matched with the size of the beam blocker to obtain the clean primary beam and reach to the lowest angle. The third slit is the guard slit that filters out the parasitic scattering from the defining slits. Not that the beam diverges from the source. To avoid air scattering, the beam path has to be under vacuum, which inherently introduces two X-ray windows. The weakly scattered X-ray from the sample has to penetrate these windows without changing in energy and direction. The advantage of the pin-hole technique is that anisotropic samples can be examined, e.g., fibers and films. By translating the sample, localized spot examination can also be carried out.

The more advanced 2D gas proportional detector system has also been used. The heart of the detector is the multi-wire grid; the spatial resolution is limited to the

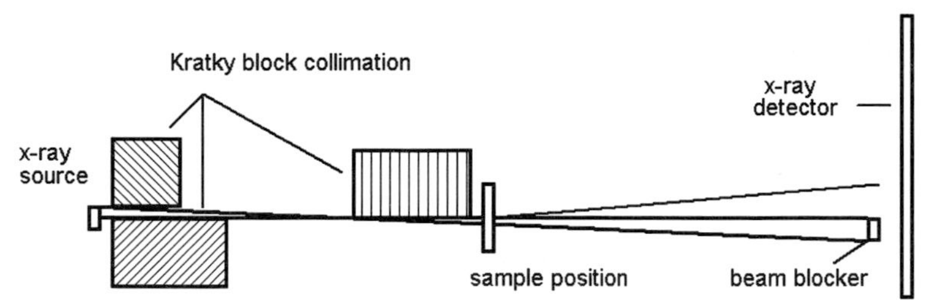

Fig. 11.8. Kratky block collimation system.

wire spacings. Normally, this is limited to about 1 mm, even though some advances have been made (0.2–0.3 mm) through electronic and software developments. As a consequence, pin-hole SAXS instruments require long path lengths to expand the image for the detector to receive.

To increase the intensity, the second method, using a line source, is often used. The optics are essentially similar to the pin-hole system shown in Fig. 11.7. This method increases the flux of the incident beam; however, only isotropic sampling can be carried out. Unfortunately, because of the fixed slit, desmearing of the data is needed. Again, three line slits properly arranged are used to minimize parasitic scattering. Adjustments to these slits are often very sensitive and difficult. Consequently, the Kratky block collimation (Fig. 11.8) was introduced and is still used today. It is easier to align and there are not too many adjustments. It can also achieve much smaller k (Å^{-1}) than the normal three-slit system.

Another method of increasing the incident beam flux and improving the spatial resolution is by using focusing mirrors and monochromators. To achieve even smaller in k, the Bonse-Hart technique is used. By taking advantage of a tighter energy filter, an even smaller angle can be achieved. This is accomplished by having the X-ray bounce several times through a channel cut crystal on the incident end as well as the diffracted end. This technique also requires significant alignments, which are normally set up at dedicated beam lines at synchrotron sources. Again, this is also a line source and the issue of desmearing will have to be considered.

11.13 Synchrotron Radiation

Undoubtedly, synchrotron radiation has played a major role in extending X-ray scattering capabilities. As pointed out in the introduction, unlike in-house X-ray sources, synchrotron radiation has unique properties. The generation of synchrotron radiation will not be covered here. The X-rays generated by a synchrotron source are highly collimated, highly polarized, tunable, and have high brightness. The probe size can be made small, down to 2 μm, which essentially provides an insight into the composition of biological cell and local structure that is not possible using in-house sources.

The high collimation enables diffraction experiments to be carried out with improved precision. The peak width spread will be extremely small. The fwhm will be essentially due to the character of the sample and not the instrument. Divergence will be extremely small. By using an incident monochromator, unlike the in-house source, only a mono-energetic beam is allowed to pass the incident monochromator. Because of the small divergence of the beam, SAXS experiments can be better set up.

To enhance atomic speciation, anomalous WAXS and SAXS have also been developed and often used in advanced laboratories. This arises from the tunability of the X-ray to the desired energies close to the absorption edges to enhance the scattering power by taking advantage of anomalous scattering factors. These techniques are covered elsewhere in this volume (see Chapter 8). Clearly, there are more examples of utilizing the advantages of performing X-ray diffraction experiments using a synchrotron source.

11.14 Concluding Remarks

In the field of materials characterization, X-ray scattering stands out as the preferred method for obtaining information on the structure and morphology of all sorts of materials. Essentially, X-rays are scattered by electrons, and all matter has electrons. The domain size that fits both wide angle and small angle X-ray scattering ranges from a fraction of a nm to 100 nm. At the low end, X-ray diffraction has provided information on atomic positions, lattices, bond lengths, atomic speciation and structure. By carefully examining the character of the scattering, secondary information, like the crystallite size, lattice distortion and occupancy, can also be extracted. The penetration depth of an X-ray is also sufficiently high compared with X-ray photoemission or electron scattering, hence X-ray scattering can be considered as bulk probing.

With the theory and technique firmly understood, the method has been used on more complex organic and polymeric systems. Unfortunately, in general, these systems are not fully crystalline. They are semi-crystalline and, more often, it is this semi-crystallinity that provides the unique properties of polymers, essentially controlling their morphology. This is one of the reasons for the development of small angle X-ray scattering.

Both the wide angle and small angle techniques are quite well developed. Basically, these characterization techniques have been extended to different aspects of biological and lifescience. The structure of cells and cell tissues, lipid membranes, proteins and protein behavior, the building blocks of many organs in our anatomy, are now being examined using both the wide angle and small angle X-ray techniques.

Structures or nanostructures in biological and lifescience are generally very complex. However, with the advancement of computing power, high energy and the spatial resolution of synchrotron sources, experiments on tissues, like cancer, tumors or sclerosis, are now possible. Clearly, X-ray scattering, together with synchrotron and neutron scattering, covers a whole range of characterization for organic and inorganic condensed matter, crystalline and semi-crystalline, and amorphous structures. These experiences have also developed into the biological field and lifesciences.

Acknowledgment

This work was performed under the auspices of the U.S. Department of Energy by the University of California, Lawrence Livermore National Laboratory under Contract W-7405-Eng-48. Thanks are given to Dr Art Nelson for reviewing the manuscript.

References

The subject presented has been reviewed in several fundamental books. Listed are some of the books used in writing this chapter.

1 Guinier, A., *X-ray Diffraction*, Freeman and Company, San Francisco, **1963**.

2 Balta-Calleja, F.J., Vonk, C.G., *X-ray Scattering of Synthetic Polymers*, Elsevier, New York, **1989**, chapter 7.

3 Wang, J.I., Harrison, I.R., *Methods of Experimental Physics: Volume 16: Polymers*, Ed. Fava, R.A., Academic Press, New York, **1980**.

4 Klug, H.P., Alexander, L.E., *X-ray Diffraction Procedures*, John Wiley & Sons, New York, **1973**.

5 Alexander, L., *X-ray Diffraction Methods in Polymer Science*, Wiley-Interscience, New York, **1969**.

6 Azaroff, L.V., *Elements of X-ray Crystallography*, McGraw-Hill Book Company, New York, **1968**.

7 Roe, R.J., *Methods of X-Ray and Neutron Scattering in Polymer Science*, Oxford University Press, New York, **1999**.

8 Kostorz, G., *Treatise on Materials Science and Technology*, Academic Press Inc., New York, **1979**, pp. 15, 227–286.

9 Richtmyer, F.K., Kennard, E.H., Cooper, J.N., *Introduction to Modern Physics*-6th Ed., McGraw-Hill, New York, **1969**.

10 Warren, B.E., *X-ray Diffraction*, Addison-Wesley Publishing Co., Reading, MA, **1969**.

11 Schwartz, L.H., Cohen, J.B., *Diffraction from Materials*, Springer, New York, **1981**.

12 Glatter, O., Kratky, O., *Small Angle X-ray Scattering*, Academic Press, London, **1982**, p. 156.

13 Balta-Calleja, F.J., Vonk, C.G., *X-ray Scattering of Synthetic Polymers*, Elsevier, New York, **1989**, chapter 7.

14 Thomson, J.J., *Conduction of Electricity Through Gases*, 2nd edn., Cambridge University Press, Cambridge, **1928–33**, p. 321, and Compton, A.H. and Allison, S.K., *X-rays in Theory and Experiment*, 2nd edn. Van Nostrand-Reinhold, Princeton, NJ, **1935**.

15 *International Table of Crystallography*, vols. 1–4, by The International Union of Crystallography, ed. N.F.M. Henry and K. Lonsdale, The Kynoch Press, Birmingham, England, **1952**.

16 Spruiell, J.E. and Clark, E.S., *Methods of Experimental Physics*, chapter 6: X-ray Diffraction, ed. Fava, R.A., Academic Press, New York, **1980**.

17 Bennett, C.H., Serially deposited amorphous aggregates of hard spheres, *J. Appl. Phys.* **1972**, *43*, 2727–2734; Boudreaux, D.S., Gregor,

J.M., Structure simulation of transition-metal-metalloid glasses. III, *J. Appl. Phys.*, **1977**, *48*, 5057–5061.

18 Saw, C.K. and Schwarz, R.B., Chemical short-range order in dense random-packed models, *J. Less-Common Metals*, **1988**, *140*, 385–393.

19 Vaughan, D. (ed), *X-ray Data Booklet*, Center for X-ray Optics, Lawrence Berkeley Laboratory, Berkeley, CA, 1986, and Dyson, N.A., *X-rays in Atomic and Nuclear Physics*, Longman, London, **1973**, chapter 3.

20 Fernandez, M., Keyrilainen, J., Serimaa, R., Torkkeli, M., Karjalainen-Lindsberg, M.-L., Tenhunen, M., Thomlinson, W., Urban, V. and Suortti, P., Small-angle x-ray scattering studies of human breast tissue samples, *Phys. Med. Biol.*, **2002**, *47*, 577–592.

21 Lewis, R.A., Rogers, K.D., Hall, C.J., Towns-Andrews, E., Slawsom, S., Evans, A., Pinder, S.E., Ellis, I.O., Boggis, C.R.M., Hufton, A., Dance, D.R., Breast caner diagnosis using scattered x-rays, *J. Synchrotron Rad.* **2000**, *7*, 348–352, James, V., Synchrotron fibre diffraction identifies and locates foetal collagenouos breast tissue associated with breast carcinoma, *J. Synchrotron Rad.*, **2002**, *9*, 71–76.

22 Jaschouz, D., Paris, O., Roschger, P., Hwang, H.S., Fratzl, P., Pole figure analysis of mineral nanoparticles orientation in individual trabecula of human vertebral bone, *J. Appl. Cryst.*, **2003**, *36*, 494–498.

23 Jager, I., Fratzl, P., Mineralized collagen fibrils: A mechanical model with a staggered arrangement of mineral particles, *Biophys. J.*, **2000**, *79*, 1737–1746.

24 Laggner, P., *Physiochemical Methods in the Study of Biomembranes: Subcellular Biochemistry*, ed. Hilderson, H.J., Ralston, G.B., Plenum Press, New York, **1994**, pp. 23, 11, 451–491.

25 Murray, C.B., Kagan, C.R., Bawendi, M.G., Synthesis and characterization of monodisperse nanocrystals and closed-packed nanocrystals assemblies, *Annu. Rev. Mater. Sci.*, **2000**, *20*, 545–610.

26 Stuhrmann, H.B., Miller, A., Small-angle scattering of biological structures, *J. Appl. Cryst.* **1978**, *11*, 325–345.

27 Fratzl, P., Cellulose and collagen: from fibres to tissues, *Curr. Opin. Colloid Interface Sci.*, **2003**, *8*, 32–39.

28 Fratzl, P., Gupta, H.S., Paschalis, E.P., Roschger, P., Structure and mechanical quality of the collagen-mineral nano-composite in bone, *J. Mater. Chem.*, **2004**, *14*, 2115–2123.

29 Abraham, T., Masakatsu, H., Hirai, M., Glycolipid based cubic nanoparticles: preparation and structural aspects, *Colloids Surf. B: Biointerfaces*, **2004**, *35*, 107–117.

Index

Nanotechnologies for the Life Sciences Vol. 3
Nanosystem Characterization Tools in the Life Sciences. Edited by Challa S. S. R. Kumar

ISBN: 3-527-31383-4

C

o

p